ORGANIC CHEMISTRY AS A SECOND LANGUAGE, 5e

First Semester Topics

DAVID KLEIN

i

Library of Congress Cataloging-in-Publication Data:

Names: Klein, David R., author.
Title: Organic chemistry as a second language : first semester topics / David Klein.
Description: 5e [Fifth edition]. | Hoboken, NJ : Wiley, [2020] | Includes bibliographical references and index.
Identifiers: LCCN 2019018766 (print) | LCCN 2019020631 (ebook) | ISBN 9798838248367 Paperback

INTRODUCTION

IS ORGANIC CHEMISTRY REALLY ALL ABOUT MEMORIZATION?

Is organic chemistry really as tough as everyone says it is? The answer is yes and no. Yes, because you will spend more time on organic chemistry than you would spend in a course on underwater basket weaving. And no, because those who say it's so tough have studied inefficiently. Ask around, and you will find that most students think of organic chemistry as a memorization game. *This is not true!* Former organic chemistry students perpetuate the false rumor that organic chemistry is the toughest class on campus, because it makes them feel better about the poor grades that they received.

If it's not about memorizing, then what is it? To answer this question, let's compare organic chemistry to a movie. Picture in your mind a movie where the plot changes every second. If you're in a movie theatre watching a movie like that, you can't leave even for a second because you would miss something important to the plot. So you try your hardest to wait until the movie is over before going to the bathroom. Sounds familiar?

Organic chemistry is very much the same. It is one long story, and the story actually makes sense if you pay attention. The plot constantly develops, and everything ties into the plot. If your attention wanders for too long, you could easily get lost.

You probably know at least one person who has seen one movie more than five times and can quote every line by heart. How can this person do that? It's *not* because he or she tried to memorize the movie. The first time you watch a movie, you learn the plot. After the second time, you understand why individual scenes are necessary to develop the plot. After the third time, you understand why the dialogue was necessary to develop each scene. After the fourth time, you are quoting many of the lines by heart. *Never at any time did you make an effort to memorize the lines.* You know them *because they make sense* in the grand scheme of the plot. If I were to give you a screenplay for a movie and ask you to memorize as much as you can in 10 hours, you would probably not get very far into it. If, instead, I put you in a room for 10 hours and played the same movie over again five times, you would know most of the movie by heart, without even trying. You would know everyone's names, the order of the scenes, much of the dialogue, and so on.

Organic chemistry is exactly the same. It's not about memorization. It's all about making sense of the plot, the scenes, and the individual concepts that make up our story. Of course you will need to remember all of the terminology, but with enough practice, the terminology will become second nature to you. So here's a brief preview of the plot.

THE PLOT

The first half of our story builds up to reactions, and we learn about the characteristics of molecules that help us understand reactions. We begin by looking at atoms, the building blocks of molecules, and what happens when they combine to form bonds. We focus on special bonds between certain atoms, and we see how the nature of bonds can affect the shape and stability of molecules. Then, we need a vocabulary to start talking about molecules, so we learn how to draw and name molecules. We see how molecules move around in space, and we explore the relationships between similar types of molecules. At this point, we know the important characteristics of molecules, and we are ready to use our knowledge to explore reactions.

Reactions take up the rest of the course, and they are typically broken down into chapters based on categories. Within each of these chapters, there is actually a subplot that fits into the grand story.

HOW TO USE THIS BOOK

This book will help you study more efficiently so that you can avoid wasting countless hours. It will point out the major scenes in the plot of organic chemistry. The book will review the critical principles and explain why they are relevant to the rest of the course. In each section, you will be given the tools to better understand your textbook and lectures, as well as plenty of opportunities to practice the key skills that you will need to solve problems on exams. In other words, you will learn the language of organic chemistry. *This book cannot replace your textbook, your lectures, or other forms of studying*. This book is not the Cliff Notes of Organic Chemistry. It focuses on the basic concepts that will empower you to do well if you go to lectures and study in addition to using this book. To best use this book, you need to know how to study in this course.

HOW TO STUDY

There are two separate aspects to this course:

1. Understanding principles
2. Solving problems

Although these two aspects are completely different, instructors will typically gauge your understanding of the principles by testing your ability to solve problems. So you must master both aspects of the course. The principles are in your lecture notes, but *you* must discover how to solve problems. Most students have a difficult time with this task. In this book, we explore some step-by-step processes for analyzing problems. There is a very simple habit that you must form immediately: *learn to ask the right questions*.

If you go to a doctor with a pain in your stomach, you will get a series of questions: How long have you had the pain? Where is the pain? Does it come and go, or is it constant? What was the last thing you ate? and so on. The doctor is doing two very important and very different things: 1) asking the right questions, and 2) arriving at a diagnosis based on the answers to those questions.

Let's imagine that you want to sue McDonald's because you spilled hot coffee in your lap. You go to an attorney who asks you a series of questions. Once again, the lawyer is doing two very important and very different things: 1) asking the right questions, and 2) formulating an opinion based on the answers to those questions. Once again, the first step is asking questions.

In fact, in any profession or trade, the first step of diagnosing a problem is always to ask questions. The same is true with solving problems in this course. Unfortunately, you are expected to learn how to do this on your own. In this book, we will look at some common types of problems and we will see what questions you should be asking in those circumstances. More importantly, we will also be developing skills that will allow you to figure out what questions you should be asking for a problem that you have never seen before.

Many students freak out on exams when they see a problem that they can't do. If you could hear what was going on in their minds, it would sound something like this: "I can't do it … I'm gonna flunk." These thoughts are counterproductive and a waste of precious time. Remember that when all else fails, there is always one question that you can ask yourself: "What questions should I be asking right now?"

The only way to truly master problem-solving is to practice problems every day, consistently. You will never learn how to solve problems by just reading a book. You must try, and fail, and try again. You must learn from your mistakes. You must get frustrated when you can't solve a problem. That's the learning process. Whenever you encounter an exercise in this book, pick up a pencil and work on it. Don't skip over the problems! They are designed to foster skills necessary for problem-solving.

The worst thing you can do is to read the solutions and think that you now know how to solve problems. It doesn't work that way. If you want an A, you will need to sweat a little (no pain, no gain). And that doesn't mean that you should spend day and night memorizing. Students who focus on memorizing will experience the pain, but few of them will get an A.

The simple formula: Review the principles until you understand how each of them fits into the plot; then *focus all of your remaining time on solving problems*. Don't worry. The course is not that bad if you approach it with the right attitude. This book will act as a road map for your studying efforts.

CONTENTS

CHAPTER 15 *SYNTHESIS*   **270**

BOND-LINE DRAWINGS

To do well in organic chemistry, you must first learn to interpret the drawings that organic chemists use. When you see a drawing of a molecule, it is absolutely critical that you can read all of the information contained in that drawing. Without this skill, it will be impossible to master even the most basic reactions and concepts.

Molecules can be drawn in many ways. For example, below are three different ways of drawing the same molecule:

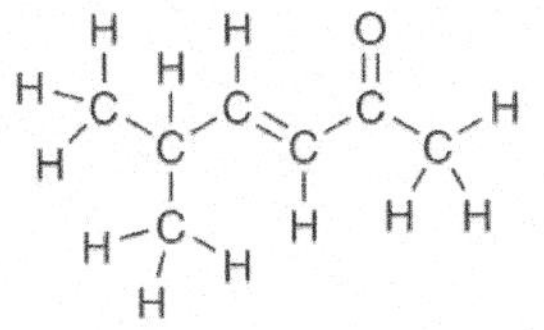

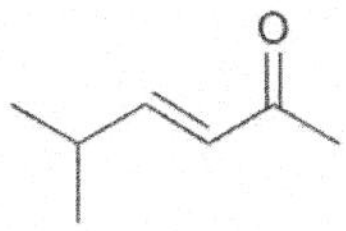

Without a doubt, the last structure (bond-line drawing) is the quickest to draw, the quickest to read, and the best way to communicate. Open to any page in the second half of your textbook and you will find that every page is plastered with bond-line drawings. It is essential that you learn to read these drawings fluently. This chapter will help you develop the skills that you will need to read these drawings quickly and fluently.

1.1 HOW TO READ BOND-LINE DRAWINGS

Bond-line drawings show the carbon skeleton (the connections of all the carbon atoms that build up the backbone, or skeleton, of the molecule) with any functional groups that are attached, such as —OH or —Br. Lines are drawn in a zigzag format, where each corner or endpoint represents a carbon atom. For example, the following compound has seven carbon atoms:

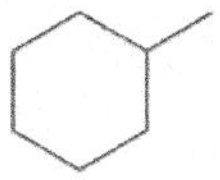

Don't forget that the ends of lines represent carbon atoms as well. For example, the following molecule has six carbon atoms (make sure you can count them):

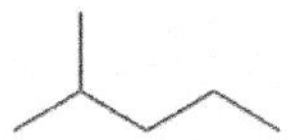

Double bonds are shown with two lines, and triple bonds are shown with three lines:

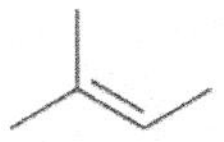

When drawing triple bonds, be sure to draw them in a straight line rather than zigzag, because triple bonds are linear (there will be more about this in the chapter on geometry). This can be quite confusing at first, because it can get hard to see just how many carbon atoms are in a triple bond, so let's make it clear:

is the same as

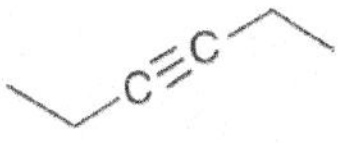

so this compound
has 6 carbon atoms

It is common to see a small gap on either side of a triple bond, like this:

is the same as

Both drawings above are commonly used, and you should train your eyes to see triple bonds either way. Don't let triple bonds confuse you. The two carbon atoms of the triple bond and the two carbons connected to them are drawn in a straight line. All other bonds are drawn as a zigzag:

$$H-\overset{\overset{H}{|}}{\underset{\underset{H}{|}}{C}}-\overset{\overset{H}{|}}{\underset{\underset{H}{|}}{C}}-\overset{\overset{H}{|}}{\underset{\underset{H}{|}}{C}}-\overset{\overset{H}{|}}{\underset{\underset{H}{|}}{C}}-H$$ is drawn like this:

BUT

$$H-\overset{\overset{H}{|}}{\underset{\underset{H}{|}}{C}}-C\equiv C-\overset{\overset{H}{|}}{\underset{\underset{H}{|}}{C}}-H$$ is drawn like this:

EXERCISE 1.1   Count the number of carbon atoms in each of the following drawings:

Answer   The first compound has six carbon atoms, and the second compound has five carbon atoms:

PROBLEMS   Count the number of carbon atoms in each of the following drawings.

1.2   Answer: _____

1.3   Answer: _____

1.4   Answer: _____

1.5   Answer: _____

1.6   Answer: _____

1.7   Answer: _____

1.8   Answer: _____

1.9   Answer: _____

1.10   Answer: _____

1.11   Answer: _____

Now that we know how to count carbon atoms, we must learn how to count the hydrogen atoms in a bond-line drawing. Most hydrogen atoms are not shown, so bond-line drawings can be drawn very quickly. Hydrogen atoms connected to atoms other than carbon (such as nitrogen or oxygen) must be drawn:

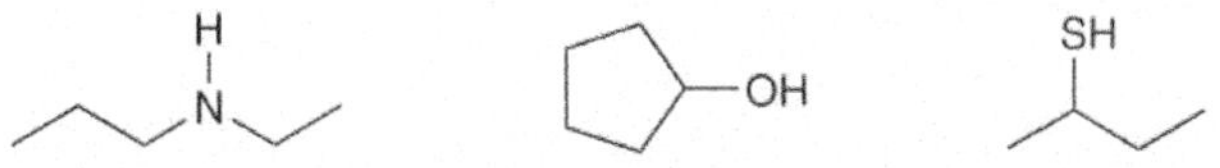

But hydrogen atoms connected to carbon are not drawn. Here is the rule for determining how many hydrogen atoms are connected to each carbon atom: *uncharged carbon atoms have a total of four bonds*. In the following drawing, the highlighted carbon atom is showing only two bonds:

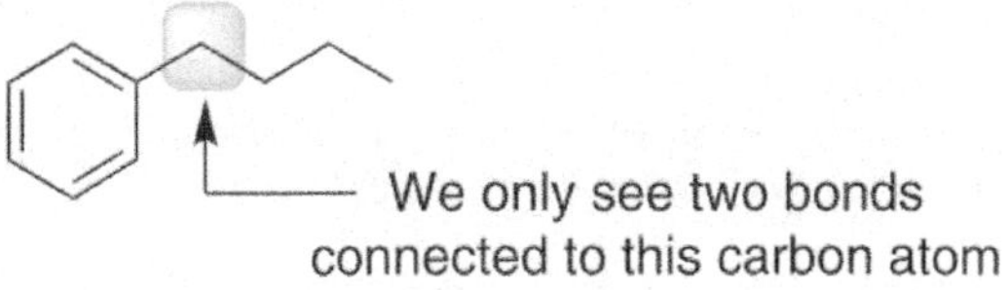

Therefore, it is assumed that there are two more bonds to hydrogen atoms (to give a total of four bonds). This is what allows us to avoid drawing the hydrogen atoms and to save so much time when drawing molecules. It is assumed that the average person knows how to count to four, and therefore is capable of determining the number of hydrogen atoms even though they are not shown.

So you only need to count the number of bonds that you can see on a carbon atom, and then you know that there should be enough hydrogen atoms to give a total of four bonds to the carbon atom. After doing this many times, you will get to a point where you do not need to count anymore. You will simply get accustomed to seeing these types of drawings, and you will be able to instantly "see" all of the hydrogen atoms without counting them. Now we will do some exercises that will help you get to that point.

EXERCISE 1.12   The following molecule has nine carbon atoms. Count the number of hydrogen atoms connected to each carbon atom.

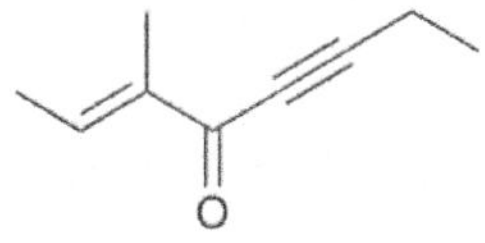

Answer

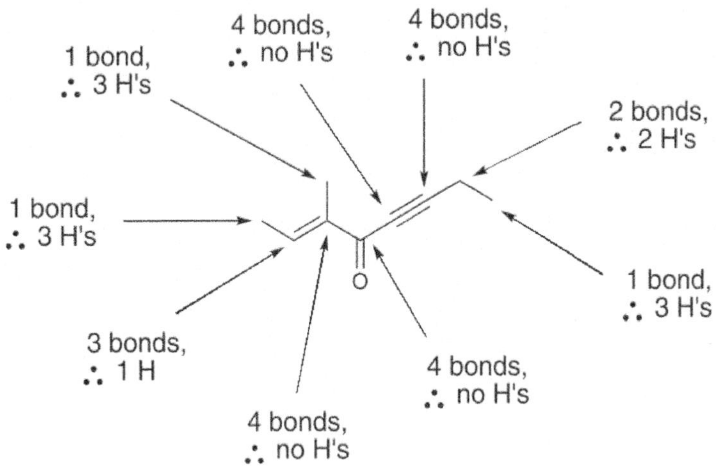

PROBLEMS For each of the following molecules, count the number of hydrogen atoms connected to each carbon atom.

1.13

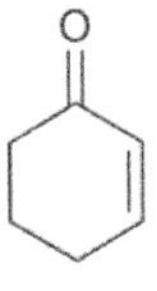

1.14

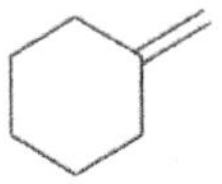

1.15

1.16

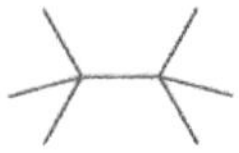

1.17

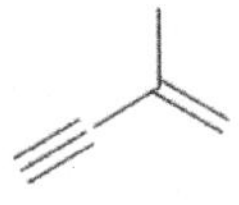

1.18

1.19

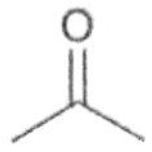

1.20

Now we can understand why we save so much time by using bond-line drawings. Of course, we save time by not drawing every C and H. But, there is an even larger benefit to using these drawings. Not only are they easier to draw, but they are easier to read as well. Take the following reaction for example:

$$(CH_3)_2C=CHCOCH_3 \quad \xrightarrow[Pt]{H_2} \quad (CH_3)_2CHCH_2COCH_3$$

It is somewhat difficult to see what is happening in the reaction. You need to stare at it for a while to see the change that took place. However, when we redraw the reaction using bond-line drawings, the reaction becomes very easy to read immediately:

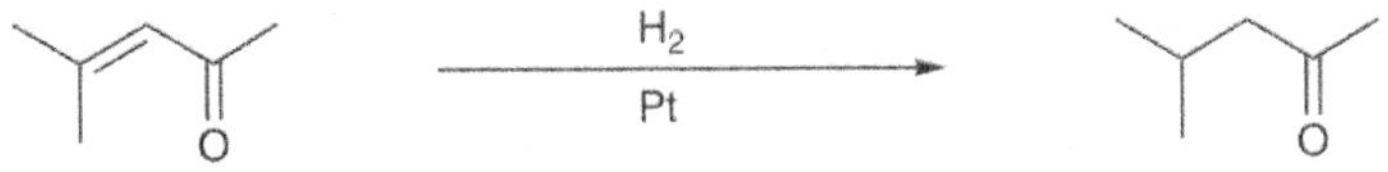

$$\xrightarrow[Pt]{H_2}$$

As soon as you see the reaction, you immediately know what is happening. In this reaction we are converting a double bond into a single bond by adding two hydrogen atoms across the double bond. Once you get comfortable reading these drawings, you will be better equipped to see the changes taking place in reactions.

1.2 HOW TO DRAW BOND-LINE DRAWINGS

Now that we know how to read these drawings, we need to learn how to draw them. Take the following molecule as an example:

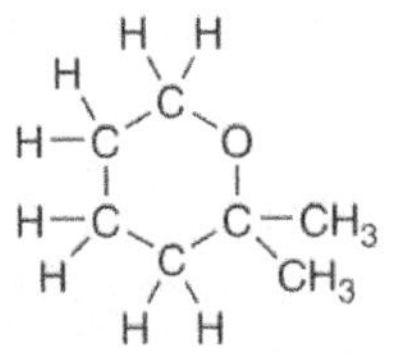

To draw this as a bond-line drawing, we focus on the carbon skeleton, making sure to draw any atoms other than C and H. All atoms other than carbon and hydrogen *must* be drawn. So the example above would look like this:

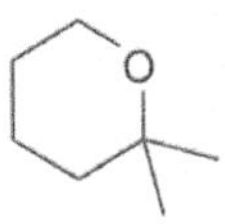

A few pointers may be helpful before you do some problems.

1. Don't forget that carbon atoms in a straight chain are drawn in a zigzag format:

 is drawn like this:

2. When drawing double bonds, try to draw the other bonds as far away from the double bond as possible:

 is much better than

BAD

3. When drawing zigzags, it does not matter in which direction you start drawing:

is the same as   is the same as

PROBLEMS   For each structure below, draw a bond-line drawing in the box provided.

1.21

1.22

1.23

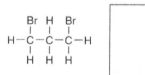

1.24

1.3 MISTAKES TO AVOID

1. *Never* draw a carbon atom with more than four bonds. This is a big no-no. Carbon atoms only have four orbitals; therefore, carbon atoms can form only four bonds (bonds are formed when orbitals of one atom overlap with orbitals of another atom). This is true of all second-row elements, and we will discuss this in more detail in the upcoming chapter.

2. When drawing a molecule, you should either show all of the H's and all of the C's, or draw a bond-line drawing where the C's and H's are not drawn. You *cannot* draw the C's without also drawing the H's:

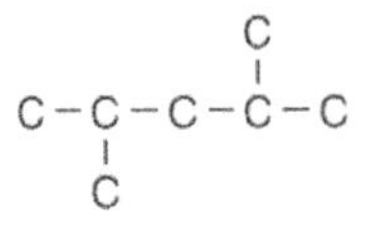

NEVER DO THIS

This drawing is insufficient. Either leave out the C's (which is preferable) or put in the H's:

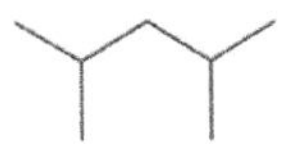

or

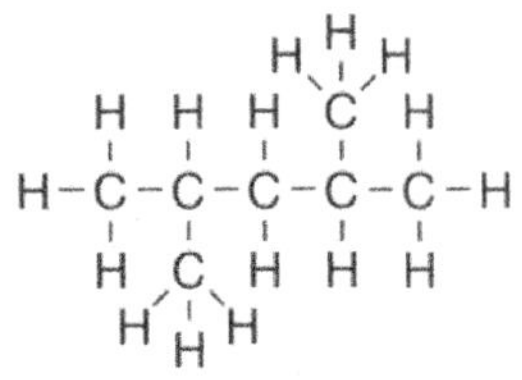

3. When drawing each carbon atom in a zigzag, try to draw all of the bonds as far apart as possible:

is better than

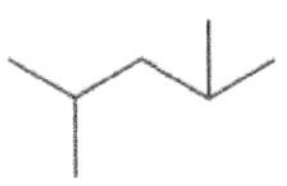

1.4 MORE EXERCISES

First, open your textbook and flip through the pages in the second half. Choose any bond-line drawing and make sure that you can say with confidence how many carbon atoms you see and how many hydrogen atoms are attached to each of those carbon atoms.

Now examine the following transformation, and think about the changes that are occurring:

Don't worry about *how* these changes occur. That will be covered much later (in Chapter 11), when we explore this type of transformation in more detail. For now, just focus on describing the changes that you see. In this case, two hydrogen atoms have been installed, and a double bond has been converted into a single bond. It is certainly clear to see that the double bond has been converted into a single bond, but you should also clearly see that two hydrogen atoms have been installed during this process.

Consider another example:

In this example, H and Br have been removed, and a single bond has been converted into a double bond (we will see in Chapter 10 that it is actually H^+ and Br^- that are removed). If you cannot see that an H was removed, then you will need to count the number of hydrogen atoms in the starting material and compare it with the product:

Now consider one more example:

In this example, a bromine atom has been replaced with a chlorine atom (as we will see in Chapter 9). Inspection of the bond-line drawings clearly indicates that no other changes occurred in this case.

PROBLEMS   For each of the following transformations, describe the changes that are occurring.

1.25
Answer: _____

1.26
Answer: _____

1.27
Answer: _____

1.28
Answer: _____

1.29
Answer: _____

1.30

Answer: _____

_____

1.31

Answer: _____

_____

1.32

Answer: _____

_____

1.5 IDENTIFYING FORMAL CHARGES

Formal charges are charges (either positive or negative) that we must often include in our drawings. They are extremely important. If you don't draw a formal charge when it is supposed to be drawn, then your drawing will be incomplete (and wrong). So you must learn how to identify when you need formal charges and how to draw them. If you cannot do this, then you will not be able to draw resonance structures (which we see in the next chapter), and if you can't do that, then you will have a very hard time passing this course.

A formal charge is a charge associated with an atom that does not exhibit the expected number of valence electrons. When calculating the formal charge on an atom, we first need to know the number of valence electrons the atom is *supposed* to have. We can get this number by inspecting the periodic table, since each column of the periodic table indicates the number of expected valence electrons (valence electrons are the electrons in the valence shell, or the outermost shell of electrons—you probably remember this from high school chemistry). For example, carbon is in Column 4A, and therefore has four valence electrons. This is the number of valence electrons that a carbon atom is supposed to have.

Next we ask how many electrons the atom *actually has* in the drawing. But how do we count this?

Let's see an example. Consider the central carbon atom in the compound below:

$$H_3C-\overset{\overset{\displaystyle :\overset{..}{O}^{\,H}}{|}}{\underset{\underset{\displaystyle H}{|}}{C}}-CH_3$$

Remember that every bond represents two electrons being shared between two atoms. Begin by splitting each bond apart, placing one electron on this atom and one electron on that atom:

$$H_3C\cdot \quad \cdot \overset{\overset{\displaystyle :\overset{..}{O}^{\,H}}{\cdot}}{\underset{\underset{\displaystyle H}{\cdot}}{C}}\cdot \quad \cdot CH_3$$

Now count the number of electrons immediately surrounding the central carbon atom:

$$H_3C\cdot \quad \boxed{\cdot \overset{\overset{\displaystyle :\overset{..}{O}^{\,H}}{\cdot}}{\underset{\underset{\displaystyle H}{\cdot}}{C}}\cdot} \quad \cdot CH_3$$

There are four electrons. This is the number of electrons that the atom actually has.

Now we are in a position to compare how many valence electrons the atom is *supposed* to have (in this case, four) with how many valence electrons it *actually* has (in this case, four). Since these numbers are the same, the carbon atom has no formal charge. This will be the case for most of the atoms in the structures you will draw in this course. But in some cases, there will be a difference between the number of electrons the atom is supposed to have and the number of electrons the atom actually has. In those cases, there will be a formal charge. So let's see an example of an atom that has a formal charge.

Consider the oxygen atom in the structure below:

Let's begin by determining the number of valence electrons that an oxygen atom is *supposed* to have. Oxygen is in Column 6A of the periodic table, so oxygen should have six valence electrons. Next, we need to look at the oxygen atom in this structure and ask how many valence electrons it *actually* has. So, we redraw the structure by splitting up the C—O bond:

In addition to the electron on the oxygen from the C—O bond, the oxygen also has three lone pairs. A lone pair is when you have two electrons that are not being used to form a bond. Lone pairs are drawn as two dots on an atom, and the oxygen above has three of these lone pairs. You must remember to count each lone pair as two electrons. So we see that the oxygen atom actually has seven valence electrons, which is one more electron than it is supposed to have. Therefore, it will have a negative charge:

EXERCISE 1.33   Consider the nitrogen atom in the structure below and determine if it has a formal charge:

$$
\begin{array}{c}
H \\
| \\
H-N-H \\
| \\
H
\end{array}
$$

Answer   Nitrogen is in Column 5A of the periodic table so it should have five valence electrons. Now we count how many it actually has:

$$
\begin{array}{c}
H \\
\cdot \\
H\cdot \ \cdot N\cdot \ \cdot H \\
\cdot \\
H
\end{array}
$$

It only has four. So, it has one less electron than it is supposed to have. Therefore, this nitrogen atom has a positive charge:

$$
\begin{array}{c}
H \\
|\oplus \\
H-N-H \\
| \\
H
\end{array}
$$

PROBLEMS For each of the structures below determine if the oxygen or nitrogen atom has a formal charge. If there is a charge, draw the charge.

1.34 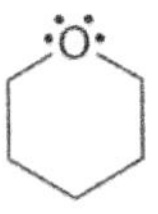   **1.35** 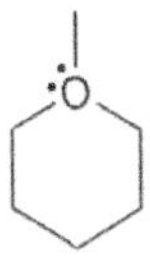   **1.36** 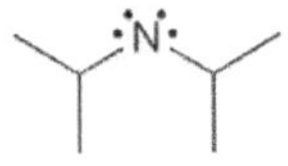   **1.37**

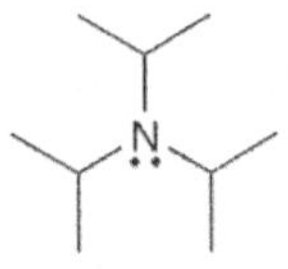

1.38 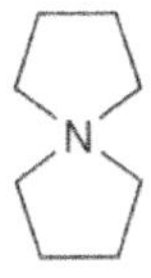   **1.39** 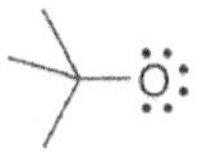   **1.40** 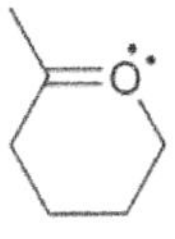   **1.41**

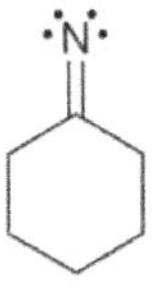

1.42 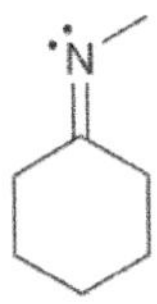   **1.43** 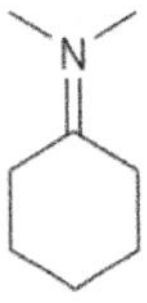   **1.44** 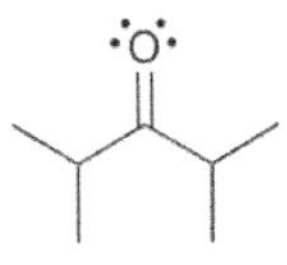   **1.45** $H_3C—C≡N:$

This brings us to the most important atom of all: carbon. We saw before that carbon forms four bonds. This allows us to ignore the hydrogen atoms when drawing bond-line structures, because it is assumed that we know how to count to four and can figure out how many hydrogen atoms are there. When we said that, we were only talking about carbon atoms without formal charges (most carbon atoms in most structures will not have formal charges). But now that we have learned what a formal charge is, let's consider what happens when carbon has a formal charge.

If carbon bears a formal charge, then we cannot just assume it will still have four bonds. In fact, it will have only three. Let's see why. Let's first consider C^+, and then we will move on to C^-.

If carbon has a *positive* formal charge, then it has only three valence electrons (it is supposed to have four valence electrons, because carbon is in Column 4A of the periodic table). Since it has only three valence electrons, it can form only three bonds. That's it. So, a carbon atom with a positive formal charge will have only three bonds, and you should keep this in mind when counting hydrogen atoms:

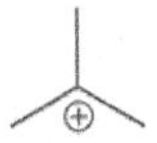

No hydrogen atoms
on this C^+

Now let's consider what happens when we have a carbon atom with a *negative* formal charge. The reason it has a negative formal charge is because it has one more electron than it is supposed to have. Therefore, it has five valence electrons. Two of these electrons form a lone pair, and the other three electrons are used to form bonds:

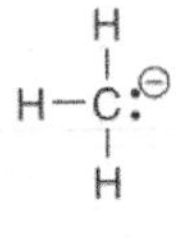

We have the lone pair, because we can't use each of the five electrons to form a bond. Carbon can *never* have five bonds. Why not? Electrons exist in regions of space called orbitals. These orbitals can overlap with orbitals from other atoms to form bonds, or the orbitals can contain two electrons (which is called a lone pair). Carbon has only four orbitals in its valence shell, so there is no way it could possibly form five bonds—it does not have five orbitals to use to form those bonds. This is why a carbon atom with a negative charge will have a lone pair (if you look at the drawing above, you will count four orbitals—one for the lone pair and then three more for the bonds).

Therefore, a carbon atom with a negative charge can also form only three bonds (just like a carbon with a positive charge). When you count hydrogen atoms, you should keep this in mind:

No hydrogen atoms
on this C⁻

1.6 FINDING LONE PAIRS THAT ARE NOT DRAWN

From all of the cases above (oxygen, nitrogen, carbon), you can see why you have to know how many lone pairs there are on an atom in order to figure out the formal charge on that atom. Similarly, you have to know the formal charge to figure out how many lone pairs there are on an atom. Take the case below with the nitrogen atom shown:

N   could either be   N⁻   or   N⁺

If the lone pairs were drawn, then we would be able to figure out the charge (two lone pairs would mean a negative charge and one lone pair would mean a positive charge). Similarly, if the formal charge was drawn, we would be able to figure out how many lone pairs there are (a negative charge would mean two lone pairs and a positive charge would mean one lone pair). So you can see that drawings must include either lone pairs or formal charges. *The convention is to always show formal charges and to leave out the lone pairs.* This is much easier to draw, because you usually won't have more than one charge on a drawing (if even that), so you get to save time by not drawing every lone pair on every atom.

Now that we have established that formal charges must *always* be drawn and that lone pairs are usually *not* drawn, we need to get practice in how to see the lone pairs when they are not drawn. This is not much different from training yourself to see all the hydrogen atoms in a bond-line drawing even though they are not drawn. If you know how to count, then you should be able to figure out how many lone pairs are on an atom where the lone pairs are not drawn.

Let's see an example to demonstrate how you do this:

O⁻

In this case, we are looking at an oxygen atom. Oxygen is in Column 6A of the periodic table, so it is supposed to have six valence electrons. Then, we need to take the formal charge into account. This oxygen atom has a negative charge, which means one extra electron. Therefore, this oxygen atom must have $6 + 1 = 7$ valence electrons. Now we can figure out how many lone pairs there are.

The oxygen atom has one bond, which means that it is using one of its seven electrons to form a bond. The other six must be in lone pairs. Since each lone pair is two electrons, this must mean that there are three lone pairs:

O⁻   is the same as   O⁻

Let's review the process:

1. Count the number of valence electrons the atom should have according to the periodic table.

2. Take the formal charge into account. A negative charge means one more electron, and a positive charge means one less electron.

3. Now you know the number of valence electrons the atom actually has. Use this number to figure out how many lone pairs there are.

Now we need to get used to the common examples. Although it is important that you know how to count and determine numbers of lone pairs, it is actually much more important to get to a point where you don't have to waste time counting. You need to get familiar with the common situations you will encounter. Let's go through them methodically.

When oxygen has no formal charge, it will have two bonds and two lone pairs:

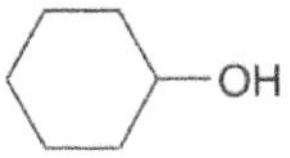

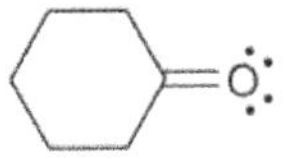

If oxygen has a negative formal charge, then it must have one bond and three lone pairs:

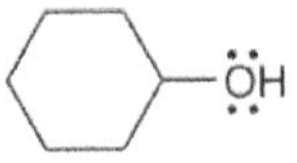

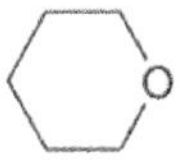

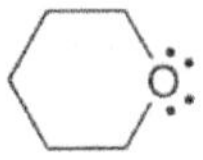

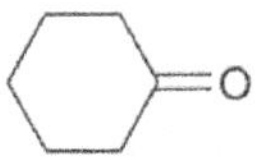

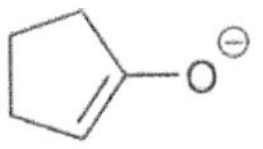

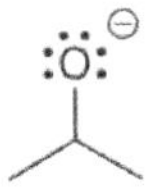

If oxygen has a positive charge, then it must have three bonds and one lone pair:

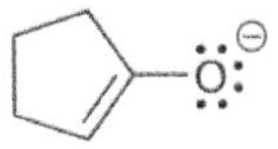

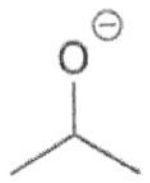

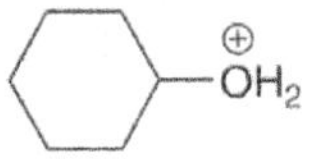

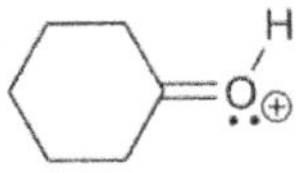

EXERCISE 1.46 Draw all lone pairs in the following structure:

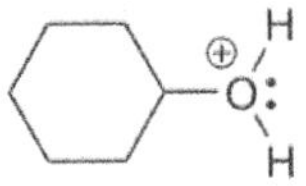

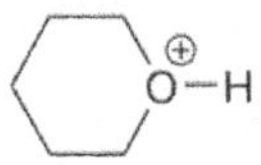

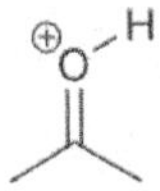

Answer The oxygen atom has a positive formal charge and three bonds. You should try to get to a point where you recognize that this must mean that the oxygen atom has one lone pair:

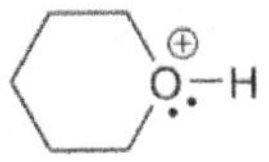

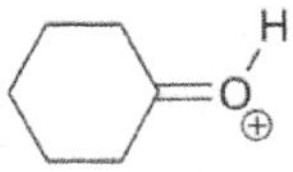

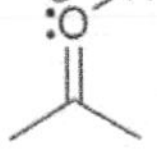

Until you get to the point where you can recognize this, you should be able to figure out the answer by counting.

Oxygen is supposed to have six valence electrons. This oxygen atom has a positive charge, which means it is missing an electron. Therefore, this oxygen atom must have $6 - 1 = 5$ valence electrons. Now, we can figure out how many lone pairs there are.

The oxygen atom has three bonds, which means that it is using three of its five valence electrons to form bonds. The other two must be in a lone pair. So there is only one lone pair.

PROBLEMS Review the common situations above, and then come back to these problems. For each of the following structures, draw all lone pairs. Try to recognize how many lone pairs there are *without* having to count. Then count to see if you were right.

1.47

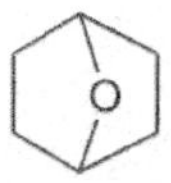

1.48

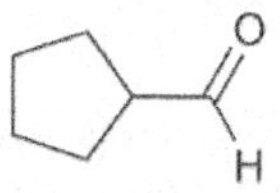

1.49

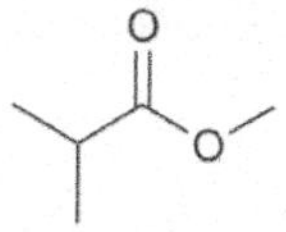

1.50

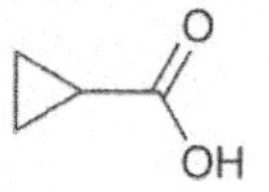

1.51

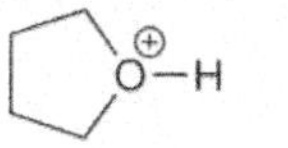

1.52

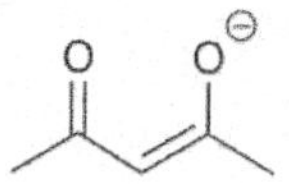

Now let's look at the common situations for nitrogen atoms. When nitrogen has no formal charge, it will have three bonds and one lone pair:

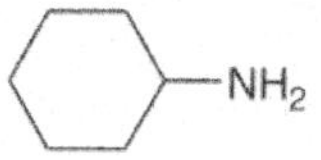

$-NH_2$ is the same as

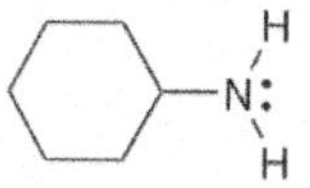

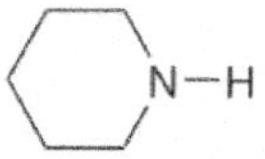

N—H   is the same as

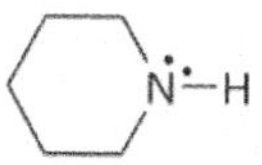

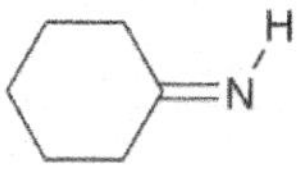

$=N$ is the same as

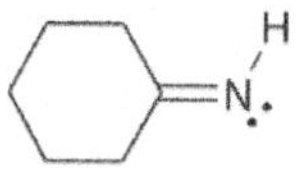

If nitrogen has a negative formal charge, then it must have two bonds and two lone pairs:

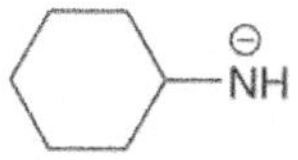

$-NH$ is the same as

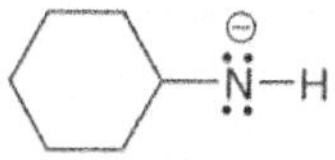

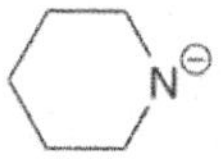

N   is the same as

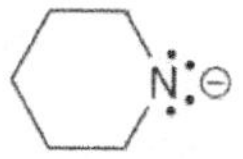

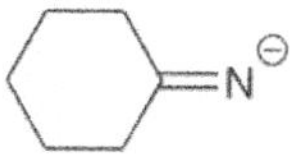

$=N$ is the same as

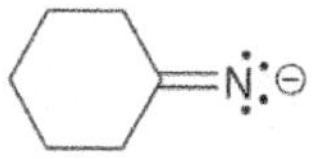

If nitrogen has a positive charge, then it must have four bonds and no lone pairs:

has no lone pairs

has no lone pairs

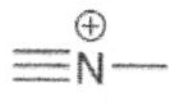

has no lone pairs

EXERCISE 1.53 Draw all lone pairs in the following structure:

$$\overset{\ominus}{N}=\overset{\oplus}{N}=\overset{\ominus}{N}$$

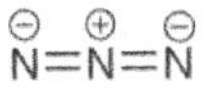

Answer The central nitrogen atom has a positive formal charge and four bonds. You should try to get to a point where you recognize that this nitrogen atom does not have any lone pairs. Each of the other nitrogen atoms has a negative formal charge and two bonds. You should try to get to a point where you recognize that each of these nitrogen atoms has two lone pairs:

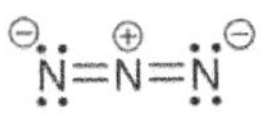

$$\overset{\ominus}{\underset{..}{\overset{..}{N}}}=\overset{\oplus}{N}=\overset{\ominus}{\underset{..}{\overset{..}{N}}}$$

Until you get to the point where you can recognize this, you should be able to figure out the answer by counting. Nitrogen is supposed to have five valence electrons. The central nitrogen atom has a positive charge, which means it is missing an electron. In other words, this nitrogen atom must have $5 - 1 = 4$ valence electrons. Now, we can figure out how many lone pairs there are. Since it has four bonds, it is using all of its electrons to form bonds. So there is no lone pair on this nitrogen atom.

For each of the remaining nitrogen atoms, there is a negative formal charge. That means that each of those nitrogen atoms has one extra valence electron, $5 + 1 = 6$ electrons. Each nitrogen atom has two bonds, which means that each nitrogen atom has four valence electrons left over, giving two lone pairs.

PROBLEMS Review the common situations for nitrogen, and then come back to these problems. For each of the following structures, draw all lone pairs. Try to recognize how many lone pairs there are *without* having to count. Then count to see if you were right.

1.54

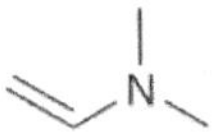

1.55

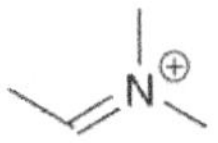

1.56

1.57

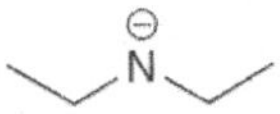

1.58

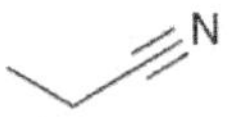

1.59

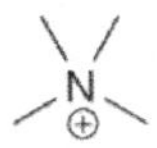

MORE PROBLEMS For each of the following structures, draw all lone pairs.

1.60

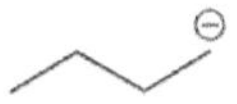

1.61

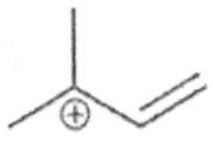

1.62 $H-C\equiv C^{\ominus}$

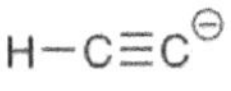

1.63

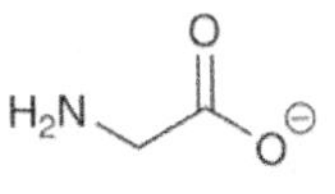

1.64

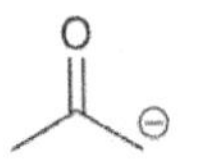

1.65

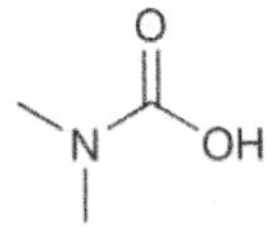

1.66

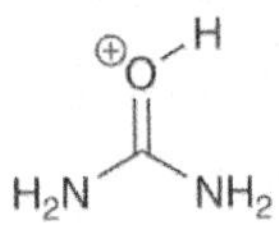

1.67

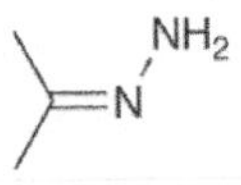

1.68

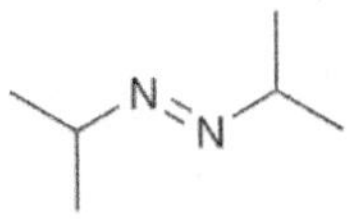

RESONANCE

In this chapter, you will learn the tools that you need to draw resonance structures with proficiency. I cannot adequately stress the importance of this skill. Resonance is the one topic that permeates the entire subject matter from start to finish. It finds its way into every chapter, into every reaction, and into your nightmares if you do not master the rules of resonance. You cannot get an A in this class without mastering resonance. So what is resonance? And why do we need it?

2.1 WHAT IS RESONANCE?

In Chapter 1, we introduced one of the best ways of drawing molecules, bond-line structures. They are fast to draw and easy to read, but they have one major deficiency: they do not describe molecules perfectly. In fact, no drawing method can completely describe a molecule using only a single drawing. Here is the problem.

Although our drawings are very good at showing which atoms are connected to each other, our drawings are not good at showing where all of the electrons are, because electrons aren't really solid particles that can be in one place at one time. All of our drawing methods treat electrons as particles that can be placed in specific locations. Instead, it is best to think of electrons as *clouds of electron density*. We don't mean that electrons fly around in clouds; we mean that electrons *are* clouds. These clouds often spread themselves across large regions of a molecule.

So how do we represent molecules if we can't draw where the electrons are? The answer is resonance. We use the term *resonance* to describe our solution to the problem: we use more than one drawing to represent a single molecule. We draw several drawings, and we call these drawings *resonance structures*. We meld these drawings into one image in our minds. To better understand how this works, consider the following analogy.

Your friend asks you to describe what a nectarine looks like, because he has never seen one. You aren't a very good artist so you say the following:

> *Picture a peach in your mind, and now picture a plum in your mind. Well, a nectarine has features of both: the inside tastes like a peach, but the outside is smooth like a plum. So take your image of a peach together with your image of a plum and meld them together in your mind into one image. That's a nectarine.*

It is important to realize that a nectarine does not switch back and forth every second from being a peach to being a plum. A nectarine is a nectarine all of the time. The image of a peach is not adequate to describe a nectarine. Neither is the image of a plum. But by imagining both together at the same time, you can get a sense of what a nectarine looks like.

The problem with drawing molecules is similar to the problem above with the nectarine. No single drawing adequately describes the nature of the electron density spread out over the molecule. To solve this problem, we draw several drawings and then meld them together in our mind into one image. Just like the nectarine.

Let's see an example:

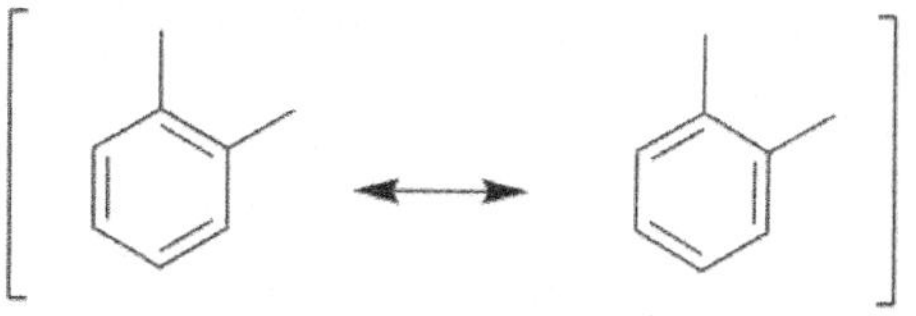

The compound above has two resonance structures. Notice that we separate resonance structures with a straight, two-headed arrow, and we place brackets around the structures. The arrow and brackets indicate that they are resonance structures *of one molecule*. The molecule is not flipping back and forth between the different resonance structures.

Now that we know why we need resonance, we can begin to understand why resonance structures are so important. Ninety-five percent of the reactions that you will see in this course occur because one molecule has a region of low electron density and the other molecule has a region of high electron density. They attract each other in space, which causes a reaction. So, to predict how and when two molecules will react with each other, we must first predict where there is low electron density and where there is high electron density. We need to have a firm grasp of resonance to do this. In this chapter, we will see many examples of how to predict the regions of low or high electron density by applying the rules of drawing resonance structures.

2.2 CURVED ARROWS: THE TOOLS FOR DRAWING RESONANCE STRUCTURES

In the beginning of the course, you might encounter problems like this: here is a drawing; now draw the other resonance structures. But later on in the course, it will be assumed and expected that you can draw all of the resonance structures of a compound. If you cannot actually do this, you will be in big trouble later on in the course. So how do you draw all of the resonance structures of a compound? To do this, you need to learn the tools that help you: curved arrows.

Here is where it can be confusing as to what is exactly going on. These arrows do NOT represent an actual process (such as electrons moving). This is an important point, because you will learn later about curved arrows used in drawing reaction mechanisms. Those arrows look exactly the same, but they actually do refer to the flow of electron density. In contrast, curved arrows here are used only as tools to help us draw all resonance structures of a molecule. The electrons are not actually moving. It can be tricky because we will say things like: "this arrow shows the electrons coming from here and going to there." But we don't actually mean that the electrons are moving; they are *not* moving. Since each drawing treats the electrons as particles stuck in one place, we will need to "move" the electrons to get from one drawing to another. Arrows are the tools that we use to make sure that we know how to draw all resonance structures for a compound. So, let's look at the features of these important curved arrows.

Every curved arrow has a *head* and a *tail*. It is essential that the head and tail of every arrow be drawn in precisely the proper place. *The tail shows where the electrons are coming from, and the head shows where the electrons are going* (remember that the electrons aren't really going anywhere, but we treat them as if they were so we can make sure to draw all resonance structures):

Tail ⟶ Head

Therefore, there are only two things that you have to get right when drawing an arrow: the tail needs to be in the right place and the head needs to be in the right place. So we need to see rules about where you can and where you cannot draw arrows. But first we need to talk a little bit about electrons, since the arrows are describing the electrons.

Atomic orbitals can hold a maximum of two electrons. So, there are only three options for any atomic orbital:

# of electrons in atomic orbital	Comments	Outcome
0	Nothing to talk about (no electrons)	-
1	Can overlap with another atomic orbital (also housing one electron) to form a bond with another atom	*bond*
2	The atomic orbital is filled and is called a lone pair	*lone pair*

So we see that electrons can be found in two places: in bonds or in lone pairs. Therefore, electrons can only come from either a bond or a lone pair. Similarly, electrons can only go to form either a bond or a lone pair.

Let's focus on tails of arrows first. Remember that the tail of an arrow indicates where the electrons are coming from. So the tail has to come from a place that has electrons: either from a bond or from a lone pair. As an example, consider the following resonance structures:

How do we get from the first structure to the second one? Notice that the electrons that make up the double bond have been "moved." This is an example of electrons coming from a bond. Let's see the arrow showing the electrons coming from the bond and going to form another bond:

Now let's see an example where electrons come from a lone pair:

Never draw an arrow that comes from a positive charge. The tail of an arrow must come from a spot that has electrons.

Heads of arrows are just as simple as tails. The head of an arrow shows where the electrons are going. So the head of an arrow must either point directly in between two atoms to form a bond, like this:

or it must point to an atom to form a lone pair, like this:

Never draw the head of an arrow going off into space, like this:

Bad arrow

Remember that the head of an arrow shows where the electrons are going. So the head of an arrow must point to a place where the electrons can go—either to form a bond or to form a lone pair.

2.3   THE TWO COMMANDMENTS

Now we know what curved arrows are, but how do we know when to use curved arrows to push electrons and where to push them? First, we need to learn where we *cannot* push electrons. There are two important rules that you should *never* violate when pushing arrows. They are the "two commandments" of drawing resonance structures:

1. Thou shall not break a single bond.
2. Thou shall not exceed an octet for second-row elements.

Let's focus on one at a time.

1. *Never break a single bond* when drawing resonance structures. By definition, resonance structures must have all the same atoms connected in the same order.

Never break a single bond

There are very few exceptions to this rule, and only a trained organic chemist can be expected to know when it is permissible to violate this rule. Some instructors might violate this rule one or two times (about half-way through the course). If this happens, you should recognize that you are seeing a very rare exception. In virtually every situation that you will encounter, you *cannot* violate this rule. Therefore, you must get into the habit of never breaking a single bond when drawing resonance structures.

There is a simple way to ensure that you never violate this rule. When drawing resonance structures, just make sure that you never draw the tail of an arrow on a single bond.

2. *Never exceed an octet for second-row elements.* Elements in the second row (C, N, O, F) have only four orbitals in their valence shell. Each of these four orbitals can be used either to form a bond or to hold a lone pair. Each bond requires the use of one orbital, and each lone pair requires the use of one orbital. So the second-row elements can never have five or six bonds; the most is four. Similarly, they can *never* have four bonds and a lone pair, because this would also require five orbitals. For the same reason, they can never have three bonds and two lone pairs. The sum of (bonds) + (lone pairs) for a second-row element can never exceed the number four. Let's see some examples of arrow pushing that violate this second commandment:

BAD ARROW   **BAD** ARROW   **BAD** ARROW

In each of these drawings, the central atom cannot form another bond because it does not have a fifth orbital that can be used. *This is impossible.* Don't ever do this.

The examples above are clear, but with bond-line drawings, it can be more difficult to see the violation because the hydrogen atoms are not drawn (and, very often, neither are the lone pairs). You have to train yourself to see the hydrogen atoms and to recognize when you are exceeding an octet:

is the same as

At first it is difficult to see that the arrow on the left structure violates the second commandment. But when we count the hydrogen atoms, we can see that the arrow above would give a carbon atom with five bonds.

From now on, we will refer to the second commandment as "the octet rule." But be careful—for purposes of drawing resonance structures, it is only a violation if we *exceed* an octet for a second-row element. However, it is OK for carbon (which is a second-row element) to have *fewer* than an octet of electrons. For example:

This carbon atom
does not have an octet.

This drawing is perfectly acceptable, even though the central carbon atom has only six electrons surrounding it. For our purposes, we will only consider the "octet rule" to be violated if we exceed an octet.

Our two commandments (never break a single bond, and never violate "the octet rule") reflect the two parts of a curved arrow (the head and the tail). A bad tail violates the first commandment, and a bad head violates the second commandment.

EXERCISE 2.1   Look at the arrow drawn on the following structure and determine whether it violates either of the two commandments for drawing resonance structures:

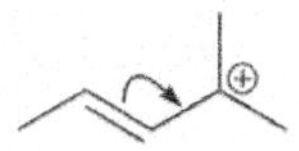

Answer   First we need to ask if the first commandment has been violated: did we break a single bond? To determine this, we look at the *tail* of the arrow. If the tail of the arrow is coming from a single bond, then that means we are breaking that single bond. If the tail is coming from a double bond, then we have not violated the first commandment. In this example, the tail is on a double bond, so we did not violate the first commandment.

Now we need to ask if the second commandment has been violated: did we violate the octet rule? To determine this, we look at the *head* of the arrow. Are we forming a fifth bond? Remember that C⁺ only has three bonds, not four. When we push the electrons as shown above, the carbon atom will now get four bonds, and the second commandment has not been violated.

The arrow above is valid, because the two commandments were not violated.

PROBLEMS   For each of the problems below, determine which arrows violate either one of the two commandments, and explain why. (Don't forget to count all hydrogen atoms and all lone pairs.)

2.2   _____

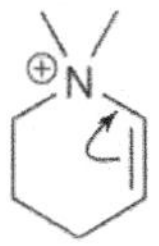

2.3   _____

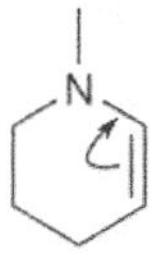

2.4   _____

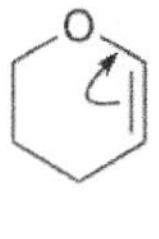

2.5   _____

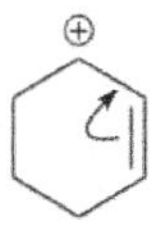

2.6   _____

2.7   _____

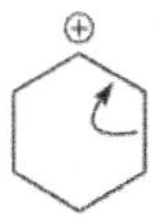

2.8

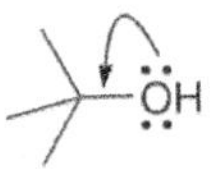

2.9

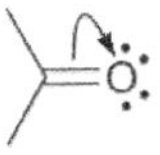

2.10

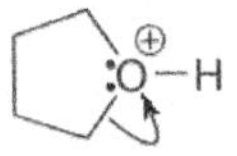

2.11

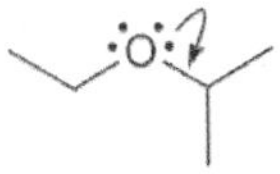

2.12

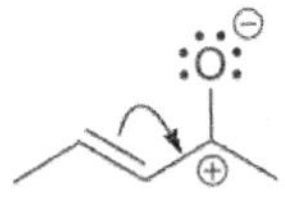

2.4 DRAWING GOOD ARROWS

Now that we know how to identify good arrows and bad arrows, we need to get some practice drawing arrows. We know that the tail of an arrow must come either from a bond or a lone pair, and that the head of an arrow must go to form a bond or a lone pair. If we are given two resonance structures and are asked to show the arrow(s) that get us from one resonance structure to the other, it makes sense that we need to look for any bonds or lone pairs that are appearing or disappearing when going from one structure to another. For example, consider the following resonance structures:

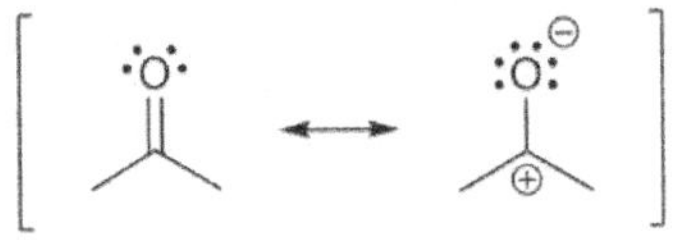

How would we figure out what curved arrow to draw to get us from the drawing on the left to the drawing on the right? We must look at the difference between the two structures and ask, "How should we push the electrons to get from the first structure to the second structure?" Begin by looking for any double bonds or lone pairs that are disappearing. That will tell us where to put the tail of our arrow. In this example, there are no lone pairs disappearing, but there is a double bond disappearing. So we know that we need to put the tail of our arrow on the double bond.

Now, we need to know where to put the head of the arrow. We look for any lone pairs or double bonds that are appearing. We see that there is a new lone pair appearing on the oxygen atom. This tells us where to put the head of the arrow:

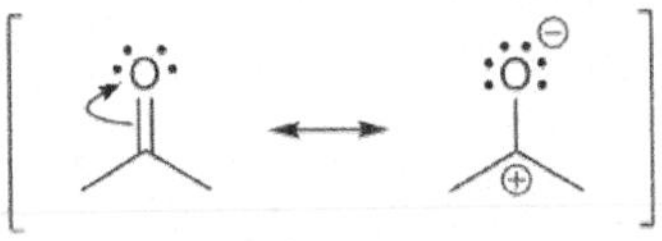

Notice that when we move a double bond up onto an atom to form a lone pair, it creates two formal charges: a positive charge on the carbon atom that lost its bond, and a negative charge on the oxygen atom that got a lone pair. This is a very important issue. Formal charges were introduced in the previous chapter, and now they will become instrumental in drawing resonance structures. For the moment, let's just focus on pushing arrows, and in the next section of this chapter, we will come back to focus on these formal charges.

It is pretty straightforward to see how to push only one curved arrow that gets us from one resonance structure to another. But what about when we need to push more than one arrow to get from one resonance structure to another? Let's do an example.

EXERCISE 2.13   Draw curved arrows that convert the first resonance structure into the second resonance structure below:

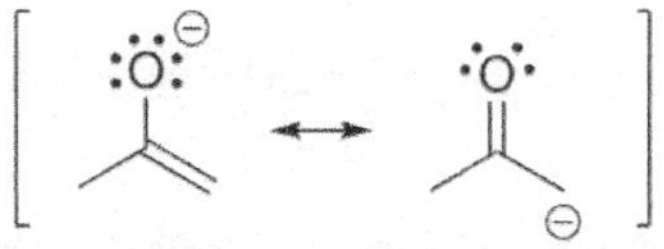

Answer   Let's analyze the difference between these two drawings. We begin by looking for any double bonds or lone pairs that are disappearing. We see that oxygen is losing a lone pair, and the C=C bond on the bottom is also disappearing. This should automatically tell us that we need two curved arrows. To lose a lone pair and a double bond, we will need two tails.

Now let's look for any double bonds or lone pairs that are appearing. We see that a C=O bond is appearing and a C with a negative charge is appearing (remember that a C⁻ means a C with a lone pair). This tells us that we need two heads, which confirms that we need two curved arrows.

So we know we need two curved arrows. Let's start at the top. We lose a lone pair from the oxygen atom and form a C=O bond. Let's draw that curved arrow:

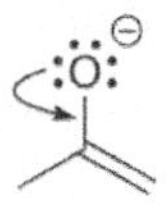

CAN'T STOP HERE

Notice that if we stopped here, we would be violating the second commandment. The central carbon atom is getting five bonds. To avoid this problem, we must also draw the second curved arrow. The C=C bond disappears (which solves our octet problem) and becomes a lone pair on carbon.

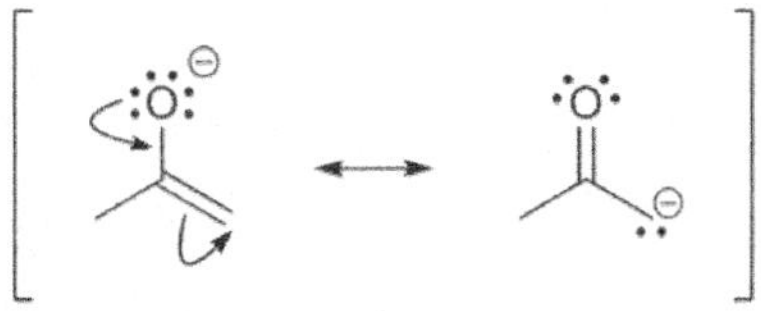

Arrow pushing is much like riding a bike. If you have never done it before, watching someone else will not make you an expert. You have to learn how to balance yourself. Watching someone else is a good start, but you have to get on the bike if you want to learn. You will probably fall a few times, but that's part of the learning process. The same is true with arrow pushing. The only way to learn is with practice.

Now it's time for you to get on the arrow-pushing bike. You would never be stupid enough to try riding a bike for the first time next to a steep cliff. Similarly, do not let your first arrow-pushing experience be during your exam. Practice right now!

PROBLEMS   Draw the curved arrows that convert one drawing to the next. In many cases you will need to draw more than one curved arrow.

2.14

2.15

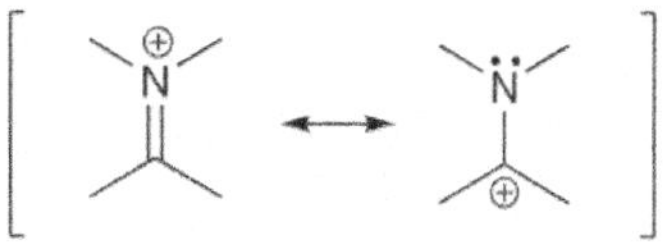

2.16

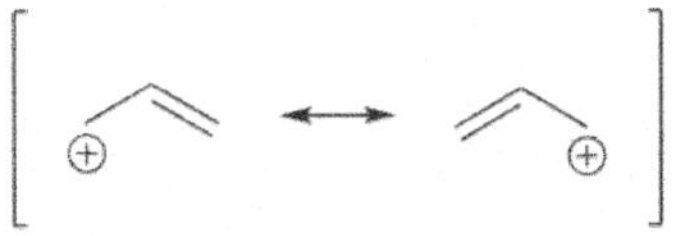

2.17

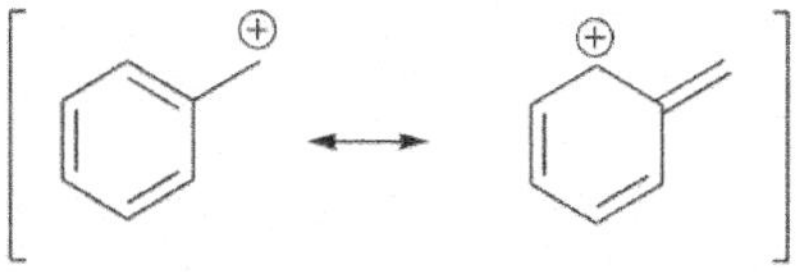

2.18 **2.19**

2.5 FORMAL CHARGES IN RESONANCE STRUCTURES

Now we know how to draw good arrows (and how to avoid drawing bad arrows). In the last section, we were given the resonance structures and just had to draw the arrows. Now we need to take this to the next level. We need to get practice drawing the resonance structures when they are not given. To ease into it, we will still show the arrows, and we will focus on drawing the resonance structures with proper formal charges. Consider the following example:

In this example, we can see that one of the lone pairs on oxygen is coming down to form a bond, and the C=C double bond is being pushed to form a lone pair on a carbon atom. When both arrows are pushed at the same time, we are not violating either of the two commandments. So, let's focus on how to draw the resonance structure. Since we know what arrows mean, it is easy to follow the arrows. We just get rid of one lone pair on oxygen, place a double bond between carbon and oxygen, get rid of the carbon–carbon double bond, and place a lone pair on carbon:

The arrows are really a language, and they tell us what to do. But here comes the tricky part: we cannot forget to put formal charges on the new drawing. If we apply the rules of assigning formal charges, we see that oxygen gets a positive charge and carbon gets a negative charge. As long as we draw these charges, it is not necessary to draw in the lone pairs:

It is absolutely critical to draw these formal charges. Structures drawn without them are *wrong*. In fact, if you forget to draw the formal charges, then you are missing the whole point of resonance. Let's see why. Look at the resonance structure we just drew. Notice that there is a negative charge on a carbon atom. This tells us that this carbon atom is a site of high electron density. We would not know this by looking only at the first drawing of the molecule:

This is why we need resonance—it shows us where there are regions of high and low electron density in the resonance hybrid. If we draw resonance structures without formal charges, then what is the point in drawing the resonance structures at all?

Now that we see that proper formal charges are essential, we should make sure that we know how to draw them when drawing resonance structures. If you are a little bit shaky when it comes to formal charges, you can go back and review formal charges in the previous chapter. More importantly, you should be able to draw formal charges without having to count each time. We saw the common situations for oxygen, nitrogen, and carbon. It is important to remember those (go back and review those if you need to).

Another way to assign formal charges is to read the arrows properly. Let's look at our example again:

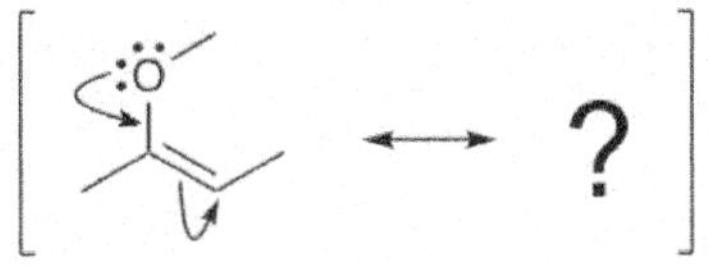

Notice what the arrows are telling us: oxygen is giving up a lone pair (two electrons entirely on oxygen) to form a bond (two electrons being shared: one for oxygen and one for carbon). So oxygen is losing an electron. This tells us that it must get a positive charge in the resonance structure. A similar analysis for the carbon atom on the bottom right shows that it will get a negative charge. Remember that the electrons are not really moving anywhere. Arrows are just tools that help us draw resonance structures. To use these tools properly, we imagine that the electrons are moving, but they are not.

Now let's practice.

EXERCISE 2.20   Draw the resonance structure that you get when you push the arrows shown below. Be sure to include formal charges.

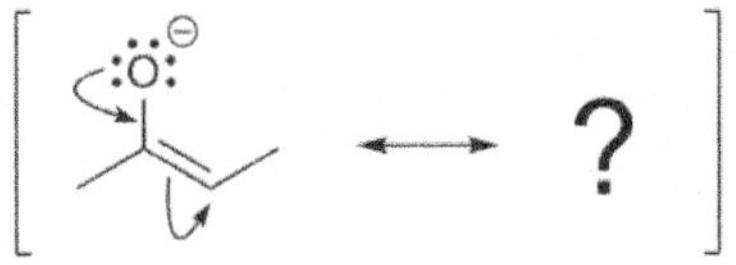

Answer   We read the curved arrows carefully. One of the lone pairs on oxygen is coming down to form a bond, and the C=C double bond is being pushed to form a lone pair on a carbon atom. This is very similar to the example we just saw. We just get rid of one lone pair on oxygen, place a double bond between carbon and oxygen, get rid of the carbon–carbon double bond, and place a lone pair on carbon. Finally, we must draw any formal charges:

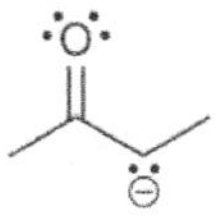

There is one subtle point that must be mentioned. We said that you do not need to draw lone pairs—you only need to draw formal charges. There will be times when you will see arrows being pushed on structures that do not have the lone pairs drawn. When this happens, you might see an arrow coming from a negative charge:

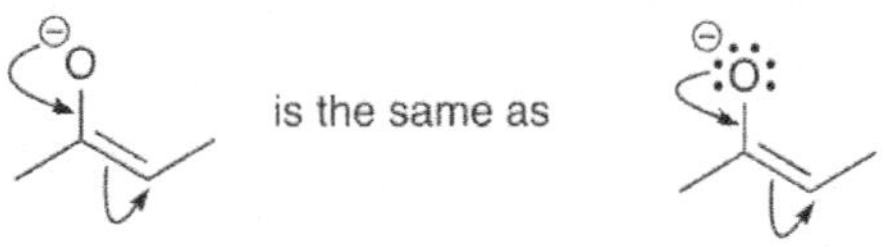

is the same as

The drawing on the right is more accurate and complete, but the drawing on the left is often used by organic chemists. So, if you are reviewing your lecture notes and you find that your instructor drew the tail of a curved arrow on a negative charge, just don't forget that the electrons are really coming from a lone pair (as seen in the drawing on the right).

One way to double check your drawing when you are done is to count the total charge on the resonance structure that you draw. This total charge should be the same as the structure you started with. So if the first structure has a negative charge, then the resonance structure you draw should also have a negative charge. If it doesn't, then you know you did something wrong (this is known as *conservation of charge*). You cannot change the total charge when drawing resonance structures.

PROBLEMS   For each of the structures below, draw the resonance structure that you get when you push the arrows shown. Be sure to include formal charges. (*Hint:* In some cases the lone pairs are drawn and in other cases they are not drawn. Be sure to take them into account even if they are not drawn—you need to train yourself to see lone pairs when they are not drawn.)

2.21

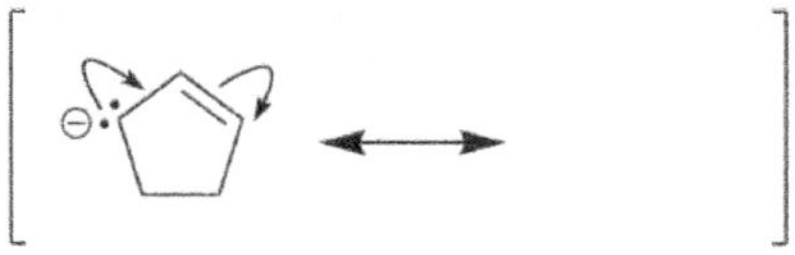

2.22

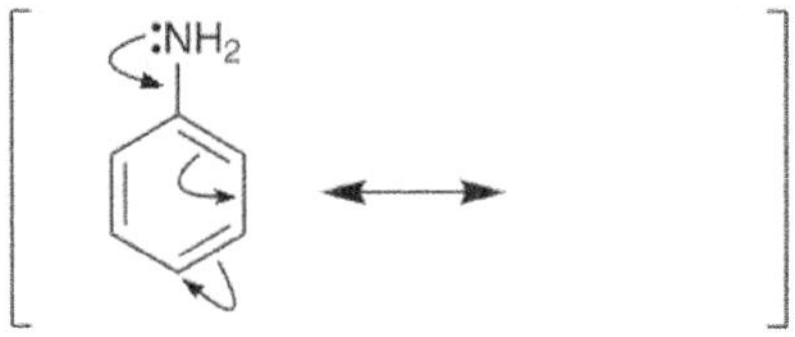

2.23

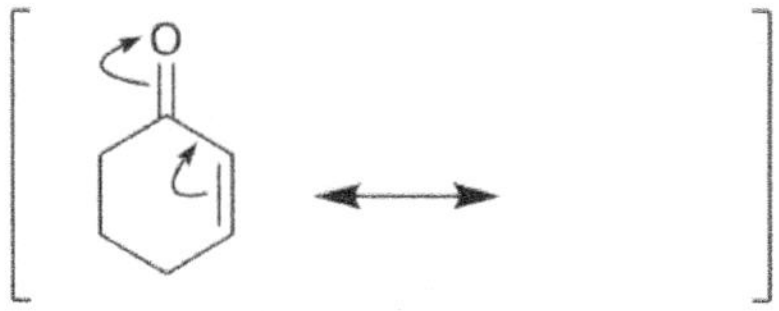

2.24

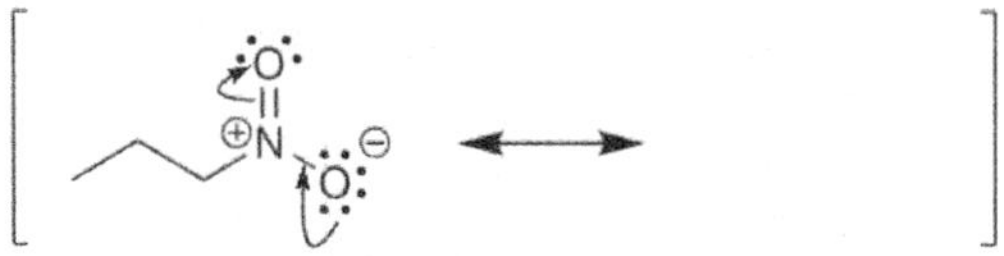

2.25

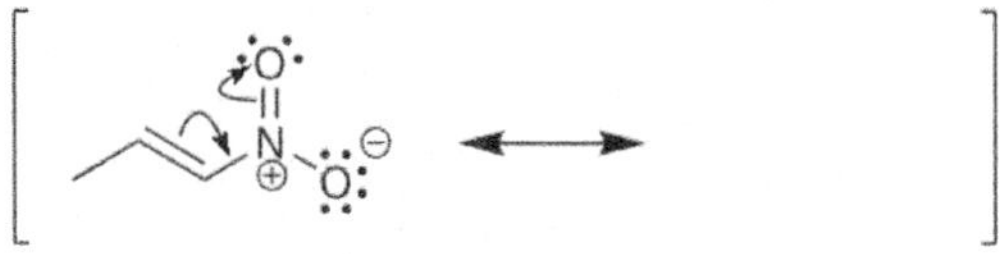

2.26

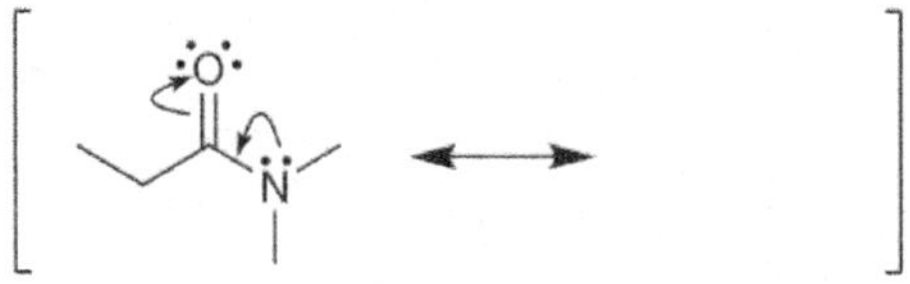

2.27

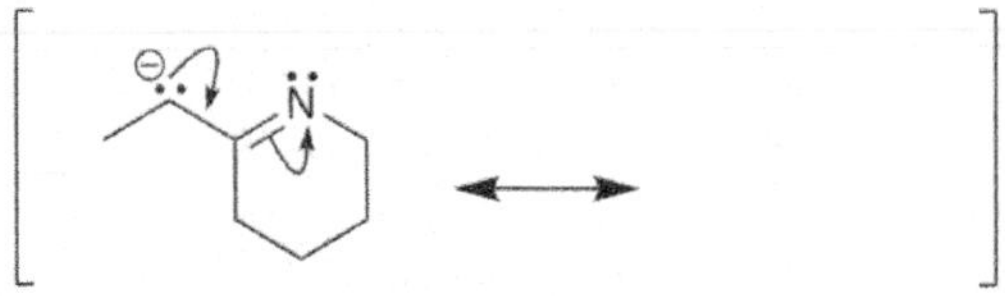

2.28

2.6   DRAWING RESONANCE STRUCTURES—STEP BY STEP

Now we have all the tools we need. We know why we need resonance structures and what they represent. We know what curved arrows represent. We know how to recognize bad arrows that violate the two commandments. We know how to draw arrows that get you from one structure to another, and we know how to draw formal charges. We are now ready to begin using curved arrows to draw resonance structures.

First we need to locate the part of the molecule where resonance is an issue. Remember that we can push electrons only from lone pairs or bonds. We don't need to worry about all bonds, because we can't push an arrow from a single bond (that would violate the first commandment). So we only care about double or triple bonds. Double and triple bonds are called *pi bonds*. So we need to look for lone pairs and pi bonds.

Once we have located the regions where resonance is an issue, now we need to ask if there is any way to push the electrons without violating the two commandments. Let's be methodical, and break this up into three questions:

1. Can we convert any *lone pairs into pi bonds* without violating the two commandments?
2. Can we convert any *pi bonds into lone pairs* without violating the two commandments?
3. Can we convert any *pi bonds into pi bonds* without violating the two commandments?

We do not need to worry about the fourth possibility (converting a lone pair into a lone pair) because electrons cannot jump from one atom to another. Only the three possibilities above are acceptable.

Let's go through these three steps, one at a time, starting with step 1, converting lone pairs into pi bonds. Consider the following example:

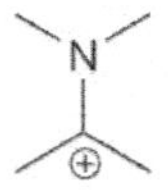

We ask if there are any lone pairs that we can move to form a pi bond. So we draw an arrow that brings the lone pair down to form a pi bond:

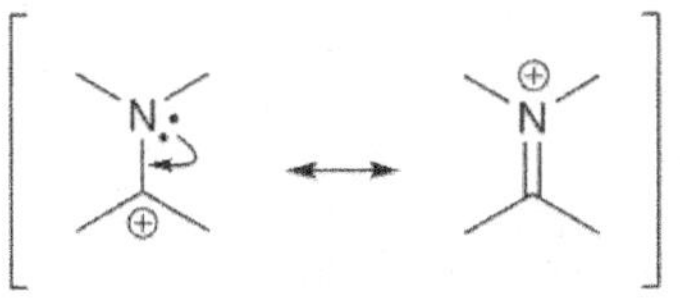

This does not violate either of the two commandments. We did not break any single bonds and we did not violate the octet rule. So this is a valid structure. Notice that we cannot move the lone pair in another direction, because then we would be violating the octet rule:

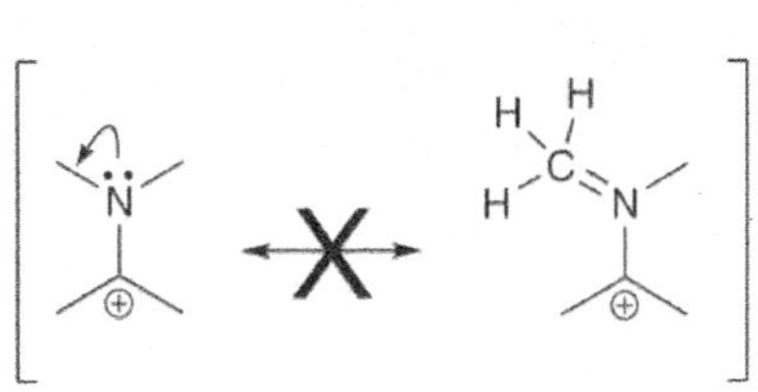

Let's try again with the following example:

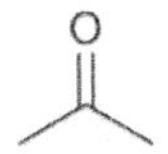

We ask if we can move one of the lone pairs down to form a pi bond, so we try to draw it:

This violates the octet rule—the carbon atom would end up with five bonds. So we cannot push the arrows that way. There is no way to turn a lone pair into a pi bond in this example.

Now let's move on to step 2, converting pi bonds into lone pairs. We try to move the double bond to form a lone pair and we see that we can move the bond in either direction:

or

Neither of these structures violates the two commandments, so both structures above are valid resonance structures. (However, the bottom structure, although valid, is not a significant resonance structure. In the next section, we will see how to determine which resonance structures are significant and which are not.)

For step 3, converting pi bonds into pi bonds, let's consider the following examples:

If we try to push the pi bonds to form other pi bonds, we find

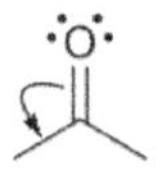

NO. This violates the octet rule

YES. Does not violate the octet rule

The arrow on the top structure violates the octet rule (giving carbon five bonds), and the arrow on the bottom structure does not violate the octet rule. The arrow on the bottom structure will therefore provide a valid resonance structure:

Now that we have learned all three steps, we need to consider that these steps can be combined. Sometimes we cannot do a step without violating the octet rule, but by doing two steps at the same time, we can avoid violating the octet rule. For example, if we try to turn a lone pair into a bond in the following structure, we see that this would violate the octet rule:

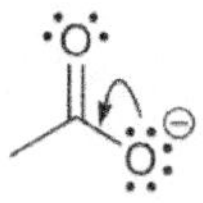

If, at the same time, we also do step 2 (push a pi bond to become a lone pair), then it works:

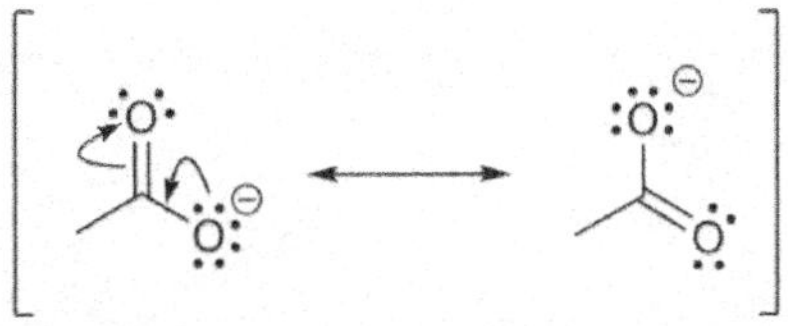

In other words, you should not always jump to the conclusion that pushing an arrow will violate the octet rule. You should first look to see if you can push another arrow that will eliminate the problem.

As another example, consider the following structure. We cannot move the C=C bond to become another bond unless we also move the C=O bond to become a lone pair:

NO   YES

In this way, we truly are "pushing" the electrons around.

Now we are ready to get some practice drawing resonance structures.

EXERCISE 2.29   Draw all resonance structures for the following compound:

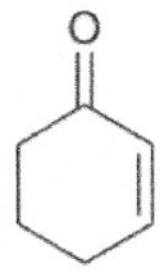

Answer   Let's start by finding all of the lone pairs and redrawing the molecule. Oxygen has two bonds here, so it must have two lone pairs (so that it will be using all four orbitals):

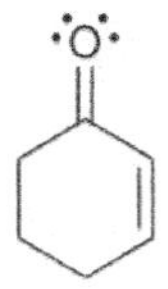

Now let's do step 1: can we convert any lone pairs into pi bonds? If we try to bring down the lone pairs, we will violate the octet rule by forming a carbon atom with five bonds:

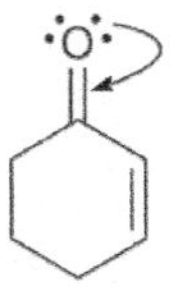

Violates second commandment

The only way to avoid forming a fifth bond for carbon would be to push an arrow that takes electrons away from that carbon. If we try to do this, we will break a single bond and we will be violating the first commandment:

or

Violates first commandment

We cannot move a lone pair to form a pi bond, so we move on to step 2: can we convert any pi bonds into lone pairs? Yes:

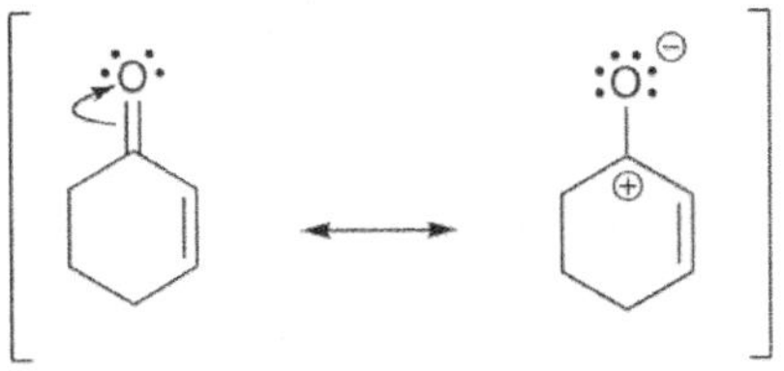

Now we move to step 3: can we convert pi bonds into pi bonds? There is only one move that will not violate the two commandments:

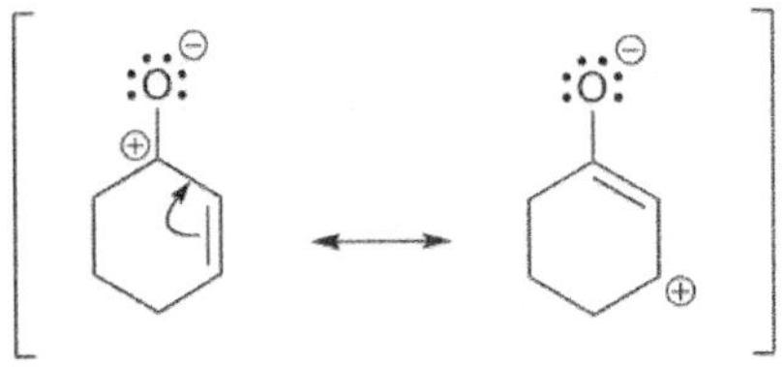

So the resonance structures are

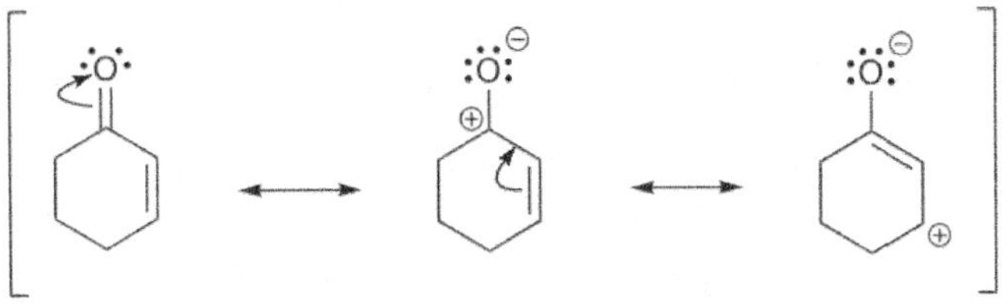

PROBLEM 2.30   For the following compound, go through all three steps (making sure not to violate the two commandments) and draw the resonance structures.

While working through this problem, you probably found that it took a very long time to think through every possibility, to count lone pairs, to worry about violating the octet rule for each atom, to assign formal charges, and so on. Fortunately, there is a way to avoid all of this tedious work. You can learn how to become very quick and efficient at drawing resonance structures if you learn certain patterns and train yourself to recognize those patterns. We will now develop this skill.

2.7 DRAWING RESONANCE STRUCTURES— BY RECOGNIZING PATTERNS

There are five patterns that you should learn to recognize to become proficient at drawing resonance structures. First we list them, and then we will go through each pattern in detail, with examples and exercises. Here they are:

1. A lone pair next to a pi bond.
2. A lone pair next to C⁺.
3. A pi bond next to C⁺.
4. A pi bond between two atoms, where one of those atoms is electronegative.
5. Pi bonds going all the way around a ring.

A Lone Pair Next to a Pi Bond

Consider the following two examples:

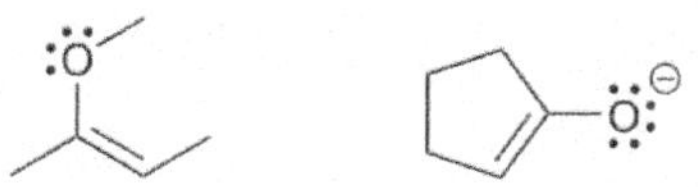

Both examples exhibit a lone pair "next to" the pi bond. "Next to" means that the lone pair is separated from the double bond by exactly one single bond—no more and no less. You can see this in all of the examples below:

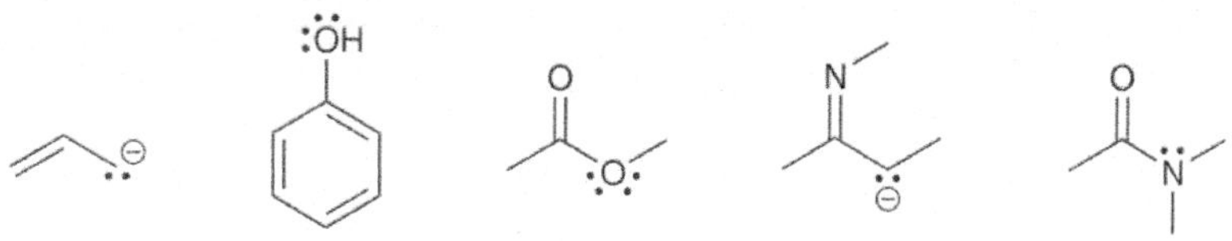

In each of these cases, you can bring down the lone pair to form a pi bond, and kick up the pi bond to form a lone pair:

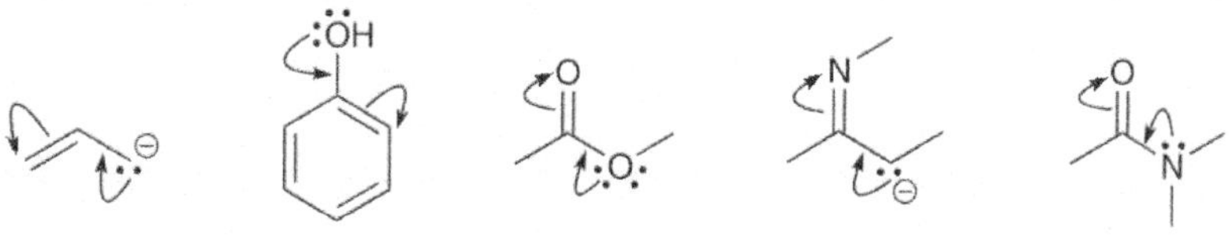

Notice what happens with the formal charges. When the atom with the lone pair has a negative charge, then it transfers its negative charge to the atom that will get a lone pair in the end:

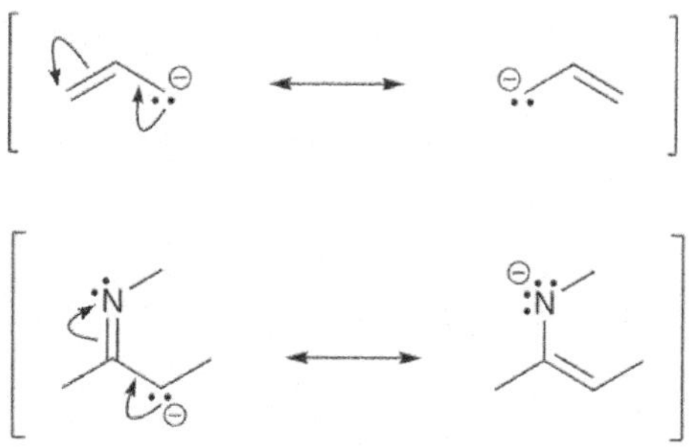

When the atom with the lone pair does not have a negative charge to begin with, then it will end up with a positive charge in the end, while a negative charge will go on the atom getting the lone pair in the end (remember conservation of charge):

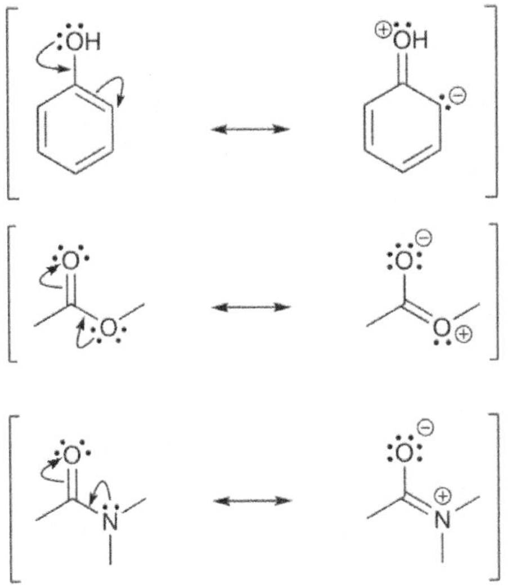

Once you learn to recognize this pattern (a lone pair next to a pi bond), you will be able to save time in calculating formal charges and determining if the octet rule is being violated. You will be able to push the arrows and draw the new resonance structure without thinking about it.

EXERCISE 2.31   Draw a resonance structure of the compound below:

Answer   We notice that this is a lone pair next to a pi bond. Therefore, we draw two curved arrows: one from the lone pair to form a pi bond, and one from the pi bond to form a lone pair.

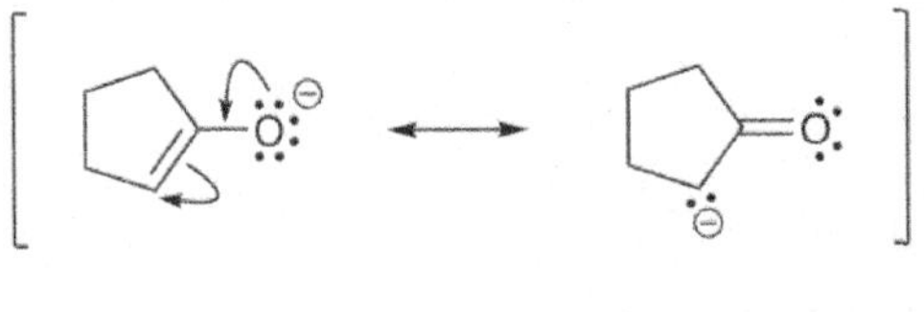

Look carefully at the formal charges. The negative charge used to be on oxygen, but now it moved to carbon.

PROBLEMS   For each of the following structures, locate the pattern we just learned and draw the resonance structure.

2.32

2.33

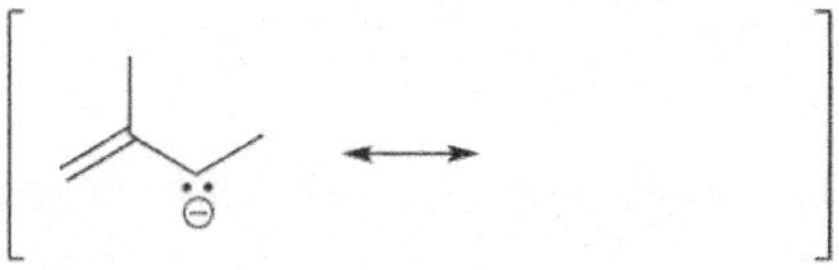

2.34

2.35

2.36

2.37

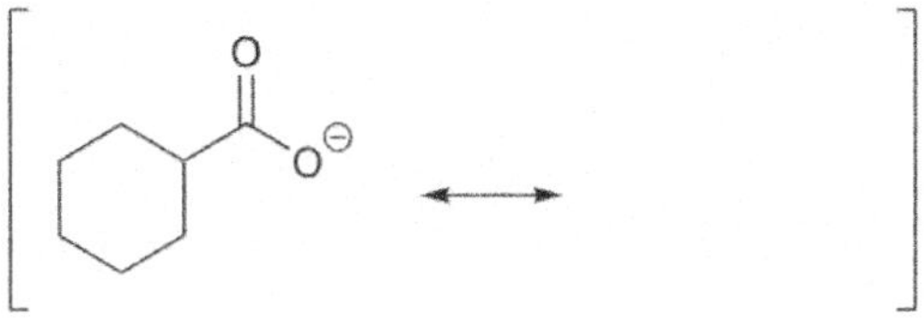

2.38

2.39

Notice that the lone pair needs to be directly next to the pi bond. If we move the lone pair one atom away, this does not work anymore:

GOOD   BAD

A Lone Pair Next to C⁺

Consider the following two examples:

Both examples exhibit a lone pair next to C⁺. In each case, we can bring down the lone pair to form a pi bond:

Notice what happens with the formal charges. When the atom with the lone pair has a negative charge, then the charges end up canceling each other:

When the atom with the lone pair does not have a negative charge to begin with, then it will end up with the positive charge in the end (remember conservation of charge):

PROBLEMS   For each of the following structures, locate the pattern we just learned and draw the resonance structure.

2.40

2.41

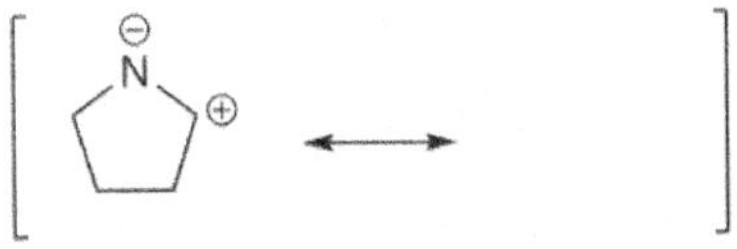

2.42

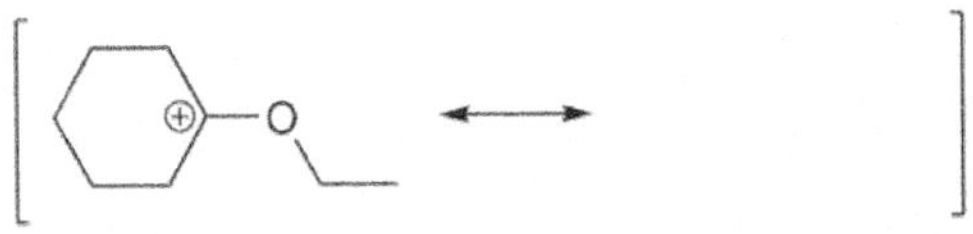

2.43

Notice that in this problem, a negative and positive charge cancel each other to become a double bond. There is one situation when we cannot combine charges to give a double bond: the nitro group. The structure of the nitro group looks like this:

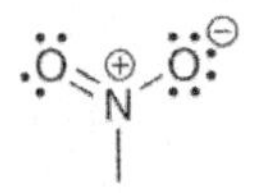

We cannot draw a resonance structure where there are no charges:

Violates
octet rule

This might seem better at first, because we get rid of the charges, but it cannot be drawn like this: the nitrogen atom would have five bonds, which would violate the octet rule.

A Pi Bond Next to C+

These cases are very easy to see:

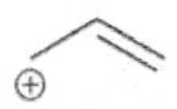

We need only one arrow going from the pi bond to form a new pi bond:

Notice what happens to the formal charge in the process. It gets moved to the other end:

It is possible to have many double bonds in conjugation (this means that we have many double bonds that are each separated by only one single bond) next to a positive charge:

When this happens, we push each of the double bonds over, one at a time:

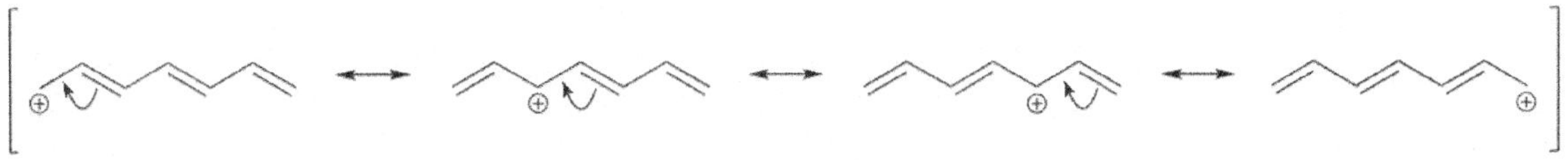

It is not necessary to waste time recalculating formal charges for each resonance structure, because the arrows indicate what is happening. Think of a positive charge as a hole (a place that is missing an electron). When we push electrons to plug up the hole, a new hole is created nearby. In this way, the hole is simply moved from one location to another. Notice that the tails of the curved arrows are placed on the pi bonds, not on the positive charge. *Never place the tail of a curved arrow on a positive charge.*

PROBLEMS   For each of the following structures, locate the pattern we just learned and draw the resonance structures.

2.44

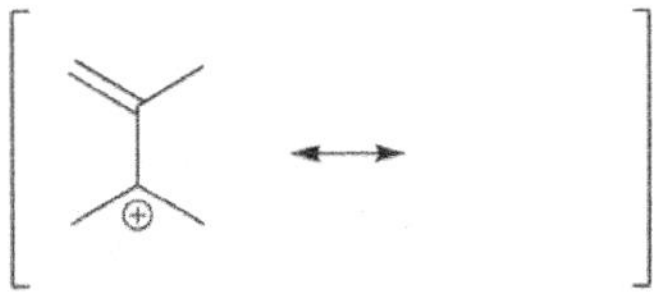

2.45

2.46

A Pi Bond Between Two Atoms, Where One of Those Atoms Is Electronegative (N, O, etc.)

Let's see an example:

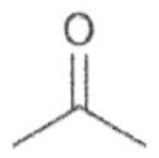

In cases like this, we move the pi bond up onto the electronegative atom to become a lone pair:

Notice what happens with the formal charges. A double bond is being separated into a positive and negative charge (this is the opposite of what we saw in the second pattern we looked at, where the charges came together to form a double bond).

PROBLEMS   For each of the compounds below, locate the pattern we just learned and draw the resonance structure:

2.47

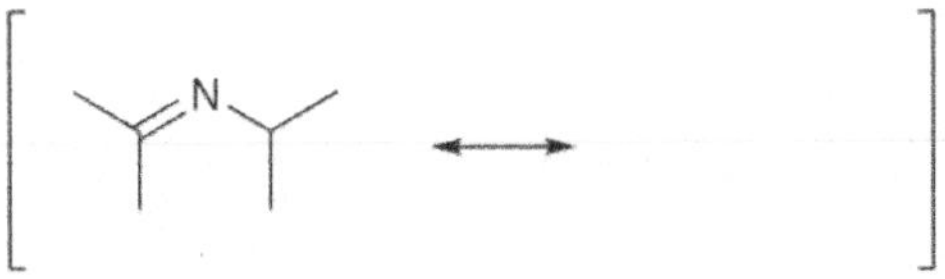

2.48

2.49

Pi Bonds Going All the Way Around a Ring

Whenever we have alternating double and single bonds, we refer to the alternating bond system as *conjugated*:

Conjugated double bonds

When we have a conjugated system that wraps around in a circle, then we can always move the electrons around in a circle:

It does not matter whether we push our arrows clockwise or counterclockwise (either way gives us the same result, and remember that the electrons are not really moving anyway).

In summary, we have seen the following five arrow-pushing patterns:

1. A lone pair next to a pi bond.
2. A lone pair next to C^+.
3. A pi bond next to C^+.
4. A pi bond between two atoms, where one of those atoms is electronegative.
5. Pi bonds going all the way around a ring.

Let's get some practice with all five patterns.

PROBLEMS   For each of the following structures, draw the additional resonance structure(s).

2.50

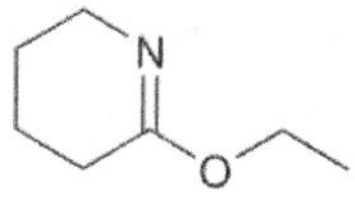

2.51

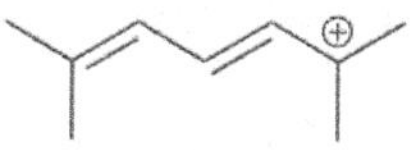

2.52

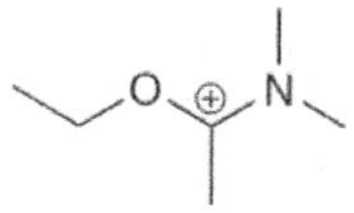

2.53

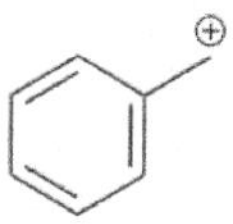

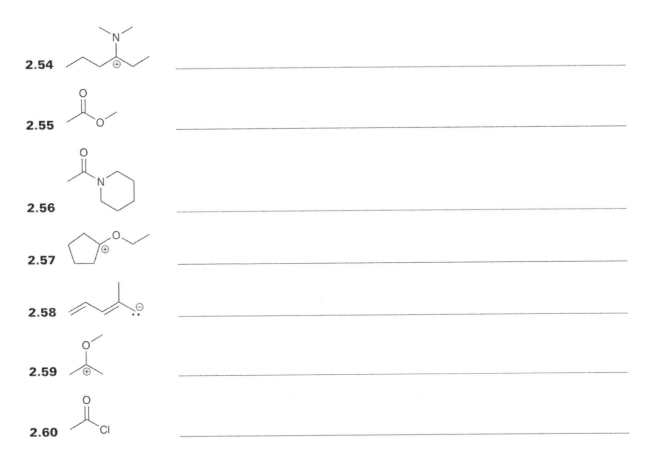

2.54

2.55

2.56

2.57

2.58

2.59

2.60

2.8 ASSESSING THE RELATIVE IMPORTANCE OF RESONANCE STRUCTURES

Not all resonance structures are equally significant. A compound might have *many* valid resonance structures, but some of those resonance structures may be more significant than others. To understand what we mean when we say "significant," let's revisit the analogy we used in the beginning of the chapter.

Recall that we used the analogy of a nectarine (being a hybrid between a peach and plum) to explain the concept of resonance. Now, imagine that we create a new type of fruit that is a hybrid between *three* fruits: a peach, a plum, and a kiwi. Suppose that the hybrid fruit that we produce has the following character: 65% peach character, 30% plum character, and 5% kiwi character. Since the kiwi contributes such a small amount to the hybrid, the hybrid fruit is still expected to look like a nectarine, but this contribution may still be significant since it might be adding something unique to the complex flavor (and even appearance) of the fruit.

A similar concept exists when comparing resonance structures. Although some resonance structures may be only minor contributors to the resonance hybrid, they are still significant and can often help explain or even predict the reactivity of a given compound. In order to understand the true nature of the compound, we must be able to compare the resonance structures and determine which structures are major contributors, which are minor contributors, and which are insignificant.

There are four rules to follow when comparing resonance structures. At this point, you are probably thinking that it is hard enough to keep track of everything we have seen so far—there are two commandments for arrow-pushing, there are three steps for determining valid resonance structures, there are five patterns, and now there are four rules for determining which resonance structures are significant. The good news is that this is the end of the line. There are no more rules or steps. We are almost done with resonance structures. More good news—drawing resonance structures really is very much like riding a bike. When you first learn to ride a bike, you need to concentrate on every movement to avoid from falling. And you have to remember a lot of rules, such as which way to lean your body and which way to turn the handlebars when you feel you are falling to the left. But eventually, you get the hang of it, and then you can ride the bike with no hands. The same is true here. It will take a lot of practice. But before you know it, you will be the resonance guru, and that is where you need to be to do well in this class.

The following rules, listed in order of importance, can be used to evaluate the relative significance of resonance contributors.

Rule 1 *The most important resonance contributors have the greatest number of filled octets.* Consider the following example:

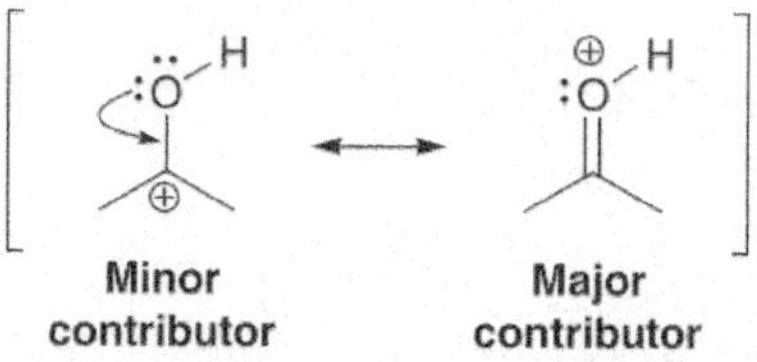

Minor	**Major**
contributor	**contributor**

The first resonance structure has a carbon atom that lacks an octet (C^+), while all of the atoms in the second resonance structure have filled octets. Therefore, the second resonance structure is the major contributor. You may also notice that the second resonance structure has more covalent bonds than the first resonance structure. So another way to state this rule is as follows: *The most important resonance contributors have the greatest number of covalent bonds.* This is because a resonance structure with more covalent bonds will have a greater number of filled octets.

As seen in the first resonance structure above, it is common to encounter a carbon atom with a positive charge, even though it lacks an octet. In contrast, oxygen is much more electronegative than carbon, so you should never draw a resonance structure in which an oxygen atom lacks an octet. This can be illustrated with the following example:

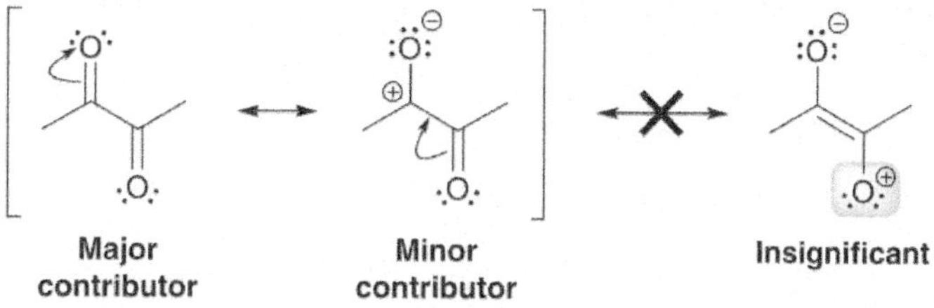

Major	**Minor**	**Insignificant**
contributor	**contributor**	

In the first resonance structure, all of the atoms have filled octets, so this resonance structure is the major contributor. The second resonance structure is a minor contributor because the carbon atom lacks an octet. The third resonance structure shown is *insignificant* because the positively charged oxygen atom lacks an octet. *Avoid drawing insignificant resonance structures.*

Rule 2 *The structure with fewer formal charges is more important.* Consider the following example:

Most significant	**Major**	**Minor**
contributor	**contributor**	**contributor**

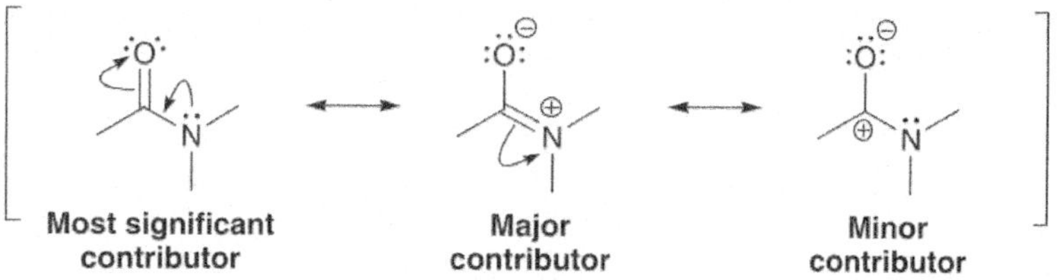

In this example, the first resonance structure is the most significant resonance structure (it is the greatest contributor to the resonance hybrid) because it has filled octets and no formal charges. The second resonance structure is still a major contributor since it has filled octets, but it is less important than the first because it has formal charges. The third resonance structure is a minor contributor because it has a carbon atom that lacks an octet.

In cases where there is an overall net charge, as seen in the example below, the creation of new charges is not favorable. For such charged compounds, the goal in drawing resonance structures is to *delocalize the charge*—relocate it to as many different positions as possible.

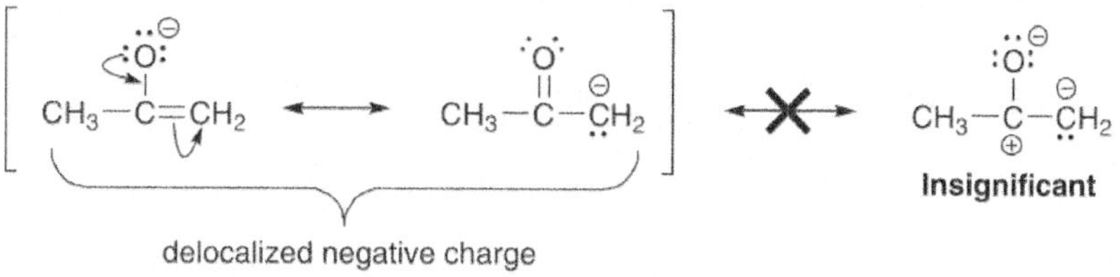

delocalized negative charge

In this case, the third resonance structure above is insignificant (and should not be drawn).

Rule 3 Other things being equal, *a structure with a negative charge on the more electronegative element will be more important*. To illustrate this, let's revisit the previous example, in which there are two significant resonance structures. The first resonance structure has a negative charge on oxygen, while the second resonance structure has a negative charge on carbon. Since oxygen is more electronegative than carbon, the first resonance structure is the major contributor:

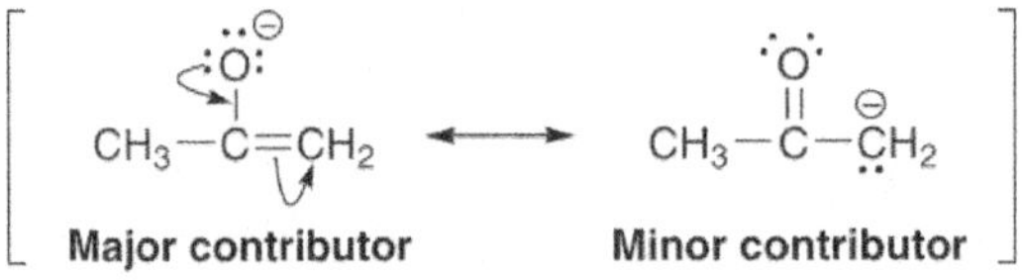

Similarly, a positive charge will be more stable on the less electronegative element. In the following example, both resonance structures have filled octets, so we consider the location of the positive charge. Nitrogen is less electronegative than oxygen, so the resonance structure with N$^+$ is the major contributor:

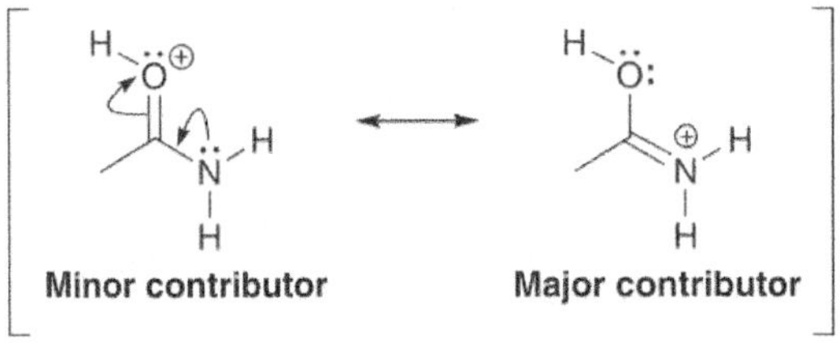

Rule 4 Resonance structures that have equally good Lewis structures are described as *equivalent* and contribute equally to the resonance hybrid. As an example, consider the carbonate ion (CO$_3^{2-}$), shown here:

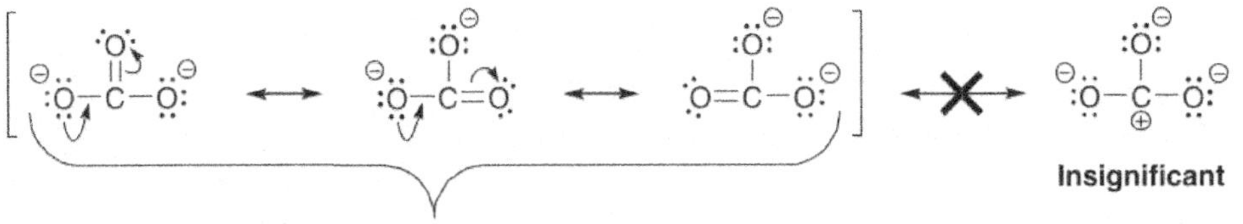

This ion has a net charge, so recall from Rule 2 that the goal is to delocalize the charges as much as possible and to avoid creating new charges. In the actual structure of the carbonate ion (the resonance hybrid), the two negative charges are shared equally among all three oxygen atoms.

EXERCISE 2.61 Draw all significant resonance structures for the following compound.

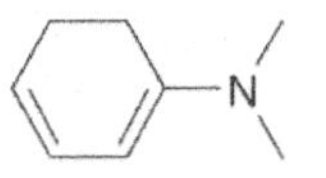

Answer This structure has a lone pair next to a pi bond, so we draw the following two curved arrows, giving a resonance structure with charge separation:

As we inspect this resonance structure, we find the same pattern again—a lone pair next to a pi bond. So, once again, we draw the following two curved arrows, giving another resonance structure:

In total, there are three significant resonance structures:

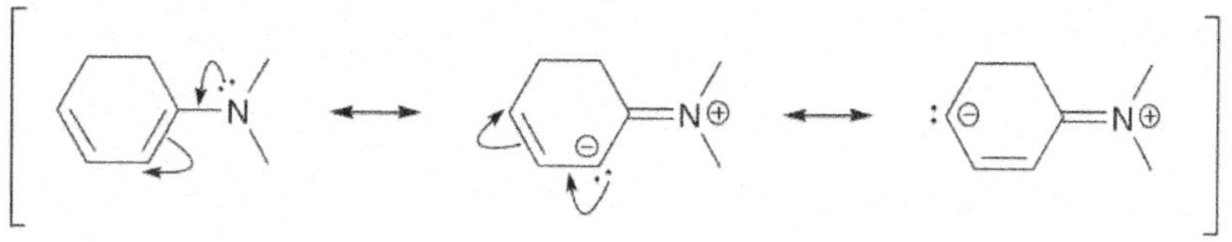

In each of the three resonance structures, all of the atoms have filled octets. The first is the most significant because it is has no formal charges, but the other two resonance structures do certainly contribute some character to the resonance hybrid.

Be careful—DON'T draw the third structure shown below, which has both C^+ and C^-, and is therefore insignificant:

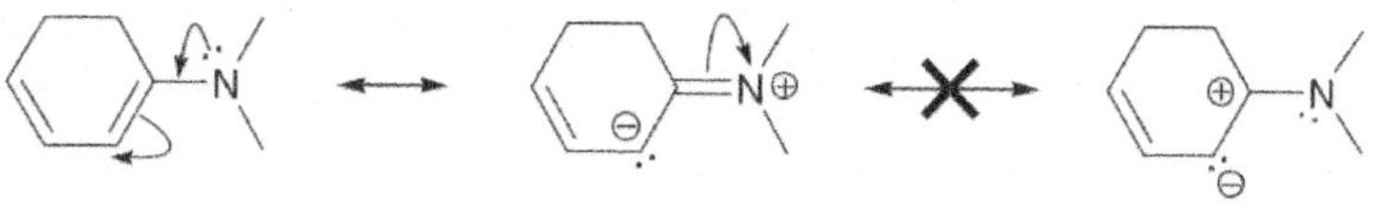

Insignificant

This resonance structure suffers from TWO major deficiencies: 1) it does not have filled octets, while the other resonance structures shown above all have filled octets, and 2) it has a negative charge on a carbon atom (which is not an electronegative atom). Either of these deficiencies alone would render the resonance structure a minor contributor. But with both deficiencies together (C^+ and C^-), this resonance structure is insignificant. The same is true for any resonance structure that has both C^+ and C^-. Such a resonance structure will generally be insignificant.

PROBLEMS   For each of the following, draw all of the *significant* resonance structures, and identify the major contributor(s) in each case.

2.62

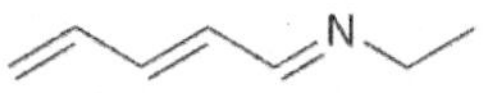

2.63

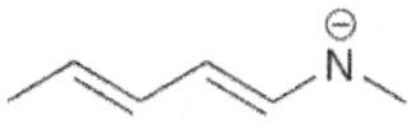

2.64

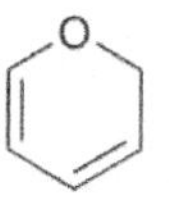

2.65

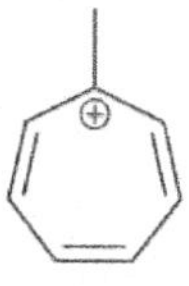

2.66

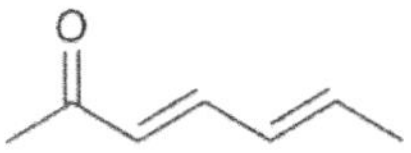

2.67

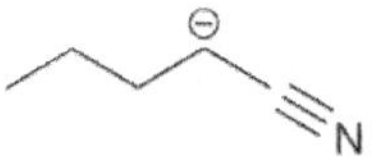

2.68

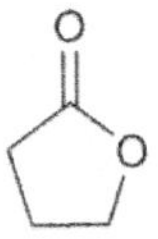

2.69

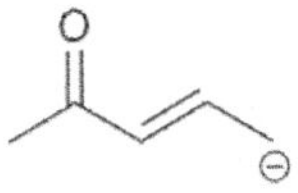

2.70

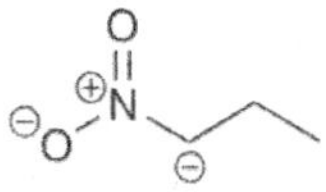

2.71

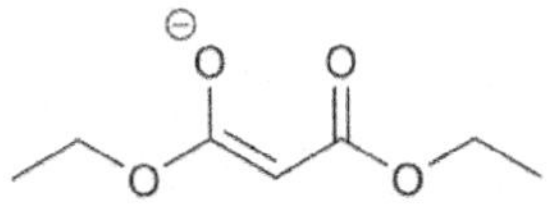

2.72

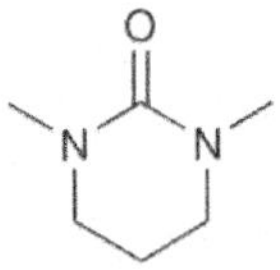

2.73

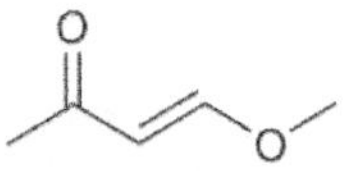

2.74

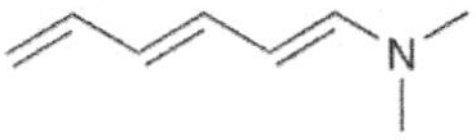

ACID–BASE REACTIONS

The first several chapters of any organic chemistry textbook focus on the structure of molecules: how atoms connect to form bonds, how we draw those connections, the inadequacies of our drawing methods, how we name molecules, what molecules look like in 3D, how molecules twist and bend in space, and so on. Only after gaining a clear understanding of structure do we move on to reactions. But there seems to be one exception: acid–base chemistry.

Acid–base chemistry is typically covered in one of the first few chapters of an organic chemistry textbook, yet it might seem to belong better in the later chapters on reactions. There is an important reason why acid–base chemistry is taught so early on in your course. By understanding this reason, you will have a better perspective of why acid–base chemistry is so incredibly important.

To appreciate the reason for teaching acid–base chemistry early in the course, we need to first have a very simple understanding of what acid–base chemistry is all about. Consider the following general reaction, in which an acid (HA) loses a proton to give the corresponding conjugate base (A^-).

$$H-A \quad \underset{\longleftarrow}{\overset{-H^+}{\rightleftharpoons}} \quad A^{\ominus}$$

In this type of reaction, we see an acid (HA) on the left side of the equilibrium, and the conjugate base (A^-) on the right side. HA is an acid by virtue of the fact that it has a proton (H^+) to donate. A^- is a base by virtue of the fact that it wants to accept its proton back (acids donate protons, and bases accept protons). Since A^- is the base that is obtained when HA is deprotonated, we call A^- the *conjugate base* of HA.

So the question is: how much is HA willing to give up its proton? If HA is very willing to give up the proton, then HA is a strong acid. However, if HA is not willing to give up its proton, then HA is a weak acid. So, how can we tell whether or not HA is willing to give up its proton? *We can figure it out by looking at the conjugate base.*

Notice that the conjugate base has a negative charge. The real question is: how stable is that negative charge? If that charge is stable, then HA will be willing to give up the proton, and therefore HA will be a strong acid. If that charge is not stable, then HA will not be willing to give up its proton, and HA will be a weak acid.

So mastering acid–base chemistry requires the following skill: you need to be able to look at a negative charge and determine how stable that negative charge is. If you can do that, then acid–base chemistry will be a breeze for you. If you cannot determine charge stability, then you will have problems even after you finish acid–base chemistry. To predict reactions, you need to know what kind of charges are stable and what kind of charges are not stable.

Now you can understand why acid–base chemistry is taught so early in the course. Charge stability is a vital part of understanding the structure of molecules. It is so incredibly important because reactions are all about how charges interact with one another. You cannot begin to discuss reactions until you have an excellent understanding of what factors stabilize charges and what factors destabilize charges. This chapter will focus on four important factors, one at a time.

3.1 FACTOR 1—WHAT ATOM IS THE CHARGE ON?

The most important factor for determining charge stability is to ask what atom the charge is on. For example, consider the following two structures:

The one on the left has a negative charge on oxygen, and the one on the right has the charge on sulfur. How do we compare these? We look at the periodic table, and we need to consider two trends: comparing atoms *in the same row* and comparing atoms *in the same column:*

Let's start with comparing atoms *in the same row*. For example, let's compare carbon and oxygen:

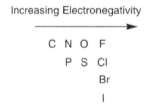

The structure on the left has the charge on carbon, and the structure on the right has the charge on oxygen. Which one is more stable? Recall that electronegativity increases as we move to the right on the periodic table:

Since electronegativity is the measure of an element's affinity for electrons (how willing the atom will be to accept a new electron), we can say that a negative charge on oxygen will be more stable than a negative charge on carbon.

Now let's compare atoms *in the same column*, for example, iodide (I^-) and fluoride (F^-). Here is where it gets a little bit tricky, because the trend is the opposite of the electronegativity trend:

It is true that fluorine is more electronegative than iodine, but there is another more important trend when comparing atoms in the same column: the *size* of the atom. Iodine is *huge* compared to fluorine. So when a charge is placed on iodine, the charge is spread out over a very large volume. When a charge is placed on fluorine, the charge is stuck in a very small volume of space:

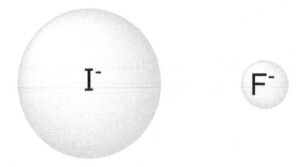

Even though fluorine is more electronegative than iodine, nevertheless, iodine can better stabilize a negative charge. If I^- is more stable than F^-, then HI must be a stronger acid than HF, because HI will be more willing to give up its proton than HF.

To summarize, there are two important trends: *electronegativity* (for comparing atoms in the same row) and *size* (for comparing atoms in the same column). The first factor (comparing atoms in the same row) is a much stronger effect. In other words, the difference in stability between C^- and F^- is much greater than the difference in stability between I^- and F^-.

Now we have all of the information we need to answer the question presented earlier in this section: Which charge below is more stable?

When comparing these two ions, we see an oxygen atom bearing the negative charge (on the left) and a sulfur atom bearing the negative charge (on the right). Oxygen and sulfur are in the same column of the periodic table, so size is the important trend to look at. Sulfur is larger than oxygen, so sulfur can better stabilize the negative charge.

EXERCISE 3.1   Compare the two protons highlighted below and determine which one is more acidic.

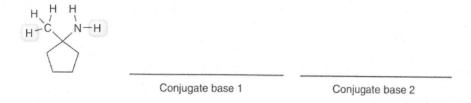

Answer   For each compound, we remove the highlighted proton and draw the resulting conjugate base.

Now we need to compare these conjugate bases and ask which one is more stable. In other words, which negative charge is more stable? We are comparing a negative charge on nitrogen with a negative charge on oxygen. So we are comparing two atoms in the same row of the periodic table, and the important trend is electronegativity. Oxygen can better stabilize the negative charge, because oxygen is more electronegative than nitrogen. Therefore, we conclude that the proton on oxygen will be more acidic, because removing this proton generates the more stable conjugate base.

PROBLEMS

3.2   Compare the two highlighted protons in the following compound and determine which is more acidic. Remember to begin by drawing the two conjugate bases, and then compare them.

Conjugate base 1   Conjugate base 2

3.3   Compare the two highlighted protons in the following compound and determine which is more acidic.

Conjugate base 1   Conjugate base 2

3.4 Compare the two highlighted protons in the following compound and determine which is more acidic.

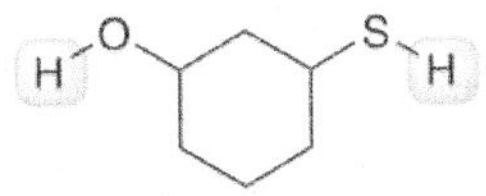

_____   _____
Conjugate base 1   Conjugate base 2

3.5 Compare the two highlighted protons in the following compound and determine which is more acidic.

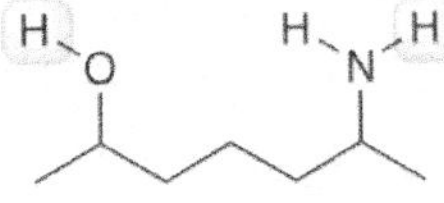

_____   _____
Conjugate base 1   Conjugate base 2

3.2   FACTOR 2—RESONANCE

The previous chapter was devoted solely to drawing resonance structures. If you have not yet completed that chapter, do so before you begin this section. We said in the previous chapter that resonance would find its way into every single topic in organic chemistry. And here it is in acid–base chemistry.

To see how resonance plays a role here, let's compare the acidity of the following two compounds:

In both cases, we remove a proton, resulting in a charge on oxygen:

Now we compare the stability of these conjugate bases. We cannot use factor 1 (what atom is the charge on) to determine which proton is more acidic. In both cases, we are dealing with a negative charge on oxygen. But there is a critical difference between these two negative charges. The first one is stabilized by resonance, as shown here:

Remember what resonance means. It does not mean that we have two structures that are in equilibrium. Rather, it means that there is only one chemical entity and we cannot use one drawing to adequately describe where the charge is. In reality, the charge is spread out equally over both oxygen atoms. To see this, we need to draw both drawings.

So what does this do in terms of stabilizing the negative charge? Imagine that you have a hot potato in your hand (too hot to hold for long). If you could grab another potato that is cold and transfer half of the warmth to the second potato, then you would have two potatoes, each of which is not too hot to hold. It's the same concept here. When we spread a charge over more than one atom, we call the charge "delocalized." A delocalized negative charge is more stable than a localized negative charge (stuck on one atom):

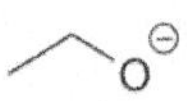

charge is stuck on one atom ("localized")

This factor is very important and explains why carboxylic acids are acidic:

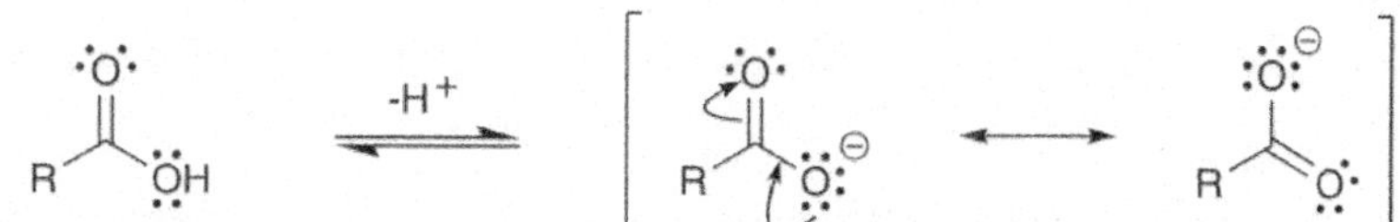

A carboxylic acid
(R = carbon chain)

They are acidic because the conjugate base is stabilized by resonance. It is worth noting that carboxylic acids are not strongly acidic. They are acidic when compared with other organic compounds, such as alcohols and amines, but not very acidic when compared with inorganic acids, such as sulfuric acid or nitric acid. In the equilibrium above showing a carboxylic acid losing a proton, we have one molecule losing its proton for every 10,000 molecules that do not give up their proton. In the world of acidity, this is not very acidic, but everything is relative.

So we have learned that resonance (which delocalizes a negative charge) is a stabilizing factor. The question now is how to roughly determine how stabilizing this factor is. Consider, for example, the following case:

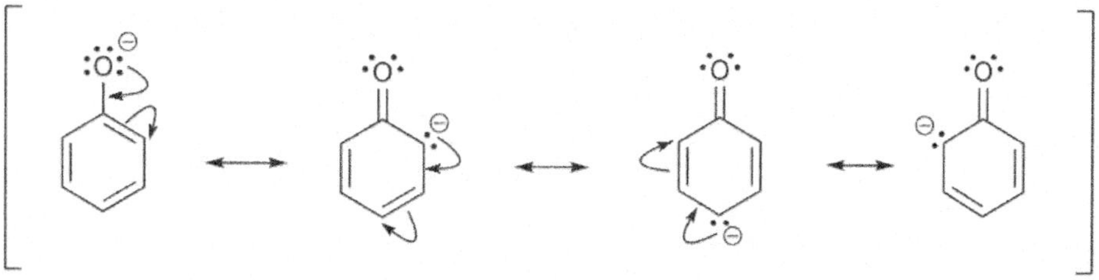

The negative charge is stabilized over four atoms: one oxygen atom and three carbon atoms. Even though carbon is not as happy with a negative charge as oxygen is, nevertheless, it is better to spread the charge over one oxygen and three carbon atoms than to leave the negative charge stuck on one oxygen. Spreading the charge around helps to stabilize that charge.

But the number of atoms sharing the charge isn't everything. For example, it is better to have the charge spread over two oxygen atoms than to have the charge spread over one oxygen and three carbon atoms:

More stable

So now we have the basic framework to compare two anionic (negatively charged) bases that are each resonance stabilized. We need to compare the bases, keeping in mind the rules we just learned:

1. The more delocalized the better. A charge spread over four atoms is more stable than a charge spread over two atoms, *but*

2. One oxygen is better than many carbon atoms.

Now let's do some problems.

EXERCISE 3.6   Compare the two protons highlighted below and determine which one is more acidic.

Answer   We begin by removing one proton and drawing the resulting conjugate base. Then we do the same thing for the other proton:

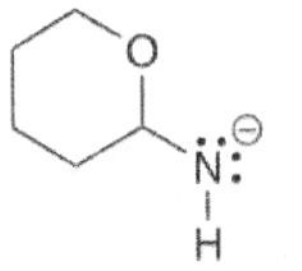

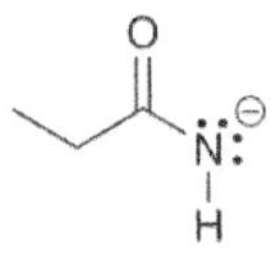

Now we need to compare these conjugate bases and ask which one is more stable. In the structure on the left, we are looking at a charge that is localized on a nitrogen atom. For the structure on the right, the negative charge is delocalized over a nitrogen atom and an oxygen atom (draw resonance structures). It is more stable for the charge to be delocalized, so the second structure is more stable.

   We therefore conclude that the compound below is the more acidic compound, because removing a proton from this compound generates the more stable conjugate base.

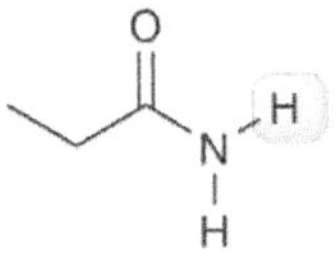

PROBLEMS

3.7 Compare the two highlighted protons and determine which one is more acidic.

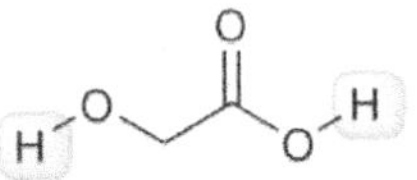

| Conjugate base 1 | Conjugate base 2 |

3.8 Compare the two highlighted protons and determine which one is more acidic.

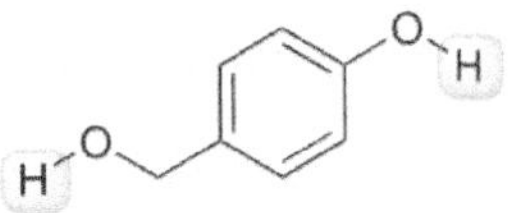

| Conjugate base 1 | Conjugate base 2 |

3.9 Compare the two highlighted protons and determine which one is more acidic.

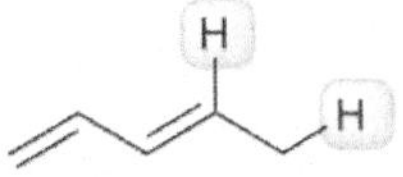

| Conjugate base 1 | Conjugate base 2 |

3.10 Compare the two highlighted protons and determine which one is more acidic.

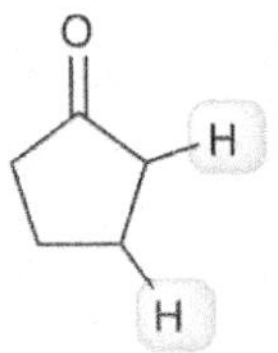

| Conjugate base 1 | Conjugate base 2 |

3.11 Compare the two highlighted protons and determine which one is more acidic.

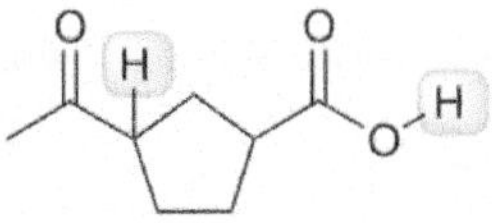

| Conjugate base 1 | Conjugate base 2 |

3.12 Compare the two highlighted protons and determine which one is more acidic.

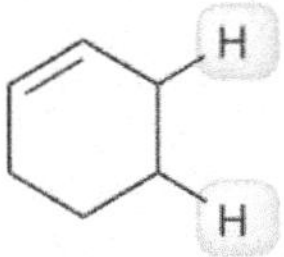

| Conjugate base 1 | Conjugate base 2 |

3.3   FACTOR 3—INDUCTION

Let's compare the following compounds:

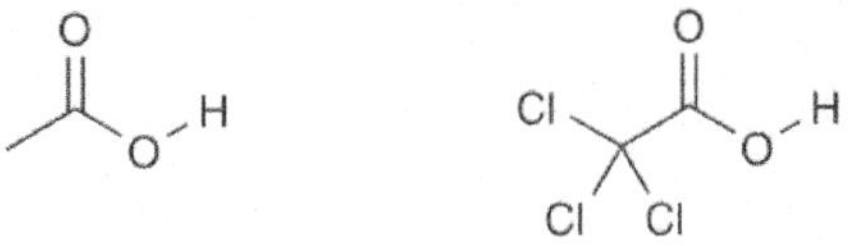

Which compound is more acidic? To answer this question, we remove the protons and draw the conjugate bases:

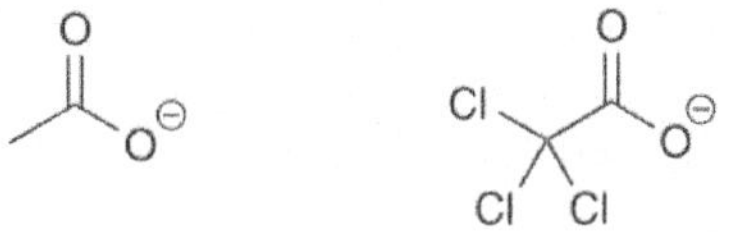

Let's go through the factors we learned so far. Factor 1 does not answer the problem: in both cases, the negative charge is on oxygen. Factor 2 also does not answer the problem: in both cases, there is resonance that delocalizes the charge over two oxygen atoms. Now we need factor 3.

The difference between the compounds is clearly the placement of the chlorine atoms. What effect will this have? For this, we need to understand a concept called induction.

We know that electronegativity measures the affinity of an atom for electrons, so what happens when you have two atoms of different electronegativity connected to each other? For example, consider a carbon–oxygen bond (C—O). Oxygen is more electronegative, so the two electrons that are shared between carbon and oxygen (the two electrons that form the bond between them) are pulled more strongly by the oxygen atom. This creates a difference in the electron density on the two atoms—the oxygen atom becomes electron rich and the carbon atom becomes electron poor. This is usually shown with the symbols $\delta+$ and $\delta-$, which indicate "partial" positive and "partial" negative charges:

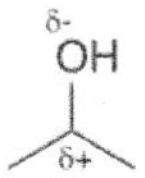

This "pulling" of electron density is called induction. Going back to our first example, the three chlorine atoms withdraw electron density from the carbon atom to which they are attached, rendering the carbon atom electron poor (δ+). This carbon atom can then withdraw electron density from the region that has the negative charge, and this effect will stabilize the negative charge:

More stable

Inductive effects fall off rapidly with distance, so there is a large difference between the following structures:

More stable

EXERCISE 3.13   Compare the two protons highlighted below and determine which one is more acidic.

Answer   Begin by drawing the conjugate bases:

In the structure on the left, the charge is somewhat stabilized by the inductive effects of the neighboring chlorine atoms. In contrast, the structure on the right lacks this stabilizing effect. Therefore, the structure on the left is more stable.

The more acidic proton is the one that will leave to give the more stable negative charge. So the following proton is more acidic:

PROBLEMS

3.14   Compare the two highlighted protons and determine which one is more acidic.

Conjugate base 1   Conjugate base 2

3.15 Compare the two highlighted protons and determine which one is more acidic.

_____   _____
Conjugate base 1   Conjugate base 2

3.16 Compare the two highlighted protons and determine which one is more acidic.

_____   _____
Conjugate base 1   Conjugate base 2

3.4   FACTOR 4—ORBITALS

The three factors we have learned so far will not explain the difference in acidity between the two highlighted protons in the compound below:

In order to determine which proton is more acidic, we remove each proton and compare the resulting conjugate bases:

In both cases, the negative charge is on a carbon, so factor 1 does not help. In both cases, the charge is not stabilized by resonance, so factor 2 does not help. In both cases, there are no inductive effects to consider, so factor 3 does not help. The answer here comes from looking at the type of orbital that is accommodating the charge.

Let's quickly review the shape of hybridized orbitals. sp^3, sp^2, and sp orbitals all have similar shapes, but they are different in size:

sp^3 sp^2 sp

Notice that the sp orbital is smaller and tighter than the other orbitals. It is closer to the nucleus of the atom, which is located at the point where the front lobe (white) meets the back lobe (gray). Therefore, a lone pair of electrons residing in an sp orbital will be held closer to the positively charged nucleus and will be *stabilized* by being close to the nucleus.

So a negative charge on an *sp* hybridized carbon is more stable than a negative charge on an *sp³* or *sp²* hybridized carbon:

More stable

Determining which carbon atoms are *sp*, *sp²*, or *sp³* is very simple: a carbon with a triple bond is *sp*, a carbon with a double bond is *sp²*, and a carbon with all single bonds is *sp³*. For more on this topic, turn to the next chapter (covering geometry).

EXERCISE 3.17 Locate the most acidic proton in the following compound:

Answer It is important to recognize where all of the protons (hydrogen atoms) are. If you cannot do this, then you should review Chapter 1, which covers bond-line drawings. Only one proton can leave behind a negative charge in an *sp* orbital. All of the other protons would leave behind a negative charge in either *sp²* or *sp³* hybridized orbitals. So the most acidic proton is:

3.5 RANKING THE FOUR FACTORS

Now that we have seen each of the four factors individually, we need to consider what order of importance to place them in. In other words, what should we look for first? And what should we do if two factors are competing with each other?

In general, the order of importance is the order in which the factors were presented in this chapter.

1. What atom is the charge on? (Remember the difference between comparing atoms in the same row and comparing atoms in the same column.)

2. Are there any resonance effects making one conjugate base more stable than the others?

3. Are there any inductive effects (electronegative atoms) that stabilize any of the conjugate bases?

4. In what orbital do we find the negative charge for each conjugate base that we are comparing?

There is an important exception to this order. Compare the following two compounds:

$$H - \!\!\!\equiv\!\!\! - H \qquad\qquad NH_3$$

In order to predict which compound is more acidic, we remove a proton from each compound and compare the conjugate bases:

$$H - \!\!\!\equiv\!\!\! :^\ominus \qquad\qquad {}^\ominus\!\!:\ddot{N}H_2$$

When comparing these two negative charges, we find two competing factors: the first factor (what atom is the charge on?) and the fourth factor (what orbital is the charge in?). The first factor says that a negative charge on nitrogen is more stable than a negative charge on carbon. However, the fourth factor says that a negative charge in an *sp* orbital is more stable than a negative charge in an *sp³* orbital (the negative charge on the nitrogen atom occupies an sp^3 hybridized orbital). In general, we would say that factor 1 wins over the others. But this case is an exception, and factor 4 (orbitals) actually wins here, so the negative charge on carbon is more stable in this case:

More stable

This example illustrates that a negative charge on an sp hybridized carbon atom is more stable than a negative charge on an sp^3 hybridized nitrogen atom. For this reason, NH_2^- can be used as a base to deprotonate a triple bond.

There are, of course, other exceptions, but the one explained above is the most common. In most cases, you should be able to apply the four factors and provide a qualitative assessment of acidity.

EXERCISE 3.18   Compare the two protons highlighted below and determine which one is more acidic.

Answer   The first thing we need to do is draw the conjugate bases:

Now we can compare them and ask which negative charge is more stable, using our four factors:

1. *Atom*   The first conjugate base has a negative charge on a nitrogen atom, while the second conjugate base has a negative charge on a carbon atom. Based on this factor alone, we predict that the first conjugate base should be more stable.

2. *Resonance*   Neither of the conjugate bases is stabilized by resonance.

3. *Induction*   Neither of the conjugate bases is stabilized by induction.

4. *Orbital*   The first conjugate base has a negative charge on an sp^3 hybridized atom, while the second conjugate base has a negative charge on an sp hybridized atom. Based on this factor alone, we predict that the second conjugate base should be more stable.

So, we have a competition between the first factor (atom) and the fourth factor (orbital). In general, the first factor takes precedence over the fourth factor. But this is the one exception that we saw. In this case, the fourth factor is the dominating factor because a negative charge on an sp hybridized carbon atom is more stable than a negative charge on an sp^3 hybridized nitrogen atom. Therefore, the proton highlighted below is the more acidic proton:

Remember the four factors, and what order they come in:

1. Atom

2. Resonance

3. Induction

4. Orbital

If you have trouble remembering the order, try remembering this acronym: ARIO.

PROBLEMS For each of the following compounds, two protons have been highlighted. In each case, determine which of the two protons is more acidic.

3.19

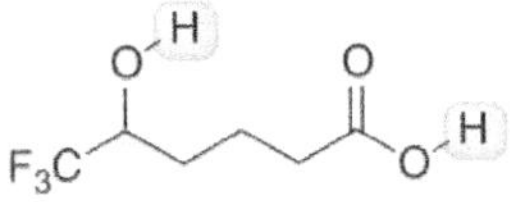

Conjugate base 1   Conjugate base 2

3.20

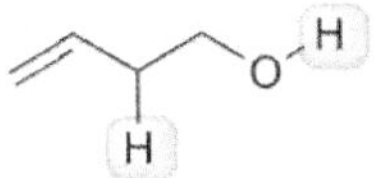

Conjugate base 1   Conjugate base 2

3.21

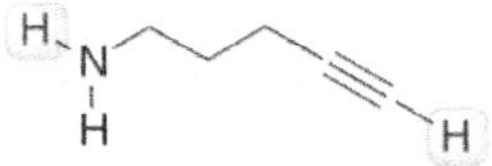

Conjugate base 1   Conjugate base 2

3.22

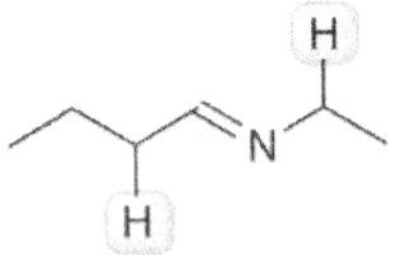

Conjugate base 1   Conjugate base 2

3.23

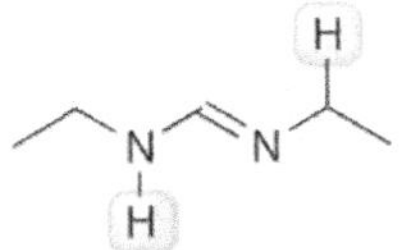

Conjugate base 1   Conjugate base 2

3.24

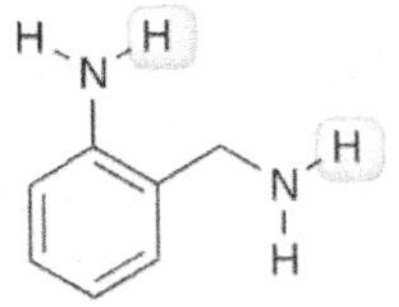

Conjugate base 1   Conjugate base 2

3.25

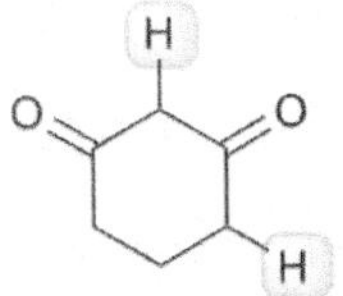

Conjugate base 1   Conjugate base 2

3.26

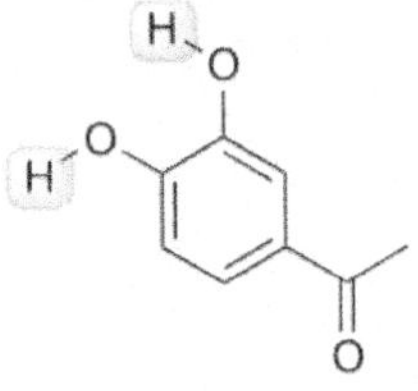

Conjugate base 1   Conjugate base 2

PROBLEMS   For each pair of compounds below, predict which will be more acidic (*Hint*: To solve 3.31, you should carefully consider your answers for 3.28 and 3.29).

3.27   HI   HBr

3.28 CH_3SH CH_3OH

3.29 NH_3 H_2O

3.30

3.31

3.32

3.6   OTHER FACTORS

In the previous sections, we presented four factors for comparing the stability of conjugate bases. But you shouldn't think that there are no other factors. There are indeed other factors that can affect the stability of conjugate bases. One such factor is best illustrated if we compare the following two compounds:

tert-Butanol   Ethanol

In order to compare the acidity of these compounds, we must compare the stability of their conjugate bases:

tert-Butoxide   Ethoxide

If we apply our four factors (ARIO) in this case, we would predict that these two conjugate bases should have similar stability. Yet, one of them (ethoxide) turns out to be more stable than the other. Why? The difference in stability is best explained by considering the interactions between each conjugate base and the surrounding solvent molecules:

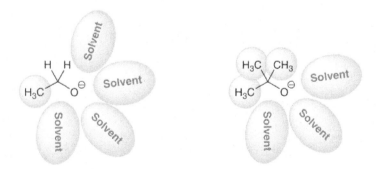

The *tert*-butoxide ion is very bulky, or sterically hindered, compared to the ethoxide ion, so *tert*-butoxide will experience fewer interactions with the solvent molecules. These interactions stabilize the conjugate bases, so ethoxide (which has more of these stabilizing interactions) will be more stable than *tert*-butoxide. And this explains why ethanol is found to be more acidic than *tert*-butanol. This effect, called a *solvating effect*, is generally a weaker factor than all of the four factors we have learned thus far (ARIO).

3.7   QUANTITATIVE MEASUREMENT (pK_a VALUES)

Everything we have mentioned so far has been the *qualitative* method for comparing acidity of different protons. In other words, we never said how *much* more acidic one proton is over another, and we never said *exactly* how acidic each proton is. We have talked only about relative acidities: which proton is *more* acidic?

There is also a *quantitative* method of measuring acidities. All protons can be given a number that quantifies exactly how acidic they are. This value is called pK_a. It is impossible to predict the exact pK_a by just looking at a structure. The pK_a must be determined empirically through experimentation. Many professors require that you know some general pK_a's for certain classes of compounds (for instance, all alcoholic protons, RO—H, will have the same ballpark pK_a). Most textbooks will have a chart that you can memorize. Your instructor will tell you if you are expected to memorize this chart. Either way, you should know what the numbers mean.

The smaller the pK_a, the more acidic the proton is. This probably seems strange, but that's the way it is. A compound with a pK_a of 4 is more acidic than a compound with a pK_a of 7. Next, we need to know what the difference is between 4 and 7. These numbers measure orders of magnitude. So the compound with a pK_a of 4 is 10^3 times more acidic (1000 times more acidic) than a compound with a pK_a of 7. If we compare a compound with a pK_a of 10 to a compound with a pK_a of 25, we find that the first compound is 10^{15} times more acidic than the second compound (1,000,000,000,000,000 times more acidic).

3.8   PREDICTING THE POSITION OF EQUILIBRIUM

Now that we know how to compare stability of charge, we can begin to predict which side of an equilibrium will be favored. Consider the following scenario:

$$H-A \ + \ B^{\ominus} \ \rightleftharpoons \ A^{\ominus} \ + \ H-B$$

This equilibrium represents the struggle between two bases competing for a proton (H$^+$). A$^-$ and B$^-$ are competing with each other. Sometimes A$^-$ captures the proton and sometimes B$^-$ captures the proton. If we have a very large amount of A$^-$ and B$^-$ and not enough H$^+$ to protonate both of them, then at any given moment in time, there will be a certain number of A's that have a proton (HA) and a certain number of B's that have a proton (HB). These numbers are controlled by the equilibrium, which is controlled by (you guessed it) *stability of the conjugate bases*. If A$^-$ is more stable than B$^-$, then A will be happy to have the negative charge and B$^-$ will capture most of the protons. However, if B$^-$ is more stable than A$^-$, then we will have the reverse effect.

Another way of looking at this is the following. In the equilibrium above, we see an A$^-$ on one side and a B$^-$ on the other side. The equilibrium will favor whichever side has the more stable negative charge. If A$^-$ is more stable, then the equilibrium will lean so as to favor the formation of A$^-$:

$$H-A \ + \ B^{\ominus} \ \rightleftharpoons \ \boxed{A^{\ominus}} \ + \ H-B$$

If B$^-$ is more stable, then the equilibrium will lean so as to favor the formation of B$^-$:

$$H-A \ + \ \boxed{B^{\ominus}} \ \rightleftharpoons \ A^{\ominus} \ + \ H-B$$

The position of equilibrium can be easily predicted by comparing the relative stability of negative charges.

EXERCISE 3.33   Predict the position of equilibrium for the following reaction:

$$H_2S \ + \ CH_3O^{\ominus} \ \rightleftharpoons \ HS^{\ominus} \ + \ CH_3OH$$

Answer   We look at both sides of the equilibrium and compare the negative charge on either side. Then we ask which one is more stable. We use the four factors:

1. *Atom*   The negative charge on the left is on oxygen, and the negative charge on the right is on sulfur. A negative charge is more stable on a sulfur atom, because sulfur is larger.

2. *Resonance*   Neither one is resonance stabilized.

3. *Induction* Neither structure exhibits inductive effects. So this factor does not help us.

4. *Orbital* No difference between the right and left.

Based on factor 1, we conclude that the one on the right is more stable, and therefore the equilibrium lies to the right. We show this in the following way:

$$H_2S \quad + \quad CH_3O^{\ominus} \quad \rightleftharpoons \quad HS^{\ominus} \quad + \quad CH_3OH$$

PROBLEMS

3.34 Predict the position of equilibrium for the following reaction:

3.35 Predict the position of equilibrium for the following reaction:

3.36 Predict the position of equilibrium for the following reaction:

3.9 SHOWING A MECHANISM

Later on in the course, you will spend a lot of time drawing mechanisms of reactions. A mechanism shows how the electrons move during a reaction to form the products. Sometimes many steps are required, and sometimes only one step is required. In acid–base reactions, mechanisms are very straightforward because there is only one step. We use curved arrows (just like we did when drawing resonance structures) to show how the electrons flow. The only difference is that here we are allowed to break single bonds, because we are using arrows to show how a reaction happened (a reaction that involved the breaking of a single bond). With resonance drawings, we can never break a single bond (remember the first commandment). The second commandment—never violate the octet rule—is still true, even when we are drawing mechanisms. Second-row elements can never have more than four bonds, as described in Section 2.3.

From an arrow-pushing point of view, all acid–base reactions are the same. It goes like this:

There are always two curved arrows. One is drawn coming from the base and attacking the proton. The second curved arrow is drawn coming from the bond (between the proton and whatever atom is connected to the proton) and going to the atom currently connected to the proton. That's it. There are always two curved arrows. Each curved arrow has a head and a tail, so there are four possible mistakes you can make. You might accidentally draw either of the heads incorrectly, or you might draw either of the tails incorrectly. With a little bit of practice you will see just how easy it is, and you will realize that acid–base reactions always follow the same mechanism.

EXERCISE 3.37 Show a mechanism for the following acid–base reaction:

Answer Remember—2 curved arrows. One from the base to the proton and the other from the bond (that is losing the proton) to the atom (currently connected to the proton):

PROBLEMS

3.38 Show a mechanism for the following acid–base reaction:

3.39 Show a mechanism for the following acid–base reaction:

PROBLEMS Show a mechanism for the reaction that takes place when you mix hydroxide (HO⁻) with each of the following compounds (remember that you need to look for the most acidic proton in each case).

3.40 CH_3SH

3.41

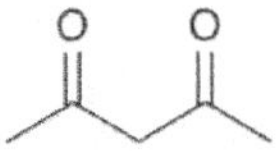

3.42

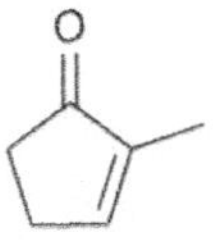

PROBLEMS Show a mechanism for the reaction that takes place when you mix the amide ion (H_2N^-) with each of the following compounds (remember that you need to look for the most acidic proton in each case).

3.43

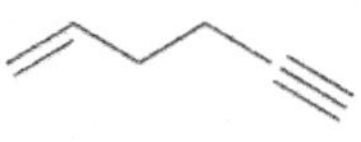

3.44

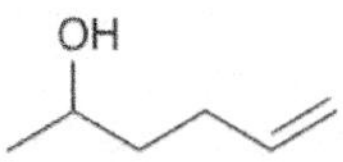

CHAPTER 4

GEOMETRY

In this chapter, we will see how to predict the 3D shape of molecules. This is important because it limits much of the reactivity that you will see in the second half of this course. For molecules to react with each other, the reactive parts of the molecules must be able to get close in space. If the geometry of the molecules prevents them from getting close, then there cannot be a reaction. This concept is called *sterics*.

Let's use an analogy to help us see the importance of geometry. Imagine that you are stuffing a turkey for Thanksgiving dinner and your hand gets stuck inside the turkey. Just at that moment, someone wants to shake your hand. You can't shake the person's hand because your hand is unavailable at the moment. It's kind of the same way with molecules. When two molecules react with each other, there are specific sites on the molecules that are reacting with each other. If those sites cannot get close to each other, the reaction won't happen.

There will be many times in the second half of this course when you will be trying to determine which way a reaction will proceed from two possible outcomes. Many times, you will choose one outcome, because the other outcome has steric problems to overcome (the geometry of the molecules does not permit the reactive sites to get close together). In fact, you will learn to make decisions like this as soon as you learn your first reactions: S_N2 versus S_N1 reactions. Now that we know why geometry is so important, we need to brush up on some basic concepts.

To determine the geometry of an entire molecule, we need to be able to determine the geometry of each atom. This is accomplished by analyzing how it is connected to the atoms around it. After all, that is what defines the geometry—how the atoms are connected in 3D space. Since atoms are connected to each other with bonds, it makes sense that we need to take a close look at bonds. In particular, we need to know the exact locations and angles of every bond to every atom. This might sound difficult, but it is actually straightforward, and with a little bit of practice, you can get to the point where you know the geometry of a molecule as soon as you look at it (without even needing to think about it). That is the point that we need to get to, and that is what this chapter is all about.

4.1 ORBITALS AND HYBRIDIZATION STATES

To determine the geometry of a molecule, we need to know how atoms bond with each other three dimensionally, so it makes sense for our discussion to start with orbitals. After all, bonds come from overlapping orbitals.

A bond is formed when an orbital of one atom overlaps with an orbital of another atom. If each orbital contains one electron, then the two electrons can be shared between both atoms, and we call that a bond. Since electrons exist in regions of space called orbitals, then what we really need to know is; what are the locations and angles of the atomic orbitals around every atom? It is not so complicated, because the number of possible arrangements of atomic orbitals is very small. You need to learn the possibilities, and how to identify them when you see them.

There are two simple atomic orbitals: *s* and *p* orbitals (we don't really deal much with *d* and *f* orbitals in organic chemistry). *s* orbitals are spherical and *p* orbitals have two lobes (one front lobe and one back lobe):

s orbital   *p* orbital

Atoms in the second row (such as C, N, O, and F) have one *s* orbital and three *p* orbitals in the valence shell. These orbitals are usually mixed together to give us hybridized orbitals (sp^3, sp^2, and sp). We get these orbitals by mathematically mixing the *properties* of *s* and *p* orbitals. What do we mean by mixing?

Imagine one swimming pool shaped like a triangle and another shaped like a pentagon; now we put them next to each other. We wave a magic wand and they magically turn into two rectangular pools. That would be a neat trick. That's what *sp* orbitals are: we take one *s* orbital and one *p* orbital, then wave a magic wand (mathematics), and poof—we now have two equivalent orbitals that look the same. The two new orbitals have a different shape from the original two orbitals. This new shape is somewhat of an average of the two original shapes.

If we mix two *p* orbitals and one *s* orbital, then we get three equivalent sp^2 orbitals. Let's go back to the pool analogy. Imagine two pools shaped like octagons and one shaped like a triangle. We wave our magic wand and get three pools shaped like hexagons. We started with three pools and we ended with three pools. But the three pools in the end all look the same. The same thing is true here with orbitals. We start with three orbitals (two *p* orbitals and one *s* orbital). Then we mix them together and end up with three orbitals that all look the same. The three new orbitals (since they came from one *s* orbital and two *p* orbitals) are called sp^2 orbitals. Similarly, when you mathematically combine three *p* orbitals and one *s* orbital, you get four equivalent sp^3 orbitals.

To truly understand the geometry of bonds, we need to understand the geometry of these three different hybridization states. The hybridization state of an atom describes the type of hybridized atomic orbitals (sp^3, sp^2, or sp) that contain the valence electrons. Each hybridized orbital can be used either to form a bond with another atom or to hold a lone pair.

It is not difficult to determine hybridization states. If you can add, then you should have no trouble determining the hybridization state of an atom. Just count how many other atoms are bonded to your atom, and count how many lone pairs your atom has. Add these numbers. Now you have the total number of hybridized orbitals that contain the valence electrons. This number is all you need to determine the hybridization state of the atom. Let's look at an example.

Consider the molecule below:

$$\underset{\text{H} \qquad \text{H}}{\overset{\displaystyle \text{O}}{\underset{\displaystyle \|}{\text{C}}}}$$

Let's try to determine the hybridization state of the carbon atom. We begin by counting the number of atoms connected to this carbon atom. There are 3 atoms (O, H, and H). *The oxygen atom only counts once.*

Next we count the number of lone pairs on the carbon atom. There are no lone pairs on the carbon atom. (If you are not sure how to tell that there are no lone pairs there, go back to Chapter 1 and review the section on counting lone pairs.) Now we take the sum of the attached atoms and the number of lone pairs—in this case, $3 + 0 = 3$. Therefore, three hybridized orbitals are being used here. That means that we have mixed two *p* orbitals and one *s* orbital (a total of three orbitals) to get three equivalent sp^2 orbitals. Thus, the hybridization is sp^2. Let's take a closer look at how this works.

Recall that the second row elements have three *p* orbitals and one *s* orbital that can be hybridized in one of three ways: sp^3, sp^2, or sp. If we are using three hybridized orbitals, then we must have mixed two *p* orbitals with one *s* orbital:

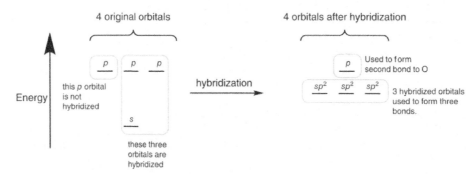

So here's the rule: Just count the number of bonded atoms and the number of lone pairs. The total indicates how many hybridized orbitals you need according to the following:

If the sum is 4, then you have 4 sp^3 orbitals.

If the sum is 3, then you have 3 sp^2 orbitals and one *p* orbital (as in our example).

If the sum is 2, then you have 2 sp orbitals and two *p* orbitals.

There are several exceptions. For now, don't worry. We will focus on simple cases.

Once you get used to looking at drawings of molecules, you should not have to count anymore. There are certain arrangements that are always *sp³* hybridized, and the same is true for *sp²* and *sp*. Here are some common examples:

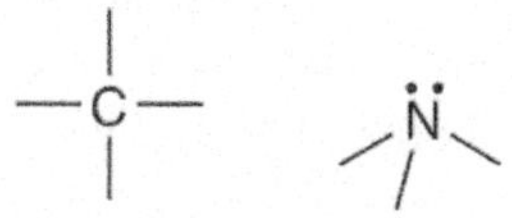

If you can determine the hybridization state of any atom, you will be able to easily determine the geometry of that atom. Let's do another example.

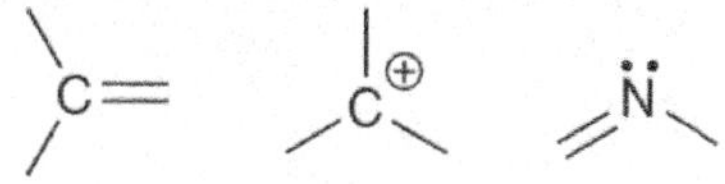

EXERCISE 4.1 Identify the hybridization state for the nitrogen atom in ammonia (NH_3).

Answer First we need to ask how many atoms are connected to this nitrogen atom. There are three hydrogen atoms. Next we need to ask how many lone pairs the nitrogen atom has. It has 1 lone pair. Now, we take the sum $3 + 1 = 4$. If we need to have four hybridized orbitals, then the hybridization state must be sp^3.

PROBLEMS For each compound below, identify the hybridization state for the central carbon atom.

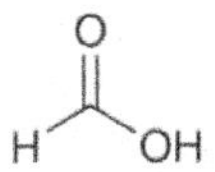

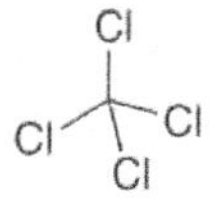

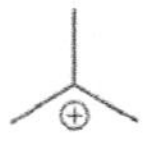

4.8 For each carbon atom in the following molecule, identify the hybridization state. Do not forget to count the hydrogen atoms (they are not shown).

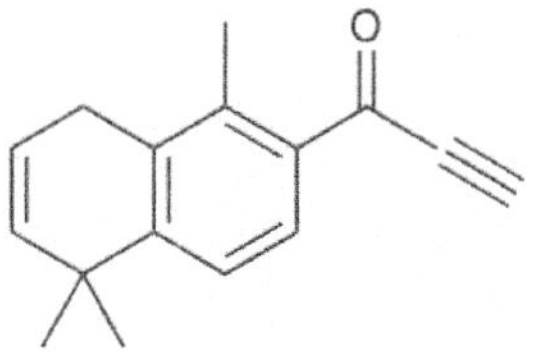

Once you get used to it, you do not need to count anymore—just look at the number of bonds. If carbon has only single bonds, then it is sp^3 hybridized. If the carbon atom has a double bond, then it is sp^2 hybridized. If the carbon atom has a triple bond, then it is sp hybridized. Consult the chart of common examples on the previous page.

4.2 GEOMETRY

Now that we know how to determine hybridization states, we need to know the geometry of each of the three hybridization states. We can predict the geometry of each hybridization state using a theory called the *valence shell electron pair repulsion theory* (VSEPR). Stated simply, all orbitals containing electrons in the outermost shell (the valence shell) want to get as far apart from each other as possible. This simple idea enables us to predict the geometry around most atoms. We will now apply VSEPR theory to the three types of hybridized orbitals.

1. Four equivalent sp^3-hybridized orbitals achieve maximum distance from one another when they arrange in a tetrahedral structure:

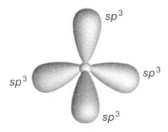

Think of this as a tripod with an additional leg sticking straight up in the air. In this arrangement, each of the four orbitals is exactly 109.5° from each of the other three orbitals.

2. Three equivalent sp^2-hybridized orbitals achieve maximum distance from one another when they arrange in a trigonal planar structure:

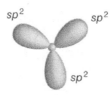

All three orbitals are in the same plane, and each one is 120° from each of the other orbitals. The remaining p orbital is orthogonal to (perpendicular to the plane of) the three hybridized orbitals.

3. Two equivalent sp-hybridized orbitals achieve maximum distance from one another when they arrange in a linear structure:

Both orbitals are 180° from each other. The remaining two p orbitals are 90° from each other and from each of the hybridized orbitals.

So far it's very simple:

1. sp^3 = tetrahedral

2. sp^2 = trigonal planar

3. sp = linear

But here's where students usually get confused. What happens when a hybridized orbital holds a lone pair? What does that do to the geometry? The answer is that the geometry of the orbitals does not change much, but the geometry of the molecule is affected. Why?

Let's look at an example. In ammonia (NH₃), the nitrogen atom is sp^3 hybridized, so *all four orbitals arrange in a tetrahedral structure*, just as we would expect. But only three of the orbitals in this arrangement are responsible for bonds. So, if we look just at the atoms that are connected, we do not see a tetrahedron. Rather, we see a trigonal pyramidal arrangement:

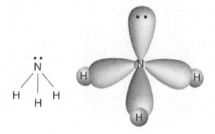

Trigonal, because there are three bonds pointing away from the central nitrogen atom, and pyramidal because it's shaped like a pyramid.

Similarly, VSEPR theory predicts that the oxygen atom in H₂O should have four orbitals in a tetrahedral arrangement. Two of these orbitals are being used for bonds, while the other two orbitals are occupied by lone pairs. If we focus only on the atoms that are connected, we do not see a tetrahedron. Rather, we see a bent arrangement:

Let's now put all of this information together:

sp^3	with	0 lone pairs = tetrahedral
sp^3	with	1 lone pair = trigonal pyramidal
sp^3	with	2 lone pairs = bent
sp^2	with	0 lone pairs = trigonal planar
sp^2	with	1 lone pair = bent
sp	with	0 lone pairs = linear

That's it. There are only six different types of geometry that we need to know. First we determine the hybridization state. Then, using the number of lone pairs, we can figure out which of the six different types of geometry we are dealing with. Let's try it out on a problem.

EXERCISE 4.9   Identify the geometry of the carbon atom below:

$$\underset{H \qquad\quad H}{\overset{\displaystyle O}{\Vert}}$$

Answer   First, we need to determine the hybridization state. We analyzed this molecule earlier in this chapter and found that the hybridization state is sp^2 (there are 3 atoms connected and no lone pairs, so we need three hybridized orbitals; therefore, it is sp^2).

Next we remind ourselves how many lone pairs there are; in this case, there are none. So the geometry must be trigonal planar.

Once you can determine the geometry around an atom, you should have no problem determining the geometry, or shape, of a molecule. Simply repeat your analysis for each and every atom in the molecule. This may seem like a large task at first, but once you get the hang of it, you will be able to determine the geometry of an atom immediately upon seeing it.

For the next set of problems, you should get to the point where you can do these problems very quickly. The first few will take you longer than the last ones. If the last problem is still taking you a long time, then you have not mastered the process and you will need more practice. If this is the case, open to any page in the second half of your textbook. You will probably see drawings of structures. Point to any atom in a structure and try to determine what the geometry is. Use the list above to help you. Go from one drawing to the next until you can do it without the list. That is the important part—doing it without needing the list.

PROBLEMS   Identify the geometry of each atom in the following compounds. Do not worry about the geometry of atoms connected to only one other atom. For example, do not worry about the geometry of any hydrogen atoms or about the geometry of the oxygen atoms in problems 4.12, 4.13, 4.15, and 4.17.

4.10

4.11

4.12

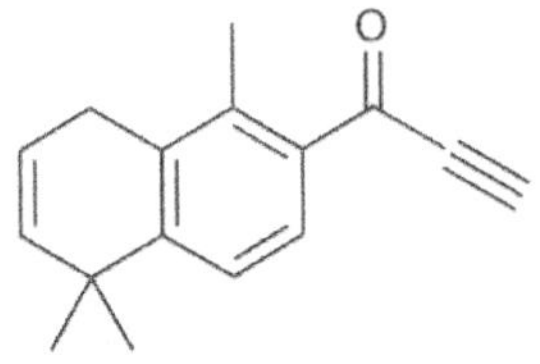

4.13

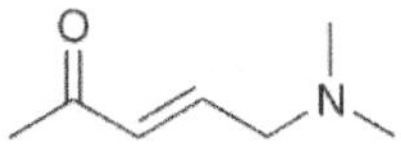

4.14

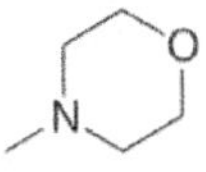

4.15

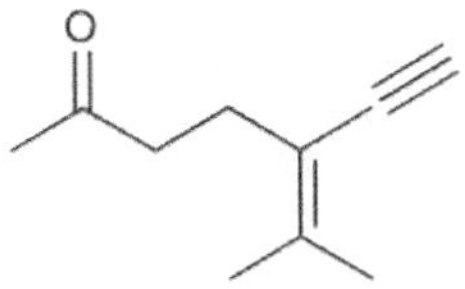

4.16

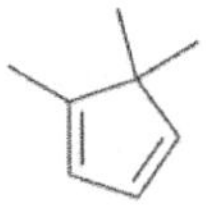

4.17

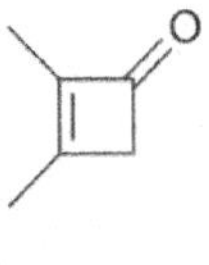

4.3   LONE PAIRS

In general, lone pairs occupy hybridized orbitals. For example, consider the lone pair on the nitrogen atom in the following compound:

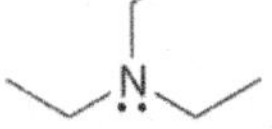

This nitrogen atom has three bonds and one lone pair, so it is sp^3 hybridized, just as we would expect. The lone pair occupies an sp^3 hybridized orbital, and the nitrogen atom has trigonal pyramidal geometry, just as we saw in the previous section. But now consider the nitrogen atom in the following compound:

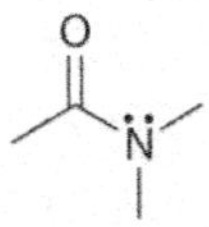

The lone pair on this nitrogen atom does NOT occupy a hybridized orbital. Why not? Because this lone pair is participating in resonance:

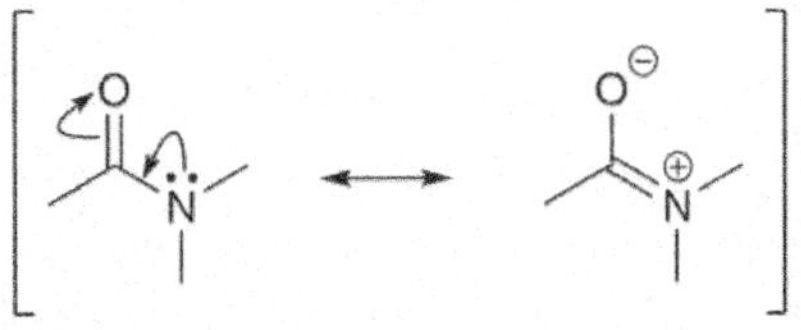

Inspection of the second resonance structure reveals that this nitrogen atom is actually sp^2 hybridized, not sp^3. It might look like it is sp^3 hybridized in the first resonance structure, but it isn't. Here is the general rule: a lone pair that participates in resonance must occupy a p orbital. In other words, the nitrogen atom in the compound above is sp^2 hybridized. And as a result, this nitrogen atom is trigonal planar rather than trigonal pyramidal.

Now let's get some practice identifying atoms that possess a lone pair participating in resonance:

PROBLEMS   Using VSEPR theory, predict the geometry of the nitrogen atom in each of the following compounds.

4.18

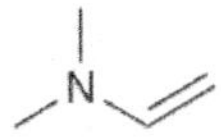

4.19

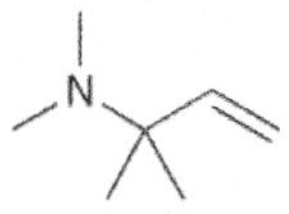

4.20

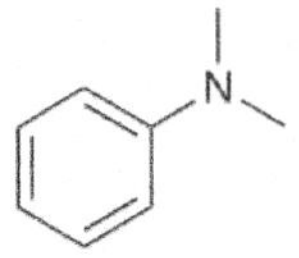

4.21

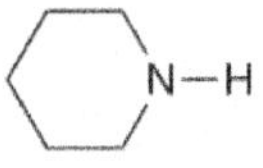

4.22

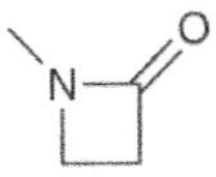

4.23

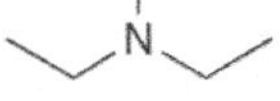

NOMENCLATURE

All molecules have names, and we need to know their names to communicate. Consider the molecule below:

Clearly, it would be inadequate to refer to this compound as "you know, that thing with five carbons and an OH coming off the side with a chlorine on a double bond." First of all, there are too many other compounds that fit that fuzzy description. And even if we could come up with a very adequate description that could only be this one compound, it would take way too long (probably an entire paragraph) to describe. By following the rules of nomenclature, we can unambiguously describe this molecule with just a few letters and numbers: Z-2-chloropent-2-en-1-ol.

It would be impossible to memorize the names of every molecule, because there are too many to even count. Instead, we have a very systematic way of naming molecules. What you need to learn are the rules for naming molecules (these rules are referred to as IUPAC nomenclature). This is a much more manageable task than memorizing names, but even these rules can become challenging to master. There are so many of them, that you could study only these rules for an entire semester and still not finish all of them. The larger the molecules get, the more rules you need to account for every kind of possibility. In fact, the list of rules is regularly updated and refined.

Fortunately, you do not need to learn all of these rules, because we deal with very simple molecules in this course. You need to learn only the rules that allow you to name small molecules. This chapter focuses on most of the rules you need to name simple molecules.

There are five parts to every name:

Stereoisomerism	Substituents	Parent	Unsaturation	Functional group

1. *Stereoisomerism* indicates whether double bonds are *cis/trans*, and indicates chiral centers (*R, S*), which we will cover in Chapter 7.
2. *Substituents* are groups connected to the main chain.
3. *Parent* is the main chain.
4. *Unsaturation* identifies if there are any double or triple bonds.
5. *Functional group* is the group after which the compound is named.

Let's use the compound above as an example:

Stereoisomerism	Substituents	Parent	Unsaturation	Functional group
(Z)	2-chloro	pent	2-en	1-ol

We will systematically go through all five parts to every name, starting at the end (functional group) and working our way backward to the first part of the name (stereoisomerism). It is important to do it backward like this, because the position of the functional group affects which parent chain you choose.

5.1   FUNCTIONAL GROUP

| Stereoisomerism | Substituents | Parent | Unsaturation | Functional group |

The term *functional group* refers to specific arrangements of atoms that have certain characteristics for reactivity. For example, when an —OH group is connected to a compound, we call the molecule an alcohol. Alcohols will display similar reactivity, because alcohols all have the same functional group, the —OH group. In fact, most textbooks have chapters arranged according to functional groups (one chapter on alcohols, one chapter on amines, etc.). Accordingly, many textbooks treat nomenclature as an ongoing learning process: As you work through the course, you slowly add to your list of functional group names. Here we focus on six common functional groups, because you will certainly learn at least these six throughout your course. All six appear in the table below, and in each case, "R" represents a carbon chain, called an alkyl group.

When a compound has one of these six functional groups, we show it in the name of the compound by adding a suffix to the name:

Functional group	Class of compound	Suffix
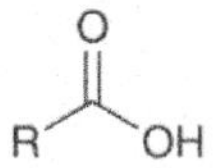 ![carboxylic acid structure] R—C(=O)—OH	Carboxylic acid	-oic acid
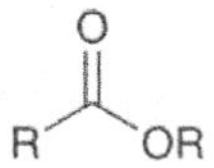 ![ester structure] R—C(=O)—OR	Ester	-oate
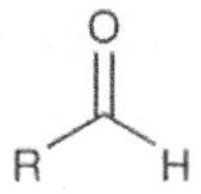 ![aldehyde structure] R—C(=O)—H	Aldehyde	-al
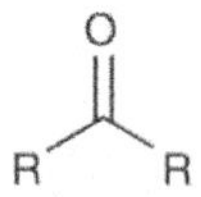 ![ketone structure] R—C(=O)—R	Ketone	-one
R—O—H	Alcohol	-ol
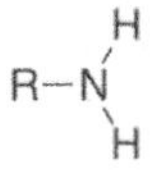 R—N(H)(H)	Amine	-amine

Halogens (F, Cl, Br, I) are usually not named in the suffix of a compound. They get named as substituents, which we will see later on.

Notice that the carboxylic acid is like a ketone and an alcohol placed next to each other. But be careful, because carboxylic acids are very different from ketones or alcohols. So don't make the mistake of thinking that a carboxylic acid is a ketone and an alcohol:

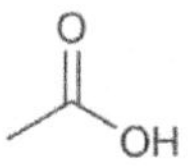

a carboxylic acid

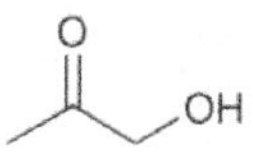

a ketone
and an alcohol

The second compound above raises an important issue: how do you name a compound with two functional groups? One will go in the suffix of the name and the other will be a prefix in the substituent part of the name. But how do we choose which one goes as the suffix of the name? There is a hierarchy that needs to be followed. The six groups shown above are listed according to their hierarchy, so a ketone takes precedence over an alcohol. A compound with both of these groups is named as a ketone and we put the term "hydroxy" in the substituent part of the name, to indicate the presence of the OH group.

EXERCISE 5.1 Identify what suffix you would use in naming the following compound:

HO~~~NH₂

Answer There are two functional groups in this compound, so we have to decide between calling this compound an amine or calling it an alcohol. If we look at the hierarchy above, we see that an alcohol outranks an amine. Therefore, we use the suffix -ol in naming this compound.

PROBLEMS Identify what suffix you would use in naming each of the following compounds.

5.2 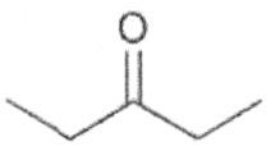 Suffix: _____

5.3 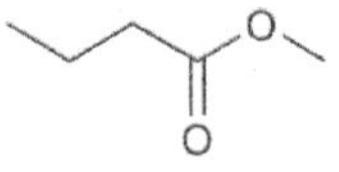 Suffix: _____

5.4 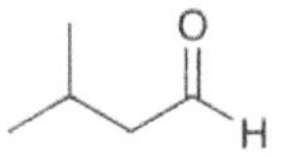 Suffix: _____

5.5 H₂N~~~Cl Suffix: _____

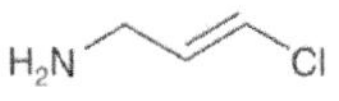

5.6 Br~~~OH Suffix: _____

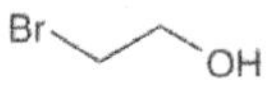

5.7 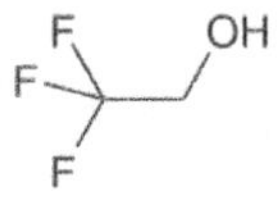 Suffix: _____

5.8 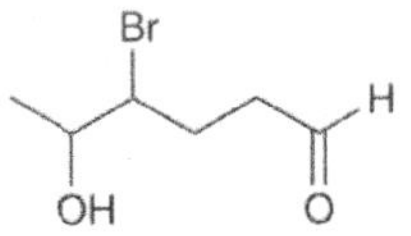 Suffix: _____

5.9 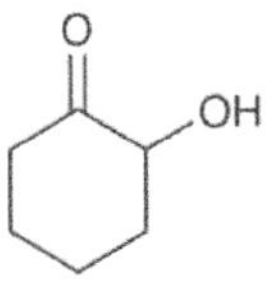 Suffix: _____

5.10 HO~~~NH₂ Suffix: _____

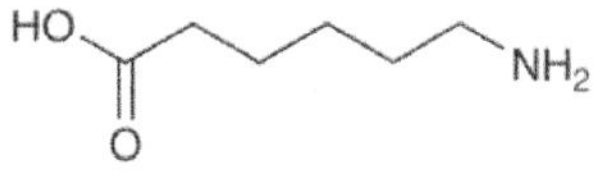

If there is no functional group in the compound, then an "e" is placed at the end of the name:

pentan**ol**

pentan**e**

5.2 UNSATURATION

Stereoisomerism	Substituents	Parent	Unsaturation	Functional group

Many compounds have double or triple bonds, and are said to be "unsaturated" because a compound with a double or triple bond has fewer hydrogen atoms than it would have without the double or triple bond. These double and triple bonds are very easy to see in bond-line drawings:

Triple bond

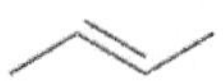

Double bond

The presence of a double bond or triple bond is indicated in the following way, where "en" (pronounced *een*) represents a double bond, and "yn" (pronounced *ine*) represents a triple bond:

pent**ane**

pent**ene**

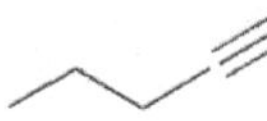

pent**yne**

Notice that "an" is used to indicate the absence of a double bond or triple bond, such as in "pentane" above. If there are two double bonds in a compound, then the unsaturation is "-dien-." Three double bonds is "-trien-." Similarly, two triple bonds is "-diyn-," and three triple bonds is "-triyn-." For multiple double and triple bonds, we use the following terms:

di = 2   penta = 5

tri = 3   hexa = 6

tetra = 4

You will rarely ever see this many double or triple bonds in one compound, but it is possible to see both double and triple bonds in the same compound. For example,

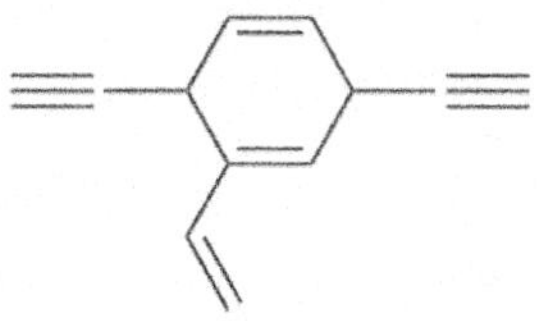

The compound shown here has three double bonds and two triple bonds. So it is a triendiyne. Double bonds always get listed first.

EXERCISE 5.11   Locate the unsaturation in the following compound, and identify how this unsaturation will be represented in the name of the compound.

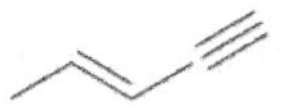

Answer   This compound has one double bond and one triple bond. For the double bond, we use the term "en." For the triple bond, we use the term "yn." Double bonds get listed first, so the unsaturation for this compound is –enyn-.

PROBLEMS   For each of the following compounds, locate the unsaturation and identify how this unsaturation will be represented in the name of the compound.

5.12

5.13

5.14

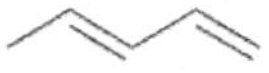

5.15

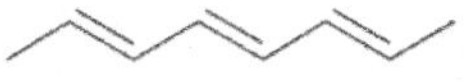

5.16

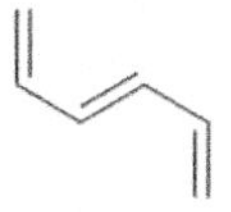

5.17

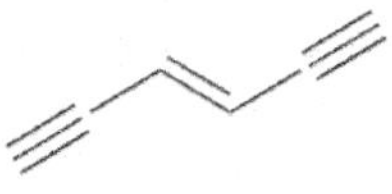

5.3 NAMING THE PARENT CHAIN

Stereoisomerism	Substituents	Parent	Unsaturation	Functional group

When naming the parent of the compound, we are looking for the chain of carbon atoms that is going to be the root of our name. Everything else in the compound is connected to that chain at a specific location, designated by numbers. So we need to know how to choose the parent carbon chain and how to number it correctly.

The first step is learning how to say "a chain of three carbon atoms" or "a chain of seven carbon atoms." Here is a table showing the appropriate names:

Number of carbon atoms in the chain	Parent
1	meth
2	eth
3	prop
4	but
5	pent
6	hex
7	hept
8	oct
9	non
10	dec

For example, pentane is a chain of five carbon atoms. If we have carbon atoms in a ring, we add the term cyclo, so a ring of six carbon atoms is called cyclohex- as the parent and a ring of five carbon atoms is cyclopent-.

You must commit these terms to memory. I am not a big advocate of memorization, but for now, you must memorize these terms. After a while, it will become habitual, and you won't have to think about it anymore.

The tricky part comes when you need to figure out which carbon chain to use. Consider the following example, which has three different possibilities for the parent chain:

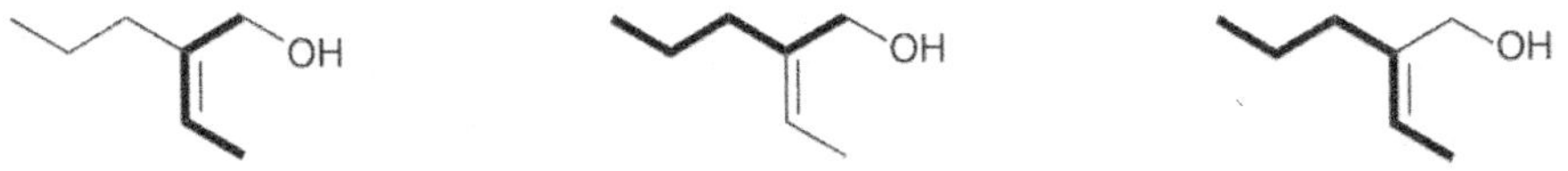

4-carbon chain   5-carbon chain   6-carbon chain

So how do we know whether to call this –but- (which is 4) or –pent- (which is 5) or –hex- (which is 6)? There is a hierarchy for this as well. The chain should be as long as possible, making sure to include the following groups, in this order:

Functional group

Double bond

Triple bond

First we need to find the functional group and make sure that the functional group is connected directly to the parent chain. Remember from the last section that if there are two functional groups, one of them gets priority. The functional group that gets priority is the one that needs to be connected to our parent chain. Of the three possibilities shown above, this rule eliminates the last possibility, because the functional group (OH) is not connected directly to the parent chain.

If there are still more choices of possible parent chains (as there are in this case), then we look for the chain that also includes both carbon atoms of the double bond (if there is a double bond in the compound). In our case, there is a double bond, and this rule determines for us which parent to use:

CORRECT   INCORRECT

Of the remaining two possibilities, we must choose a parent that includes both carbon atoms of the double bond. Only one parent chain contains both the functional group and the two carbon atoms of the double bond. "Containing the functional group" means that the OH is connected to a carbon that is part of the chain. We do not count the oxygen atom itself as part of the chain. It is only attached to the chain. So the chain above is made up of four carbon atoms.

In cases where there is no functional group, then we look for the longest chain that includes the double bond. If there is no double bond, then we look for a triple bond, and choose the longest chain that has the triple bond in it.

If there are no functional groups, no double bonds, and no triple bonds, then we simply choose the longest chain possible.

Now you can see why we are moving our way backward through the naming process. We cannot name the parent correctly unless we can pick out the highest ranking functional group in the compound. So we start naming a compound by first asking which functional group has priority.

EXERCISE 5.18   Name the parent chain in the following compound:

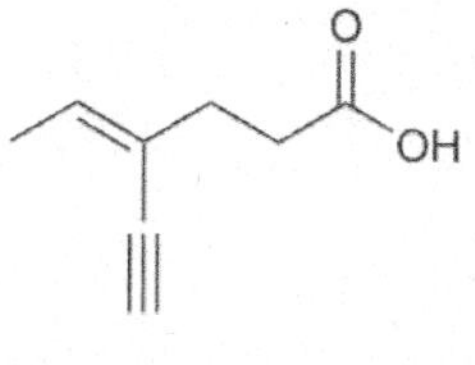

Answer   First we look for a functional group. This compound is a carboxylic acid, so we know the parent chain must include the carbon atom of the carboxylic acid group. Next we look for a double bond. The parent chain should include that as well. This gives us our answer. The triple bond will not be included in the parent chain, because the functional group and the double bond are a higher priority than a triple bond.

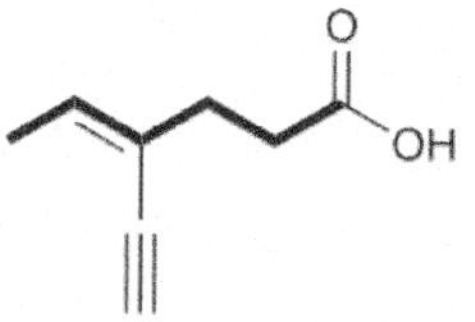

So we count the number of carbon atoms in this chain. There are six (notice that we include the carbon atom of the carboxylic acid group). Therefore, the parent will be called "hex."

PROBLEMS   Name the parent chain in each of the following compounds.

5.19 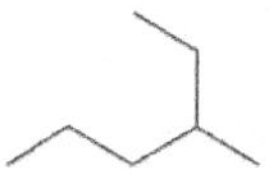 Parent: _____

5.20 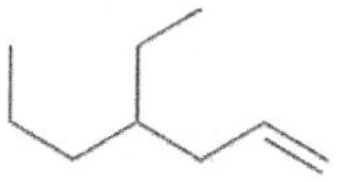 Parent: _____

5.21 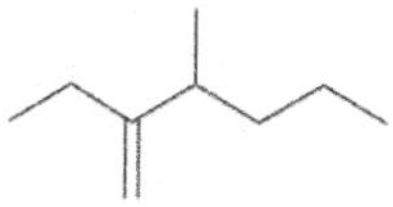 Parent: _____

5.22 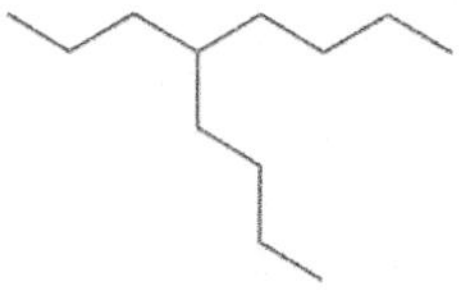 Parent: _____

5.23 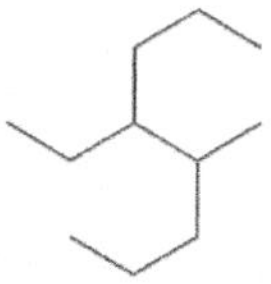 Parent: _____

5.24 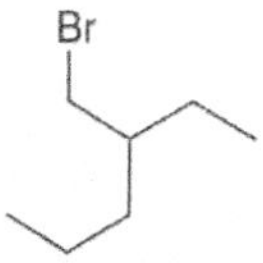 Parent: _____

5.25 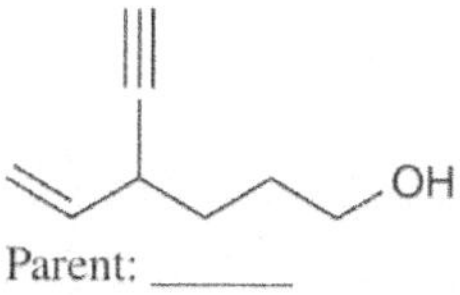 Parent: _____

5.26 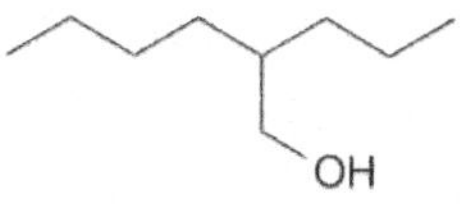 Parent: _____

5.27 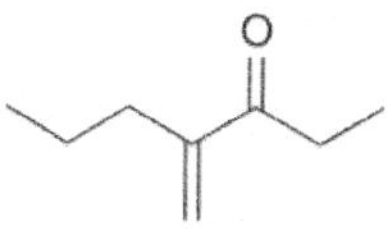 Parent: _____

5.4   NAMING SUBSTITUENTS

Stereoisomerism	Substituents	Parent	Unsaturation	Functional group

Once we have identified the functional group and the parent chain, then everything else connected to the parent chain is called a substituent. In the following example, all of the highlighted groups are substituents, because they are not part of the parent chain:

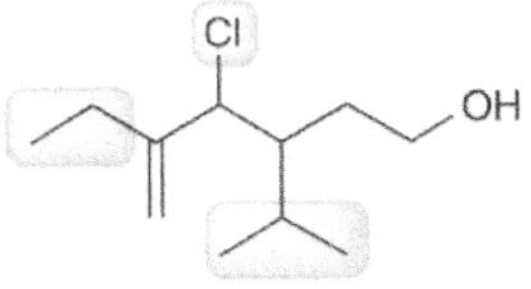

We start by learning how to name the alkyl substituents. These groups are named with the same terminology that we used for naming parent chains, but we add "yl" to the end to indicate that it is a substituent:

Number of carbon atoms in the substituent	Substituent
1	methyl
2	ethyl
3	propyl
4	butyl
5	pentyl
6	hexyl
7	heptyl
8	octyl
9	nonyl
10	decyl

Methyl groups can be shown in a number of ways, and all of them are acceptable:

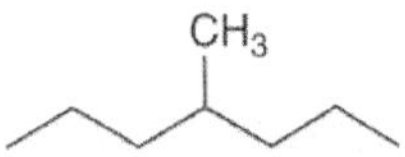

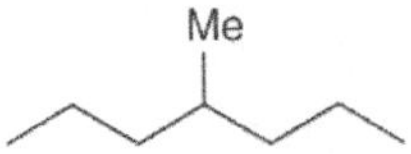

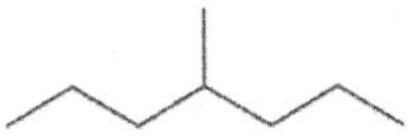

Ethyl groups can also be shown in a number of ways:

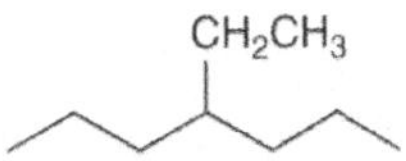

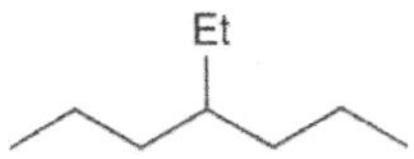

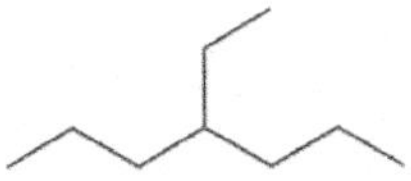

Propyl groups have three carbon atoms and can either be linear or branched, as shown in the following examples:

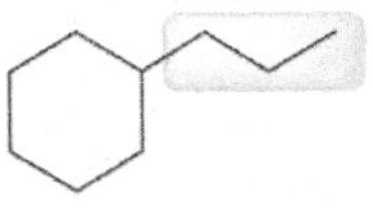

Linear substituent
n-propyl

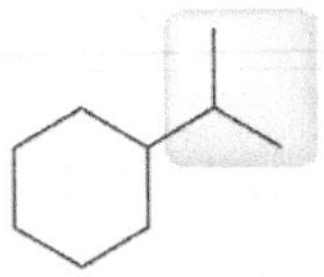

Branched substituent
isopropyl

The first compound has a linear substituent, called an *n*-propyl group (where the "*n*" stands for "normal"), while the second compound has a branched substituent, called an isopropyl group.

Butyl groups can also be linear or branched, although now that we have four carbon atoms, there are more possible ways to connect those atoms. Below is a linear butyl group, called an *n*-butyl group, as well as three different ways for a butyl group to be branched. These branched substituents are called an isobutyl group, a *sec*-butyl group, and a *tert*-butyl group:

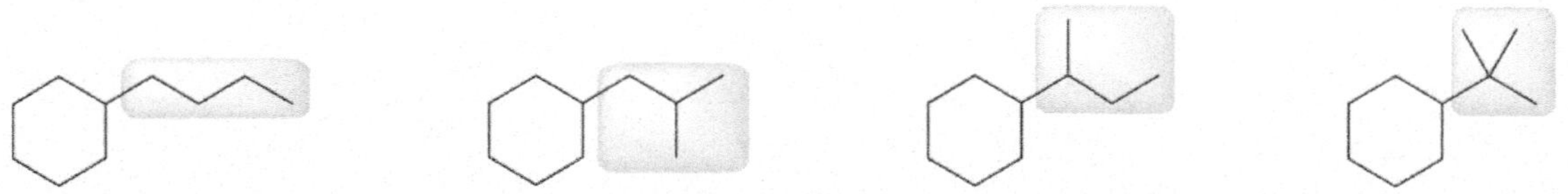

n-butyl   **isobutyl**   **sec-butyl**   **tert-butyl**

Once again, the "*n*" (of an *n*-butyl group) indicates that the substituent is linear. Note that throughout your organic course, you will regularly encounter all of the substituents shown above (*n*-propyl, isopropyl, *n*-butyl, isobutyl, *sec*-butyl and *tert*-butyl groups), so it would be wise to be familiar with them.

There is another important type of substituent that we need to consider (other than alkyl groups). When we learned about functional groups, we saw that some compounds can have two functional groups. When this happens, we choose one of the functional groups to be named as the suffix, and the other functional group must be named as a substituent. To choose the functional group that gets the priority, go back to the section on functional groups and you will see the list of functional groups. We need to know how to name these functional groups as substituents. The OH group is named –hydroxy- as a substituent. The NH$_2$ group is called –amino- if it is named as a substituent. A ketone is called –keto- as a substituent, and an aldehyde is called –aldo- as a substituent. Knowing how to name those four functional groups as substituents will probably cover you for most situations that you will encounter in your course.

Halogens are named as substituents in the following way: fluoro, chloro, bromo, and iodo. Essentially, we add the letter "o" at the end to say that they are substituents. If there are multiple substituents of the same kind (for example, if there are five chlorine atoms in the compound), we use the same prefixes that we used earlier when classifying the number of double and triple bonds:

di = 2   penta = 5
tri = 3   hexa = 6
tetra = 4

So a compound with five chlorine atoms would be "pentachloro." Each and every substituent needs to be numbered so that we know where it is connected to the parent chain, but we will learn about this after we have finished going through the five parts of the name. At that time, we will also discuss in what order to place substituents in the name.

EXERCISE 5.28 In the following compound, identify all groups that would be considered substituents, and then indicate how you would name each substituent:

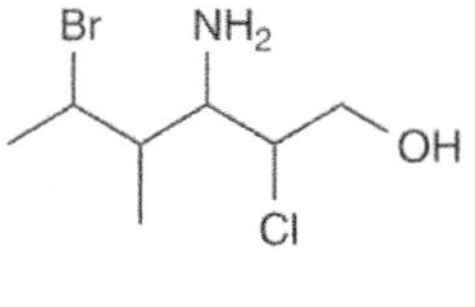

Answer First we must locate the functional group that gets the priority. Alcohols outrank amines, so the OH group is the priority functional group. Then, we need to locate the parent chain. There are no double or triple bonds, so we choose the longest chain containing the carbon atom connected to the OH group:

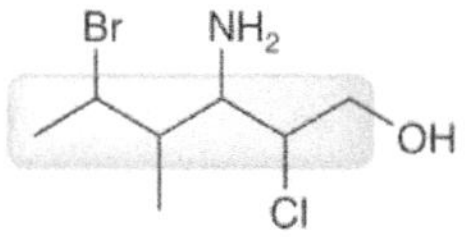

Now we know which groups must be substituents, and we name them accordingly:

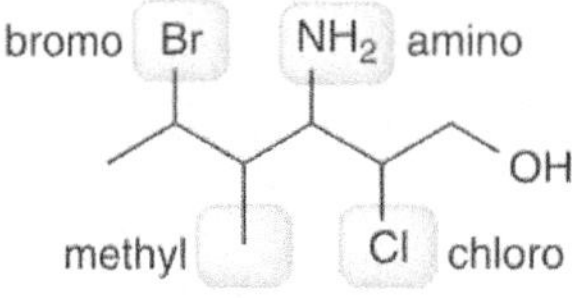

PROBLEMS For each of the following compounds, identify all groups that would be considered substituents, and then indicate how you would name each substituent.

5.29

5.30

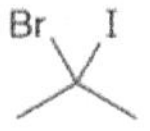

5.31

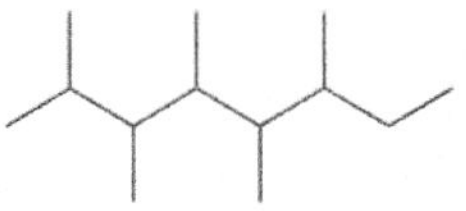

5.32

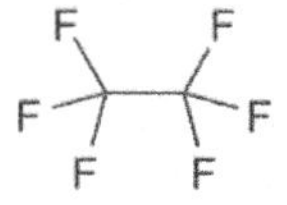

5.33

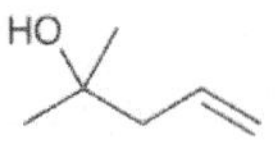

5.34

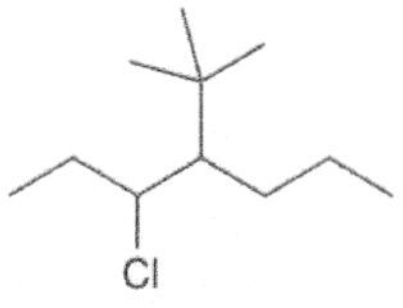

5.35

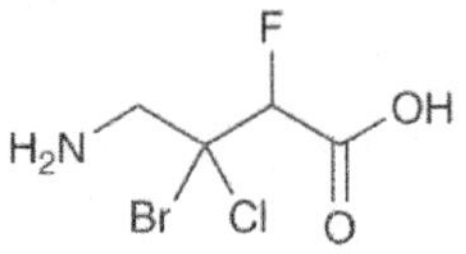

5.36

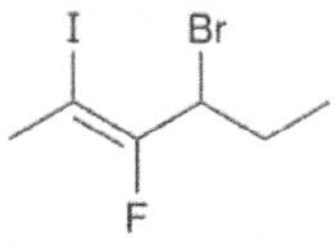

5.37

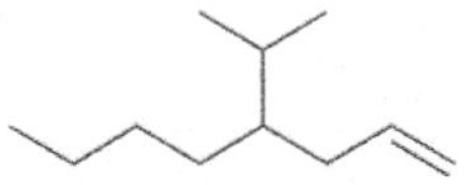

5.38

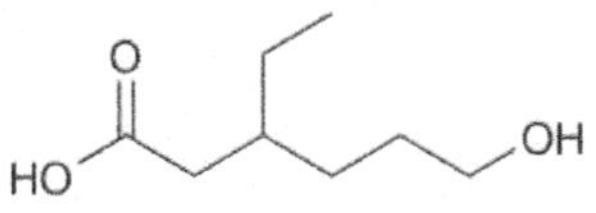

5.5 STEREOISOMERISM

Stereoisomerism	Substituents	Parent	Unsaturation	Functional group

Stereoisomerism is the first part of every name. It identifies the configuration of any double bonds or chiral centers (chiral centers will be covered in Chapter 7). If there are no double bonds or chiral centers in the molecule, then you don't need to worry about this part of the name. If there are, you must learn how to identify the configuration of each. Identifying the configuration of a chiral center requires a chapter to itself. You will need to learn what chiral centers are, how to locate them in molecules, how to draw them, and how to assign a configuration (*R* or *S*). These topics will all be covered in detail in Chapter 7. At that time, we will revisit how to appropriately identify the configuration in the name of the molecule. For now, you should know that configurations are indicated in the first part of a name.

Here we will focus on double bonds, which can often be arranged in two ways:

This is very different from the case with single bonds, which are freely rotating all of the time. A double bond is the result of overlapping *p* orbitals, and double bonds *cannot* freely rotate at room temperature (if you had trouble with this concept when you first learned it, you should review the bonding structure of a double bond in your textbook or notes). So there are two ways to arrange the atoms in space: *cis* and *trans*. If you compare which atoms are connected to each other in each of the two possibilities, you will notice that all of the atoms are connected in the same order. The difference is how they are connected *in 3D space*. This is why they are called stereoisomers (this type of isomerism stems from a difference of orientation in space—"stereo").

To name a double bond as being *cis* or *trans*, you need to have identical groups that can be compared to each other. If these identical groups are on the same side of the double bond, we call them *cis*. If they are on opposite sides, we call them *trans*:

The two groups that we compare can even be hydrogen atoms. For example,

But what do you do if you don't have two identical groups to compare? For example,

These compounds are clearly not the same. We cannot use *cis/trans* terminology to differentiate them, because we don't have two identical groups to compare. In situations when all four groups on the double bond are different, we have to use another method for naming them.

The other way of naming double bonds uses rules similar to those used in determining the configuration of a chiral center (*R* versus *S*), so we will wait until Chapter 7 (when we learn about *R* and *S*), and then we will cover this alternative way of naming double bonds. The alternative method is far superior, because it can be used to name any double bond. In contrast, *cis/trans* nomenclature is generally only used to name disubstituted alkenes. The reason that we do not drop the *cis/trans* terminology altogether is probably based in deep-rooted tradition and usage of these terms.

There is one situation when we don't have to worry about *cis/trans* because there aren't two ways to arrange the double bond. If we have two identical groups connected *to the same atom*, then we cannot have stereoisomers. For example, because there are two chlorine atoms connected to one side of the double bond. Why are the two drawings the same? Remember that the carbon atoms of the double bond are sp^2 hybridized, and therefore trigonal *planar*. So, if we flip over the first drawing, we get the other drawing. They are the same thing. To see this, take two pieces of paper. Draw one of these compounds on one piece of paper, and draw the other compound on the second piece of paper. Then flip over one of the pieces of paper, and hold it up to the light so that you can see the drawing through the back side of the paper. Compare it to the other drawing and you will see that they are the same. If you try to do the same thing with some of the previous examples (that did have *cis* and *trans* stereoisomers), you will find that flipping the page over does not make the two drawings the same.

PROBLEMS   For each of the compounds below, identify whether the double bond is *cis* or *trans*.

5.39

5.40 F   F

5.41

5.42

5.43

5.6   NUMBERING

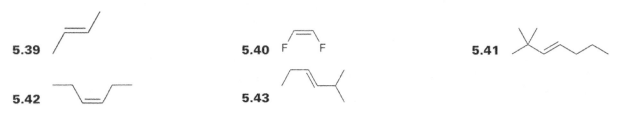

| Stereoisomerism | Substituents | Parent | Unsaturation | Functional group |

Numbering applies to all parts of the name

We're almost ready to start naming molecules. We finished learning about the individual parts of a name, and now we need to know how to identify how the pieces are connected. For example, let's say you determine that the functional group is OH (therefore, the suffix is –ol), there is one double bond (-en-), the parent chain is six carbon atoms long (hex), there are four methyl groups attached to the parent chain (tetramethyl), and the double bond is *cis*. Now you know all of the pieces, but we must find a way to identify where all of the pieces are on the parent chain. Where are all of those methyl groups? (and so on). This is where the numbering system comes in. First we will learn how to number the parent chain, and then we will learn the rules of how to apply those numbers in each part of the name.

Once we have chosen the parent chain, there are only two ways to number it: right to left or left to right. But how do we choose? To number the parent chain properly, we begin with the same hierarchy that we used when choosing the parent in the first place:

Functional group

Double bond

Triple bond

If there is a functional group, then number the parent chain so that the functional group gets the lower number:

OH gets the number 2 instead of 5

If there is no functional group, then number the chain so that the double bond gets the lower number:

the double bond is 1 instead of 5

If there is no functional group or double bond, then number the chain so that the triple bond gets the lower number:

the triple bond is 1 instead of 5

If there is no functional group, double bond, or triple bond, then we should number the chain so that the substituent has the lower number:

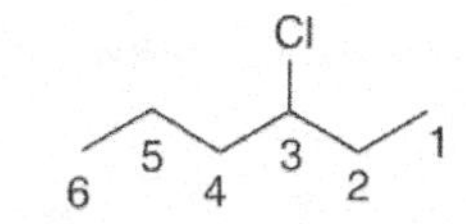

Cl gets the number 3 instead of 4

If there is more than one substituent on the parent chain, then we should number the chain in the direction that gives the lower number to the first substituent. In the following example, whether we number from left-to-right or from right-to-left, either way, the first substituent will be at position 3. So in this case, we consider the position of the second substituent. In this case, we will number from right-to-left, so that the second substituent is at position 3, rather than at position 4:

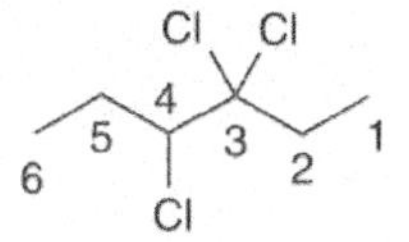

3,3,4-trichloro instead of 3,4,4-trichloro

EXERCISE 5.44   For the compound below, choose the parent chain and then number it correctly:

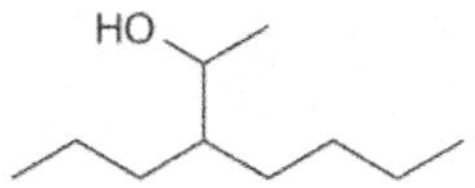

Answer   To choose the parent chain, remember that we need to choose the longest chain that contains the carbon atom connected to the OH group:

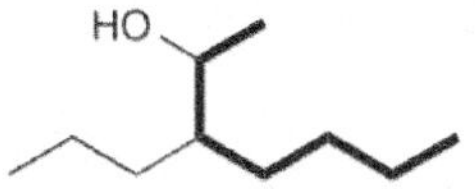

To number the parent chain correctly, we need to go in the direction that gives the functional group the lowest number:

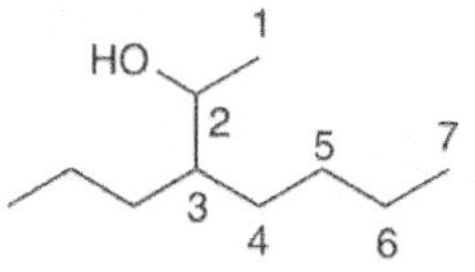

PROBLEMS   For each of the compounds below, choose the parent chain and number it correctly.

5.45

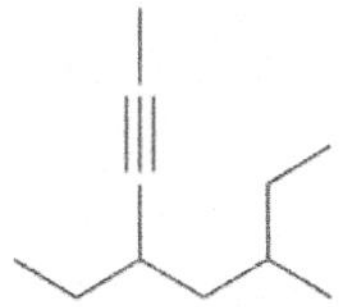

5.46

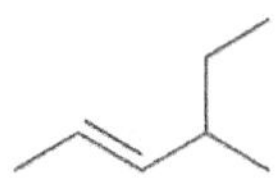

5.47

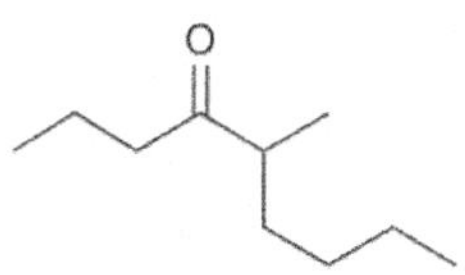

5.48

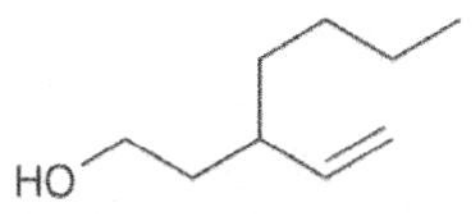

5.49

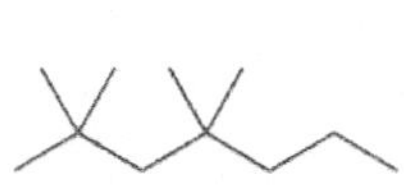

5.50

5.51 [structure] **5.52** [structure with OH] **5.53** [structure]

Now that we know how to number the parent chain, we need to see how to apply those numbers to the various parts of the name.

Functional Group The number generally gets placed directly in front of the suffix (for example, hexan-2-ol). When placing a number, it is OK to place the number at the beginning of the parent. For example, 2-hexanol is the same as hexan-2-ol. Both names are acceptable.

Unsaturation For double and triple bonds, the number indicates the lower number of the two carbon atoms. For example,

[structure with numbers 2 4 6 on top and 1 3 5 on bottom] We use the number 2

The double bond is between C2 and C3, so we use C2 to number the double bond. So the example above is hex-2-ene (or 2-hexene). We treat triple bonds in a similar way.

If there are two double bonds in the molecule, then we must indicate both numbers; for example, hexa-2,4-diene, or 2,4-hexadiene. Every double and triple bond must be numbered.

Substituents The number of the substituent goes immediately in front of the substituent. Examples:

[structure with Cl] **2-chloro**hexane   [structure] **3-methyl**pentane

If there are multiple substituents, then every substituent must be numbered:

[structure with Cl, Cl] **2,3-dichloro**hexane   [structure] **2,2,4-trimethyl**pentane

If there are multiple substituents of different types, then we must alphabetize the substituents in the name. Consider the following example:

[structure with Cl F F CH₃]

There are four types of substituents in the example above (chloro, fluoro, ethyl, and methyl). They must be alphabetized (c, e, f, m) (we do not count di, tri, tetra, etc. as part of the alphabetization system). So the compound above is called

2-chloro-3-ethyl-2,4-difluoro-4-methylnonane

Note that two numbers are always separated by commas (2,4 in example above) but letters and numbers are separated by dashes.

Stereoisomerism If there are any stereoisomeric double bonds, we place the term *cis* or *trans* at the beginning of the name. If there is more than one double bond, then we need to indicate *cis* or *trans* for each double bond, and we must number accordingly (for example, 2-*cis*-4-*trans* ...). If there are any chiral centers, here is where we would indicate them; for example, (2R,4S). We will see more of this when we learn about chiral centers in the upcoming chapters.

There it is. A lot of rules. No one ever said nomenclature would take 10 minutes to learn, but with enough practice, you should get the hang of it. Let's now take everything we have learned and practice naming compounds:

EXERCISE 5.54   Name the following compound:

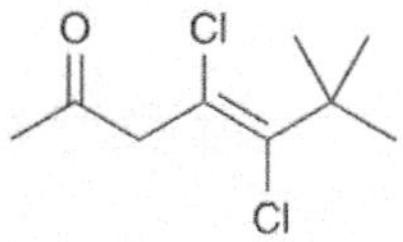

Answer   We go through the five parts of the name backward. So we start by looking for the functional group. We see that this compound is a ketone. So we know the end of the name will be –one.

Next, we look for unsaturation. There is a double bond here, so there will be –en- in the name.

Next we need to name the parent. We locate the longest chain that includes the functional group and double bond. In this example, it is an obvious choice. The parent has seven carbon atoms, so the parent is –hept-.

Next we look for substituents. There are two methyl groups and two chlorine atoms. We need to alphabetize, and c comes before m, so it will be dichlorodimethyl.

Next we look for stereoisomerism. The double bond in this molecule has two chlorine atoms on opposite sides, so it is *trans*. This part of the name (*trans*) is generally italicized and surrounded by parentheses. So far, we have

(*trans*)-dichlorodimethylheptenone

Now that we have figured out all of the pieces, we must assign numbers. We need to number the parent to give the functional group the lowest number, so the numbering, in this example, will go from left to right. This puts the functional group at the number 2 position, the double bond at the 4 position, the chlorine atoms at 4 and 5, and the methyl groups at 6 (both of them). So the name is

(*trans*)-4,5-dichloro-6,6-dimethylhept-4-en-2-one

Note: In Chapter 7, we will learn an alternate method for assigning the configuration of the double bond in this example. To be accurate, *cis-trans* terminology is typically reserved for cases in which the alkene is disubstituted, and the alternate naming method (that we will soon see) is the preferred method for naming.

PROBLEMS   Name each of the following compounds. (Ignore chiral centers for now. We will focus on chiral centers in the upcoming chapters.)

5.55 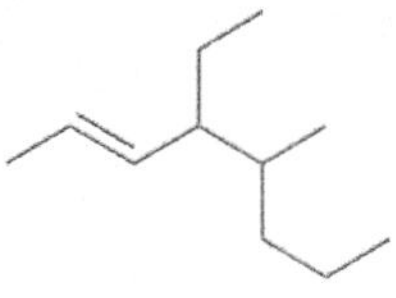   Name: _____

5.56 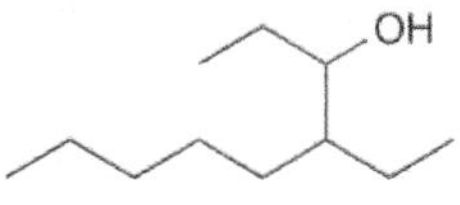   Name: _____

5.57 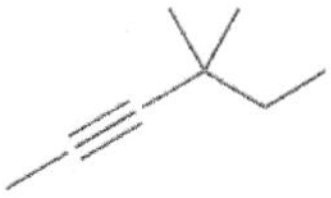   Name: _____

5.58 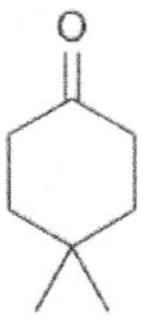   Name: _____

5.59   Name: _____

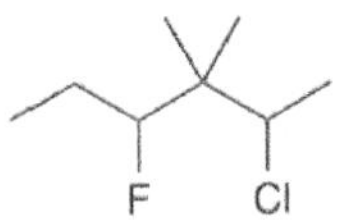

5.60   Name: _____

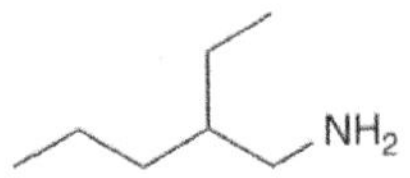

5.61   Name: _____

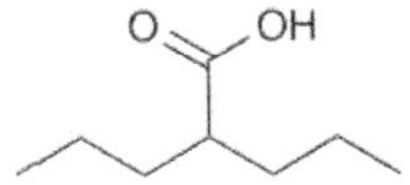

5.62   Name: _____

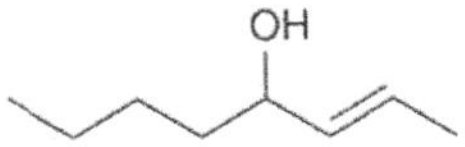

5.63   Name: _____

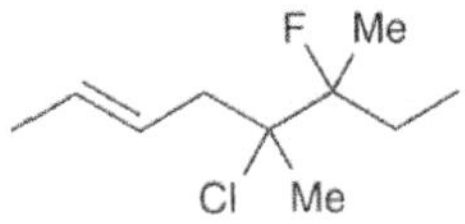

5.7 COMMON NAMES

In addition to the rules for naming compounds, there are also some common names for some simple and common organic compounds. You should be aware of these names to the extent that your course demands this of you. Each course will be different in terms of how many of these common names you should be familiar with. Here are some examples:

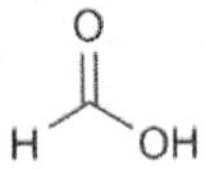

IUPAC Name = methanoic acid
Common Name = formic acid

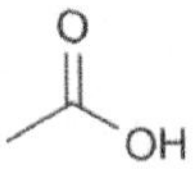

IUPAC Name = ethanoic acid
Common Name = acetic acid

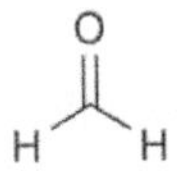

IUPAC Name = methanal
Common Name = formaldehyde

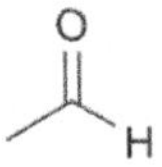

IUPAC Name = ethanal
Common Name = acetaldehyde

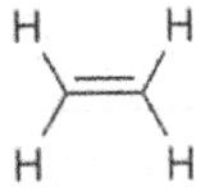

IUPAC Name = ethene
Common Name = ethylene

IUPAC Name = ethyne
Common Name = acetylene

Most of these examples are so common that it is quite rare to hear someone refer to these compounds by their IUPAC names. Their common names are much more "common," which is why we call them common names.

Ethers are typically called by their common names. The group on either side of the oxygen atom is named as a substituent before the term ether. Examples:

Diethyl ether   Dimethyl ether

It is not a bad idea to familiarize yourself with all of the common names listed in whatever textbook you are using.

5.8 GOING FROM A NAME TO A STRUCTURE

Once you have completed all of the problems in this chapter, you will find that it is much easier to draw a compound when you are given the name than it is to name a compound that is drawn in front of you. It is easier for the following reason: when naming a compound, there are a lot of decisions you need to make (which functional group has priority, what is the parent chain, how the chain should be numbered, in what order to put the substituents in the name, etc.). But when you have a name in front of you and you need to draw a structure, you do not need to make any of these decisions. Just draw the parent and then start adding everything else to it according to the numbering system provided in the name.

For practice, make a list of the answers to problems 5.55–5.63. This list should just be names. Wait a few days until you cannot remember what the structures looked like and then try to draw them based on the names. You can also use your textbook for more examples.

From this point on, I will assume that I can say names like 2-hexanol and you will know what I mean. That is what your textbook will do as well, so now is the time to master nomenclature.

CHAPTER 6

CONFORMATIONS

Molecules are not inanimate objects. Unlike rocks, they can twist and bend into all kinds of shapes, very much like people do. We have limbs and joints (elbows, knees, etc.) that give us flexibility. Although our bones might be very rigid, nevertheless, we can achieve a great range of movement by twisting our joints in different ways. Molecules behave the same way. Once you learn about the general types of joints that molecules have and the ranges of motion available to those joints, you will then be able to predict how individual molecules can twist around in space. Why is this important?

Think about how many common positions you assume every day. You can sit, you can stand, you can lean against something, you can lie down, you can even stand on your head, and so on. Some of these positions are comfortable (such as lying down), while other positions are very uncomfortable (such as standing on your head). There are some activities, like drinking a glass of water, which can be done only in certain positions. You cannot drink a glass of water while lying down or standing on your head, not easily anyway.

Molecules are very similar. They have many positions into which they can twist and bend. There are some comfortable positions (low in energy), while other positions are uncomfortable (high in energy). The various positions available to a molecule are called *conformations*. It is important to be able to predict the conformations available to molecules, because there are certain activities that can be performed only in specific conformations. Just as a person standing on her head cannot drink a glass of water, so, too, molecules cannot undergo some reactions from certain conformations. Just as you can run only when you are in a standing position, molecules can often undergo certain reactions from only one specific conformation. If the molecule cannot twist into the conformation necessary for a reaction to take place, then the reaction will not happen. So you can understand why you will need to be able to predict what conformations are available to molecules—so that you can predict when reactions can and cannot happen.

There are two very important drawing styles that show conformations and give us the power to predict what conformations are available to different types of molecules: Newman projections and chair conformations:

Newman projection 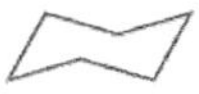   Chair

We begin this chapter with Newman projections.

6.1 HOW TO DRAW A NEWMAN PROJECTION

Before we can talk about drawing Newman projections, we need to first review one aspect of drawing bond-line structures that we did not cover in Chapter 1. To show how groups are positioned in 3D space, we often use wedges and dashes:

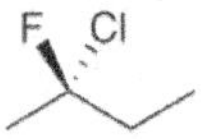

In the bond-line drawing above, the fluorine atom is on a wedge and the chlorine atom is on a dash. The wedge means that the fluorine is coming out toward us in 3D space, and the dash means that the chlorine is going away from us in 3D space. Imagine that all four carbon atoms in the molecule above are positioned in the plane of the page. If you look at this page from the left side (so the page looks two-dimensional), you would see the fluorine sticking out of the page to your right and the chlorine sticking out of the page to the left.

Do not be confused by whether the dash is drawn on the right or left:

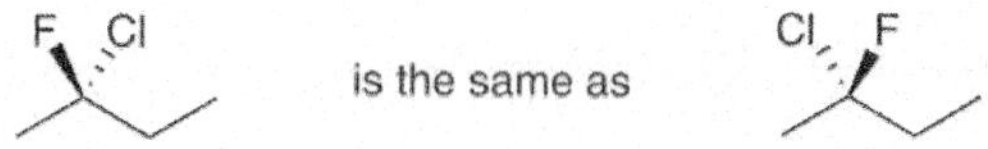

is the same as

In both drawings above, the chlorine is on a dash and the fluorine is on a wedge, so these drawings are the same. In reality, both the chlorine and the fluorine should be drawn straight up—the chlorine goes straight up and behind the page, while the fluorine goes straight up and in front of the page. If we drew it like that, we would not be able to see the chlorine because the fluorine would be blocking it (the way the moon blocks the sun in a solar eclipse). To clearly see both groups, we move one of the groups slightly to the left and the other slightly to the right. It does not matter which is on the right and which is on the left. All that matters is which one is on the wedge and which one is on the dash.

Now that we understand what the dash and wedge mean, let's consider what the molecule would look like from a slightly different angle:

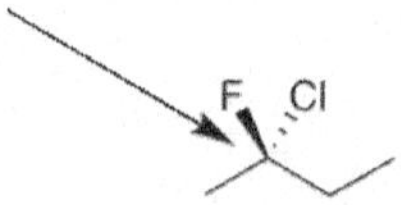

Imagine looking at the molecule from the angle shown by the arrow above. If you are not sure what angle we are talking about, try doing the following: Turn your book so that it is facing your stomach instead of your head. Now turn the page so you are looking at it from the side, like we did before. You should be looking down the path of this arrow at the molecule. In this view, you are looking directly down a carbon–carbon bond, where one carbon is in front and one is in back:

Look from here

front
carbon   back
carbon

In this view, you will see three groups connected to the front carbon atom. You should expect to see a fluorine atom sticking out of the page to the right and a chlorine atom sticking out of the page to the left. You would also see a methyl group pointing straight down. This is what it would look like from that view:

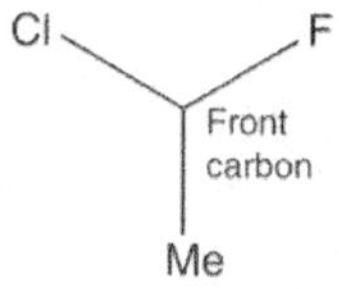

Cl   F
Front
carbon
Me

You would see the three groups like this, and you would not see the back carbon atom, because the carbon atom in front would be covering it up (again, like the moon covers the sun during a solar eclipse). Let's try to draw that back carbon that we cannot see, and by convention, we will draw it as a big circle:

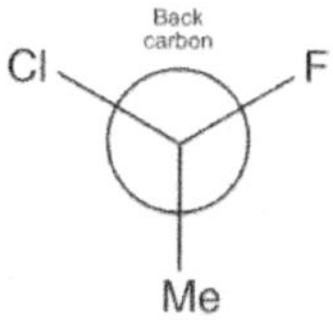

Back
carbon
Cl   F
Me

Now we draw the three groups that are connected to the back carbon atom. There is one methyl group and two hydrogen atoms. If we put them into our drawing, then we get our Newman projection. It looks like this:

It is important that you can see what this drawing represents, because we cannot move forward until you see it very clearly. We are essentially viewing down a carbon–carbon bond, focusing on the three groups attached to each carbon atom. The central point in our Newman projection (where the lines to the Cl, F, and Me meet each other) is the first carbon. The big circle in the back is our back carbon. All at once you can see all six groups (the three connected to the front carbon atom and the three connected to the back carbon atom). So a Newman projection is another way of drawing the compound we showed earlier:

is the same as

Let's use one more analogy to help us understand this. Imagine that you are looking at a fan that has three blades. Behind this fan, there is another fan that also has three blades. So you see a total of six blades. If both fans were spinning, and you started taking photographs, you might find some photos where you can clearly see all six blades, and other photos where the three blades in the front are blocking our view of the three blades in the back. In this last case, the three blades in front would be *eclipsing* the three blades in back.

This is where the analogy helps us understand why Newman projections are useful. The bond connecting the two carbon atoms is a single bond that can freely rotate. As such, the three groups in the front and the three groups in the back can be staggered or eclipsed:

Staggered

Eclipsed

Think of the front carbon atom and its three groups as one fan, and the back carbon atom and its three groups as a different fan. These two fans can spin independently of each other, which gives rise to many different possible conformations. This is why Newman projections are so incredibly powerful at showing conformations. They are drawn in a way that is perfect for showing the various conformations that arise as an individual single bond rotates.

It gets a little more complicated when we realize that most single bonds experience free rotation, giving rise to a very large number of conformations. We can avoid that kind of complexity by focusing just on one particular bond, and the various conformations that arise from free rotation of just that bond. If we can learn how to do that, then we can do that for the part of the molecule that is undergoing a reaction.

EXERCISE 6.1 Draw a Newman projection of the following compound looking from the perspective of the arrow:

Answer   The first thing to realize is that there are groups on dashes and wedges that have not been shown. They are the hydrogen atoms. We did not focus on drawing the dashes and wedges in Chapter 1, but the hydrogen atoms are, in fact, on dashes and wedges. We learned in the chapter on geometry that these carbon atoms would be classified as sp^3 hybridized, and, therefore, their geometry is tetrahedral. That means the hydrogen atoms are coming in and out of the page:

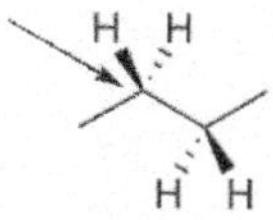

Now we draw the front carbon atom with its three groups. Looking along the direction of the arrow, we see a hydrogen atom that is up and to the left, another hydrogen atom that is up and to the right, and a methyl group pointing straight down. So we draw it like this:

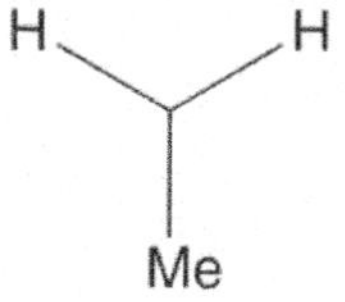

Next we draw the back carbon atom as a large circle and we look at all three groups attached to it. There is a methyl group pointing straight up and then two hydrogen atoms pointing left and right. So the answer is

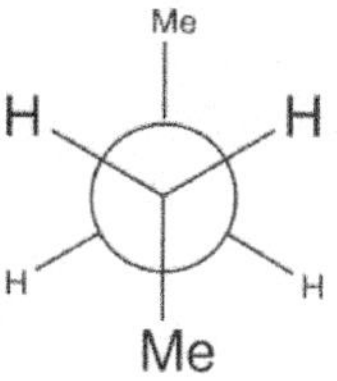

PROBLEMS   Draw Newman projections for each of the following compounds. In each case the skeleton of the Newman projection is drawn for you. You just need to fill in the six groups in their proper places.

6.2   Answer

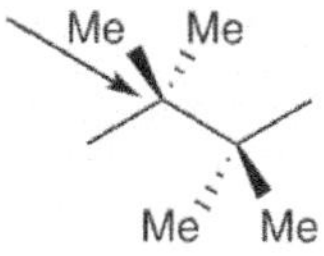

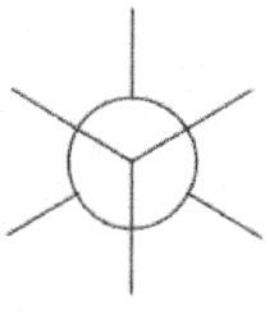

6.3   Answer

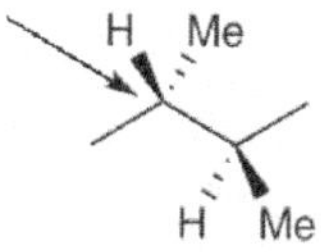

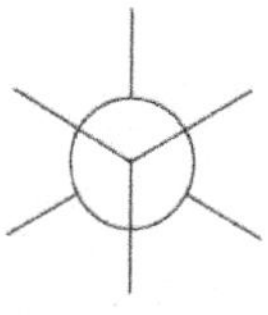

6.4   Answer

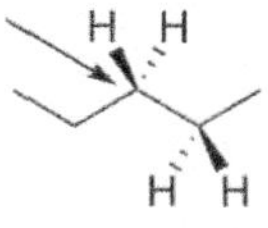

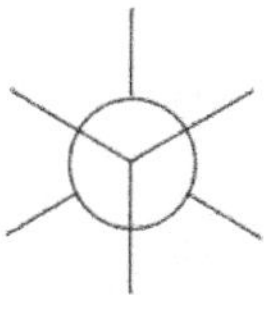

6.5   Answer

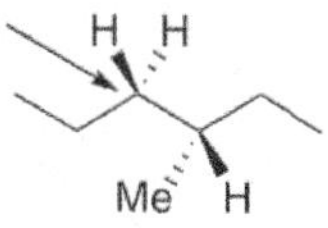

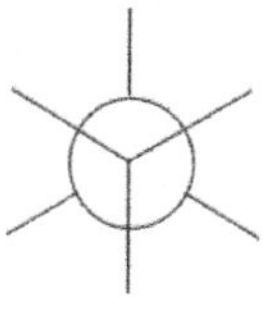

6.6   Answer

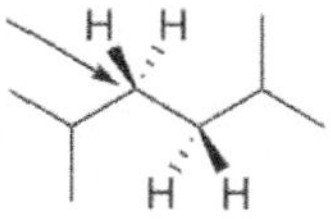

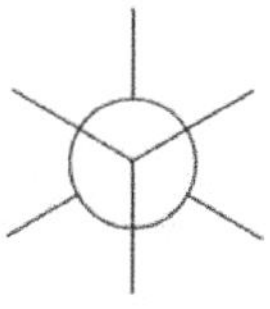

6.7   Answer

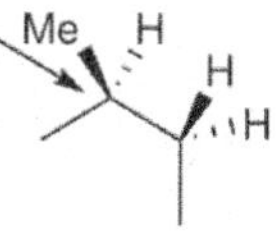

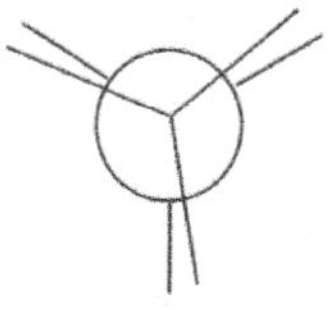

6.2 RANKING THE STABILITY OF NEWMAN PROJECTIONS

We have seen that Newman projections are a powerful way to show the different conformations of a molecule. We mentioned earlier that there are staggered conformations and eclipsed conformations. In fact, there are three staggered and three eclipsed conformations. Let's draw all three staggered conformations of butane. The best way to do this is to keep the back carbon atom motionless (so the fan in the back is not spinning), and let's slowly turn the groups in the front (only the front fan is spinning):

Look at the first drawing above and notice the placement of the methyl group at the bottom. If we rotate the front carbon atom clockwise with all three of its groups, this methyl group ends up in the top left (as seen in the second drawing). Then we rotate one more time to get the third drawing. Rotating one more time regenerates the first drawing. In the first drawing, the methyl groups are as far away from each other as possible, which is the most stable conformation, called the *anti* conformation. The other two drawings both have the methyl groups near each other in space. They feel each other and they are a bit crowded (a steric interaction). This interaction is called a *gauche* interaction, which makes these conformations a little bit less stable than the *anti* conformation.

If we were to go back to our comparison between molecules and people, we would say that the *anti* conformation would be like lying down in a bed, and the gauche conformations are both like sitting in a chair. All of these are comfortable positions, but lying down is the most comfortable. The *anti* conformation is the most stable.

Now let's look at the three eclipsed conformations of butane. Again, let's keep the back where it is, and let's just rotate the front carbon atom with its three groups:

These conformations are all high in energy relative to the staggered conformations. All of the groups are eclipsing each other, so they are very crowded. All three of these would be like standing on your head, which is extremely uncomfortable. But the middle one is the most unstable, because the two methyl groups (the two largest groups) are eclipsing each other. This would be like standing on your head without using your hands for help—now that *really* hurts. So, all of these are high in energy, but the middle one is the most unstable.

To summarize, the most stable conformation will be the staggered conformation where the large groups are as far apart as possible (*anti*), and the least stable conformation will be the eclipsed conformation where the large groups are eclipsing each other.

EXERCISE 6.8 Draw the most stable and the least stable conformations of the following compound, using Newman projections looking down the following bond:

Answer   We begin by drawing the Newman projection that we would see when we look down the bond indicated. This Newman projection will have a methyl group and two hydrogen atoms connected to the front carbon atom, and there will be an ethyl group and two hydrogen atoms connected to the back carbon atom, in the following way:

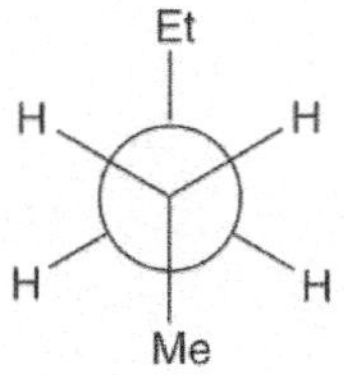

Now we decide how to rotate the front carbon atom so as to provide the most stable conformation. The most stable conformation will be a staggered conformation with the two largest groups *anti* to each other, so in this case, we do not need to rotate at all. The drawing that we just drew is already the most stable conformation, because the methyl and ethyl groups are *anti* to each other:

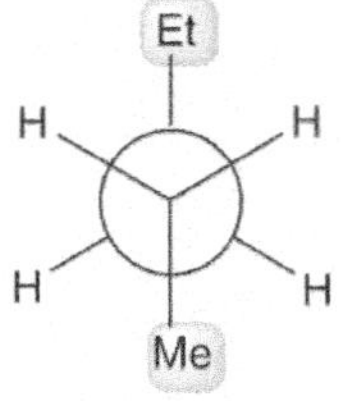

To find the least stable conformation, we need to rotate the front carbon atom and consider all three eclipsed conformations. The least stable conformation will be the one with the two largest groups eclipsing each other:

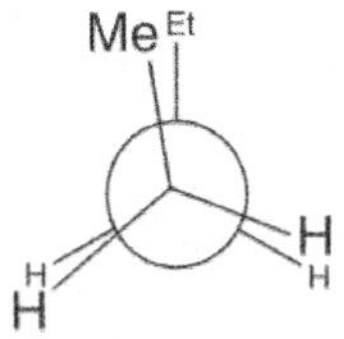

PROBLEMS   Draw the most stable and the least stable conformations for each of the following compounds. In each case, fill in the groups on the Newman projections below.

6.9

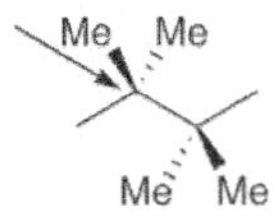

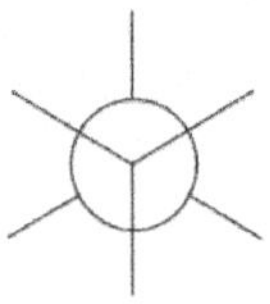

Most stable 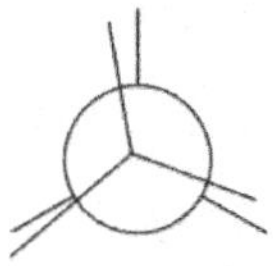   Least stable

6.10

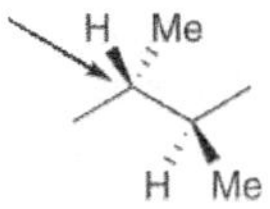

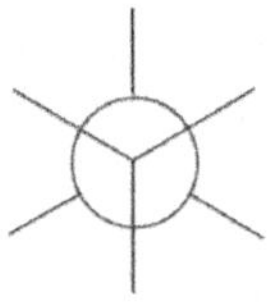

Most stable 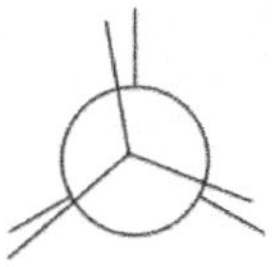   Least stable

6.11

Most stable   Least stable

6.12

Most stable   Least stable

6.13

Most stable   Least stable

6.14

Most stable   Least stable

6.3 DRAWING CHAIR CONFORMATIONS

An interesting case of conformational analysis comes to play when we consider a six-membered ring (cyclohexane). There are many conformations that this compound can adopt. You will see them all in your textbook: the chair, the boat, the twist-boat. The most stable conformation of cyclohexane is the chair. We call it a chair, because when you draw it, it looks like a chair:

You can almost imagine someone sitting on this structure, as if it were a beach chair. Most students have a difficult time drawing the chair and its substituents correctly. In this section, we will focus on learning how to draw the chair correctly. It is very important, because we cannot move on to see more about chairs until you know how to draw them.

You will need to practice this, step by step. Begin by drawing a *very wide* V:

Next, draw a line from the top right of the V, going down at a 60-degree angle, and stop a little bit to the right of an imaginary line coming straight down from the center of the V:

Next, draw a line parallel to the left side of the V and stop a little bit to the right of an imaginary line coming straight down from the top left of the V:

Then, start at the top left of the V and draw a line parallel to the line all the way on the right side, going down exactly as low as that line goes:

Finally, connect the last two lines:

Please don't *ever* draw a chair like this:

When you draw the chair sloppily (as so many students do), it makes it impossible to place the substituents correctly on the ring. And that's when you start losing silly points on your exam. So, take the time to practice and learn how to draw it properly. Practice in the space below:

Now we can begin practicing how to draw substituents connected to the ring. Start with the top right corner and draw a line going straight up:

Then go around the ring and draw a straight line at each carbon atom, alternating between up and down:

These six substituents are called *axial* substituents. They go straight up and down, in the order shown above.

Next we need to learn how to draw the *equatorial* substituents. These are the substituents pointing out toward the sides. There are also six of them. Each one is drawn so that it is parallel to the two bonds from which it is once removed:

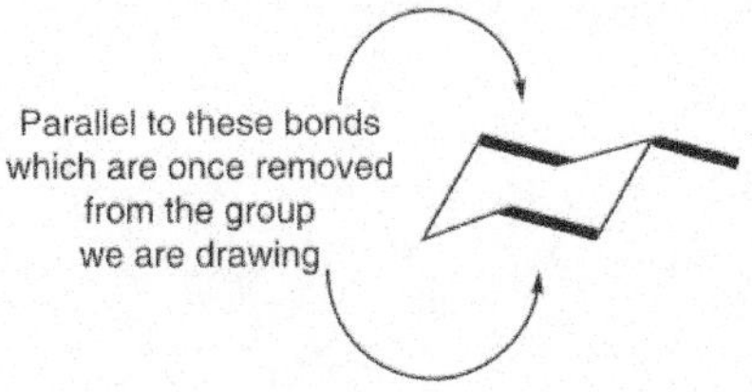

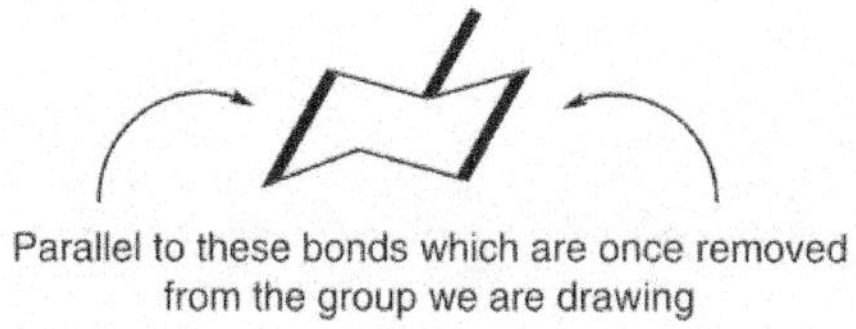

We go all the way around the ring like this, until we have drawn all of the equatorial groups:

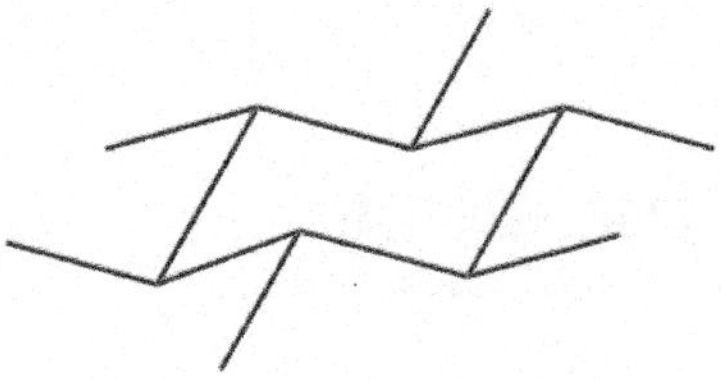

Now we know how to draw all twelve substituents, but remember that if we draw a line and don't put anything at the end of that line, then this implies a methyl group. So, unless we are referring to dodecamethylcyclohexane (that's 12 methyl groups), we must draw hydrogen atoms at the end of these lines:

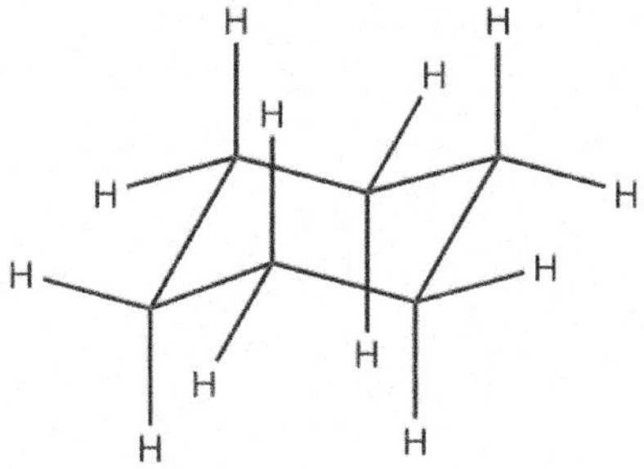

Generally, we do not have to draw all 12 lines and place hydrogen atoms there. Remember how bond-line drawings work: if we don't draw any groups at all, it is assumed that there are hydrogen atoms. We are going through this exercise because it is important to know *how* to draw all 12 substituents. We will see in the next section that in most problems you will need to draw only a few of them. Which ones you draw will depend on the problem. So the only way to be sure that you can draw whatever the problem throws at you is to master drawing *all* of them. *Never* draw groups like this:

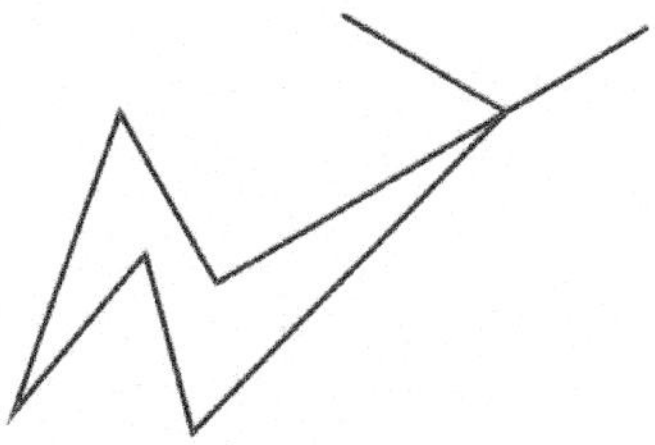

Very bad

Drawings like this will cause you to lose serious points on exams (not to mention the fact that a drawing like this defeats the whole purpose of a chair drawing—the exact positioning of the substituents is very important).

Use the space below to practice drawing a chair with all twelve substituents. When you are finished, label every substituent as axial or equatorial. Do not move on to the next section until you can do this.

6.4 PLACING GROUPS ON THE CHAIR

Now we need to see how to draw a chair with proper placement of the groups when we are given a regular hexagon-style drawing:

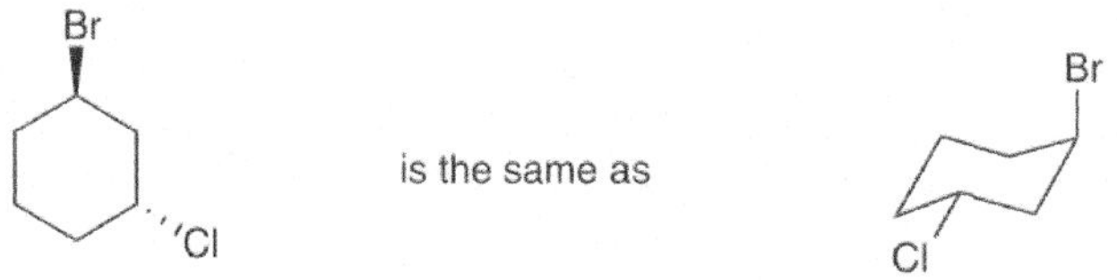

is the same as

Before we can get started, we need to remember what the dashes and wedges mean in the hexagon-style drawing. Remember that wedges are coming out toward you, and dashes are going back away from you. So, at each of the six carbon atoms of the ring, there are two groups—one coming out at you and one going away from you. If the groups are not drawn, then it means that there are two hydrogen atoms, one coming out and one going away.

Now let's introduce some new terminology. This is not scientific terminology, and you likely won't find it in your textbook, but it will help you master the task at hand. Anything coming off of the ring that is on a wedge, we will call *up*, because the group is coming up above the ring. Any group on a dash, we will call *down*, because the group is going down below the ring. So in the previous example, Br is up and Cl is down.

Now let's apply the same terminology to the groups on a chair. Each carbon atom has two groups, one pointing above the ring (up) and one pointing below the ring (down):

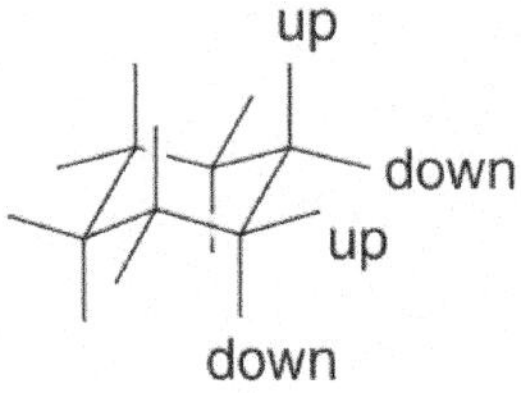

You can do this for every carbon (and you should try on the drawing above), and you will see that each carbon has two groups (up and down). It is important to realize that there is *no correlation* between up/down and axial/equatorial. Look at the drawing above. For one of the carbon atoms, the up position is axial. For the other carbon atom showing its groups, the up position is equatorial. Take a close look at the two equatorial positions shown above. One of them is up and the other is down.

Now we are ready to draw a chair when we are given a hexagon. Let's work through the example shown at the beginning of this section:

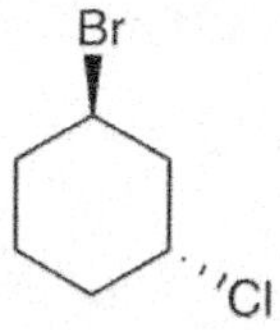

We begin by placing numbers. These do not necessarily have to be the same as the numbers that we used in naming compounds. These are just numbers that help us draw the chair with the groups in the right place. It does not matter where we start or which direction we go in, so let's just say that we will always start at the top and go clockwise:

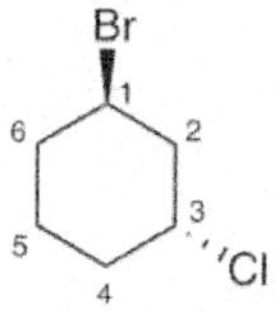

Now, we draw a chair and we put numbers on the chair also. We can start our numbers anywhere on the chair, but we *must* go in the same direction that we did in the hexagon. If we went clockwise there, then we must go clockwise here. To avoid a mistake, let's just say that we will always go clockwise from now on and we will always start at the top right corner:

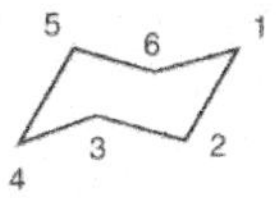

Now we know where to put in the groups. Br is connected to position number 1, and Cl is connected to position number 3. This brings us to the up/down system. Draw the chair, showing both positions (up and down) at each of the carbon atoms where we need to place a group:

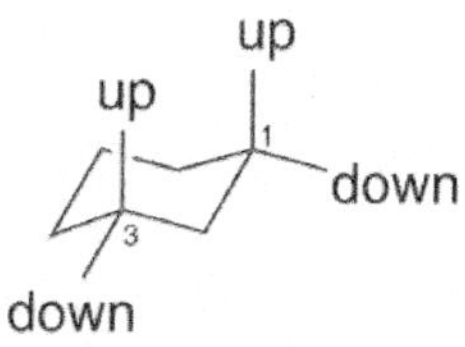

Look at the hexagon drawing again and ask whether each group is up or down. Br is on a wedge, so it is up. Cl is on a dash, so it is down. Now we are ready to put the groups into our chair drawing:

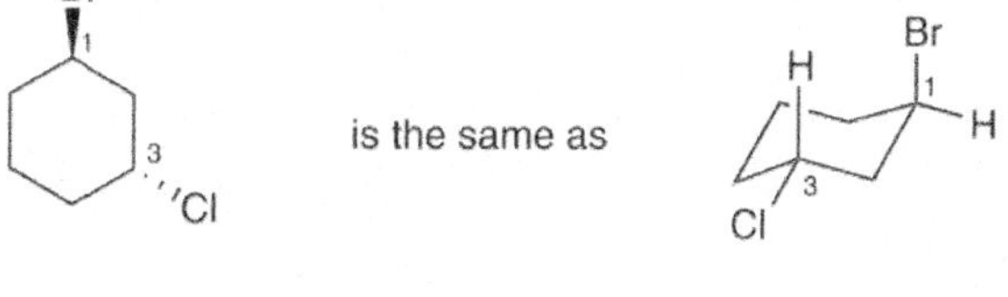

is the same as

That's all there is to it. To review, we need to draw the chair, number the chair and hexagon (both clockwise), determine where the groups go, determine whether they are up or down, and then draw them in. Let's get some practice now.

EXERCISE 6.15   Draw a chair conformation of the compound below:

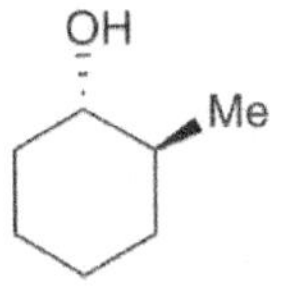

Answer   Begin by numbering the hexagon at the first group and then going clockwise. This puts the OH at the position numbered 1 and the Me at the position numbered 2:

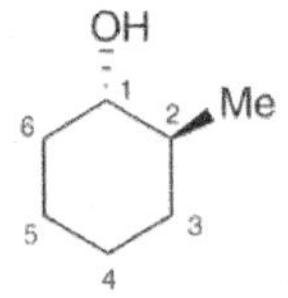

Next draw the chair, number it going clockwise, and then put in the up and down positions at the carbon atoms numbered 1 and 2:

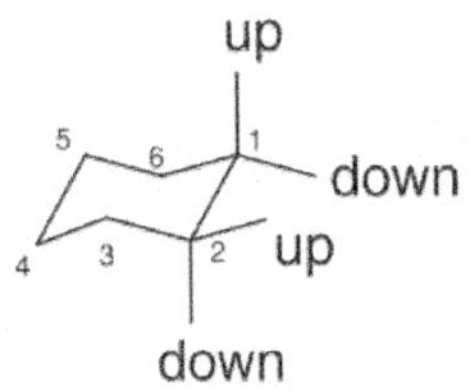

Finally, place the groups where they belong. The OH should be at the number 1 position in the down direction (because it was on a dash in the hexagon drawing), and the Me should be at the number 2 position in the up direction (because it was on a wedge):

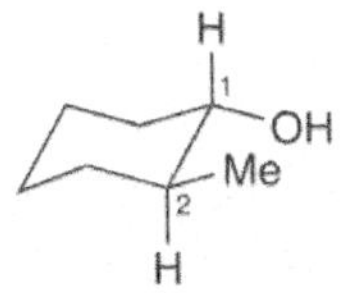

The example above illustrates an important point. Take a close look at the Me and the OH in the hexagon drawing. One is on a wedge and one is on a dash. We call this relationship *trans* (when you have two groups that are on opposite sides of the ring). If the two groups had been on the same side (either both on wedges, or both on dashes), then we would have called their relationship *cis*. So in the example above, the groups are *trans* to each other. Now take a close look at the chair drawing we just drew above. The OH and Me don't look *trans* in this drawing. In fact, they look like they are *cis*, but they are *trans*. The OH is in the down position and the Me is in the up position.

It will become very clear that these groups are *trans* to each other when we learn how to draw the chair after it has "flipped" to give us a new chair drawing. We will see this in the next section. For now, let's get some practice drawing the first chair correctly.

PROBLEMS Draw a chair conformation for each of the compounds below.

6.16

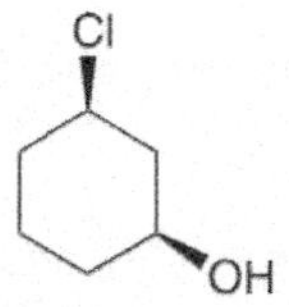

6.17

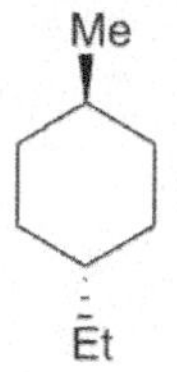

6.18

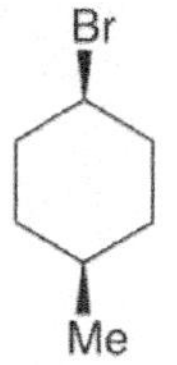

6.19

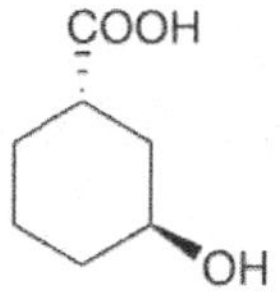

6.20

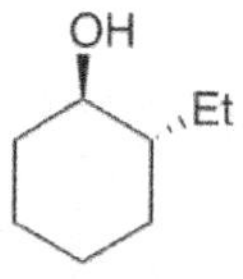

6.21

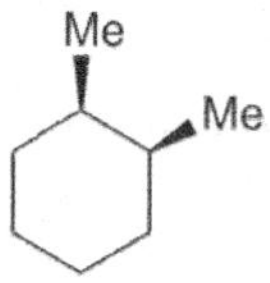

6.5 RING FLIPPING

Ring flipping is one of the most important aspects of understanding chair conformations, yet students commonly misunderstand this. Let's try to avoid the mistake by starting off with what ring flipping is not. It is *not* simply turning the ring over:

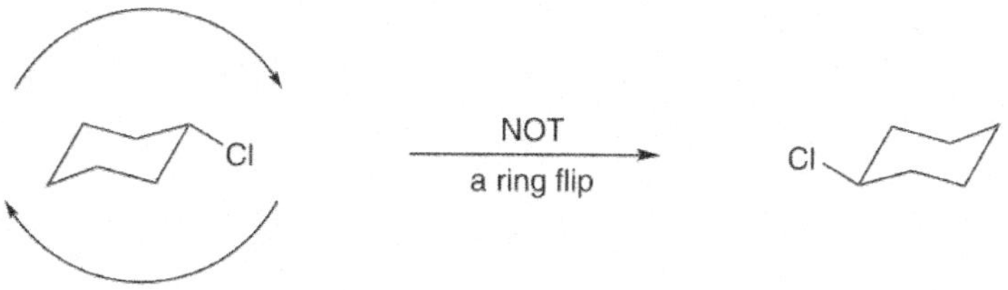

NOT
a ring flip

It makes sense why students think that this would be a flip—after all, this is the common meaning of the word "flipping." But we are talking about something very different when we say that rings can flip. Here is what we really mean:

Notice that in the drawing on the left, the left side of the chair is pointing down. In the drawing on the right, the left side of the chair is pointing up. This is a different chair. Also notice that chlorine went from being in an equatorial position to being in an axial position. This is a critical feature of ring flips. When performing a ring flip, all axial positions become equatorial, and all equatorial positions become axial.

Let's consider an analogy to help us get through this. Imagine that you are walking down a long hallway. Your hands are swaying back and forth as you walk, as most people do with their hands when they walk. One second, your left hand is in front of you and your right hand is behind you; the next second, it switches. Your hands switch back and forth with every step you take. The cyclohexane ring is doing something similar. It is moving around all of the time, flipping back and forth between two different chair conformations. So all of the substituents are constantly flipping back and forth between being axial and being equatorial.

There is one more important feature to recognize. Let's go back to the example above with chlorine. We said that the chair flip moves chlorine from an equatorial position into an axial position. But what about the up/down terminology? Let's see:

Notice that chlorine is down all of the time. In other words, up/down is *not* something that changes during a ring flip, but axial/equatorial does change during a ring flip. This illustrates that there is no relationship between up/down and axial/equatorial. If a substituent is up, then it will stay up all of the time, throughout the ring flipping process.

So now we can understand that a common hexagon-style drawing represents a molecule that is flipping back and forth between two chair conformations. The hexagon drawing shows us which substituents are up and which are down. That never changes. But whether those groups are axial or equatorial will depend on which chair you are drawing. So far, we have learned how to draw only one of these chairs. Now we will learn how to draw the other.

The process for drawing the skeleton of the chair is very similar to how we did it before. The only difference is that we draw our lines in the other direction. When we drew our first chair, we followed these steps:

| Step 1 | Step 2 | Step 3 | Step 4 | Step 5 |

Now, to draw the other chair, we follow these steps:

| Step 1 | Step 2 | Step 3 | Step 4 | Step 5 |

Compare the method for drawing the second chair to the method for drawing the first. The key is in step 2. If you compare step 2 for the first and second chair, everything else should flow from there. Use the following space to practice drawing the second chair:

Now, let's make sure you know how to draw the substituents. The rules are the same as before. All axial positions are drawn straight up and down, alternating:

and all equatorial positions are drawn parallel to the two bonds that are once removed:

Parallel to these bonds which are once removed
from the group we are drawing

PROBLEM

6.22   In the space below, practice drawing the second chair, showing all 12 substituents.

Let's now go back and review, because it is important that you understand the following points. When we are given a hexagon-style drawing, the drawing shows us which positions are up and which positions are down. No matter which chair we draw, up will always be up, and down will always be down. There are two chair conformations for this compound, and the molecule is flipping back and forth between these two conformations. With each flip, axial positions become equatorial positions and vice versa. Let's see an example.

Consider the following compound:

Notice that there are two groups on this ring. Cl is down (because it is on a dash), and Br is up (because it is on a wedge). There are two chair conformations that we can draw for this compound:

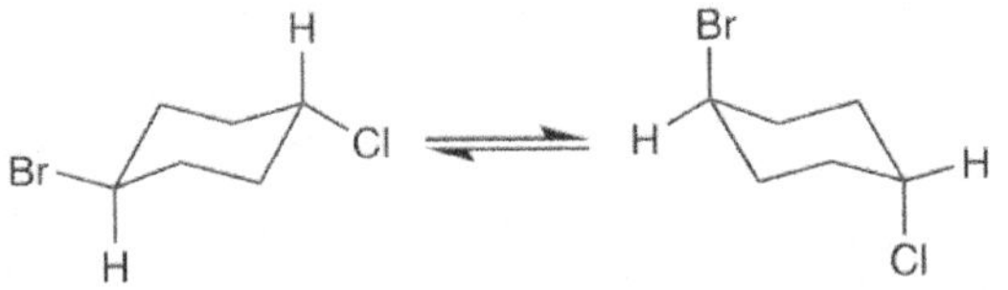

In both chair conformations, Cl is down and Br is up. The difference between these drawings is the axial/equatorial positions. In the conformation on the left, both groups are equatorial. In the conformation on the right, both groups are axial.

So any hexagon-style drawing will have two chair conformations. Now let's focus on making sure you can draw both conformations for any compound. We already saw in the last section how to draw the first one. We used a numbering system to determine where to put the groups, and we used the up/down terminology to figure exactly how to draw them (whether to draw them as equatorial or axial). To draw the second chair, we simply follow the same procedure. We begin by drawing the skeleton for the second chair (this is where you begin to see the difference between the chairs):

Skeleton for First Chair   Skeleton for Second Chair

Once we have drawn the skeleton, we number the carbon atoms going clockwise. Then we place the groups in the correct positions, making sure to draw them in the correct direction (up or down). So we can really use this method to draw both chairs at the same time. Let's do an example.

EXERCISE 6.23 Draw both chair conformations for the following compound:

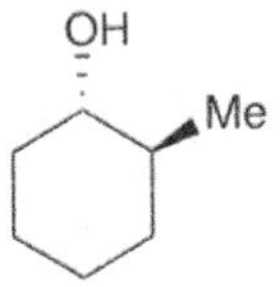

Answer Begin by numbering the hexagon clockwise. This puts the OH group at the position numbered 1 and the Me group at the position numbered 2.

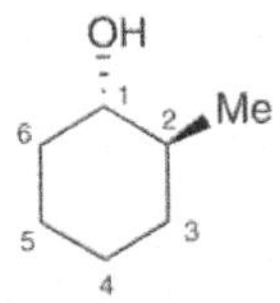

Next draw both chair skeletons and number them clockwise. Then identify the up and down positions at the carbon atoms numbered 1 and 2:

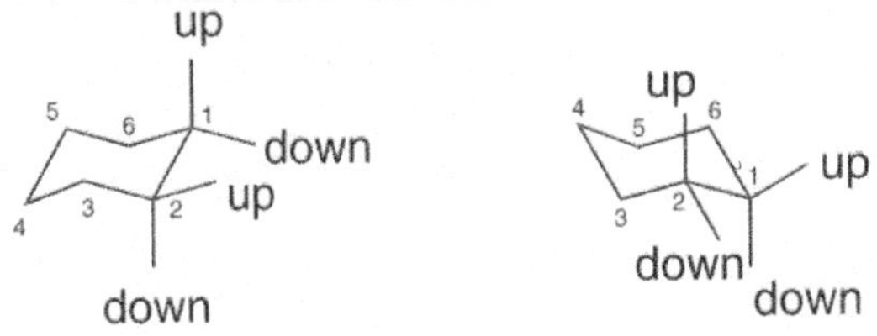

Finally, place the groups where they belong in both chairs. The OH group should be at the number 1 position in the down direction (because it was on a dash in the hexagon drawing), and the Me group should be at the number 2 position in the up direction (because it was on a wedge):

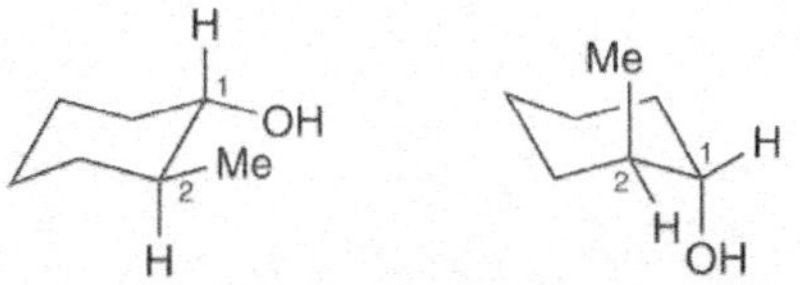

When we redraw these structures without showing any numbers or hydrogen atoms, it is clear to see that we need to go through these steps methodically because the relationship between these two conformations is not so obvious:

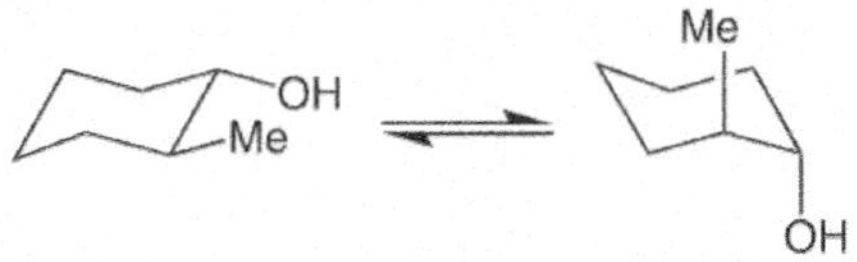

PROBLEMS   For each of the compounds below, draw both chair conformations.

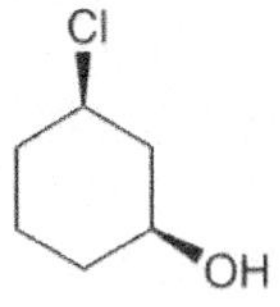

6.24

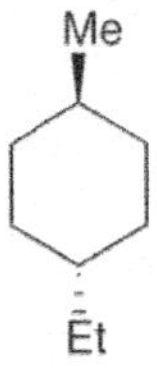

6.25

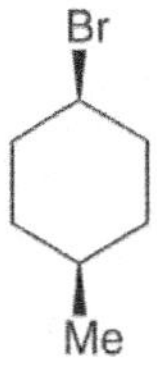

6.26

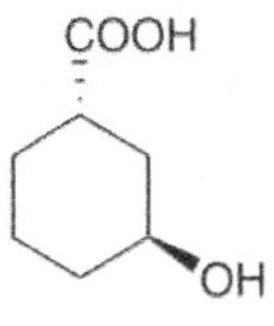

6.27

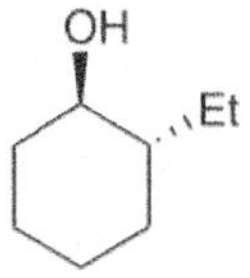

6.28

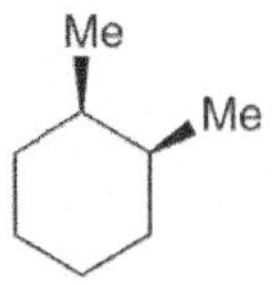

6.29

Sometimes, we might be given one chair conformation and be asked to draw the second chair conformation. Again, we use numbers to help us out. Let's see an example:

EXERCISE 6.30   Below you will see one chair conformation of a substituted cyclohexane. Draw the other chair (i.e., do a ring flip):

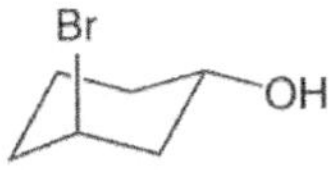

Answer   Begin by numbering the first chair. Start on the right side of the chair, and put a 1 at the first group. Then go clockwise. This puts the Br at the position numbered 3.

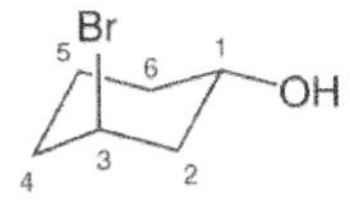

Notice that OH is down and Br is up.

    Next draw the skeleton for the second chair. Begin numbering on the right side again, making sure to go clockwise. Then draw the down position at the 1 position, and draw the up position at the 3 position:

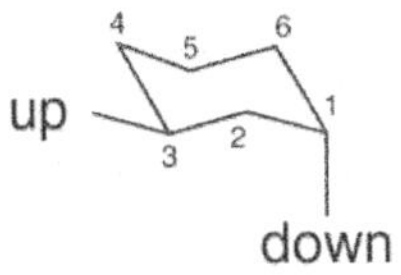

Finally, place the groups where they belong:

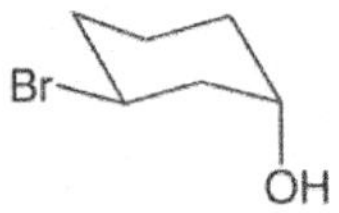

PROBLEMS   For each chair conformation shown below, do a chair flip and draw the other chair conformation.

6.31

6.32

6.33

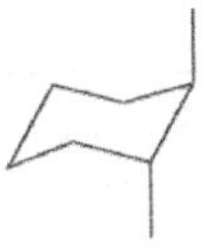

6.34

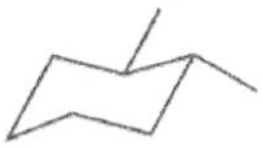

6.35

6.36

6.6 COMPARING THE STABILITY OF CHAIRS

Once you have drawn both chair conformations for a substituted cyclohexane, you should be able to predict which conformation is more stable. This is where it gets important for reactivity. Imagine that you are learning about a reaction that can proceed only if a certain group is in an axial position (you will learn about a reaction like this very soon—it is called E2). You already know that the groups are flipping back and forth between axial and equatorial positions (as it goes back and forth from one chair to the other). But what if one of the chairs is so unstable that the ring is spending 99% of its time in the other chair conformation? Then the question becomes, where is the important group in the stable chair conformation? Is it axial or equatorial? If it is equatorial, then the reaction can't happen. It could happen only during the 1% of the time that the group is in the axial position, so the reaction would be very slow. However, if the group is in an axial position 99% of the time, then the reaction will happen very quickly.

So you can see that it is important to understand what makes chair conformations unstable. There is really only one rule you need to worry about for now: a chair will be more stable with a group in an equatorial position, because it is not bumping into anything (this bumping is called a *steric interaction*). Axial positions are bumping into other axial positions, but equatorial positions are not:

The larger the group, the more it will prefer to be equatorial. So, a *tert*-butyl group will spend almost all of its time in an equatorial position. This essentially "locks" the ring in one conformation and prevents the ring from flipping (the truth is that the ring is still flipping, but the ring is spending more than 99% of its time in the more stable chair conformation):

So what happens if you also have a chlorine atom on the ring that is axial while the *tert*-butyl group is equatorial?

This will essentially lock the ring in the chair conformation that puts chlorine in an axial position. If we are trying to run a reaction where Cl needs to be axial, then this effect will speed up the reaction. However, if Cl is locked in an equatorial position, then the reaction will be too slow:

Now we understand why this can be important. So let's go step by step in determining which of two chair conformations is more stable.

If you have only one group on the ring, then the more stable chair will be the one with the group in an equatorial position:

More stable

If you have two groups, then it is best if both can occupy equatorial positions:

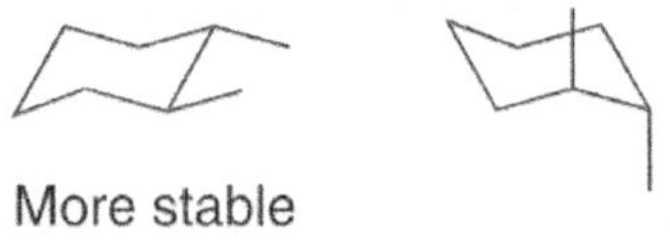

More stable

If only one can be equatorial in either chair, then the more stable chair will be the one with the larger group in the equatorial position:

More stable

In the example above, we have a choice to put the *tert*-butyl group in an equatorial position or the methyl group in an equatorial position. We can't have both. So we put the *tert*-butyl in an equatorial position.

If the ring has more than two groups, you just use the same logic that we used above to choose the more stable chair. Just try to put the largest groups in equatorial positions.

EXERCISE 6.37   For the following compound, draw the most stable chair conformation:

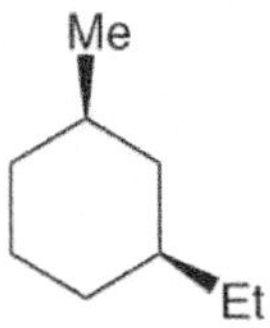

Answer   We begin by drawing both chair conformations (if you have trouble with this, review the two previous sections in this chapter):

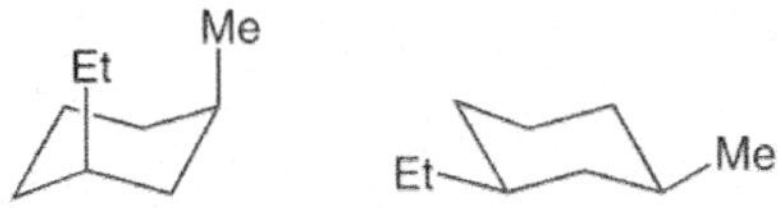

Now we select the chair that has the larger group in the equatorial position. In this case, both groups can be equatorial, so we choose this one:

PROBLEMS   For each compound below, draw the most stable chair conformation.

6.38

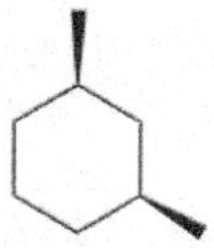

6.39

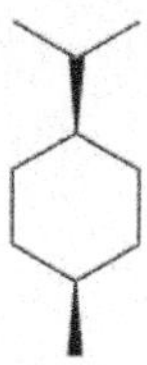

6.40

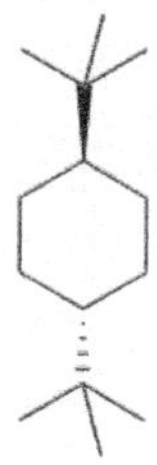

6.41

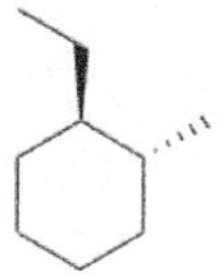

6.42

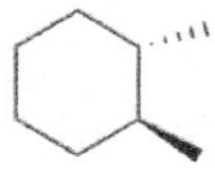

6.43

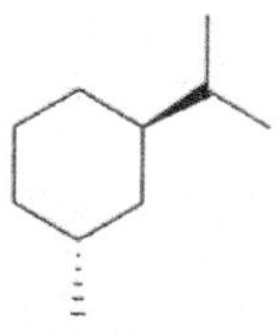

6.44

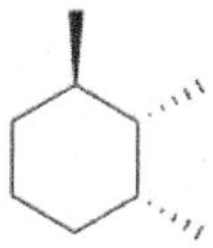

6.45

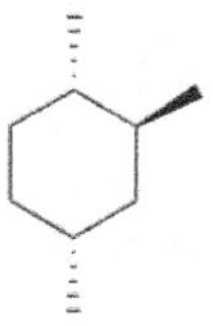

6.7 DON'T BE CONFUSED BY THE NOMENCLATURE

The terms *cis* and *trans* can be used in two different contexts, so it is worth a couple of paragraphs to clear things up. When two groups are both up or both down, we call them *cis* to each other; when one group is up and one group is down, we call them *trans* to each other:

cis   trans

Do not confuse this with *cis* and *trans* double bonds. There are no double bonds here. Don't draw any double bonds. It is amazing how many students will draw a double bond when you ask them to draw *cis*-1,2-dimethylcyclohexane. Remember that the ending –ane means that there are no double bonds anywhere in the molecule. The only comparison between double bonds and disubstituted cylohexanes is that, in both cases, *cis* means "on the same side" and *trans* means "on opposite sides."

CHAPTER 7

CONFIGURATIONS

In the previous chapter, we saw that molecules can assume many different conformations, much like a person. You can move your hands all around: hold them straight up in the air, out to the sides, straight down, and so on. In all of these positions, your right hand is still your right hand, regardless of how you move it around. There is no way to twist your right hand to turn it into a left hand. The reason it is always a right hand has nothing to do with the fact that it is connected to your right shoulder. If you were to chop off your arms and sew them on backward (*Don't try this at home!*), you would not look normal. You would look like someone with his right hand attached to his left shoulder and vice versa. You would look very strange, to say the least.

Your right hand is a right hand because it fits into a right-handed glove, and it does not fit into a left-handed glove. This will always be true no matter how you move your hand. Molecules can have this property also.

It is possible for a molecule to have a region where there are two possibilities for how the atoms can be connected in 3D space, much like the difference between a right hand and a left hand. Instead of "right hand" and "left hand," we call the two possibilities *R* and *S*. When we talk about the configuration of a compound, we are talking about whether it is *R* or *S*. If the arrangement is *S*, then it will always be *S* no matter how the arms of the molecule twist about as the molecule moves. In other words, the molecule can move into any conformation it wants, but the configuration will never change. The only way for a configuration to change would be to undergo a chemical reaction.

This explains something we saw in the previous chapter: when drawing chair conformations, we saw that up is always up regardless of which chair you draw. That is because up and down are issues of *configuration*, which does not change when the molecule twists into another conformation.

Don't confuse *conformation* with *configuration*. Students confuse these terms all of the time. *Conformations* are the different positions that a molecule can twist into, but *configuration* is a matter of right-handedness or left-handedness (*R* or *S*).

In molecules, the regions that can be *R* or *S* are called *chiral centers* (the term "chiral" comes from the Greek word for "hand," and we can understand the symbolism there). In this chapter, we will learn how to locate chiral centers, how to draw them properly, how to label them as *R* or *S*, and what happens when you have more than one chiral center in a compound.

This is all *extremely* important stuff. You will understand this as soon as you begin learning reactions. You will see that some reactions convert a chiral center from *R* into *S* (and vice versa), while other reactions will not. To predict the products of a reaction, you absolutely *must* know how to show these chiral centers.

7.1   LOCATING CHIRAL CENTERS

If a carbon atom is connected to four different groups, that carbon atom will be a chiral center, for example:

This drawing has a carbon atom in the center with four different groups: ethyl, methyl, bromine, and chlorine. Therefore, we have a chiral center. Anytime you have four different groups connected to a carbon atom, there will be two ways to arrange the groups in space (always two; never more and never less). These two arrangements are different configurations:

These two compounds are different from each other even though they have the same constitution (connectivity of atoms). The difference between them is their arrangement in 3D space. Therefore, they are called stereoisomers ("stereo" for space). More specifically, they are called *enantiomers*, because the two compounds are mirror images of each other and they are not superimposable. If we construct models of these two compounds, we see that they are not the same—i.e., they cannot be superimposed.

Notice that we are not looking at just the four *atoms* that are connected to the carbon atom in the middle (which would be Br, Cl, C, and C—and we might think that two of these are the same), but we are looking at the entire groups. In other words, whenever we look at the four groups connected to an atom, we are looking at the entire molecule, no matter how big those groups are. Consider the following example:

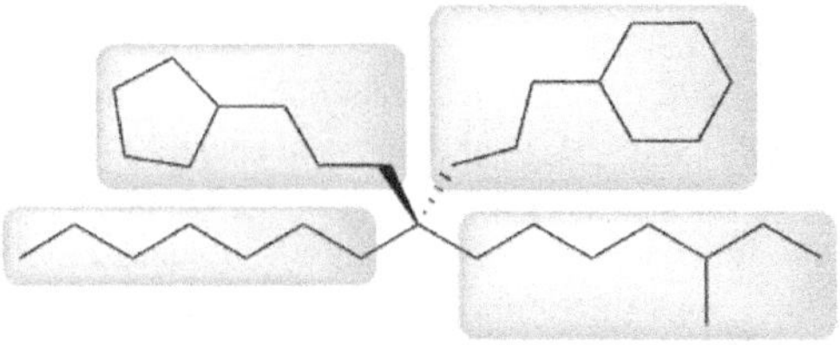

All four of these groups are different.

You must learn how to recognize when an atom has four different groups attached to it. To help you with this, let's begin with examples that are *not* chiral centers:

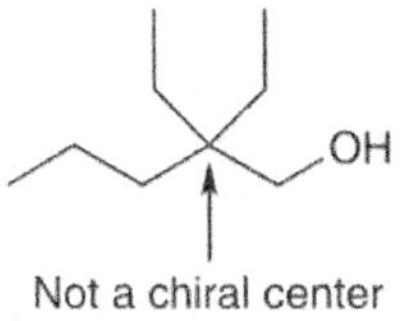

Not a chiral center

The carbon atom indicated above is not a chiral center because there are two groups that are the same (there are two ethyl groups). The same is true in the following case:

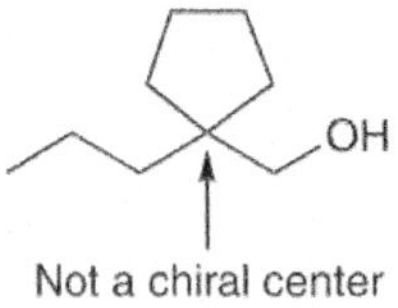

Not a chiral center

Whether you go around the ring clockwise or counterclockwise, you see the same thing, so this is not a chiral center. If we wanted to make it a chiral center, we could do so by putting a group on the ring:

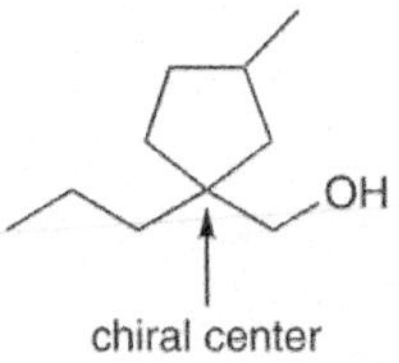

chiral center

Usually, chiral centers are drawn with dashes and wedges to show the configuration. If the dashes and wedges are not drawn, then we assume that there is a mixture of equal amounts of both configurations (which we call a *racemic* mixture). In fact, in the compound above, there is a second chiral center. Can you find it? Each of the two chiral centers in the compound above can be either *R* or *S*. Since there are two chiral centers, there will be four possibilities: *R,R* and *R,S* and *S,R* and *S,S*. Since neither chiral center has been drawn with dashes and wedges, we must assume that we have all four possible stereoisomers.

EXERCISE 7.1   In the compound below, there is one chiral center. Find it.

Answer   Let's start on the left side and work our way across the compound. The methyl group has three hydrogen atoms, so that can't be it. Then there is a CH_2 group, which also cannot be it, because two groups are the same (two H's). Then you have a carbon atom with four different groups: ethyl (on the left), methyl (on the right), OH sticking up, and H (don't forget about the H's that are not shown). This is our chiral center.

PROBLEMS   In each of the compounds below, there is one chiral center. Find it.

7.2

7.3

7.4

7.5

7.6

7.7

In the previous problems, you knew that you were looking for just one chiral center. Hopefully, you started to realize some tricks that make it faster to find the chiral center (for example, ignore CH_2 groups). So, now, we will move on to examples where you don't know how many chiral centers there are. There could be five chiral centers or there could be none.

EXERCISE 7.8   In the following compound, find all of the chiral centers, if any:

Answer   If we go around the ring, we find that there are only six carbon atoms in this compound. Four of them are CH_2 groups, so we know that they are not chiral centers. If we look at the remaining two carbon atoms, we see that each of them is connected to four different groups. They are both chiral centers.

PROBLEMS   For each of the compounds below, find all of the chiral centers, if any.

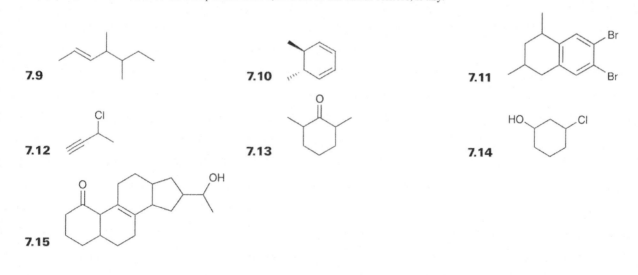

7.9

7.10

7.11

7.12

7.13

7.14

7.15

7.2 DETERMINING THE CONFIGURATION OF A CHIRAL CENTER

Now that we can find chiral centers, we must learn how to determine whether a chiral center has the *R* or *S* configuration. There are two steps involved in making the determination. First, we give each of the four groups a number (from 1 to 4). Then we use the orientation of these numbers to determine the configuration. So, how do we assign numbers to each of the groups?

We start by making a list of the four *atoms* attached to the chiral center. Let's look at the following example:

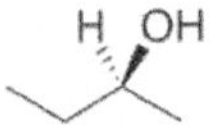

The four atoms attached to the chiral center are C, C, O, and H. We rank them from 1 to 4 based on atomic number. To do this, we must either consult a periodic table every time or commit to memory a small part of the periodic table—just those atoms that are most commonly used in organic chemistry:

$$C \quad N \quad O \quad F$$
$$P \quad S \quad Cl$$
$$Br$$
$$I$$

When comparing the four atoms in the example above, we see that oxygen has the highest atomic number, so we give it the first priority—we give it the number 1. Hydrogen is the smallest atom, so it will always get the number 4 (lowest priority) when a chiral center has a hydrogen atom. We don't have to worry about what to do if there are two hydrogen atoms, because if there were, it would not be a chiral center. But it is possible to have two carbon atoms, as in the example above. So how do we determine which one gets the number 2 priority and which gets the number 3 priority?

This is how we rank the two carbon atoms: for each carbon atom, we write a list of the three atoms to which it is attached (other than the chiral center). Let's do the example above to see how this works. The carbon atom on the left side of the chiral center has four bonds: one to the chiral center, one to another carbon atom, and then to two hydrogen atoms. So, other than the chiral center, it has three bonds (C, H, and H). Now let's look at the carbon atom on the right side of the chiral center. It has four bonds: one to the chiral center and then to three hydrogen atoms. So, other than the chiral center, it has three bonds (H, H, and H). We compare the two lists and look for the first point of difference:

$$
\begin{array}{cc}
C & H \\
H & H \\
H & H \\
\end{array}
$$

We see the first point of difference immediately: carbon beats hydrogen. So the left side of the chiral center gets priority over the right side, and the numbering turns out like this:

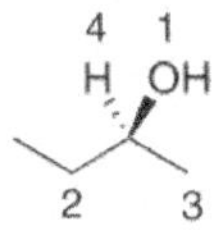

EXERCISE 7.16 In the compound below, find the chiral center, and label the four groups from 1 to 4 using the system of priorities based on atomic number.

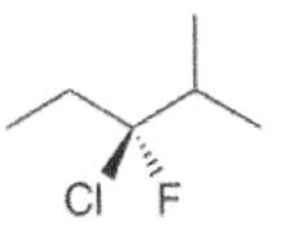

Answer   The four atoms attached to the chiral center are C, C, Cl, and F. Of these, Cl has the highest atomic number, so it gets the first priority. Then comes F as number 2. We need to decide which carbon atom gets the number 3 and which carbon atom gets the number 4. We do this by listing the three atoms attached to each of them:

Left Side	Right Side
C	C
H	C
H	H

So the right side wins. Therefore, the numbering goes like this:

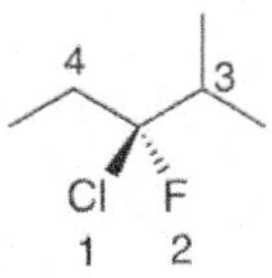

PROBLEMS   In each of the compounds below, find the chiral center, and label the four groups from 1 to 4 using the system of priorities based on atomic number.

7.17 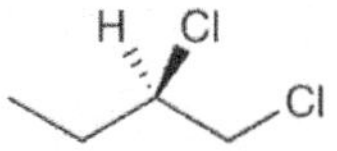   **7.18** 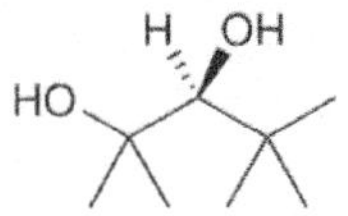   **7.19**

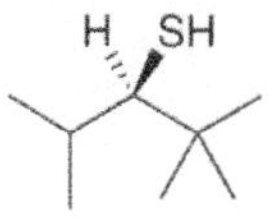

There are a few more situations you may encounter when numbering the four groups. If you are comparing two carbon atoms and you find that the three atoms on one side are the same as the three atoms on the other, then keep going farther out until you find the first difference. Below is an example:

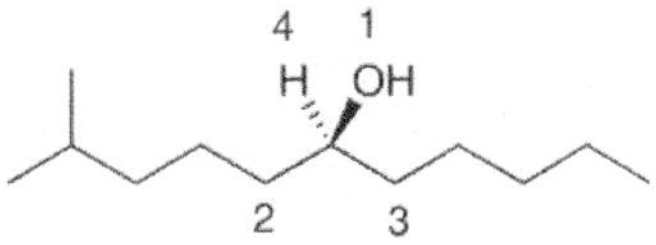

Also, you should know that we are looking for the first point of difference as we travel out, and we don't add the atomic numbers. This is best explained with an example:

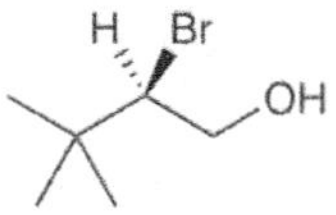

In this example, we know that the Br gets the first priority, and the H gets the number 4. When comparing the two carbon atoms, we find the following situation:

Left Side	Right Side
C	O
C	H
C	H

In this case, we do not add the atomic numbers and say that the left side wins. Rather we go down the list and compare each row. In the first row above, we have C versus O. That's it, end of story—the O wins. It doesn't matter what comes in the next two rows. Always look for the *first* point of difference. So the priorities go like this:

Finally, you should count a double bond as if the atom were connected to two carbon atoms. For example,

The group on the left gets the number 2, because we counted the following way:

Left Side	Right Side
C	C
C	H
H	H

EXERCISE 7.20   In the compound below, find the chiral center, and label the four groups from 1 to 4 using the system of priorities based on atomic number.

Answer   All four atoms connected to the chiral center are carbon atoms, so we must compare the following four lists:

The oxygen atom wins. Next comes the one with three carbon atoms. The remaining two are the same, so we need to move out one farther on the chain and compare again. Remember to count the double bond like two carbon atoms:

So the order of priorities goes like this:

PROBLEMS   In each of the compounds below, find the chiral center and label the four groups from 1 to 4 using the system of priorities based on atomic number.

7.21

7.22

7.23

7.24

7.25

7.26

Now we need to learn how to use this numbering system to determine the configuration of a chiral center. The idea is simple, but it is difficult to do if you have a hard time closing your eyes and rotating 3D objects in your mind. For those who cannot do this, don't worry. There is a trick. Let's first see how to do it without the trick.

If the number 4 group is pointing away from us (on the dash), then we ask whether 1, 2, and 3 are going clockwise or counterclockwise:

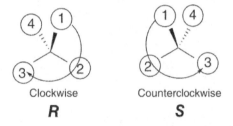

Clockwise
R

Counterclockwise
S

In the example on the left, we see that 1, 2, and 3 go clockwise, which is called *R*. In the example on the right, we see that 1, 2, and 3 go counterclockwise, which is called *S*. If the molecule is already drawn with the number 4 priority on the dash, then your life is very simple. Consider the following example:

The 4 is already on the dash, so you just look at 1, 2, and 3. In this case, they go counterclockwise, so it is *S*.

It gets a little more difficult when the number 4 is not on a dash, because then you must rotate the molecule in your mind. For example,

Let's redraw just the chiral center showing the location of the four priorities:

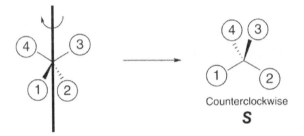

Now we need to rotate the molecule so that the fourth priority is on a dash. To do this, imagine spearing the molecule with a pencil and then rotating the pencil 90°:

Now the 4 is on a dash, so we can look at 1, 2, and 3, and we see that they go counterclockwise. Therefore, the configuration is *S*.

Let's see one more example:

We redraw just the chiral center showing the location of the four priorities, and then we spear the molecule with a pencil and rotate 180° to put the 4 on a dash:

Now, the 4 is on a dash, so we can look at 1, 2, and 3, and we see that they go clockwise. Therefore, the configuration is *R*.

And now, for the trick. If you were able to see all of that, great! But if you had trouble seeing the molecules in 3D, there is a simple trick that will help you get the answer every time. To understand how the trick works, you need to realize that if you redraw the molecule so that any two of the four groups are switched, then you have switched the configuration (*R* turns into *S*, and *S* turns into *R*):

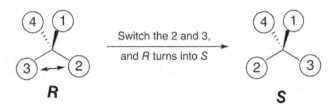

You can use this idea to your advantage. Here is the trick: Switch the number 4 with whatever group is on the dash—then your answer is the opposite of what you see. Let's do an example:

This looks tough because the 4 is on a wedge. But let's do the trick: switch the 4 with whichever group is on the dash; in this case, we switch the 4 with the 1:

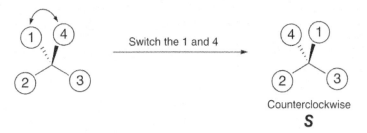

After doing the switch, the 4 is on a dash, and it becomes easy to figure out. It is counterclockwise, which means *S*. We had to do one switch to make it easy to figure out, which means that we changed the configuration. So if it became *S* after the switch, then it must have been *R* before the switch. That's the trick. *But be careful.* This trick will work every time, but you must not forget that the answer you immediately get is the opposite of the real answer, because you did one switch.

Now, let's practice determining *R* or *S* when you are given the numbers, so that we can make sure you know how to do this step. You can either visualize the molecule in 3D, or you can use the trick—whatever works best for you.

PROBLEMS   In each case, assign the correct configuration (*R* or *S*).

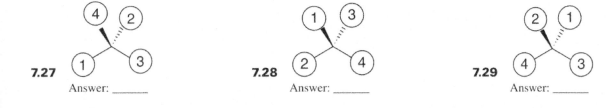

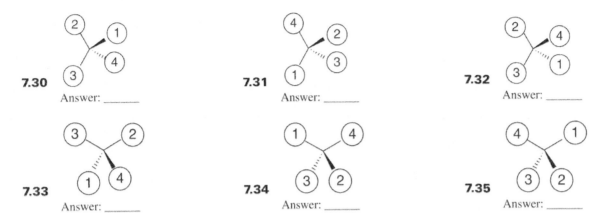

7.30 Answer: _____

7.31 Answer: _____

7.32 Answer: _____

7.33 Answer: _____

7.34 Answer: _____

7.35 Answer: _____

So we know how to assign priorities (1–4), and we know how to use those priorities to determine configuration. Now, let's get practice assigning configurations of chiral centers.

EXERCISE 7.36   The compound below has one chiral center. Find it, and determine whether it is *R* or *S*.

Answer   The carbon atom bearing the two chlorine atoms cannot be the chiral center because there are two Cl atoms (two of the same group). The chiral center is the carbon atom with the OH group attached. It has four different groups attached to it. Now that we found it, we need to assign priorities.

The O on the dash gets priority number 1, and the hydrogen atom (not shown, but it is on a wedge) gets number 4. Between the two carbon atoms, the one on the right is connected to two Cl atoms. This wins. So the numbers go like this:

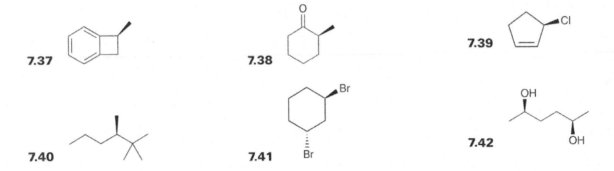

The 4 is on a wedge, so let's use the trick. If we switch the 4 and the 1, we get something that is *R*. So, it must be that it was *S* before we switched the groups and our answer is *S*.

PROBLEMS   For each compound below, find all chiral centers, and determine the configuration of each.

7.37

7.38

7.39

7.40

7.41

7.42

7.3   NOMENCLATURE

When we learned how to name compounds, we said that we would skip over the naming of chiral centers until we learned how to determine configuration. Now that we know how to determine whether a chiral center is R or S, we can now see how to include this in the name of a compound. It is actually quite simple. If there is only one chiral center, then you simply place either (R) or (S) at the beginning of the name. For example, 2-butanol has one chiral center, and can either be (R)-2-butanol or (S)-2-butanol, depending on the configuration of the chiral center. If more than one chiral center is present, then you must also use numbers to identify the location of each chiral center. Consider the following example:

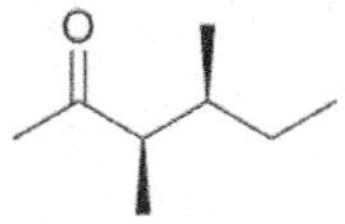

Based on everything we learned in Chapter 5 (nomenclature), we would name this compound 3,4-dimethylhexan-2-one. Now we must also add the configurations to the name. The chiral center on the left has the R configuration, and the chiral center on the right has the S configuration. We use the numbering system of the parent chain to determine where the chiral centers are. Since the parent chain was numbered left to right, we add (3R,4S) at the very beginning of the name:

Stereoisomerism	Substituents	Parent	Unsaturation	Functional group

so the name is (3R,4S)-3,4-dimethylhexan-2-one. As we saw earlier, we italicize stereoisomerism when it is part of a name.

Now let's turn to a different type of stereoisomerism, one that we already discussed in the chapter on nomenclature. Let's look at double bonds. Recall that we indicate the presence of a double bond using the term –en-:

Stereoisomerism	Substituents	Parent	Unsaturation	Functional group

And we indicate the position of the double bond with the numbering system. But then we saw that there are often two ways for the atoms of a double bond to connect to each other in 3D space. We saw a system for distinguishing these possibilities, using the terms *cis* and *trans*:

cis   trans

This was indicated in the first part of the name (stereoisomerism):

Stereoisomerism	Substituents	Parent	Unsaturation	Functional group

This system of identifying double-bond stereoisomers is very limited, because you need two groups that are the same to use the *cis/trans* system of naming. So what do you do if you have four different groups on a double bond? There are still two possible stereoisomers:

but we cannot use the *cis/trans* system here. So, we have another system that allows us to differentiate between these two compounds. This other system uses the same numbering based on priorities that we used for chiral centers (based on atomic numbers). We look at both sides of the double bond; each side has two groups:

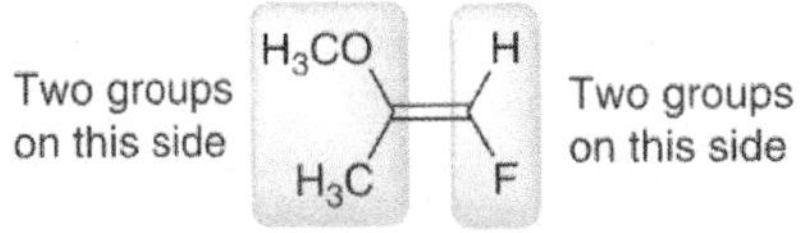

We begin with one side (let's start with the left), and we ask which of the two groups on the left has priority:

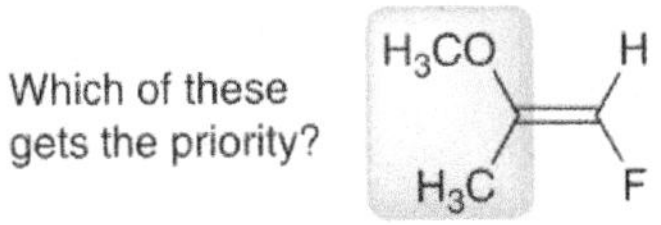

The oxygen atom gets priority over the carbon atom, based on atomic number. When comparing the two groups on the right side, the fluorine atom gets priority over the hydrogen atom, again based on atomic numbers. So now we know which group gets the priority on each side:

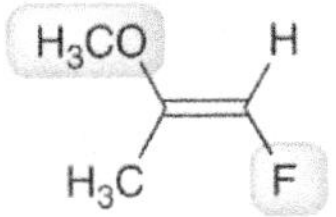

We look at the priority group on one side and the priority group on the other side, and we ask: are these groups on the same side (like *cis*) or on opposite sides (like *trans*)?

The same side is called *Z* (for the German word "zusammen" meaning together), and opposite sides are called *E* (for the German word "entgegen" meaning opposite).

In the example above, it was easy to assign priorities, but what about when it gets a little more complicated:

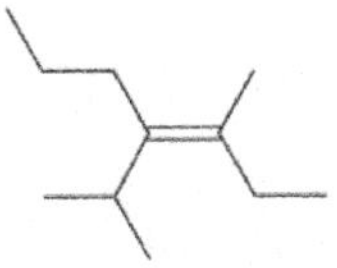

In this example, we have to compare carbon atoms to each other. The groups are all different, so we need to find a way to assign priorities. To do so, we follow the same rules that we did when assigning priorities with *R* and *S*:

1. If the atoms are the same on one side, then just move farther out and analyze again.

2. One oxygen beats three carbon atoms (remember to look for the first point of difference).

3. A double bond counts as two individual bonds.

This way of naming double bonds is far superior to *cis/trans* nomenclature because you can use this *E/Z* system for any double bond, even if all four groups are different.

We include this information in the name of a molecule, very much like we did for *R* and *S* configurations. For example, if the double bond is between carbon atoms numbered 5 and 6 on a parent chain, then we would add the term (5*E*) or (5*Z*) at the beginning of the name.

EXERCISE 7.43   Name the following compound, including stereochemistry in the beginning of the name.

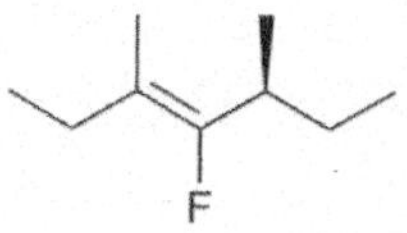

Answer   Remember that we go through all five parts of the name, starting at the end and working our way backward:

Stereoisomerism	Substituents	Parent	Unsaturation	Functional group

We begin with the functional group. There is no functional group (so the suffix is –e). Moving backward in the name, we look for any unsaturation, and there is one double bond (so the unsaturation is –en-). Then, we choose the longest parent that includes the double bond, which is seven carbon atoms long, so the parent is –hept-. There are three substituents (two methyl groups and a fluorine), so we add fluoro and dimethyl before the parent. Then we assign the numbers. We give the double bond the lowest number, so we number from left to right. This gives us

4-fluoro-3,5-dimethylhept-3-ene

If you do not remember how to do that, then you should review Chapter 5 on nomenclature.

Now let's consider stereoisomerism. The double bond has the *Z* configuration:

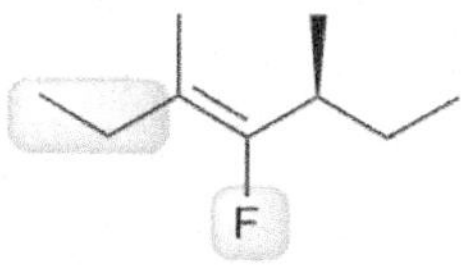

and there is a chiral center that has the *S* configuration:

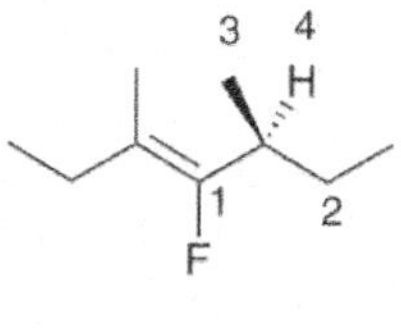

When we number the parent chain, we see that the double bond is at the third position in the parent chain, and the chiral center is at the fifth position in the parent chain:

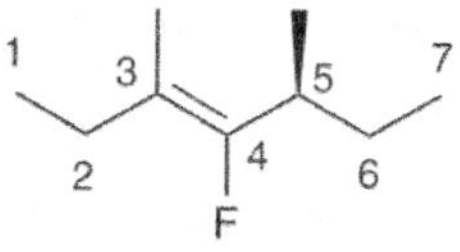

So the name is

(3*Z*,5*S*)-4-fluoro-3,5-dimethylhept-3-ene

PROBLEMS Name each of the following compounds. Be sure to include stereoisomerism at the beginning of every name:

7.44

Name: _____

7.45

Name: _____

7.46

Name: _____

7.47

Name: _____

7.48

Name: _____

7.49

Name: _____

7.4 DRAWING ENANTIOMERS

In the beginning of the chapter, we said that enantiomers are two compounds that are nonsuperimposable mirror images. Let's first clear up the term "enantiomers," since students will often use this word incorrectly in a sentence. Let's compare it to people again. If two boys are born to the same parents, those boys are called brothers. Each one is the brother of the other. Similarly, when you have two compounds that are nonsuperimposable mirror images, they are called enantiomers. Each one is the enantiomer of the other. Together, they are a pair of enantiomers. But what do we mean by "nonsuperimposable mirror images"? Let's go back to the brother analogy to explain it.

Imagine that the two brothers are twins. They are identical in every way except one. One of them has a mole on his right cheek, and the other has a mole on his left cheek. This allows you to distinguish them from each other. They are mirror images of each other, but they don't look exactly the same (one cannot be superimposed on top of the other). It is very important to be able to see the relationship between different compounds. It is important to be able to draw enantiomers. Later in the course, you will see reactions where a chiral center is created and both enantiomers are formed. To predict the products, you must be able to draw both enantiomers. In this section, we will see how to draw enantiomers.

The first thing you need to realize is that enantiomers always come in pairs. Remember that they are mirror images of each other. There are only two sides to a mirror, so there can be only two different compounds that have this relationship (nonsuperimposable mirror images). This is very much like the twin brothers. Each brother only has one twin brother, not more.

So we must learn how to draw one enantiomer when we are given the other. When we see the different ways of doing this, we will begin to recognize when compounds are enantiomers and when they are not.

The simplest way to draw an enantiomer is to redraw the carbon skeleton, but invert all chiral centers. In other words, change all dashes into wedges and change all wedges into dashes. For example,

The compound above has a chiral center (what is the configuration?). If we wanted to draw the enantiomer, we would redraw the compound, but we would turn the wedge into a dash:

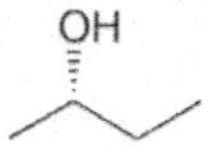

This is a pretty simple procedure for drawing enantiomers. It works for compounds with many chiral centers just as easily. For example,

The enantiomer of 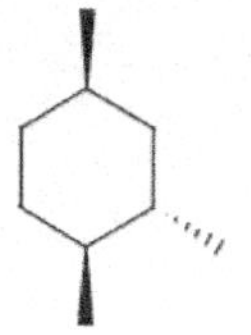 is

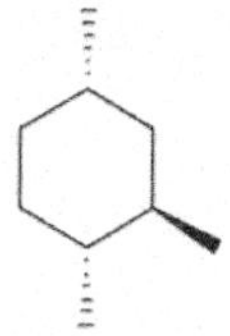

We simply invert all chiral centers. This is actually what we would see if we placed a mirror directly behind the first compound and then looked into the mirror. The carbon skeleton would look the same, but the chiral centers would all be inverted:

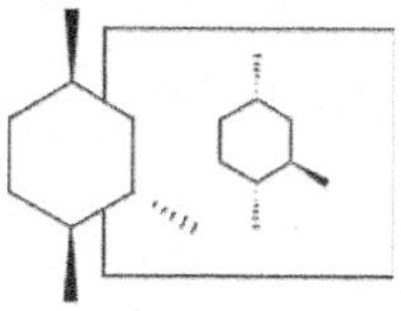

EXERCISE 7.50   Draw the enantiomer of the following compound:

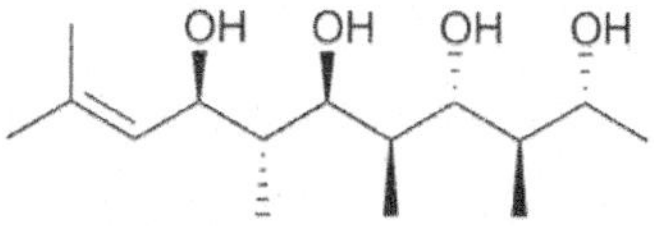

Answer   Redraw the molecule, but invert every chiral center. Convert all wedges into dashes, and convert all dashes into wedges:

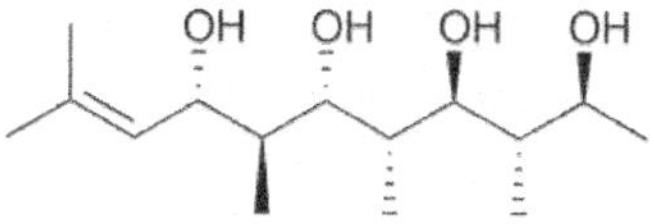

PROBLEMS   Draw the enantiomer of each of the following compounds:

7.51 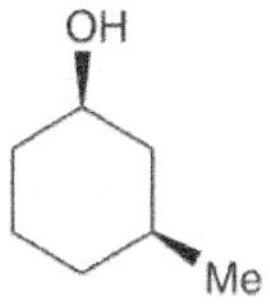   Answer:

7.52 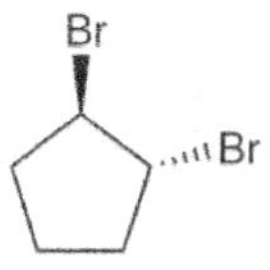   Answer:

7.53 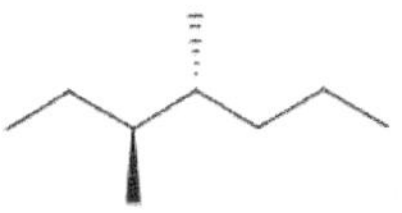   Answer:

7.54 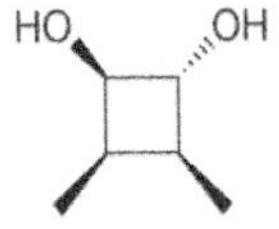   Answer:

7.55 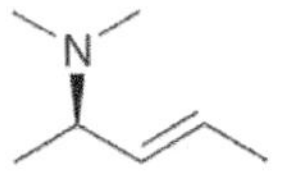   Answer:

7.56 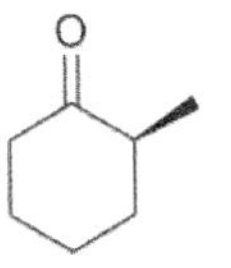   Answer:

There is another way to draw enantiomers. In the previous method, we placed an imaginary mirror *behind* the compound, and we looked into that mirror to see the reflection. In the second method for drawing enantiomers, we place the imaginary mirror *on the side of* the compound, and we look into the mirror to see the reflection. Let's see an example:

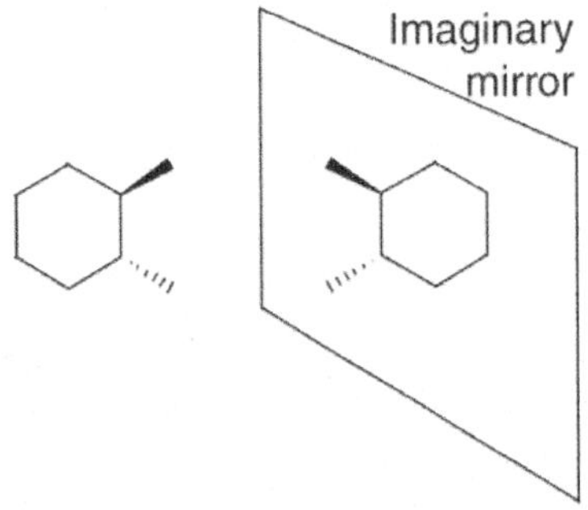

Imaginary
mirror

But why do we need a second way of drawing enantiomers? Didn't the first method seem good enough? The first method (switching all dashes with wedges) was pretty simple to do. But there are times when the first method will not work so well. There are a few examples of cyclic and bicyclic carbon skeletons where the wedges and dashes are not drawn, because they are implied. We have actually already seen an example of one of these: the chair conformation of a substituted cyclohexane.

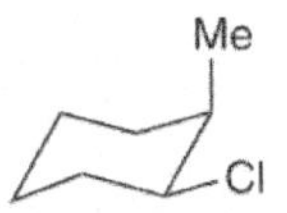

Me

Cl

In this drawing, all of the lines are drawn as straight lines (no wedges and dashes) even though we know that the bonds are not all in the plane of the page. We don't need to draw the wedges and dashes because the geometry can be understood from the drawing. We could try to draw the enantiomer by converting the drawing into a hexagon-style drawing (with wedges and dashes), then drawing the enantiomer using the first method (switching all dashes for wedges), and then redrawing the chair conformation of that enantiomer. But that is a lot of steps to go through when there is a simpler way to draw the enantiomer—just put the imaginary mirror on the side (there is no need to actually draw the mirror), and draw the enantiomer like this:

Me   Me

Cl   Cl

Whenever we have a structure where the wedges and dashes are implied but not drawn, it is much easier to use this method. There are other examples of carbon skeletons that, by convention, do not show the wedges and dashes. Most of these examples are rigid bicyclic systems. For example,

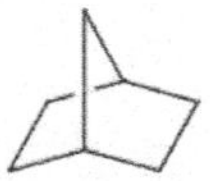

When dealing with these kinds of compounds, it is much easier to use the second method to draw enantiomers. Of course, if you like this method, you can always use this second method for all compounds (even those that show wedges and dashes).

You should get practice placing the mirror on either side (and you should notice that you get the same result whether you put the mirror on the left side or the right side).

EXERCISE 7.57   Draw the enantiomer of the following compound:

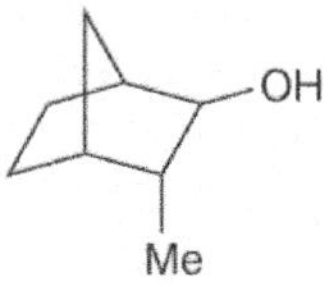

Answer   This is a rigid bicyclic system, and the dashes and wedges are not shown. Therefore, we will use the second method for drawing enantiomers. We will place the mirror on the side of the compound, and draw what would appear in the mirror:

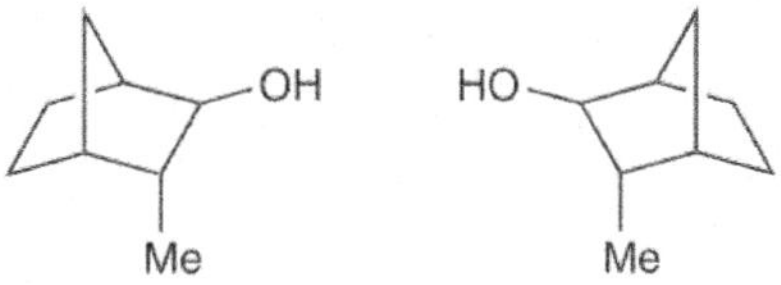

PROBLEMS   Draw the enantiomer of each of the following compounds:

7.58 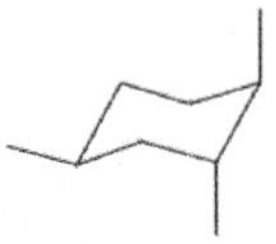   Answer: _____

7.59 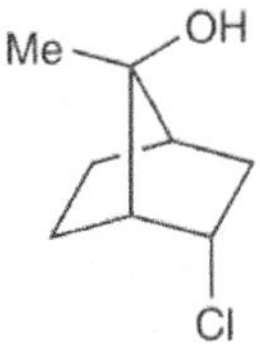   Answer: _____

7.60 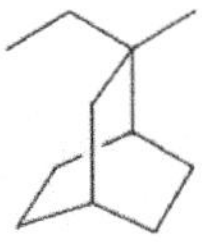   Answer: _____

7.61    Answer: _____

7.62 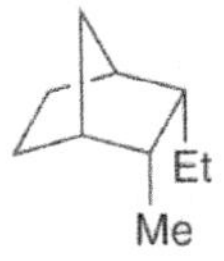   Answer: _____

7.63 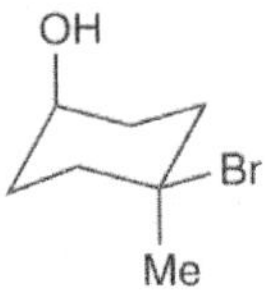   Answer: _____

7.5 DIASTEREOMERS

In all of our examples so far, we have been comparing two compounds that are mirror images. For them to be mirror images, they need to have opposite configurations for every single chiral center. Remember that our first method for drawing enantiomers was to switch all wedges with dashes. For the two compounds to be enantiomers, every chiral center had to be inverted. But what happens if we have many chiral centers and we only invert some of them?

Let's start off with a simple case where we only have two chiral centers. Consider the two compounds below:

We can clearly see that they are not the same compound. In other words, they are nonsuperimposable. But, they are *not* mirror images of each other. The top chiral center has the same configuration in both compounds. If they are not mirror images, then they are not enantiomers. So what is their relationship? They are called diastereomers. Diastereomers are any compounds that are stereoisomers that are not mirror images of each other.

We use the term "diastereomer" very much like we used the term "enantiomer" (remember the brother analogy). One compound is called the diastereomer of the other, and you can have a group of diastereomers. When we were talking about enantiomers, we saw that they always come in pairs, never more than two. But diastereomers can form a much larger family. We can have 100 compounds that are all diastereomers of each other (if there are enough chiral centers to allow for that many permutations of the chiral centers).

E/Z isomers (or *cis/trans* isomers) are called diastereomers, because they are stereoisomers that are not mirror images of each other:

If you are given two stereoisomers, you should be able to determine whether they are enantiomers or diastereomers. All you need to look at are the chiral centers. They must all be of different configuration for the compounds to be enantiomers.

EXERCISE 7.64   Identify whether the two compounds shown below are enantiomers or diastereomers:

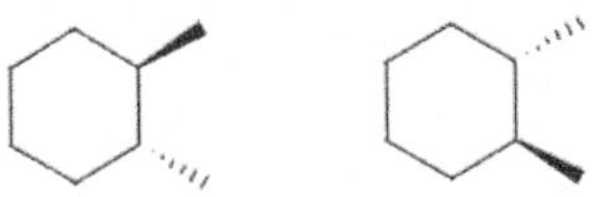

Answer   There are two chiral centers in each compound. The configurations are different for both chiral centers, so these compounds are enantiomers. In fact, if you were given the first compound only, you could have drawn the enantiomer by using the first method—switching all wedges and dashes.

PROBLEMS For each pair of compounds below, determine whether the pair are enantiomers or diastereomers.

7.65 OH   OH — Me   Me
Answer: _____

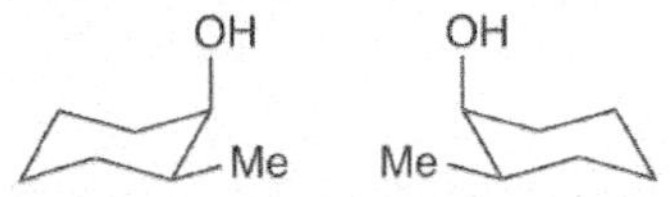

7.66 OH   OH — Me   Me
Answer: _____

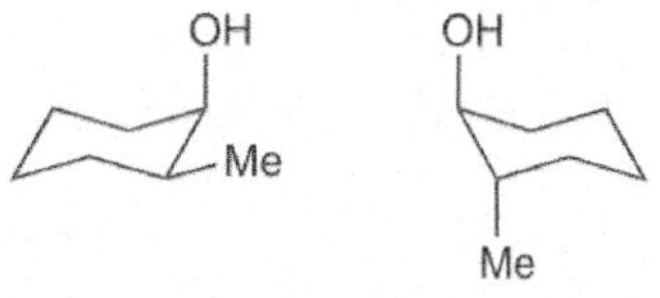

7.67 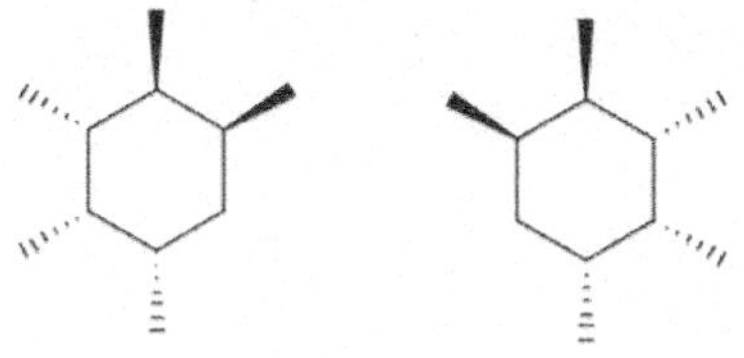
Answer: _____

7.68 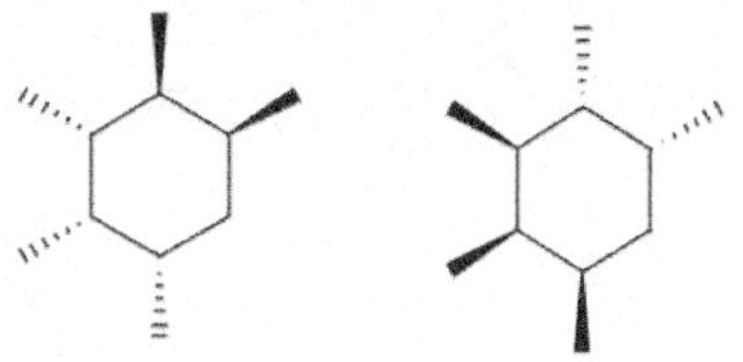
Answer: _____

7.69 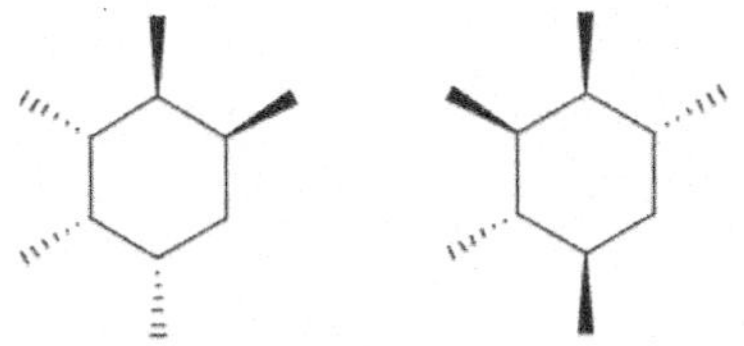
Answer: _____

7.70 F   F
Answer: _____

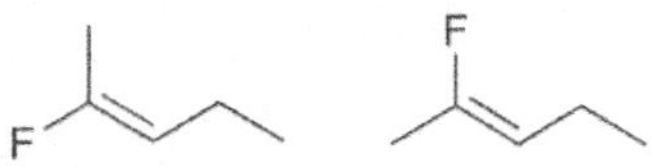

7.6 *MESO* COMPOUNDS

This is a topic that notoriously confuses students, so let's start off with an analogy. Let's use the analogy of the twin brothers who look identical except for one feature: one of them has a mole on the left side of his face, and the other has a mole on the right side of his face. You can tell them apart based on the mole, and the brothers are mirror images of each other. Imagine that their parents had other sets of twins, lots of sets of twins. So, all in all, there are a lot of siblings (who are all brothers and sisters of each other), but they are paired up, two in a group. Each child has *only one* twin sibling, who is his or her mirror image. Now imagine that the parents, out of nowhere, have one more child who is born without a twin—just a regular one-baby birth. When you look at this family, you would see lots of sets of twins, and then one child who has no twin (and has two moles—one on each side of his face). You might ask that child, where is your twin? Where is your mirror image? He would answer: I don't have a twin. I am the mirror image of myself. That's why the family has an odd number of children, instead of an even number.

The analogy goes like this: when you have a lot of chiral centers in a compound, there will be many stereoisomers (brothers and sisters). But, they will be paired up into sets of enantiomers (twins). Any one molecule will have many, many diastereomers (brothers and sisters), but it will have only one enantiomer (its mirror image twin). For example, consider the following compound:

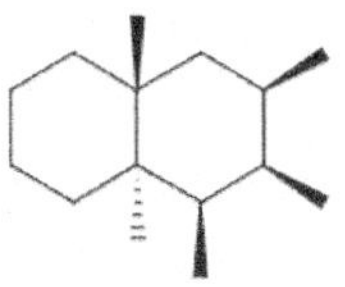

This compound has five chiral centers, so it will have many diastereomers. There are many, many possible compounds that fit that description, so this compound will have many brothers and sisters. But this compound will only have one twin—only one enantiomer (there is *only one* mirror image of the compound above):

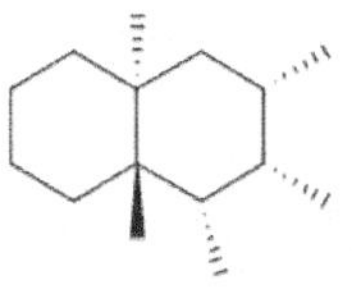

It is possible for a compound to be its own mirror image. In such a case, the compound will not have a twin, and the total number of stereoisomers will be an odd number, rather than an even number. That one lonely compound is called a *meso* compound. If you try to draw the enantiomer (using either of the methods we have seen), you will find that your drawing will be the same compound as what you started with.

So how do you know if you have a *meso* compound?

A *meso* compound has chiral centers, but the compound also has symmetry that allows it to be the mirror image of itself. Consider *cis*-1,2-dimethylcyclohexane as an example. This molecule has a plane of symmetry cutting the molecule in half. Everything on the left side of the plane is mirrored by everything on the right side:

If a molecule with chiral centers has an internal plane of symmetry, then it is a *meso* compound. If you try to draw the enantiomer (using either one of the two methods we saw), you will find that you are drawing the same thing again. This molecule does not have a twin. It is its own mirror image:

So, to be *meso*, a compound must have chiral centers, AND the compound must be superimposable on itself. We have seen that this can happen when we have an internal plane of symmetry. It can also happen when the compound has a center of inversion. For example,

This compound does not possess a plane of symmetry, but it does have a center of inversion. If we invert everything around the center of the molecule, we regenerate the same thing. Therefore, this compound will be superimposable on its mirror image, and the compound is *meso*. You will rarely see an example like this one, but it is not correct to say that the plane of symmetry is the only symmetry element that makes a compound *meso*. In fact, there is a whole class of symmetry elements (to which the plane of symmetry and center of inversion belong) called S_n axes, but we will not get into this, because it is beyond the scope of the course. For our purposes, it is enough to look for planes of symmetry.

There is one fail-safe way to tell if a compound is a *meso* compound: simply draw what you think should be the enantiomer and see if you can rotate the new drawing in any way to superimpose on the original drawing. If you can, then the compound will be *meso*. If not, then your second drawing is the enantiomer of the original compound.

EXERCISE 7.71   Is the following a *meso* compound?

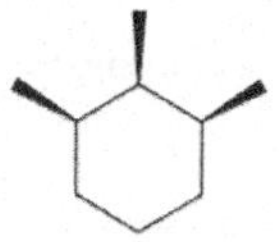

Answer   We need to try to draw the mirror image and see if it is just the same compound redrawn. If we use the second method for drawing enantiomers (placing the mirror on the side), then we will be able to see that the compound we would draw is the same thing:

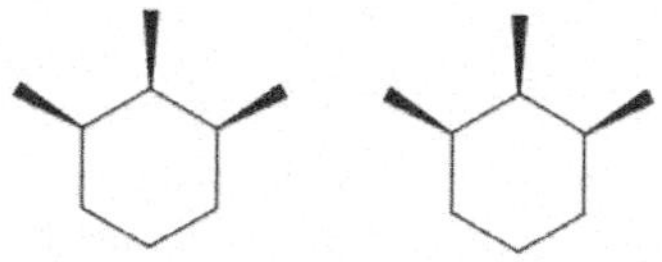

This compound has two chiral centers, AND the compound is superimposable on it's mirror image. Therefore, it is a *meso* compound.

A simpler way to draw the conclusion would be to recognize that the molecule has an internal plane of symmetry that chops through the center of one of the methyl groups, reflecting one chiral center onto another:

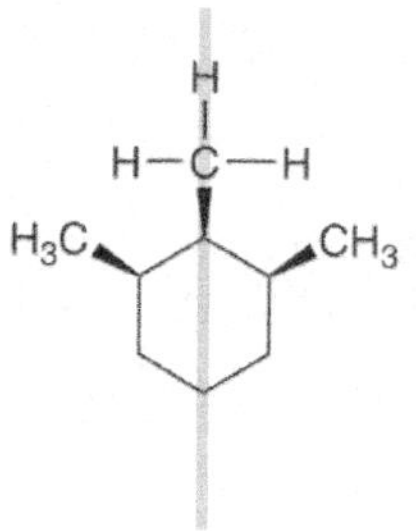

PROBLEMS   Identify whether each of the following compounds is a *meso* compound.

7.72

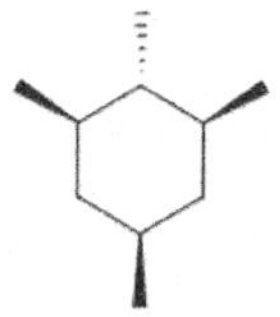

7.73

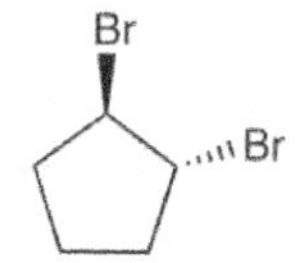

7.74

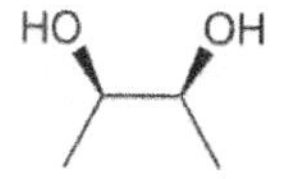

7.7   DRAWING FISCHER PROJECTIONS

There is an entirely different way to draw chiral centers (instead of using regular bond-line drawings with dashes and wedges). Fischer projections are helpful for drawing molecules that have many chiral centers, one after another. These drawings look like this:

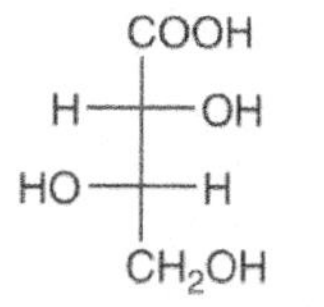

2 chiral centers

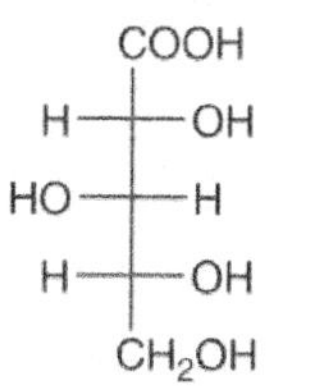

3 chiral centers

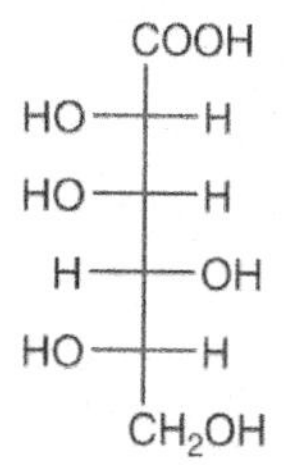

4 chiral centers

First we need to understand exactly what these drawings mean, and then we will learn a step-by-step method for drawing them properly.

Using Fischer projections saves time because we don't have to draw all of the dashes and wedges for each chiral center. Instead, we draw only straight lines, with the idea that all horizontal lines are coming out at us and all vertical lines are going away from us. Let's see exactly how this works. Consider the following molecule, which is drawn in a zigzag format (R_1 and R_2 represent groups whose identities are not being defined yet, because it does not matter for now):

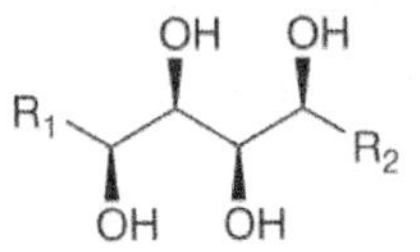

Remember that all of the single bonds are freely rotating, so there are a large number of conformations that the molecule can assume. When we rotate a single bond, the dashes and wedges change, but this is *not* because the configuration has changed. Remember that configurations do not change when a molecule twists and bends. Watch what happens when we rotate one of the single bonds:

Notice that R_1 is now pointing straight down, and the OH group is now on a dash. The configuration has *not* changed. If you need to convince yourself of this, determine the configuration of that chiral center in each of the two drawings. You will see that it has not changed.

Now let's draw another of the possible conformations for this molecule. If we rotate a couple more single bonds until the carbon skeleton is looping around like a bracelet, we get the following conformation:

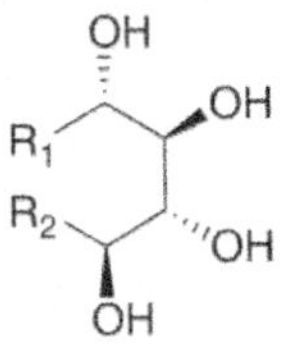

The molecule is twisting and bending around all of the time, and the conformation with the bracelet-shaped skeleton is just one of the possible conformations. The molecule probably spends very little of its time like this (it is a relatively high-energy conformation), but this is the conformation that we will use to draw our Fischer projection.

Now imagine piercing a pencil through R_1 and R_2 (the pencil is represented by the dotted line below). If you grab the ends of the pencil and rotate, you will find that R_1 and R_2 will stay in the page, but the rest of the molecule will pop out in front of the page:

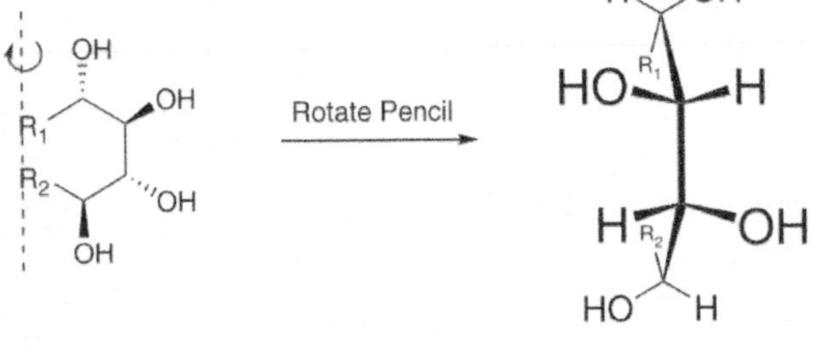

Now we imagine flattening the skeleton into a straight vertical line, and we redraw the molecule using only straight lines for the groups:

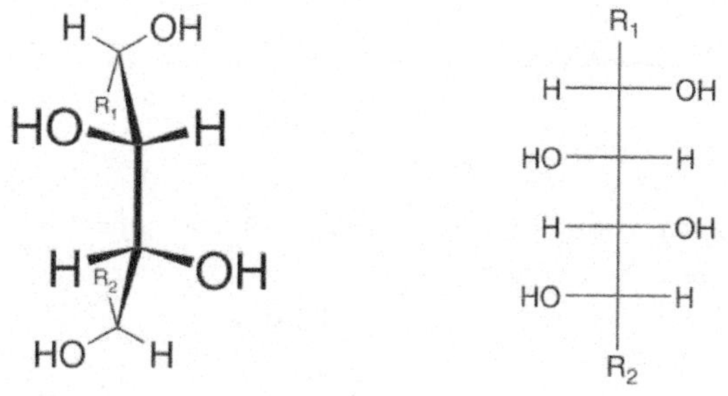

This is our Fischer projection. All of the configurations can be seen on this drawing, because we are able to picture in our minds what the 3D shape is. So the rule is that all horizontal lines are coming out at us, and all vertical lines are going away from us:

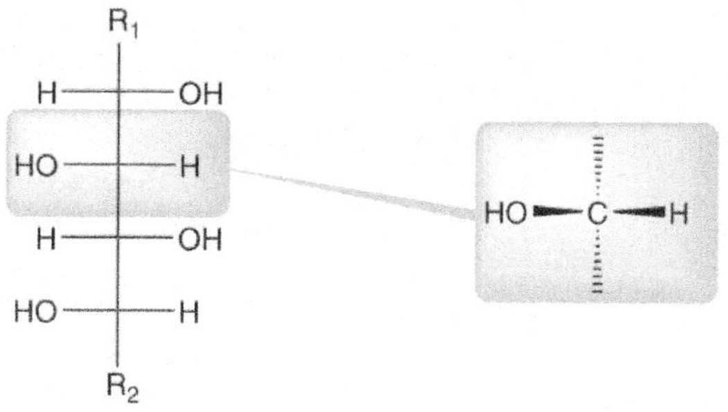

You might be wondering how you would assign the configuration of a chiral center when you are given a Fischer projection. If each chiral center is drawn as two wedges and two dashes, how do you figure out how to look at the chiral center? The answer is simple. Just choose one dash and one wedge, and draw them as straight lines. It doesn't matter which ones you choose—you will get the answer right regardless:

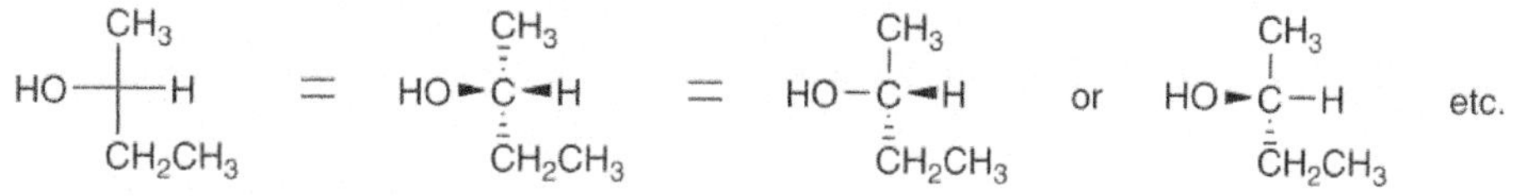

Once you have a drawing with two straight lines, one dash and one wedge, then you should be able to determine whether the chiral center is R or S. If you cannot, then you should go back and review the section on assigning configuration.

Fischer projections can also be used for compounds with just one chiral center, as above, but they are usually used to show compounds with multiple chiral centers. You will utilize Fischer projection heavily when you learn about carbohydrates at the end of the course.

Now we can understand why we cannot draw a Fischer projection sideways. If we did, we would be inverting all of the chiral centers. To draw the enantiomer of a Fischer projection, do not turn the drawing sideways. Instead, you should use the second method we saw for drawing enantiomers (place the mirror on the side of the compound and draw the reflection). Recall that this was the method that we used for drawings where wedges and dashes were implied but not shown. Fischer projections are another example of drawings that fit this criterion:

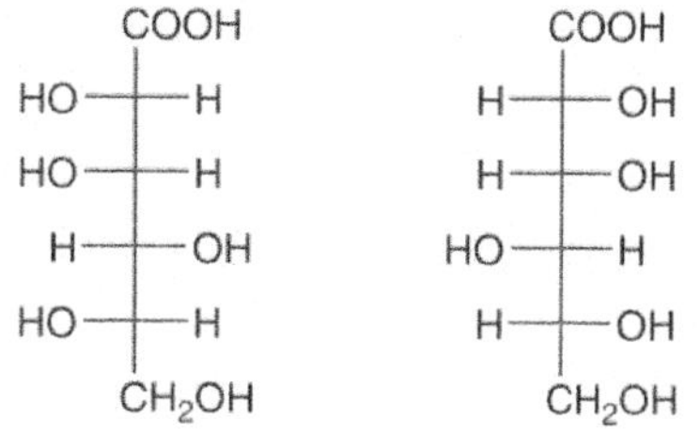

Enantiomers

EXERCISE 7.75   Determine the configuration of the chiral center below. Then draw the enantiomer.

$$\underset{\text{Me}}{\overset{\text{CH}_2\text{OH}}{\text{H}-\!\!\!-\!\!\!-\text{Cl}}}$$

Answer   We begin by drawing the chiral center as it is implied by the Fischer projection:

$$\underset{\text{Me}}{\overset{\text{CH}_2\text{OH}}{\text{H}\blacktriangleright\text{C}\blacktriangleleft\text{Cl}}}$$

Next, we choose one dash and one wedge, and we turn them into straight lines (it doesn't matter which dash or which wedge we choose):

$$\underset{\text{Me}}{\overset{\text{CH}_2\text{OH}}{\text{H}-\text{C}\blacktriangleleft\text{Cl}}}$$

Then we assign priorities based on atomic numbers:

$$\overset{\textcircled{2}}{\underset{\textcircled{3}}{\overset{\text{CH}_2\text{OH}}{\textcircled{4}\,\text{H}-\text{C}\blacktriangleleft\text{Cl}\,\textcircled{1}}}}$$

The 4 is not on a dash, so we switch it with the 3 so it can be on a dash, and we see that the configuration is *S*. Since we had to do a switch to get this, the configuration of the original chiral center (before the switch) was *R*:

$$\overset{\textcircled{2}}{\underset{\textcircled{4}\;\textbf{\textit{S}}}{\textcircled{3}-\text{C}\blacktriangleleft\textcircled{1}}}$$ therefore: $$\boxed{\overset{\textcircled{2}}{\underset{\textcircled{3}\;\textbf{\textit{R}}}{\textcircled{4}-\text{C}\blacktriangleleft\textcircled{1}}}}$$

Now, we need to draw the enantiomer. For Fischer projections, we use the method where we place a mirror on the side, and then we draw the reflection:

$$\underset{\text{Me}}{\overset{\text{CH}_2\text{OH}}{\text{H}-\!\!\!-\!\!\!-\text{Cl}}} \qquad \underset{\text{Me}}{\overset{\text{CH}_2\text{OH}}{\text{Cl}-\!\!\!-\!\!\!-\text{H}}}$$

Enantiomer

PROBLEMS   For each compound below, determine the configuration of the chiral center, and then draw the enantiomer.

7.76

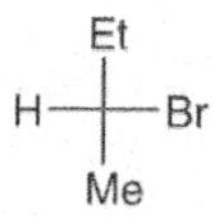

```
        Et
        |
  H ----+---- Br
        |
        Me
```

7.77

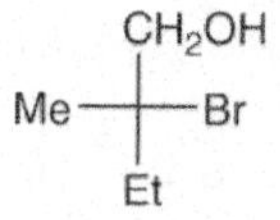

```
        CH2OH
        |
  Me ----+---- Br
        |
        Et
```

7.78

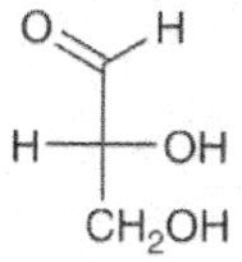

```
   O    H
    \\  /
      C
        |
  H ----+---- OH
        |
        CH2OH
```

PROBLEMS   For each compound below, determine the configuration of every chiral center. Then draw the enantiomer of each compound below (the COOH group is a carboxylic acid group).

7.79

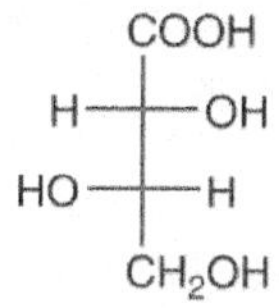

```
        COOH
        |
  H ----+---- OH
        |
  HO ---+---- H
        |
        CH2OH
```

7.80

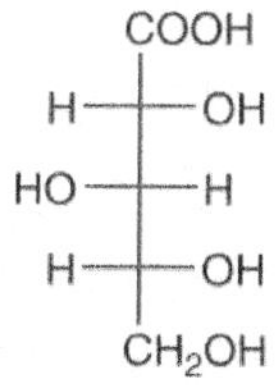

```
        COOH
        |
  H ----+---- OH
        |
  HO ---+---- H
        |
  H ----+---- OH
        |
        CH2OH
```

7.81

```
        COOH
        |
  H ----+---- Cl
        |
  Br ---+---- H
        |
  H ----+---- OH
        |
  HO ---+---- H
        |
        CH2OH
```

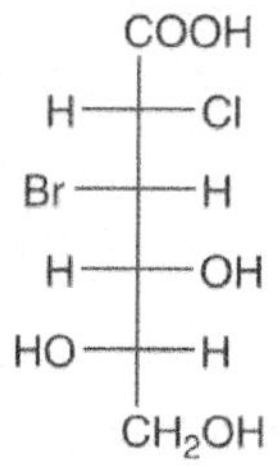

7.8   OPTICAL ACTIVITY

Students confuse *R/S* with +/− all of the time, so let's conclude our chapter by clearing up the difference. Compounds are chiral if they have chiral centers and they are not *meso* compounds. A chiral compound will have an enantiomer (a nonsuperimposable mirror image). An interesting thing happens when you take a chiral compound and subject it to plane-polarized light. The plane of the polarized light rotates as it passes through the sample. If this rotation is clockwise, then we say the rotation is +. If the rotation is counterclockwise, then we say the rotation is −. If we want to refer to a racemic mixture (an equal mixture of both enantiomers), we will often say (+/−) in the beginning of the name to mean that both enantiomers are present in solution (and the rotations cancel each other).

Do not confuse clockwise rotation of plane-polarized light with clockwise ordering of atomic numbers when determining configurations. They are not related. When determining configuration, we impose a set of man-made rules to help us distinguish between the two possible configurations. By using these rules, we will always be able to communicate which configuration we are referring to, and we only need one letter to communicate this (*R* or *S*) if we use the rules properly. However, +/− is totally different.

The rotation of plane-polarized light (either + or −) is not a man-made convention. It is a physical effect that is measured in the lab. It is impossible to predict whether a compound will be + or − without actually going into the lab and trying. If a chiral center is *R*, this does not mean that the compound will be +. It could just as easily be −. In fact, whether a compound is + or − may even depend on temperature! So a compound can be + at one temperature and − at another temperature. But clearly, temperature has nothing to do with *R* and *S*. So, don't confuse *R/S* with +/−.

You will never be expected to look at a compound that you have never seen and then predict in which direction it will rotate plane-polarized light (unless you know how the enantiomer rotates plane-polarized light, because enantiomers have opposite effects). But you will be expected to assign configurations (*R* and *S*) for chiral centers in compounds that you have never seen.

CHAPTER 8

MECHANISMS

8.1 INTRODUCTION TO MECHANISMS

Recall from Section 3.8 that a mechanism shows how we believe a reaction occurs, using curved arrows to illustrate the flow of electrons. Throughout your organic chemistry course, you will encounter a large number of reaction mechanisms, and they are all really important. One of the biggest mistakes that you can make is to underestimate the centrality of mechanisms. Mechanisms are literally the keys to your success in this course, and they MUST occupy a central focus of your study efforts. If you study mechanisms thoroughly and understand them exceptionally well, you will find that the reactions actually make sense because they follow a small handful of simple principles. You will see that electrons have predictable behavior, and the details of each reaction will become easy to remember (because a good mechanism will explain the reasons for these details), and this will make it easier for you to master the material and associated skills.

   In this chapter, we will focus exclusively on the mechanisms of ionic reactions, which are reactions that involve ions as reactants, intermediates, or products. Ionic reactions represent most (95%) of the reactions that we will encounter in this course (you will likely spend very little time on radical reactions and pericyclic reactions). As we develop the tools for drawing mechanisms, keep in mind that curved arrows take on a different meaning than they did when we were drawing resonance structures. When drawing resonance structures, the curved arrows were just tools, and they did not actually represent any physical process. That is, the electrons were not actually moving—we were just treating them as if they were moving. But when we draw mechanisms, the curved arrows actually DO represent the flow of electrons. Electrons really ARE moving during chemical reactions, and a mechanism shows how the electrons flowed during the reaction. Therefore, it is OK to break a single bond during a mechanism. Indeed, it happens all of the time (breaking a single bond was a big no-no when we drew resonance structures in Chapter 2).

8.2 NUCLEOPHILES AND ELECTROPHILES

Most of the mechanisms that you will see in upcoming chapters will have a step in which a nucleophile attacks an electrophile, so let's make sure we understand the terms. A *nucleophile* is an ion or a compound that is capable of donating a pair of electrons. The following ions can all serve as nucleophiles, because all of them have an atom bearing a lone pair of electrons:

$$:\overset{..}{\underset{..}{I}}:^{\ominus} \qquad :\overset{..}{\underset{..}{Cl}}:^{\ominus} \qquad :\overset{..}{\underset{..}{Br}}:^{\ominus} \qquad H-\overset{..}{\underset{..}{S}}:^{\ominus} \qquad R-\overset{..}{\underset{..}{O}}:^{\ominus} \qquad H-\overset{..}{\underset{..}{O}}:^{\ominus} \qquad H-\overset{\ominus}{\underset{H}{N}}$$

Similarly, the following compounds can all serve as nucleophiles as well:

$$H-\overset{..}{\underset{}{S}}-H \qquad R-\overset{..}{\underset{}{O}}-H \qquad H-\overset{..}{\underset{}{O}}-H \qquad H-\overset{\overset{H}{|}}{\underset{..}{N}}-H \qquad R-\overset{\overset{H}{|}}{\underset{..}{N}}-H$$

Despite the absence of a negative charge, each of these compounds can serve as a nucleophile, because there is an atom bearing a lone pair of electrons. In each case, the atom bearing the lone pair represents a nucleophilic center. Pi bonds can also serve as nucleophilic centers, because pi bonds are regions in space of high electron density:

$$\underset{R}{\overset{R}{\diagdown}}C=C\underset{R}{\overset{R}{\diagup}} \qquad R-C\equiv C-R$$

EXERCISE 8.1 Identify all of the nucleophilic centers in the following compound:

Answer This compound has a nitrogen atom. While the lone pair has not been drawn, we know from Section 1.6 that the nitrogen atom does indeed have one lone pair:

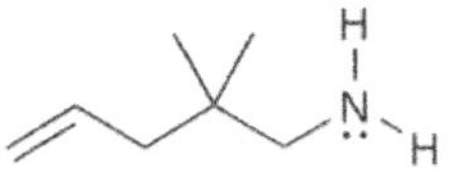

Therefore, this nitrogen atom can serve as a nucleophilic center. In addition, the pi bond can also serve as a nucleophilic center. So this compound has two nucleophilic centers, highlighted here:

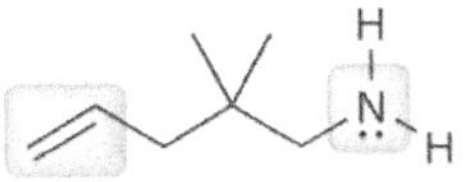

PROBLEMS Identify the nucleophilic center(s) in each of the following compounds or ions:

8.2

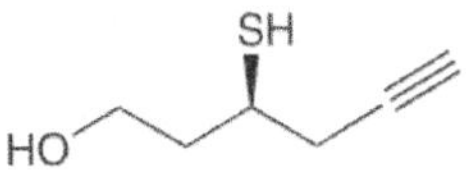

8.3

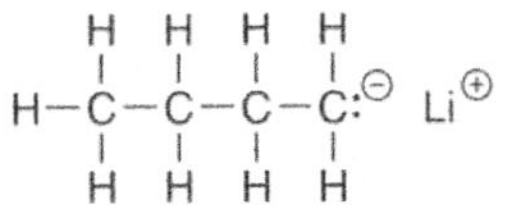

8.4

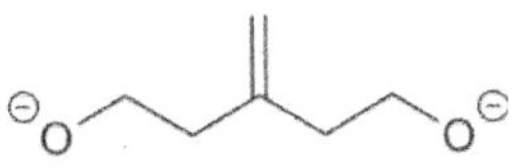

8.5

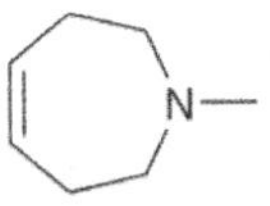

8.6

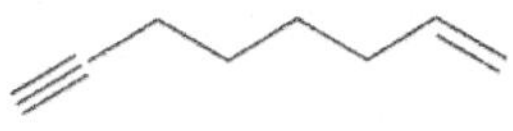

8.7

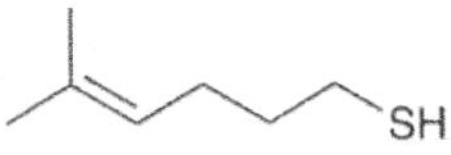

Now let's shift our attention to electrophiles. *Electrophiles* are compounds or ions that are poor in electron density, which is why they can be attacked by nucleophiles. As examples, consider the following two electrophiles:

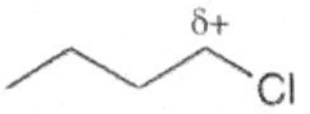

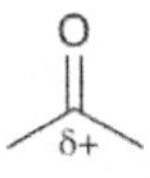

In the first example, the carbon atom connected directly to chlorine is poor in electron density ($\delta+$) as a result of induction. In the second example, the carbon atom of the C=O bond is poor in electron density ($\delta+$) because of induction, as well as resonance, shown here:

We will also see many examples of carbocation intermediates functioning as electrophiles:

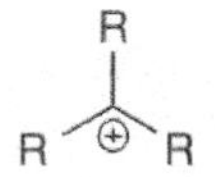

For example, consider the second step in the following mechanism:

Notice that a chloride ion functions as a nucleophile and attacks the carbocation, which functions as an electrophile. Now let's get some practice identifying electrophilic centers.

PROBLEMS   Identify the electrophilic center(s) in each of the following compounds or ions:

8.8

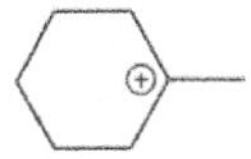

8.9

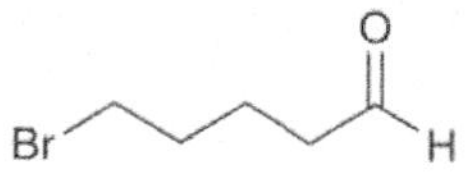

8.10

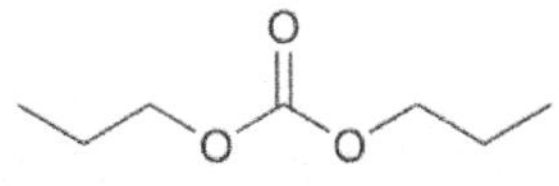

8.11

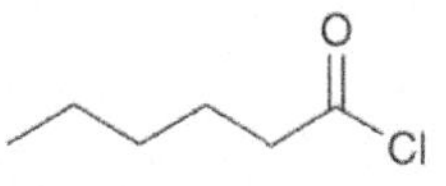

8.12

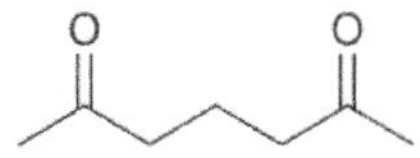

8.13

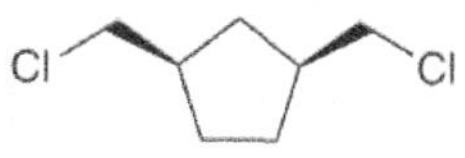

8.14   The following compound has two electrophilic centers. Identify both of them. Hint: You will need to draw all resonance structures in order to see both electrophilic centers.

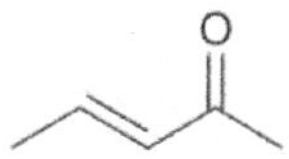

8.3 BASICITY VS. NUCLEOPHILICITY

In this chapter, we have seen that a nucleophile is a compound or ion that is capable of donating a pair of electrons (and attacking an electrophile). The term *nucleophilicity* is a measure of *how quickly* a nucleophile will attack an electrophile. A strong nucleophile will attack electrophiles rapidly, while a weak nucleophile will attack electrophiles slowly. That is, nucleophilicity is a measure of reaction *rates*.

Basicity (first encountered in Chapter 3) is an entirely different concept. Basicity is not a measure of how fast anything happens. Rather, it is a measure of the position of equilibrium for an acid–base reaction. The position of equilibrium is determined by the relative energy levels of starting materials and products, regardless of how long it takes to achieve the equilibrium. In practice, proton transfers are extremely rapid processes, often faster than most other processes, but we rarely ever talk about the rates of proton transfer steps. When we measure basicity, we are measuring the position of equilibrium.

So we see that nucleophilicity is a kinetic phenomenon (a measure of reaction rates), while basicity is a thermodynamic phenomenon (a measure of equilibrium concentrations). That is, nucleophilicity and basicity measure entirely different phenomena. As such, it is possible for a strong nucleophile to be a weak base. For example, the iodide ion (I^-) is among the weakest known bases, yet it is also one of the strongest known nucleophiles. Why? It is a weak base because the negative charge is spread over a large volume and is thereby highly stabilized (iodide is the conjugate base of a very strong acid, HI). The size of the iodide ion is also responsible for it being such a strong nucleophile. Large atoms are said to be *polarizable*, because their electron density can more easily shift and move in response to external influences.

It is also possible for a strong base to be a weak nucleophile. For example, consider the hydride ion (H^-), as in sodium hydride (NaH). This reagent is a very strong base (it is very unstable), yet it is a weak nucleophile because it is not large and polarizable. Hydrogen is the smallest atom, and so it is not polarizable. As a result, the hydride ion does not function as a nucleophile, despite the fact that it is a very strong base.

In many cases, nucleophilicity and basicity will indeed appear to parallel each other. For example, water is a weak base and a weak nucleophile, while hydroxide is a strong base and a strong nucleophile:

Weak base and
weak nucleophile

Strong base and
strong nucleophile

Make sure to keep in mind that the main difference between basicity and nucleophilicity is one of *function*. That is, a hydroxide ion can either function as a base, or it can function as a nucleophile. In some cases, we will see hydroxide functioning as a base and removing a proton, while in other cases, we will see hydroxide functioning as a nucleophile and attacking an electrophile. The following mechanism shows both occurring:

In the first step of the mechanism, hydroxide functions as a nucleophile and attacks the electrophile (the ester). Then, in the last step of the mechanism, hydroxide functions as a base and removes a proton.

Similarly, in the following mechanism (which we will learn in Chapter 11), water functions both as a nucleophile and as a base:

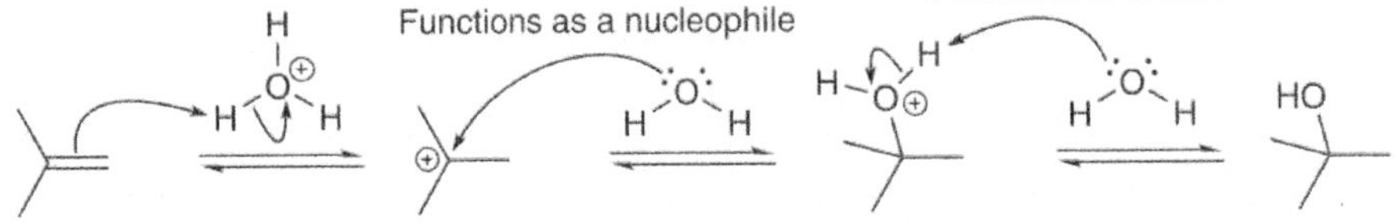

In the second step of the mechanism, water functions as a nucleophile and attacks a carbocation (which functions as an electrophile). Then, in the third (final) step of the mechanism, water functions as a base and removes a proton.

PROBLEMS Each of the following mechanistic steps will appear as part of a larger mechanism that you will encounter in an upcoming chapter. In each case, identify whether the hydroxide ion is functioning as a base or as a nucleophile.

8.15

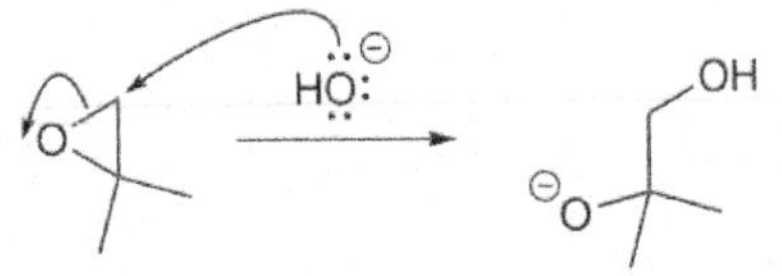

8.16

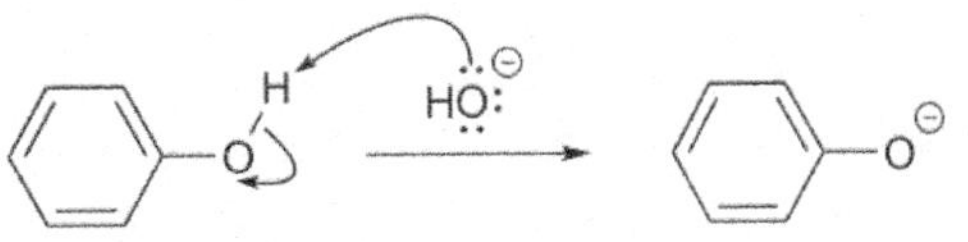

8.17

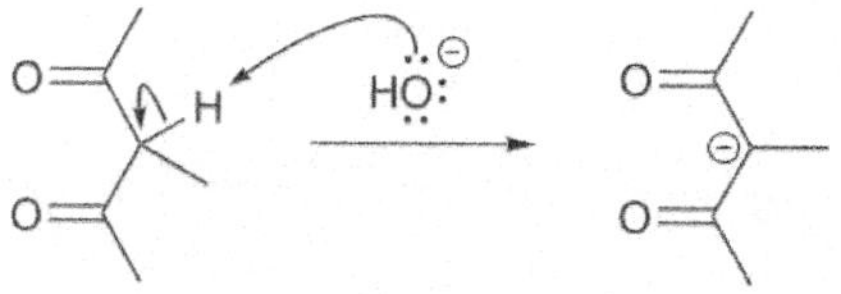

8.18

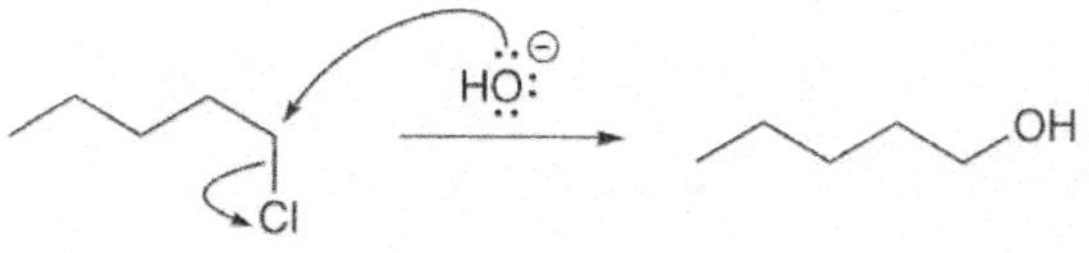

8.19

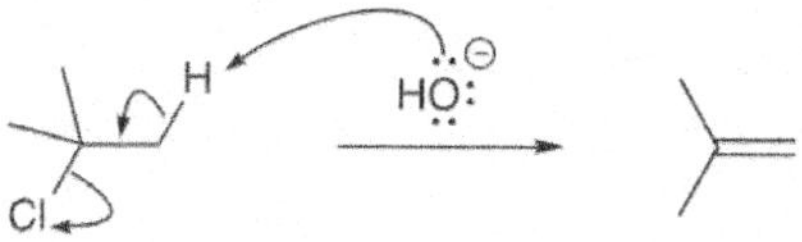

8.20

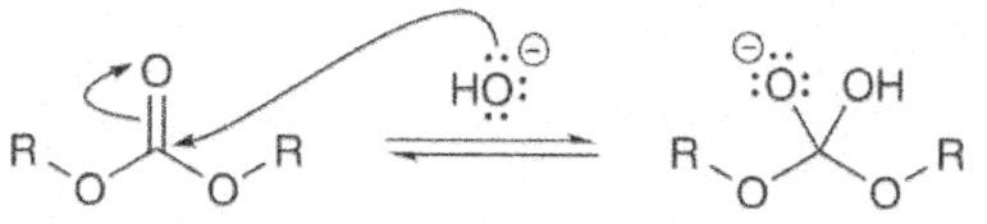

8.4 ARROW-PUSHING PATTERNS FOR IONIC MECHANISMS

In Chapter 2, we saw five patterns for pushing arrows when drawing resonance structures. With those five patterns, we were able to draw resonance structures in most common situations. Now, with ionic mechanisms, there are once again patterns for arrow pushing. But this time, there are only four patterns. And we have already seen the first of these four patterns. The first pattern is called *nucleophilic attack*, and is characterized by a nucleophile attacking an electrophile, for example:

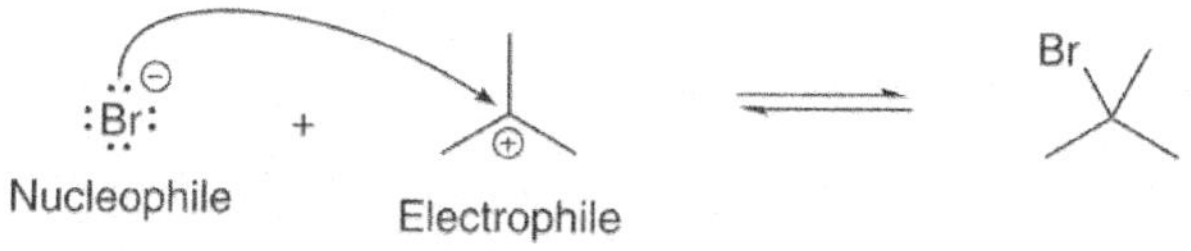

Nucleophile   Electrophile

Notice that one curved arrow is necessary to show the nucleophilic attack. The tail of this arrow is placed on the nucleophilic center, and the head of the arrow is placed on the electrophilic center. In some cases, you might see a nucleophilic attack where more than one curved arrow is used (especially when a pi bond is attacked), for example:

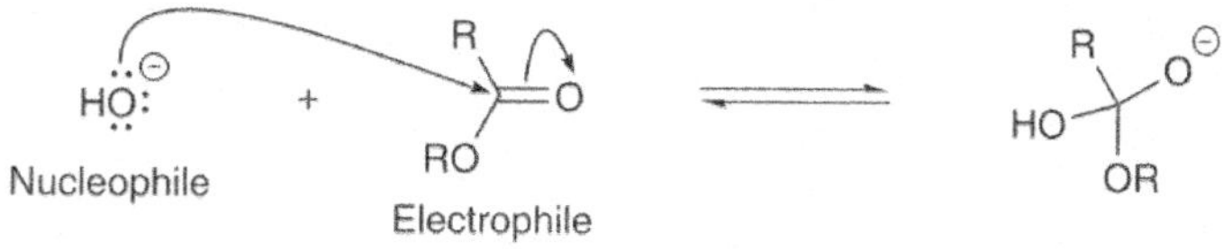

Nucleophile   Electrophile

Even though this example uses two curved arrows (rather than one), there is only one thing happening here: a nucleophile is attacking an electrophile.

Now let's move on to the second arrow-pushing pattern called *proton transfer*. This pattern should also seem familiar to you, because we already discussed it in Section 3.8. Recall that a base will deprotonate (remove a proton from) an acid, as shown in the following example:

Base   Acid

Proton transfer steps always require at least two curved arrows. One curved arrow shows the base attacking the proton, and the other curved arrow shows the acid being converted into its conjugate base. When drawing a proton transfer step, make sure that you draw BOTH of these curved arrows. It is a common mistake to forget the second curved arrow (showing formation of the conjugate base).

Before we move on to the next arrow-pushing pattern, there is one more issue to consider regarding proton transfer steps. We have seen that a pi bond can function as a nucleophile, but you should know that a pi bond can also function as a base to remove a proton from an acid, as seen in the following example:

So far, we have seen two of the four patterns: 1) nucleophilic attack and 2) proton transfer. Now let's move on to the third pattern: *loss of a leaving group*. In the following example, a chloride ion functions as a leaving group:

What is a good leaving group? We will explore this important question more fully in Section 9.4. For now, we will just say that good leaving groups are weak bases. For example, iodide is a weak base, so it is an excellent leaving group. In contrast, hydroxide is a strong base, so it is a very poor leaving group.

In the previous example, chloride is the leaving group, which is ejected from the starting alkyl chloride. During this type of step (loss of a leaving group), it is very common to see the leaving group being ejected, as above. After all, the step is called "loss" of a leaving group. But in some cases, the leaving group is not fully ejected, because it can remain tethered to the starting compound, even after it has departed. Consider the following step as an example:

In this example, the C—O single bond is broken, giving a carbocation. Notice that the leaving group is an alcohol (ROH), where R is the rest of the structure. This step would most certainly be classified as the loss of a leaving group, even though the leaving group remains tethered to the newly formed carbocation.

We have now seen three of the four patterns. The final arrow-pushing pattern is *rearrangement*. There are several types of rearrangements, but in this course, we will commonly encounter one type, called a carbocation rearrangement. The following is an example of a carbocation rearrangement:

A carbocation rearrangement is characterized by a change in the location of the electron deficient center (C^+). We will discuss carbocation rearrangements in the next section of this chapter.

To review, there are only four arrow-pushing patterns for ionic mechanisms: 1) nucleophilic attack, 2) proton transfer, 3) loss of a leaving group, and 4) carbocation rearrangements.

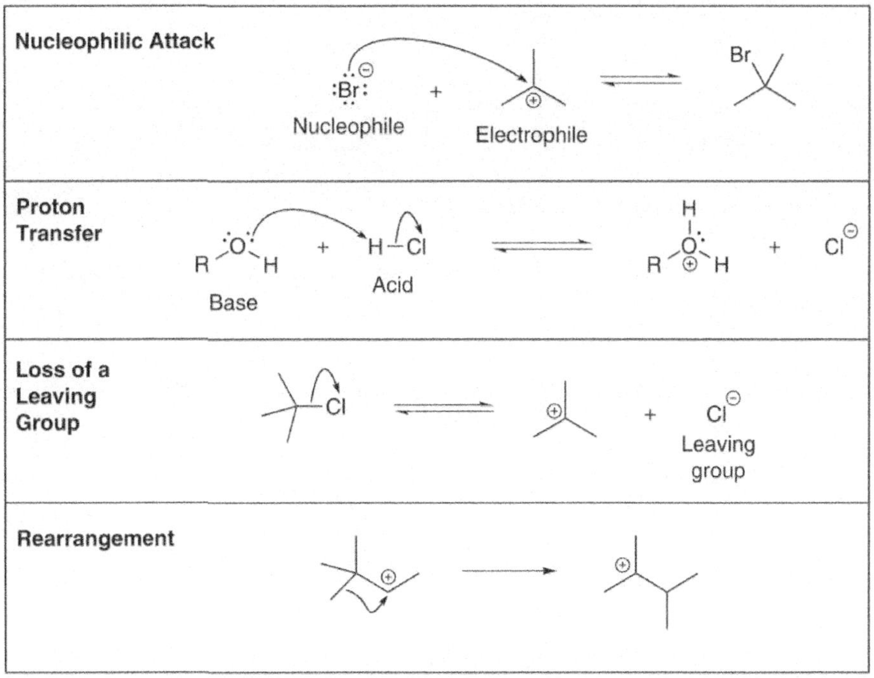

Each ionic mechanism that we encounter (moving forward) will be comprised of a particular sequence of arrow-pushing patterns. For example, consider the following mechanism, which we will encounter in Chapter 13:

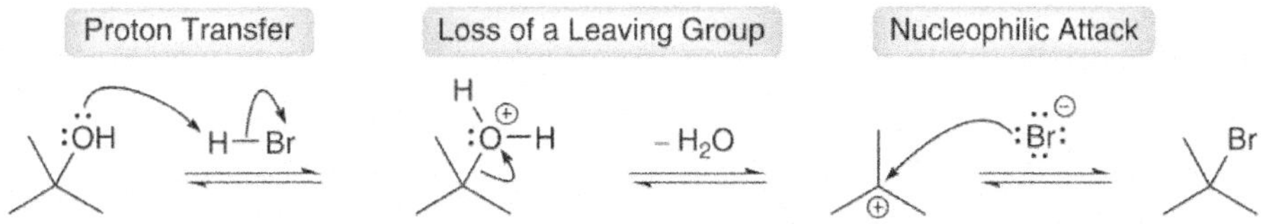

This mechanism has three steps, in the following order: 1) proton transfer, 2) loss of a leaving group, and 3) nucleophilic attack. Now let's compare this sequence of steps to the sequence of steps in the following mechanism:

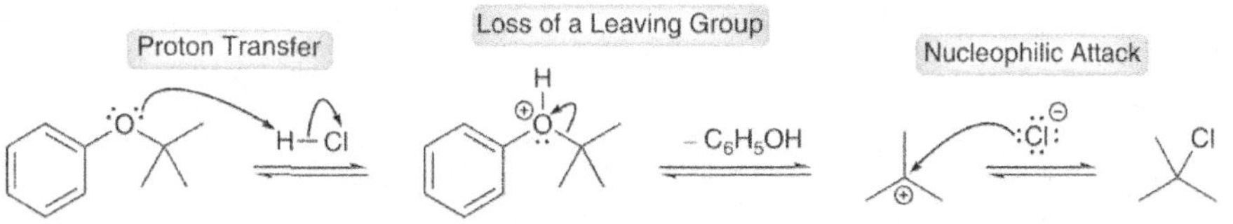

Once again, this mechanism has three steps, in the following order: 1) proton transfer, 2) loss of a leaving group, and 3) nucleophilic attack. That is, both reactions share the same sequence of events. By seeing the similarity, two apparently different reactions can be viewed under a single umbrella. Both of these reactions occur via a similar mechanism, called an S_N1 process (as we will see in Chapters 9 and 13). A proper understanding of reaction mechanisms will unify seemingly different reactions, and you will find that you will be able to predict how and when electrons will flow, using the four arrow-pushing patterns, together with a short list of rules and principles.

Sometimes, two arrow-pushing patterns are drawn in a single step. When this occurs, we call it a *concerted process*. For example, in an S_N2 process (see Chapter 9), a nucleophilic attack occurs at the same time as loss of a leaving group. Since both events occur simultaneously, an S_N2 process is said to be a concerted process.

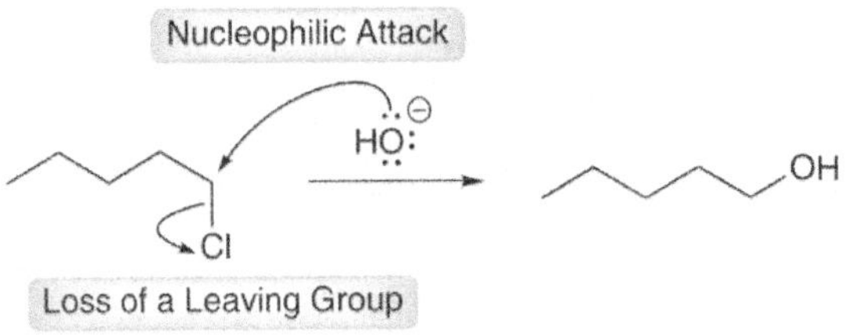

Nucleophilic Attack

Loss of a Leaving Group

We will explore S_N1 and S_N2 processes in more detail in Chapter 9. For now, let's get some practice identifying the sequence of arrow-pushing patterns in ionic mechanisms.

EXERCISE 8.21   Identify the sequence of arrow-pushing patterns in the following ionic mechanism:

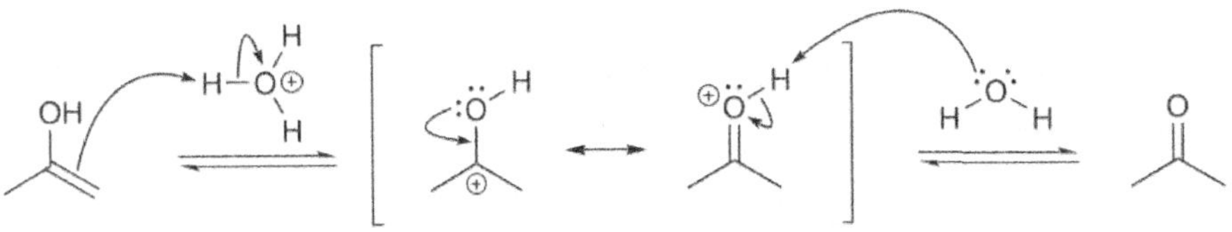

Answer   In the first step of this process, the pi bond functions as a base and removes a proton from H_3O^+. So the first step is a proton transfer step:

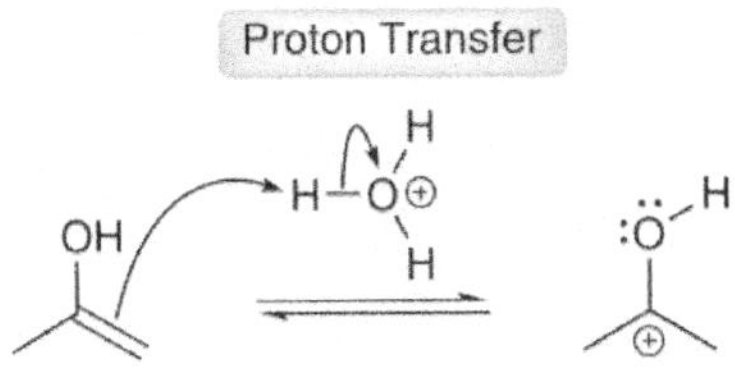

Proton Transfer

Now be careful. The next drawing does not represent a new step. It is just a resonance structure:

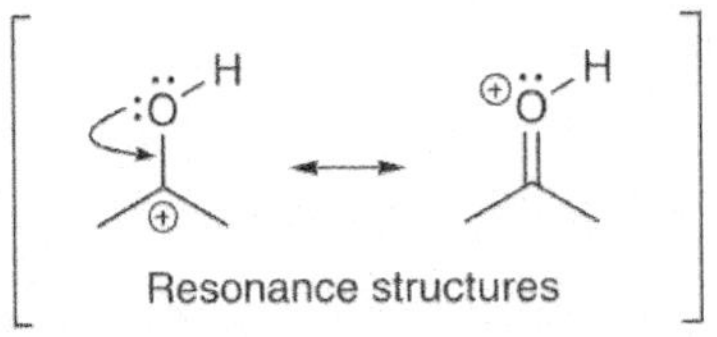

Resonance structures

Remember that resonance does not represent any physical process (see Chapter 2). Nothing is happening here. Do not attempt to label this as if it were a step of the mechanism. It is not. This mechanism has only two steps. We saw that the first step was a proton transfer. In the final step, water functions as a base and removes a proton. Therefore, this step is also a proton transfer step:

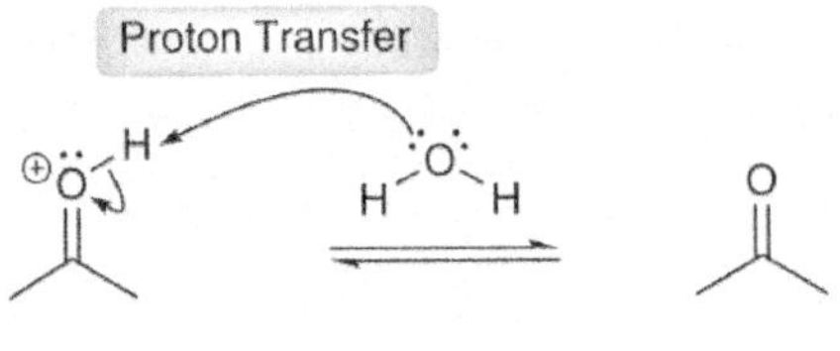

Proton Transfer

PROBLEMS Identify the sequence of arrow-pushing patterns in each of the following ionic mechanisms:

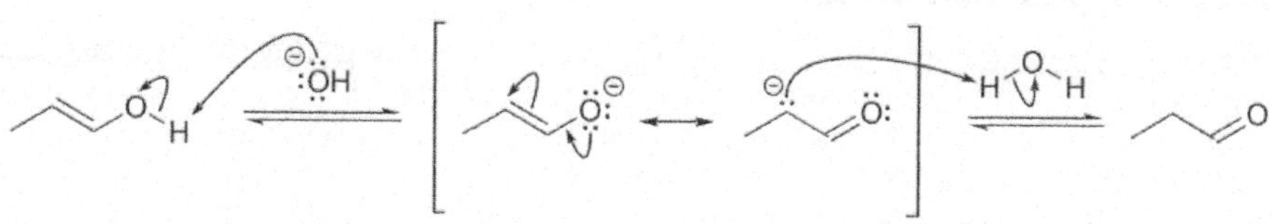

8.22

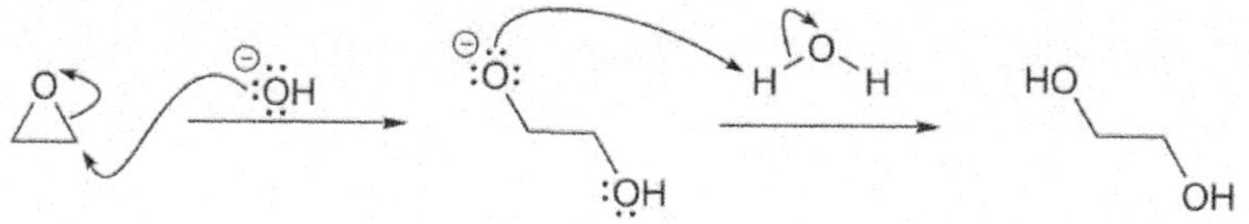

8.23

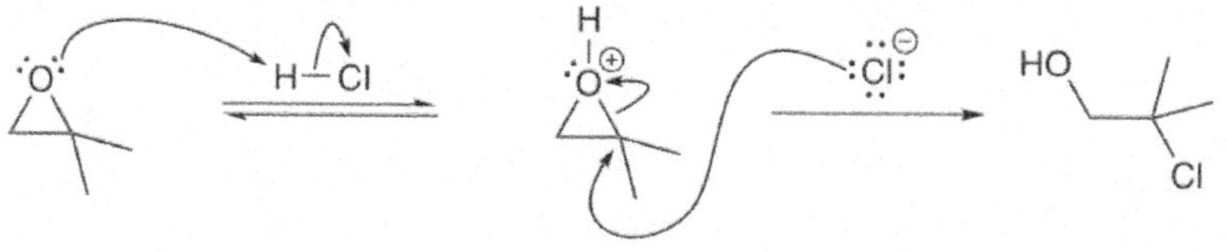

8.24

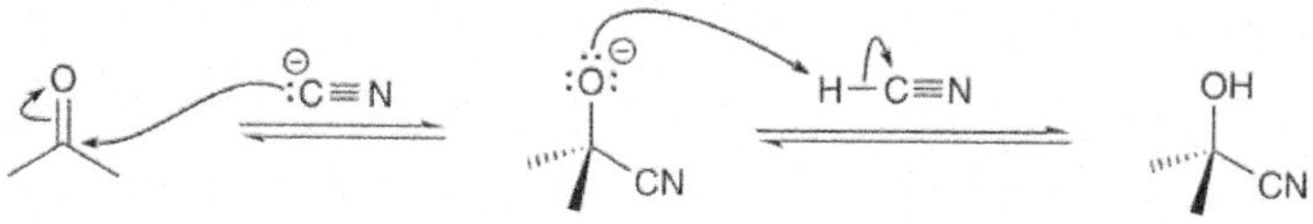

8.25

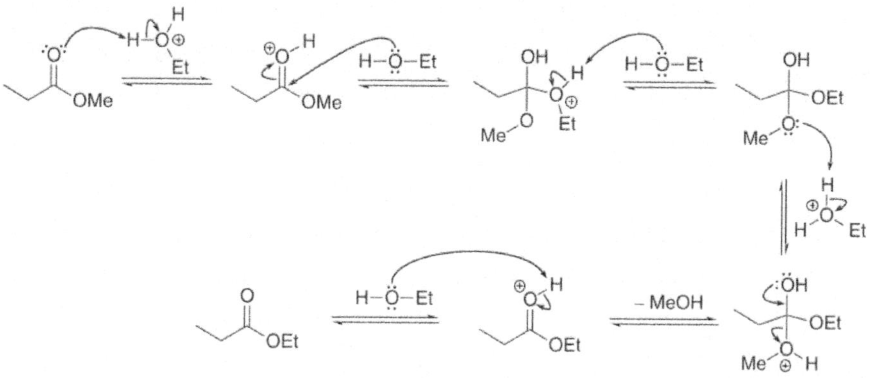

8.26

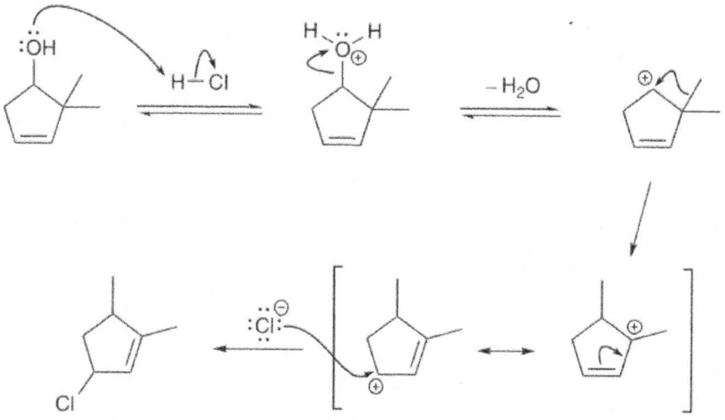

8.27

8.5 CARBOCATION REARRANGEMENTS

Carbocations are short-lived intermediates. They can't be stored in a bottle for future use in a synthesis. But they will appear in many of the mechanisms that we will encounter in upcoming chapters. Consider the following mechanism, which we saw in the previous section as an example of an S_N1 reaction (discussed further in Chapter 9). This mechanism involves a carbocation intermediate (in the last step, the carbocation is attacked by bromide):

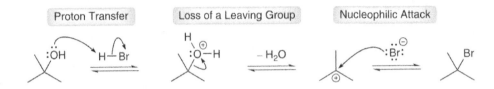

This mechanism has three steps. In the first step, the OH group is protonated (proton transfer). The second step (loss of a leaving group) generates a carbocation intermediate (C^+). This carbocation intermediate is then attacked by bromide (nucleophilic attack) in the final step of the mechanism, giving the product. These three steps are seen in the following energy diagram:

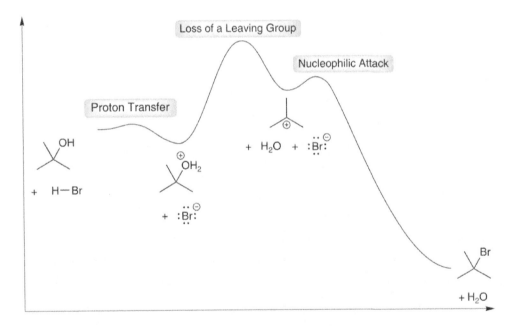

The shape of this energy diagram reveals several features that are worthy of mention:

- There are three humps, because there are three steps. Each hump corresponds to a step of the mechanism: proton transfer, loss of a leaving group, and finally, nucleophilic attack. These three steps are labeled on the energy diagram.

- The hump of highest energy represents the rate-determining step of the process. The rate of this step (formation of the carbocation) determines the rate of the entire process. Any factors which speed up the rate of the other two steps (proton transfer and nucleophilic attack) will have very little effect on the overall rate of the entire process. In contrast, the rate of the entire process will be profoundly affected by any factors that speed up the rate of the rate-determining step.

- Notice that there are two valleys in the energy diagram (in between the humps). Each of these valleys represents an intermediate. The first intermediate has a positive charge on an oxygen atom, so it is called an oxonium ion. The second intermediate has a positive charge on a carbon atom, so it is called a carbocation. Notice that a carbocation intermediate is higher in energy than an oxonium ion. Why? Because a carbocation has an atom (C^+) that lacks an octet. In contrast, an oxonium (O^+) does not have any atoms missing an octet (the positively charged oxygen atom has an octet of electrons).

- This energy diagram, together with the features mentioned above, should make it clear that carbocations are *intermediates* of reactions, not starting materials or products. They cannot be stored in a bottle for later use. Carbocations are short-lived intermediates that we will often draw as we propose mechanisms for a variety of reactions. *Carbocations are not starting materials for syntheses.*

Carbocations exhibit the following trend in stability:

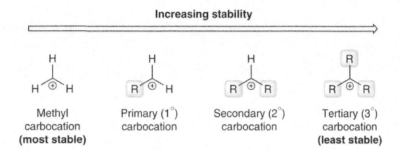

Tertiary carbocations are more stable than secondary carbocations, which are more stable than primary carbocations (methyl carbocations are very unstable). So, we see that alkyl groups stabilize a carbocation. This stabilizing effect is the result of a phenomenon called hyperconjugation. You should consult your textbook and/or lecture notes to see if you are responsible for understanding hyperconjugation.

Carbocations are also stabilized by resonance:

In this case, called an allylic carbocation, the positive charge is stabilized because it is spread over two locations (delocalized), rather than being localized on one atom. For this reason, a tertiary allylic carbocation is even more stable than a tertiary carbocation:

vs.

Tertiary allylic
(more stable)

Tertiary

Carbocations are also highly stabilized when the positive charge is located at a benzylic position (the carbon atom connected directly to an aromatic ring):

Can you draw all of the resonance structures for the benzylic carbocation above?

In some cases, a carbocation will rearrange, if doing so enables it to become more stable. For example, a secondary carbocation will rearrange if doing so will give a more stable, tertiary carbocation. And similarly, a tertiary carbocation will rearrange if doing so will give a more stable tertiary allylic carbocation. Let's take a closer look at how carbocations rearrange.

Carbocations rearrange in a variety of ways, but the two most common ways are called a *methyl shift* and a *hydride shift*. An example of a methyl shift is shown here:

Methyl shift

Notice that a methyl group migrates together with the two electrons of the bond that is connecting the methyl group to the main chain. You can think of the migrating methyl group as if it is a methyl anion (H_3C^-). When this methyl group (with its negative charge) migrates, it plugs the hole (you can think of C^+ as a "hole"—a place where electron density is missing). In doing so, a new "hole" is created where the methyl group was formerly connected. So, in a sense, the hole is now somewhere else. It is similar to the following analogy. Imagine that there is a hole in the ground, and you want to fill it with dirt. So you dig a new hole, nearby, and use that dirt to fill the original hole. After transferring the dirt, you might have filled the original hole, but you have also created a new hole. In a sense, you have simply moved the location of the hole.

The other common type of carbocation rearrangement is called a hydride shift, as seen here:

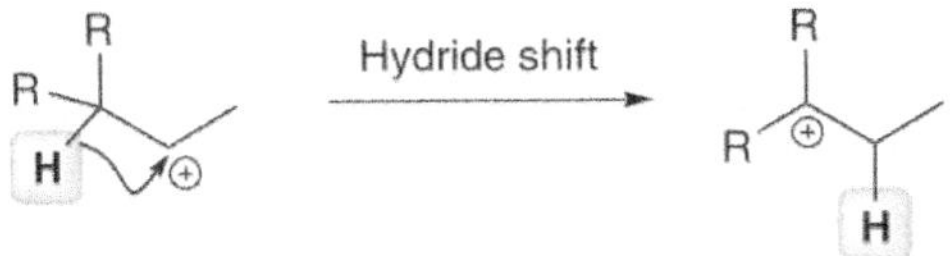

This is similar to a methyl shift, but this time, it is H^- (hydride) that is migrating, rather than H_3C^-. When this hydride (with its negative charge) migrates over, it plugs the hole. In doing so, a new hole (C^+) is created where the hydride was formerly connected.

When do carbocation rearrangements occur? Whenever they can. If a carbocation rearrangement is possible (to generate a more stable carbocation), then it will likely occur. So, whenever you propose a mechanism that involves a carbocation intermediate, you should analyze the structure of that carbocation intermediate and determine if it is susceptible to rearrangement. Let's get some practice identifying carbocations that can or cannot rearrange.

EXERCISE 8.28   Predict whether the following carbocation is expected to undergo a carbocation rearrangement, and if so, draw the rearrangement:

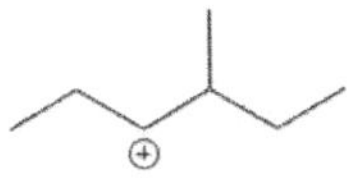

Answer   We begin by identifying the positions connected directly to the C^+. This carbocation is secondary, so it is connected to two carbon atoms, highlighted here:

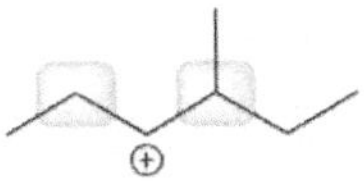

Then, we ask whether either of these highlighted positions bears an H or a methyl group. Let's start with the position on the left. This position bears both a methyl group and hydrogen atoms:

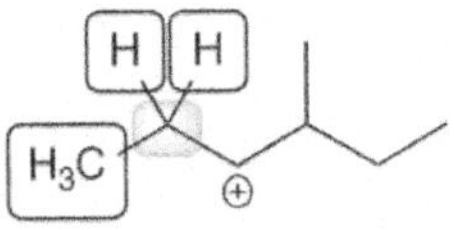

Yet, neither a hydride shift nor a methyl shift will generate a more stable carbocation. Why not? A methyl shift would give a less stable carbocation (converting a secondary carbocation into a primary carbocation):

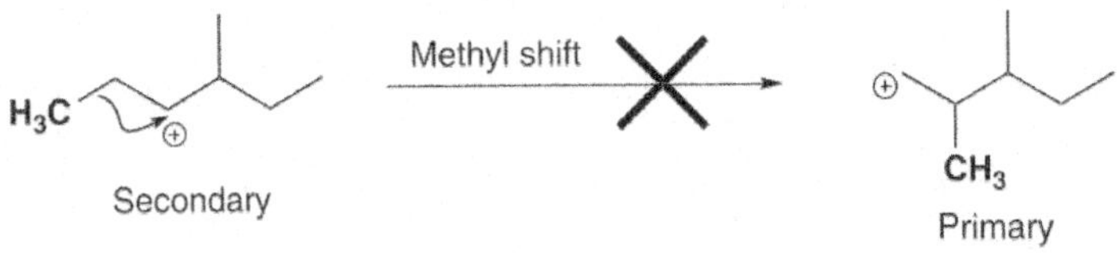

And a hydride shift would also not generate a more stable carbocation. A hydride shift would just convert a secondary carbocation into another secondary carbocation:

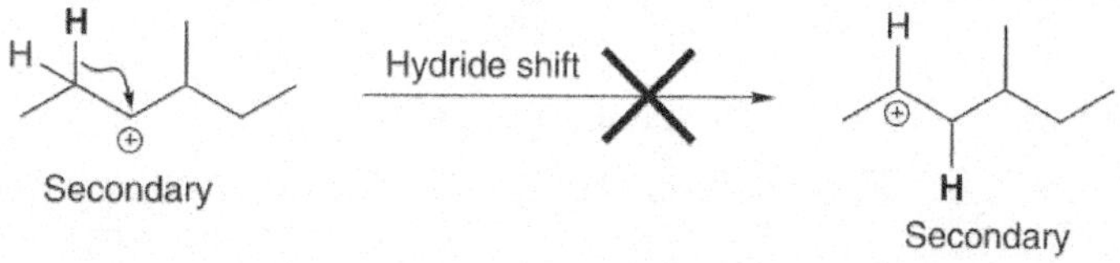

This does not represent a decrease in energy, so we keep looking. We move on to the other side of C⁺:

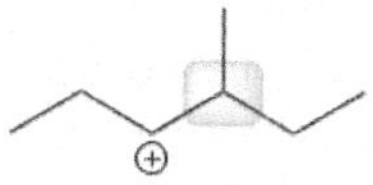

This position has both a methyl group and a hydride:

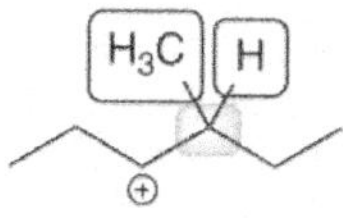

A methyl shift would not generate a more stable carbocation:

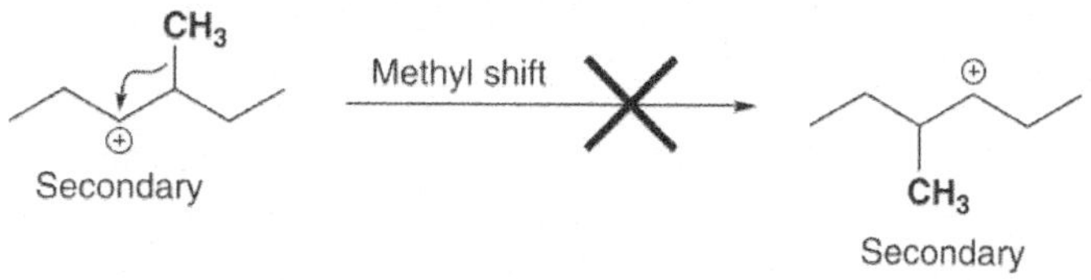

But a hydride shift WILL generate a more stable, tertiary carbocation:

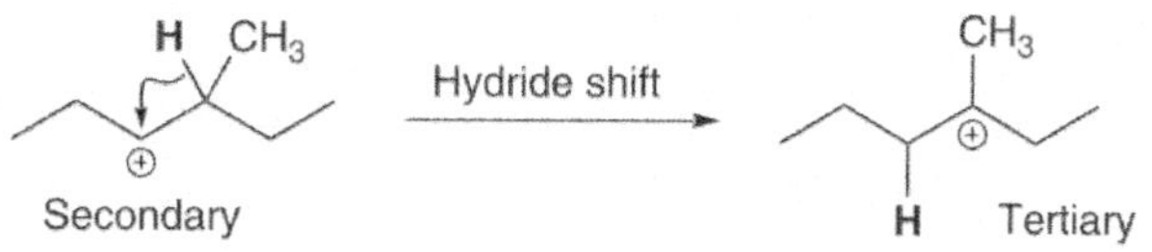

So, this carbocation can (and is therefore likely to) rearrange.

PROBLEMS   Predict whether each of the following carbocations is expected to undergo a carbocation rearrangement, and if so, draw the rearrangement:

8.29

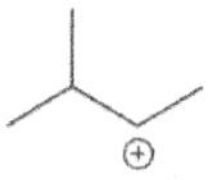

8.30

8.31

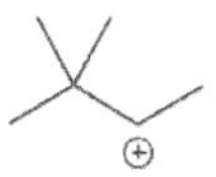

8.32

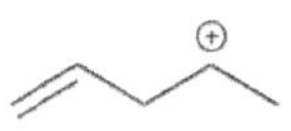

8.33

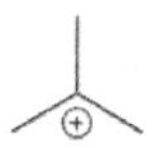

8.34

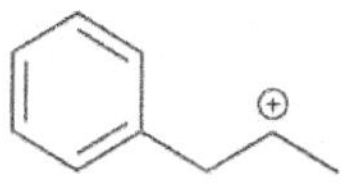

8.6 INFORMATION CONTAINED IN A MECHANISM

For any reaction, the proposed mechanism must always explain the experimental observations, including the *stereochemical outcome* and the *regiochemical outcome*. Let's explore each of these terms, one at a time. We will begin with the stereochemical outcome.

In Chapter 5, we learned to locate chiral centers and assign their configurations (*R* or *S*). When we talk about the stereochemical outcome of a reaction, we are talking about issues such as the following:

- If a chiral center is already present in the starting material, what happens to the configuration of that chiral center during the course of the reaction?
- If a new chiral center is formed, do we expect one configuration to predominate, or do we expect a 50-50 mixture of *R* and *S* (a racemic mixture)?
- If multiple diastereomeric products are possible, which diastereomers are obtained, and which are not?

These are all issues of stereochemistry. Let's see a few specific examples that you will soon encounter in the next few chapters. In Chapter 9, we will cover substitution reactions, in which one group (X) is replaced for another (Y):

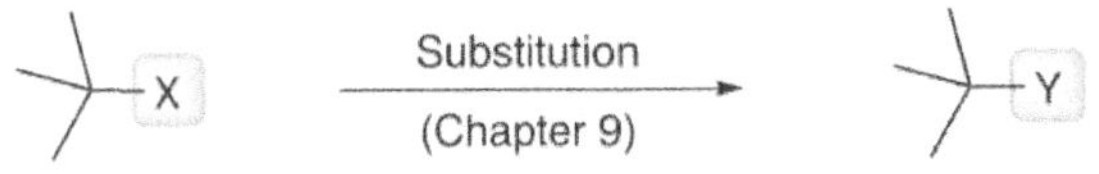

For example, Br can be replaced with OH. In the following reaction, Br is replaced with OH, but take special notice of the stereochemical outcome in this case:

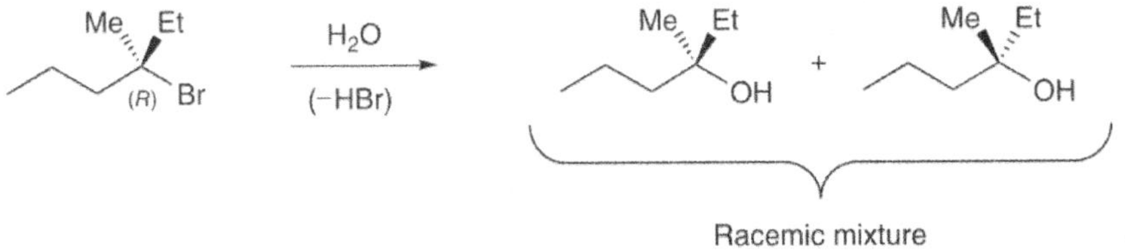

Racemic mixture

We are starting with a single enantiomer (only the *R* configuration), but the product is a racemic mixture. Why? Any proposed mechanism must account for this observation. When we cover substitution reactions, we will see that not all substitution reactions proceed with racemization. We will see that some substitution reactions proceed with inversion of configuration (*R* is converted into *S*). Why? Once again, the mechanism must account for this observation, and it will certainly be a different mechanism from the mechanism for the reaction shown above. We will learn two different mechanisms for substitution reactions, called S_N1 and S_N2. We will see that one of these mechanisms justifies racemization, while the other justifies inversion of configuration.

Issues of stereochemistry are not limited to substitution reactions. In Chapter 10, we will learn about elimination reactions, in which a proton (H^+) and a leaving group (LG) are removed to form a double bond:

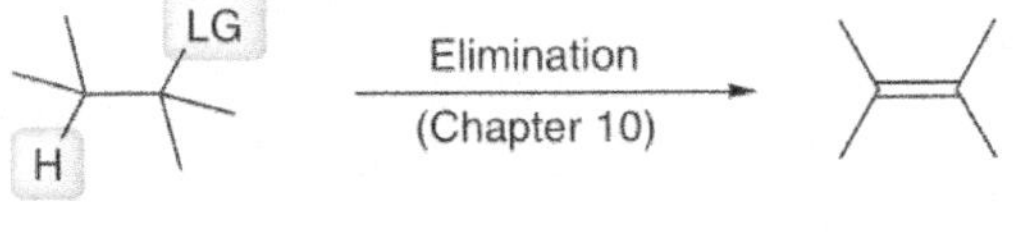

In some cases, stereoisomeric alkenes are possible, as seen in the following example:

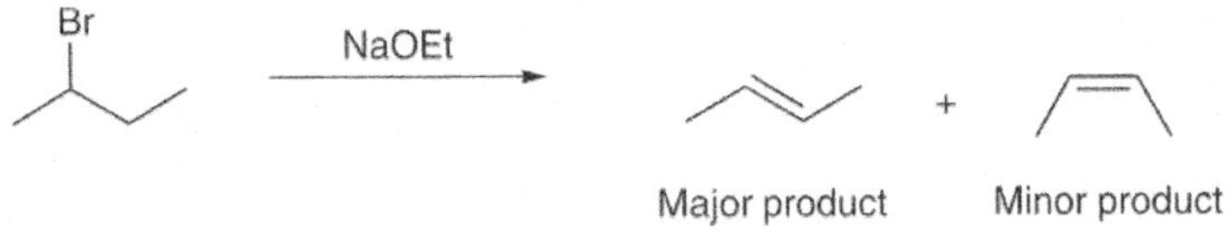

Major product   Minor product

Notice that two diastereomers are possible (*cis* and *trans*). The *trans* isomer is observed to predominate, so our proposed mechanism must somehow justify this observation.

If you think that you can avoid stereochemistry, you can't. As we move from chapter to chapter, stereochemistry will continue to be a theme that is revisited. This is also true of addition reactions, as we will see in Chapter 11. Alkenes are observed to undergo a variety of addition reactions, in which the pi bond is destroyed, and two new groups (X and Y) are installed:

We will learn many different kinds of addition reactions, and we will analyze the stereochemical outcome in each case. Consider, for example, the following process, called hydroboration–oxidation:

Notice that two new chiral centers are formed in the process, so we might expect four possible stereoisomeric products. Yet, we only observe two of these stereoisomers (the pair of enantiomers shown). H and OH were added across the same side of the alkene, which is called a *syn* addition. The following products are not obtained:

In order for these products to be observed, the H and OH would have to be added on opposite sides of the alkene, called *anti* addition. For this particular process, *anti* addition is not observed. Only *syn* addition is observed. And once again, any proposed mechanism must justify this observed stereochemical outcome (*syn* addition).

So far, in this section, we have focused on the stereochemical outcome of reactions. Now let's move on to analyzing the *regiochemical* outcome of reactions. What does this mean? It means that more than one region of a molecule can react. Regiochemistry is important when a reaction generates constitutional isomers, rather than stereoisomers. For example, consider the following elimination reaction:

The double bond can be formed in one of two locations, depending on which proton is removed in the process (as we will see in Chapter 10 when we explore the mechanism in more detail). For now, notice that the two products are not stereoisomers. Rather they are constitutional isomers. If we ask which product predominates, this is an issue of regiochemistry. In the reaction above, one alkene is the major product, and any proposed mechanism must justify this regiochemical observation.

As another example, let's go back to the reaction sequence that we just saw a moment ago:

Notice that the OH group is ultimately installed at the position that does not bear the R group. We do not obtain a product in which the OH group is installed at the position bearing the R group. Such a product would be a constitutional isomer of the products obtained (above). This is an issue of regiochemistry. Once again, any proposed mechanism must justify this regiochemical observation.

We have seen that a mechanism must justify both the stereochemical and regiochemical outcomes of a reaction. Students often try to memorize the stereochemical and regiochemical outcomes of every single reaction. This method of studying will prove to be very difficult as you move through the course, because there are many, many reactions. It is not easy to memorize all of the details. However, if you focus on understanding the proposed mechanism for each reaction, with a complete understanding of how the mechanism justifies both the stereochemical and regiochemical outcomes, then you will find that you will remember the details of each reaction more easily (without much need for memorization). Each mechanism will make sense, and that will help you remember all of the details for each reaction. This is true because a mechanism must explain all of the experimental observations.

For each mechanism that you encounter, you should be able to draw the entire mechanism on a blank piece of paper. Make sure to do that for every new mechanism. Then, wait a day, and do it again, when it is not fresh in your mind. For each mechanism that you learn, you need to get to a point where you can draw the entire mechanism, without mistakes, on a blank sheet of paper. If you discipline yourself to perform this exercise, you will find that your understanding of organic chemistry will be greatly enhanced (as will your grade in the course!). Good luck!

CHAPTER 9

SUBSTITUTION REACTIONS

In the last chapter we saw the importance of understanding mechanisms. We said that mechanisms are the keys to understanding everything else. In this chapter, we will see a very special case of this. We will see that there are two common types of substitution reactions, called S_N2 and S_N1 reactions, which have very different mechanisms. By focusing on the differences in their mechanisms, we can understand why we get S_N2 in some cases and S_N1 in other cases.

9.1 THE MECHANISMS

The vast majority of the reactions that we will encounter involve nucleophiles and electrophiles. A nucleophile has a region of high electron density (a lone pair or a pi bond), while an electrophile has a region of low electron density. When a nucleophile encounters an electrophile, a reaction can occur.

In both S_N2 and S_N1 reactions, a *nucleophile* is attacking an electrophile, giving a *substitution* reaction. That explains the S_N part of the name. But what do the "1" and "2" stand for? To understand these labels, we need to look at the mechanisms. Let's start with S_N2:

On the left, we see a nucleophile. It is attacking a compound that has an electrophilic carbon atom that is attached to a leaving group (LG). A *leaving group* is any group that can be expelled (we will see examples of this very soon). The leaving group serves two important functions: 1) it withdraws electron density from the carbon atom to which it is attached, rendering the carbon atom electrophilic and 2) it is capable of stabilizing the negative charge after being expelled.

An S_N2 mechanism employs two curved arrows: one going from a lone pair on the nucleophile to form a bond between the nucleophile and carbon, and the other going from the bond between the carbon atom and the LG to form a lone pair on the LG. Notice that the configuration at the carbon atom gets inverted in this reaction. So the reaction is said to proceed with inversion of configuration. Why does this happen? It is kind of like an umbrella flipping in a strong wind. It takes a good force to do it, but it is possible to flip the umbrella. The same is true here. If the nucleophile is good enough, and if all of the other conditions are just right, a reaction can take place in which the configuration of the chiral center is inverted (by bringing the nucleophile in on one side, and kicking off the LG on the other side).

Now we get to the meaning of "2" in S_N2. Remember from the last chapter that nucleophilicity is a measure of kinetics (how fast something happens). Since this is a *nucleophilic* substitution reaction, we care about how fast the reaction is happening. In other words, what is the rate of the reaction? This mechanism has only one step, and in that step, two chemical entities need to find each other: the nucleophile and the electrophile. So it makes sense that the rate of the reaction will be dependent on how much electrophile is around *and* how much nucleophile is around. In other words, the rate of the reaction is dependent on the concentrations of two entities. The nucleophilic attack is said to be "second order," and we signify this by placing a "2" in the name of the reaction.

Now let's look at the mechanism for an S_N1 reaction:

In this reaction, there are two steps. The first step has the LG leaving all by itself, without any help from an attacking nucleophile. This generates a carbocation, which then gets attacked by the nucleophile in step 2. This is the major difference between S_N2 and S_N1 reactions. In S_N2 reactions, everything happens in one step. In S_N1 reactions, it happens in two steps, via a carbocation intermediate. The existence of the carbocation as an intermediate in *only* the S_N1 mechanism is the key. By understanding this, we can understand everything else.

For example, let's look at the stereochemistry of S_N1 reactions. We already saw that S_N2 reactions proceed via inversion of configuration. But S_N1 reactions are very different, because a carbocation is sp^2 hybridized, so its geometry is trigonal planar. When the nucleophile attacks, there is no preference as to which face of the carbocation can be attacked, leading to two possible configurations (R or S). Half of the molecules would have one configuration and the other half would have the other configuration. We learned before that this is called a racemic mixture. Notice that we can explain the stereochemical outcome of this reaction by understanding the nature of the carbocation intermediate that is formed.

To understand the meaning of the number "1" in S_N1, we must focus on the rate-determining step of an S_N1 process. As seen above, an S_N1 process has two steps, so the energy diagram of an S_N1 process is expected to have two humps, one for each step:

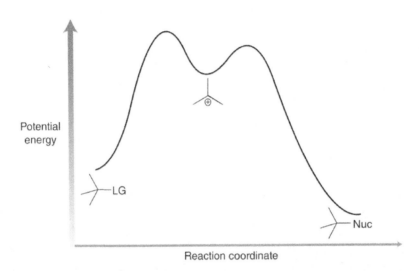

Whenever you are looking at an energy diagram for any process (whether it is S_N1 or whether it is any other process), the tallest hump will always indicate the rate-determining step. For an S_N1 process, the tallest hump is the first step, because that step involves loss of the leaving group, which has a large energy of activation. Therefore, the first step of an S_N1 process is the rate-determining step. In other words, the rate of the entire process is determined by how fast this step occurs. But notice that this step (loss of a leaving group) does not involve the nucleophile at all. This step involves only one chemical entity (the electrophile), rather than two entities, so this step is called "first order." This is signified by placing a "1" in the name of the reaction (S_N1). Of course, this does not mean that you only need the electrophile. You still need the nucleophile for the reaction to occur. You still need two different chemical entities (the nucleophile and the electrophile). The "1" simply means that the rate is not dependent on the concentration of both of them. The rate is dependent only on the concentration of one of them—the electrophile.

The mechanisms of S_N1 and S_N2 reactions helped us justify the observed stereochemical outcome in each case, and we were also able to see why we call them S_N1 and S_N2 reactions (based on reaction rates that are justified by the mechanisms). So, the mechanisms really do explain a lot. This should make sense, because a proposed mechanism must successfully explain the experimental observations. So, of course the mechanism explains the reason for racemization in an S_N1 process. That is what makes the mechanism plausible.

We need to consider four factors when choosing whether a reaction will proceed via an S_N1 or S_N2 mechanism. These four factors are: electrophile, nucleophile, leaving group, and solvent. We will go through each factor one at a time, and we will see that the difference between the two mechanisms is the key to understanding each of these four factors. Before we move on, it is very important that you understand the two mechanisms. For practice, try to draw them in the space provided without looking back to see them again.

Remember, an S_N2 mechanism has one step: the nucleophile attacks the electrophile, expelling the leaving group. An S_N1 mechanism has two steps: first the leaving group leaves to form a carbocation, and then the nucleophile attacks that carbocation. Also remember that S_N2 involves inversion of configuration, while S_N1 involves racemization. Now, try to draw them.

S_N2:

S_N1:

9.2   FACTOR 1—THE ELECTROPHILE (SUBSTRATE)

The electrophile is the compound being attacked by the nucleophile. In substitution and elimination reactions (which we will see in the next chapter), we generally refer to the electrophile as the *substrate*.

Remember that carbon has four bonds. So, other than the bond to the leaving group, the carbon atom that we are attacking has three other bonds:

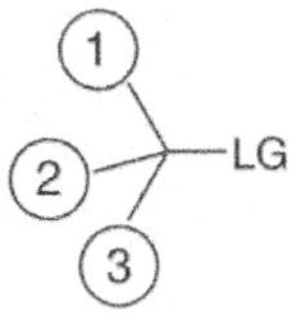

The question is, how many of these groups are alkyl groups (methyl, ethyl, propyl, etc.)? We represent alkyl groups with the letter "R." If there is one alkyl group, we call the substrate "primary" (1°). If there are two alkyl groups, we call the substrate "secondary" (2°). And if there are three alkyl groups, we call the substrate "tertiary" (3°):

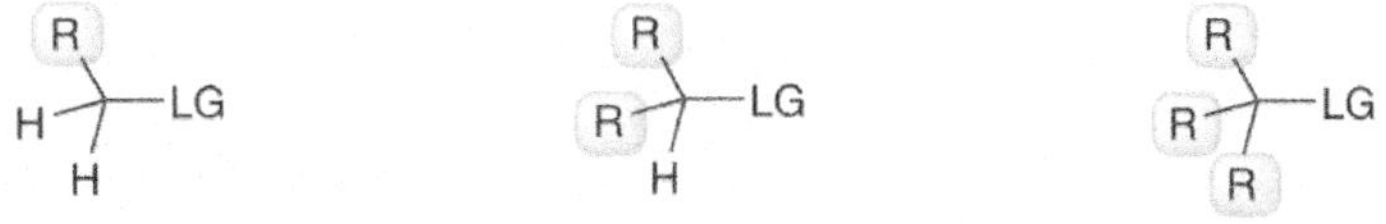

In an S_N2 reaction, alkyl groups make it very crowded at the electrophilic center where the nucleophile needs to attack. If there are three alkyl groups, then it is virtually impossible for the nucleophile to get in and attack (sterically hindered):

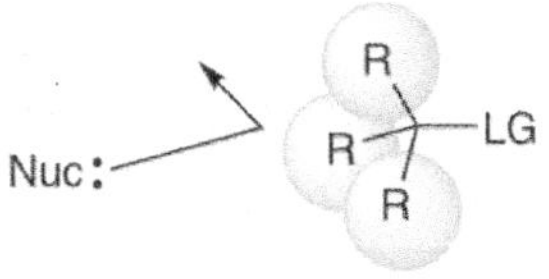

So, for S_N2 reactions, a primary substrate is better, while a tertiary substrate is unreactive.

But S_N1 reactions are totally different. The first step is not attack of the nucleophile. The first step is loss of the leaving group to form the carbocation. Then the nucleophile attacks the carbocation. Remember that carbocations are trigonal planar, so it doesn't matter how big the groups are. The groups go out into the plane, so it is easy for the nucleophile to attack. There is no steric hindrance.

In S_N1 reactions, the stability of the carbocation (formed in the rate-determining step) is the paramount issue. Recall that alkyl groups are electron donating. Therefore, 3° substrates are best because the three alkyl groups stabilize the carbocation. 1° substrates are generally unreactive toward S_N1 because there is only one alkyl group to stabilize the carbocation. This is not a steric argument; this is an electronic argument (stability of charge). So we have two opposite trends, for completely different reasons: If you have a 1° or 2° substrate, then the reaction will generally proceed via an S_N2 mechanism. If you have a 3° substrate, then the reaction will proceed via an S_N1 mechanism.

EXERCISE 9.1   Identify whether the following substrate is more likely to participate in an S_N2 or S_N1 reaction:

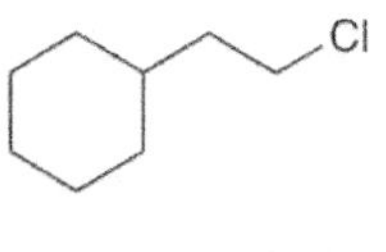

Answer   The substrate is primary, so we predict an S_N2 reaction.

PROBLEMS   Identify whether each of the following substrates is more likely to participate in an S_N2 or S_N1 reaction:

9.2   _____

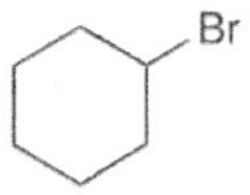

9.3   _____

9.4 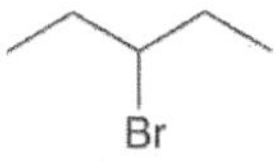 S_N2 _____

9.5   _____

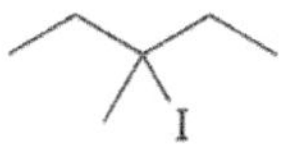

There is one other way to stabilize a carbocation (other than alkyl groups)—resonance. If a carbocation is resonance stabilized, then it will form more readily:

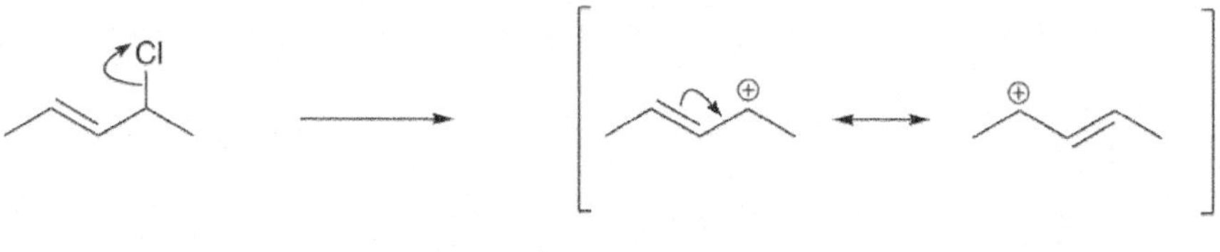

The carbocation above is stabilized by resonance. Therefore, the LG is willing to leave, and we can have an S_N1 reaction.

There are two kinds of systems that you should learn to recognize: a LG in a benzylic position and a LG in an allylic position. Compounds like this will generate resonance-stabilized carbocation intermediates upon loss of the leaving group.

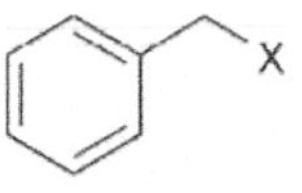

Benzylic    Allylic

If you see a double bond near the LG and you are not sure if it is a benzylic or allylic system, just draw the carbocation that would form upon loss of the leaving group, and determine if that carbocation is resonance stabilized.

EXERCISE 9.6   In the compound below, circle the LGs that occupy a benzylic or allylic position:

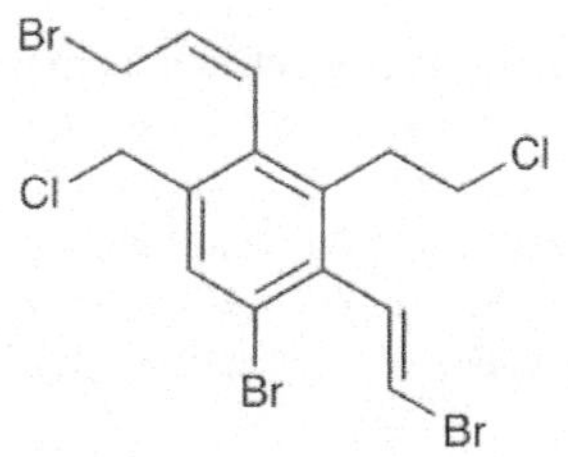

Answer

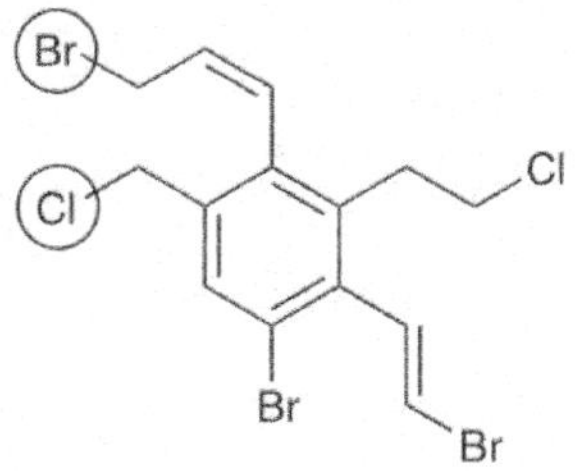

PROBLEMS   For each compound below, determine whether loss of the leaving group would result in a resonance-stabilized carbocation. If you are not sure, try to draw resonance structures of the carbocation that would form upon loss of the leaving group.

9.7

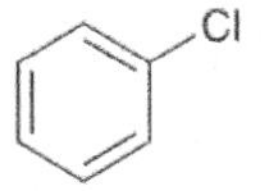

9.8

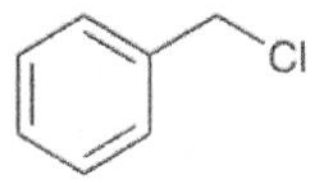

9.9

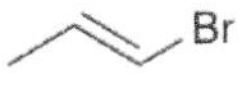

9.10

9.3   FACTOR 2—THE NUCLEOPHILE

The rate of an S_N2 process is dependent on the strength of the nucleophile. A strong nucleophile will speed up the rate of an S_N2 reaction, while a weak nucleophile will slow down the rate of an S_N2 reaction. In contrast, an S_N1 process is not affected by the strength of the nucleophile. Why not? Recall that the "1" in S_N1 means that the rate of reaction is dependent only on the concentration of the substrate, *not* on the concentration of the nucleophile. The concentration of the nucleophile is not relevant in determining the rate of reaction. Similarly, the *strength* of the nucleophile is also not relevant.

In summary, the nucleophile has the following effect on the competition between S_N2 and S_N1:

- A strong nucleophile favors S_N2.
- A weak nucleophile disfavors S_N2 (and thereby allows S_N1 to compete successfully).

We must therefore learn to identify nucleophiles as strong or weak. The strength of a nucleophile is determined by many factors, such as the presence or absence of a negative charge. For example, hydroxide (HO^-) and water (H_2O) are both nucleophiles, because in both cases, the oxygen atom has lone pairs. But hydroxide is a stronger nucleophile since it has a negative charge.

Charge is not the only factor that determines the strength of a nucleophile. In fact, there is a more important factor, called polarizability, which describes the ability of an atom or molecule to distribute its electron density unevenly in response to external influences. For example, sulfur is highly polarizable, because its electron density can be unevenly distributed when it comes near an electrophile. Polarizability is directly related to the size of the atom (and more specifically, the number of electrons that are distant from the nucleus). Sulfur is very large and has many electrons that are distant from the nucleus, and it is therefore highly polarizable. Iodine shares the same feature. As a result, I^- and HS^- are particularly strong nucleophiles. For similar reasons, H_2S is also a strong nucleophile, despite the fact that it lacks a negative charge.

Below are some strong and weak nucleophiles that we will encounter often:

Common Nucleophiles

Strong			Weak
:Ï:⊖	HṠ:⊖	HÖ:⊖	:F̈:⊖
:B̈r:⊖	H₂S̈:	RÖ:⊖	H₂Ö:
:C̈l:⊖	RṠH	N≡C:⊖	RÖH

EXERCISE 9.11 Identify whether the following nucleophile will favor S_N2 or S_N1:

Answer This compound has a sulfur atom with lone pairs. A lone pair on a sulfur atom will be strongly nucleophilic, even without a negative charge, because sulfur is large and highly polarizable. Strong nucleophiles favor S_N2 reactions.

PROBLEMS Identify whether each of the following nucleophiles will favor S_N2 or S_N1:

9.12   Answer: _____

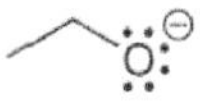

9.13   Answer: _____

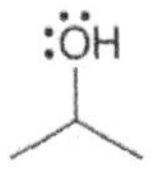

9.14   Answer: _____

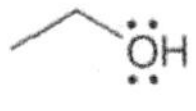

9.15 :B̈r:⊖   Answer: _____

9.16 HÖ:⊖   Answer: _____

9.17 ⊖:C≡N   Answer: _____

9.4 FACTOR 3—THE LEAVING GROUP

Both S_N1 and S_N2 mechanisms are sensitive to the identity of the leaving group. If the leaving group is bad, then neither mechanism can operate, but S_N1 reactions are generally more sensitive to the leaving group than S_N2 reactions. Why? Recall that the rate-determining step of an S_N1 process is loss of a leaving group to form a carbocation and a leaving group:

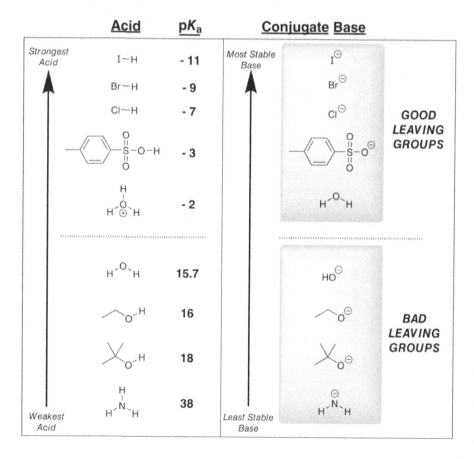

We have already seen that the rate of this step is very sensitive to the stability of the carbocation, so it should make sense that it is also sensitive to the stability of the leaving group. The leaving group must be highly stabilized in order for an S_N1 process to be effective.

What determines the stability of a leaving group? As a general rule, good leaving groups are the conjugate bases of strong acids. For example, iodide (I^-) is the conjugate base of a very strong acid (HI):

Iodide is a very weak base because it is highly stabilized. As a result, iodide can function as a good leaving group. In fact, iodide is one of the best leaving groups. The following figure shows a list of good leaving groups, all of which are the conjugate bases of strong acids:

In contrast, hydroxide is a bad leaving group, because it is not a stabilized base. In fact, hydroxide is a relatively strong base, and, therefore, it rarely functions as a leaving group. It is a bad leaving group. But under certain circumstances, it is possible to convert a bad leaving group into a good leaving group. For example, when treated with a strong acid, an OH group is protonated, converting it into a good leaving group:

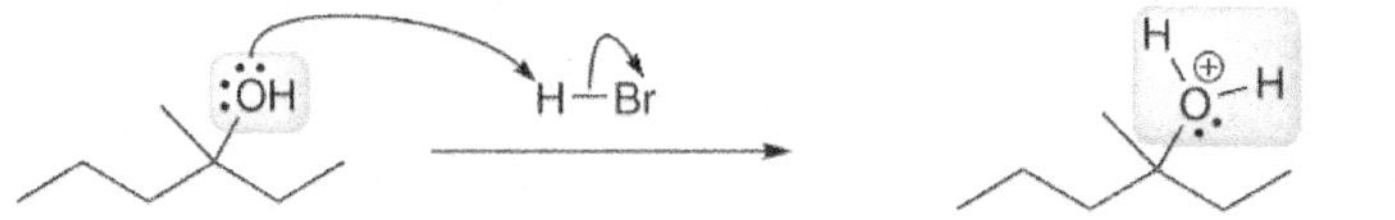

Bad leaving group (HO⁻)   Good leaving group (H_2O)

The most commonly used leaving groups are halides and sulfonate ions:

halides			sulfonate ions		
iodide	bromide	chloride	tosylate	mesylate	triflate

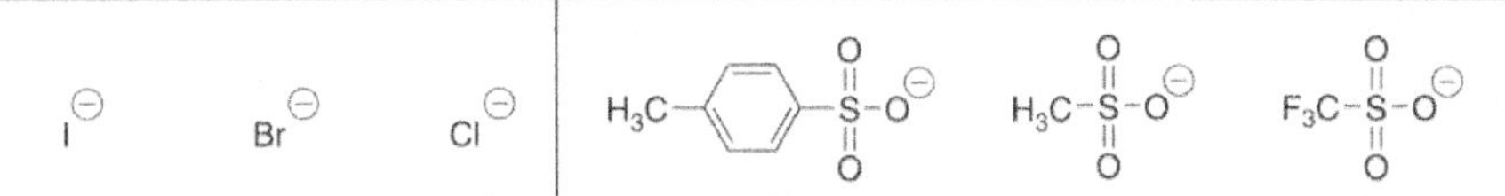

Among the halides, iodide is the best leaving group because it is a weaker base (more stable) than bromide or chloride. Among the sulfonate ions, the best leaving group is the triflate group, but the most commonly used is the tosylate group. It is abbreviated as OTs. When you see OTs connected to a compound, you should recognize the presence of a good leaving group.

EXERCISE 9.18   Identify the leaving group in the following compound:

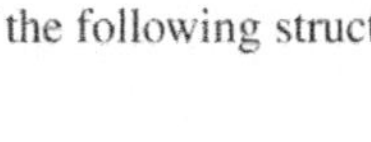

Answer   We have seen that hydroxide is not a good leaving group, because its conjugate acid (H_2O) is not a strong acid. As a result, hydroxide is not a weak base, so it does not function as a leaving group. In contrast, chloride is a good leaving group because its conjugate acid (HCl) is a strong acid. Therefore, chloride is a weak base, so it can serve as a leaving group.

PROBLEMS   Identify the best leaving group in each of the following structures:

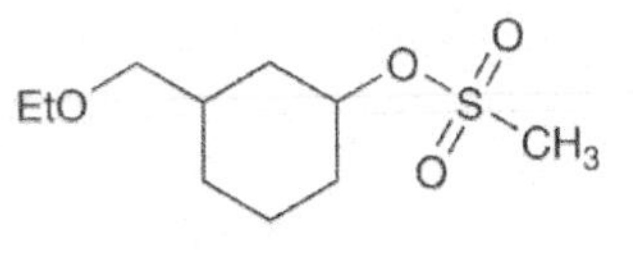

9.19   Answer: _____

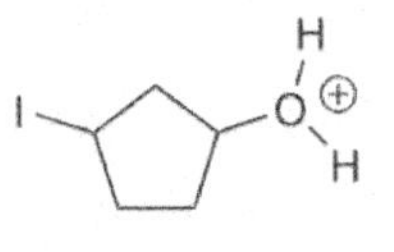

9.20   Answer: _____

9.21 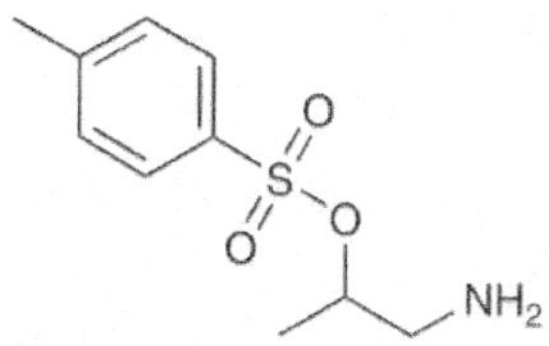   Answer: _____

9.22 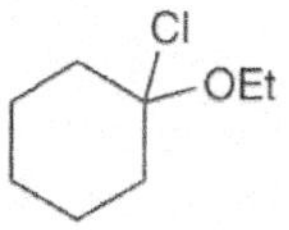   Answer: _____

9.23 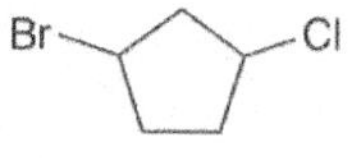   Answer: _____

9.24 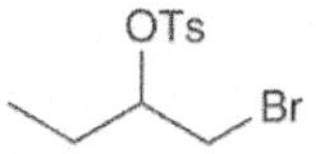   Answer: _____

9.25   Compare the structures of 3-methoxy-3-methylpentane and 3-iodo-3-methylpentane, and identify which compound is more likely to undergo an S_N1 reaction when treated with a nucleophile in non-acidic conditions.

9.26   When 3-ethyl-3-pentanol is treated with excess chloride, no substitution reaction is observed, because hydroxide is a bad leaving group. If you wanted to force an S_N1 reaction, using 3-ethyl-3-pentanol as the substrate, what reagent would you use to change the leaving group into a better leaving group and provide chloride ions at the same time?

9.5   FACTOR 4—THE SOLVENT

So far, we have explored the substrate, the nucleophile, and the leaving group. This takes care of all of the parts of the compounds that are reacting with each other. Let's summarize substitution reactions in a way that allows us to see this:

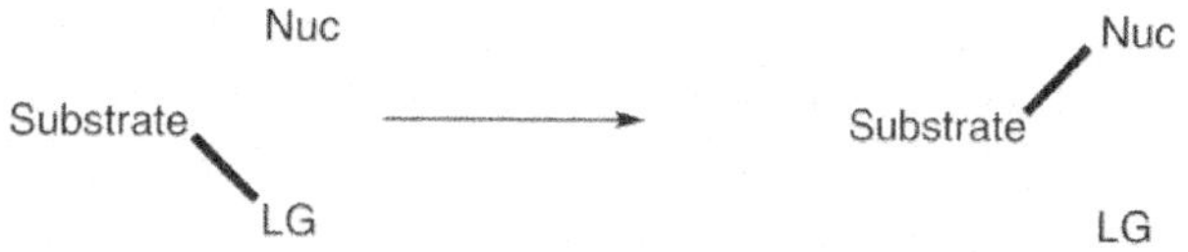

So, by talking about the substrate, the nucleophile, and the leaving group, we have covered almost everything. But there is one more factor to take into account. What solvent are these compounds dissolved in? It can make a difference. Let's see how.

There is a really strong solvent effect that *greatly* affects the competition between S_N1 and S_N2, and here it is: polar aprotic solvents favor S_N2 reactions. So, what are polar aprotic solvents, and why do they favor S_N2 reactions?

Let's break it down into two parts: *polar* and *aprotic*. Hopefully, you remember from general chemistry what the term "polar" means, and you should also remember that "like dissolves like" (so polar solvents dissolve polar compounds, and nonpolar solvents dissolve nonpolar compounds). Therefore, we really need a polar solvent to run substitution reactions. S_N1 desperately needs the polar solvent to stabilize the carbocation, and S_N2 needs a polar solvent to dissolve the nucleophile. S_N1 certainly needs the polar solvent more than S_N2 does, but you will rarely see a substitution reaction in a nonpolar solvent. So, let's focus on the term aprotic.

Let's begin by defining a protic solvent. We will need to jog our memories about acid–base chemistry. Recall that in Chapter 3 we talked about the acidity of protons (these are hydrogen atoms without the electrons, symbolized by H^+), and we saw that a proton can be removed from a compound if the compound can stabilize the negative charge that develops when H^+ is removed. A protic solvent is a solvent that has a proton connected to an electronegative atom (for example, H_2O or EtOH). It is called protic because the solvent can serve as a source of protons. So what is an aprotic solvent?

Aprotic means that the solvent does *not* have a proton on an electronegative atom. The solvent can still have hydrogen atoms, but none of them are connected to electronegative atoms. Common examples of polar aprotic solvents include acetone, DMSO, DME, and DMF:

Acetone

Dimethylsulfoxide (DMSO)

Dimethoxyethane (DME)

Dimethylformamide (DMF)

There are, of course, other polar aprotic solvents. You should look through your textbook and your class notes to determine if there are any other polar aprotic solvents that you will be expected to know. If there are any more, you can add them to the list above. You should learn to recognize these solvents when you see them.

So why do these solvents speed up the rate of S_N2 reactions? To answer this question, we need to talk about a solvent effect that is usually present when we dissolve a nucleophile in a solvent. A nucleophile with a negative charge, when dissolved in a polar solvent, will be surrounded by solvent molecules in what is called a *solvent shell:*

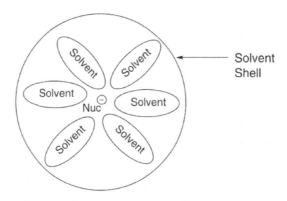

This solvent shell is in the way, holding back the nucleophile from doing what it is supposed to do (go attack something). For the nucleophile to do its job, the nucleophile must first shed this solvent shell. This is always the case when you dissolve a nucleophile in a polar solvent, *except* when you use a polar aprotic solvent.

Polar aprotic solvents are not very good at forming solvent shells around negative charges. So if you dissolve a nucleophile in a polar aprotic solvent, the nucleophile is said to be a "naked" nucleophile, because it does not have a solvent shell. Therefore, it does not need to first shed a solvent shell before it can react with something. It never had a solvent shell to begin with. This effect is drastic. As you can imagine, a nucleophile with a solvent shell is going to spend most of its existence with the solvent shell, and there will be only brief moments every now and then when it is free to react. By allowing the nucleophile to react all of the time, we are greatly speeding up the reaction. S_N2 reactions performed with nucleophiles in polar aprotic solvents can occur thousands (or even millions) of times faster than those in regular protic solvents.

Bottom line: Whenever a solvent is indicated, you should look to see if it is one of the polar aprotic solvents listed above. If it is, it is a safe bet that the reaction is going to be S_N2.

EXERCISE 9.27   Predict whether the reaction below will occur via an S_N2 or an S_N1 mechanism:

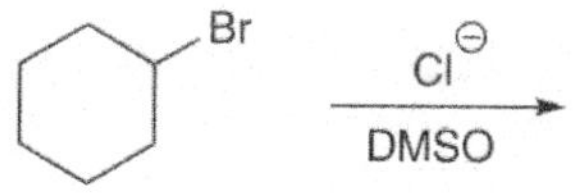

Answer   This reaction utilizes DMSO, which is a polar aprotic solvent, so we expect an S_N2 reaction even though the substrate is secondary (although, if the substrate was tertiary, a polar aprotic solvent would not help).

PROBLEM 9.28   Go back to the list of polar aprotic solvents, study the list, and then try to copy the list on a blank piece of paper (without looking back).

9.6 USING ALL FOUR FACTORS

Now that we have seen all four factors individually, we need to see how to put them all together. When analyzing a reaction, we need to look at all four factors and make a determination of which mechanism, S_N1 or S_N2, is predominating. For example, it is clear that a reaction will be S_N2 if we have a primary substrate with a strong nucleophile in a polar aprotic solvent. On the flipside, a reaction will clearly be S_N1 if we have a tertiary substrate with a weak nucleophile and an excellent leaving group.

EXERCISE 9.29   For the reaction below, look at all of the reagents and conditions, and determine if the reaction will proceed via an S_N2 or an S_N1 mechanism.

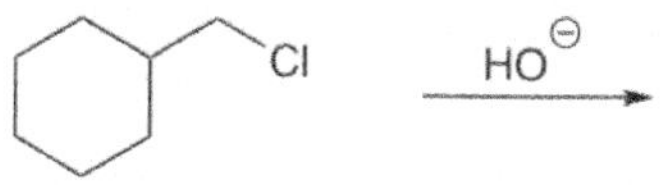

Answer   The substrate is primary, which immediately tells us that it needs to be an S_N2 pathway. On top of that, we see that we have a strong nucleophile, which also favors S_N2. The LG is good, which doesn't tell us much. The solvent is not indicated. So, taking everything into account, we predict that the reaction follows an S_N2 mechanism.

PROBLEMS   For each reaction below, look at all of the reagents and conditions, and determine if the reaction will proceed via an S_N2 or an S_N1 pathway, or neither. (HINT: The answer is "neither" for at least one of the problems below.)

9.30

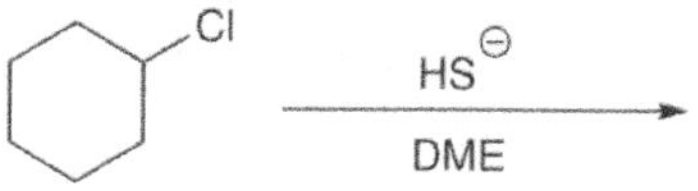

9.31

9.32

9.33

9.34

$$Br\text{-}\diagdown\diagup\diagdown \quad \xrightarrow[\text{DMSO}]{\text{Cl}^{\ominus}}$$

9.35

$$\xrightarrow{\text{CH}_3\text{OH}}$$

9.7 SUBSTITUTION REACTIONS TEACH US SOME IMPORTANT LESSONS

S_N1 and S_N2 reactions produce almost the same products. In both reactions, a leaving group is replaced by a nucleophile. One major difference arises when the leaving group is attached to a chiral center. In this situation, the S_N2 mechanism will invert the chiral center, while the S_N1 mechanism will produce a racemic mixture.

There is another major difference between the outcome of S_N1 and S_N2 processes. We have seen that an S_N1 pathway will involve a carbocation intermediate, while an S_N2 pathway will not involve a carbocation intermediate. This difference can be significant when carbocation rearrangements are possible (as seen in Section 8.5). That is, S_N1 reactions are susceptible to carbocation rearrangements, while S_N2 reactions are not. You should consult your textbook and/or lecture notes to see if you are responsible for any examples of carbocation rearrangements during S_N1 processes.

We spent a lot of time on the competition between S_N1 and S_N2 mechanisms, and we learned many valuable lessons that will be important as we move forward and learn more reactions. First we learned the important concept that all of the relevant information is contained within the mechanisms. By understanding the mechanisms completely, everything else can be justified based on the mechanisms. All of the factors that influence the reaction are summarized in the mechanism. This is true for every reaction you will see from now on. Now you have had some practice thinking this way.

Next we learned that there are multiple factors at play when analyzing a reaction. Sometimes the factors can all be pointing in the same direction, while at other times the factors can be in conflict. When they are in conflict, we need to weigh them and decide which factors win out in determining the path of the reaction. This concept of competing factors is a theme in organic chemistry. The experience of going through S_N1 and S_N2 mechanisms has prepared you for thinking this way for all reactions from now on.

Finally we learned that if we analyze the first factor (substrate), we will find two effects at play: steric and electronic considerations. We saw that S_N2 reactions require primary or secondary substrates because of steric considerations—it is too crowded for the nucleophile to attack a tertiary substrate. On the other hand, for S_N1 reactions, electronic considerations were paramount. Tertiary substrates were best, because the alkyl groups were needed to stabilize the carbocation.

These two effects (steric and electronic effects) are major themes in organic chemistry. Much of what you learn in the rest of the course can be explained with either an electronic or a steric argument. The sooner you learn to consider these two effects in every problem you encounter, the better off you will be. Electronic effects are usually more complicated than steric effects. In fact, the other three factors that we saw (nucleophile, leaving group, and solvent effects) were all electronic arguments. Once you get the hang of the kinds of electronic arguments that are generally made, you will begin to see common threads in all of the reactions that you will encounter in this course.

CHAPTER *10*

ELIMINATION REACTIONS

In the previous chapter, we saw that a substitution reaction can occur when a compound possesses a leaving group. In this chapter, we will explore another type of reaction, called *elimination*, which can also occur for compounds with leaving groups. In fact, substitution and elimination reactions frequently compete with each other, giving a mixture of products. At the end of this chapter, we will learn how to predict the products of these competing reactions. For now, let's consider the different outcomes for substitution and elimination reactions:

In a substitution reaction, the leaving group is replaced with a nucleophile. In an elimination reaction, a beta (β) proton is removed together with the leaving group, forming a double bond. In the previous chapter, we saw two mechanisms for substitution reactions (S_N1 and S_N2). In a similar way, we will now explore two mechanisms for elimination reactions, called E1 and E2. Let's begin with the E2 mechanism.

10.1   THE E2 MECHANISM

In an E2 process, a base removes a proton, causing the simultaneous expulsion of a leaving group:

Notice that there is only one mechanistic step (no intermediates are formed), and that step involves both the substrate and the base. Because that step involves two chemical entities, it is said to be bimolecular. Bimolecular elimination reactions are called E2 reactions, where the "2" stands for "bimolecular."

Now let's consider the effect of the substrate on the rate of an E2 process. Recall from the previous chapter that S_N2 reactions generally do not occur with tertiary substrates, because of steric considerations. But E2 reactions are different than S_N2 reactions, and in fact, tertiary substrates often undergo E2 reactions quite rapidly. To explain why tertiary substrates will undergo E2 but not S_N2 reactions, we must recognize that the key difference between substitution and elimination is the role played by the reagent. In a substitution reaction, the reagent functions as a nucleophile and attacks an electrophilic position. In an elimination reaction, the reagent functions as a base and removes a proton, which is easily achieved even with a tertiary substrate. In fact, tertiary substrates react even more rapidly than primary substrates.

157

10.2   THE REGIOCHEMICAL OUTCOME OF AN E2 REACTION

Recall from Chapter 8 that the term "regiochemistry" refers to *where* the reaction takes place. In other words, in what *region* of the molecule is the reaction taking place? When H and X are eliminated (where X is some leaving group), it is sometimes possible for more than one alkene to form. Consider the following example, in which two possible alkenes can be formed:

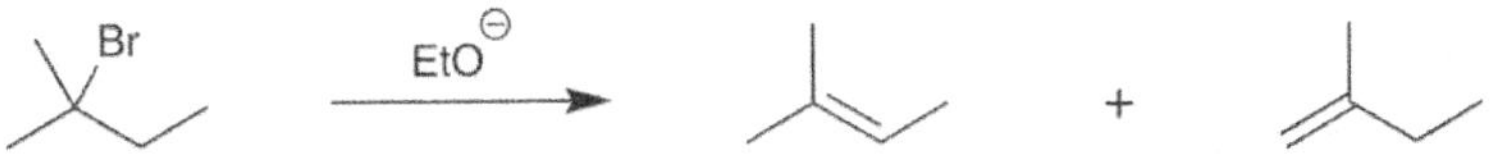

Where does the double bond form? This is a question of regiochemistry. The way we distinguish between these two possibilities is by considering how many groups are attached to each double bond. Double bonds can have anywhere from 1 to 4 alkyl groups attached to them:

Monosubstituted   **Di**substituted   **Tri**substituted   **Tetra**substituted

So, if we look back at the reaction above, we find that the two possible products are trisubstituted and disubstituted:

Trisubstituted   **Di**substituted

For an elimination reaction where there is more than one possible alkene that can be formed, we have names for the different products based on which alkene is more substituted and which is less substituted. The more-substituted alkene is called the Zaitsev product, and the less-substituted alkene is called the Hofmann product. Usually, the Zaitsev product is the major product:

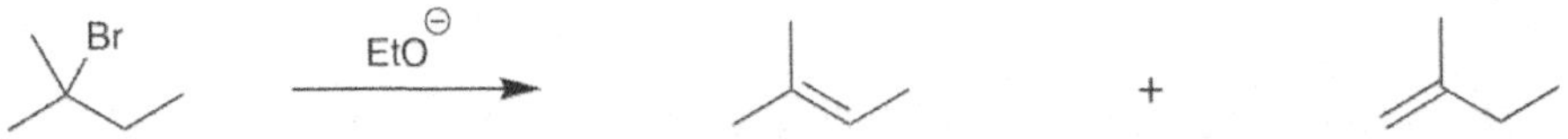

Major   **Minor**

However, there are many exceptions in which the Zaitsev product (the more-substituted alkene) is not the major product. For example, if the reaction above is performed with a sterically hindered base (rather than using ethoxide as the base), then the major product will be the less-substituted alkene:

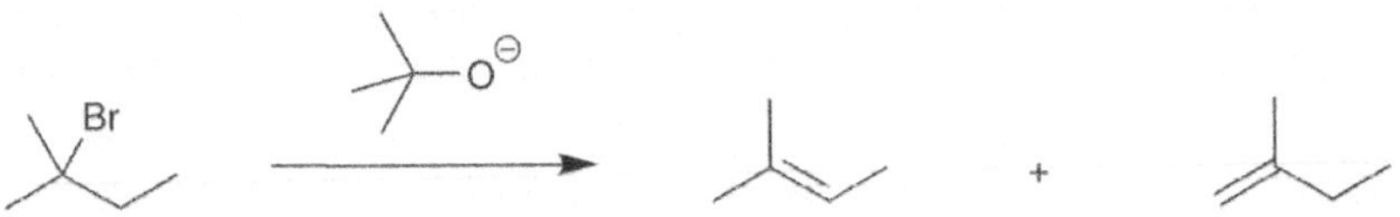

Minor   **Major**

In this case, the Hofmann product is the major product, because a sterically hindered base was used. This case illustrates an important concept: *The regiochemical outcome of an E2 reaction can often be controlled by carefully choosing the base*. Below are two examples of sterically hindered bases that will be encountered frequently throughout your organic chemistry course:

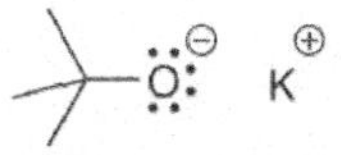

Potassium *tert*-butoxide
(*t*-BuOK)

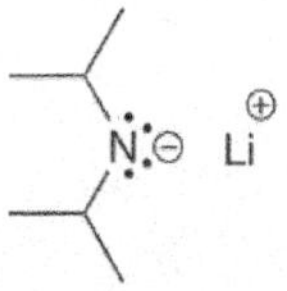

Lithium diisopropylamide
(LDA)

PROBLEMS   Draw the Zaitsev and Hofmann products that are expected when each of the following compounds is treated with a strong base to give an E2 reaction. For the following problems, don't worry about identifying which product is major and which is minor, since the identity of the base has not been indicated. Just draw both possible products:

10.1

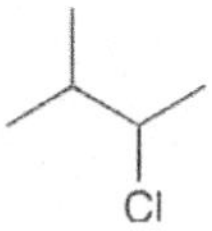

_____   _____
Zaitsev   Hofmann

10.2

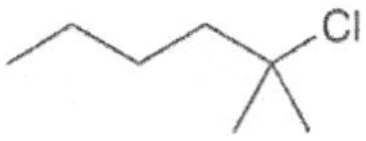

_____   _____
Zaitsev   Hofmann

10.3

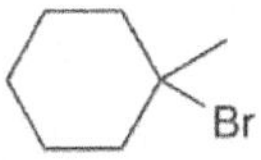

_____   _____
Zaitsev   Hofmann

10.3   THE STEREOCHEMICAL OUTCOME OF AN E2 REACTION

The examples in the previous section focused on regiochemistry. We will now focus our attention on stereochemistry. For example, consider performing an E2 reaction with the following substrate:

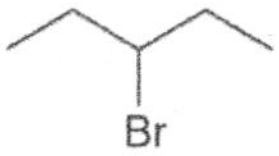

This substrate has two identical β positions so regiochemistry is not an issue in this case. Deprotonation of either β position produces the same result. But in this case, stereochemistry is relevant, because two stereoisomeric alkenes are possible:

Major   *Minor*

Both stereoisomers (*cis* and *trans*) are produced, but the *trans* product predominates. This specific example is said to be *stereoselective*, because the substrate produces two stereoisomers in unequal amounts.

In the previous example, the β position had two different protons:

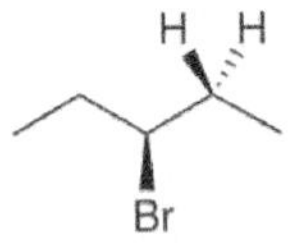

In such a case, we saw that both the *cis* and the *trans* isomers were produced, with the *trans* isomer being favored. Now let's consider a case where the β position contains only one proton. In such a case, only one product is formed. The reaction is said to be *stereospecific* (rather than stereoselective), because the proton and the leaving group must be *antiperiplanar* during the reaction. This is best illustrated using Newman projections, which allow us to draw the compound in a conformation in which the proton and the leaving group are antiperiplanar. This conformation then shows you which stereoisomer is expected. The following example will illustrate how this is done.

EXERCISE 10.4   Draw the major product that is expected when the following substrate is treated with ethoxide to give an E2 reaction:

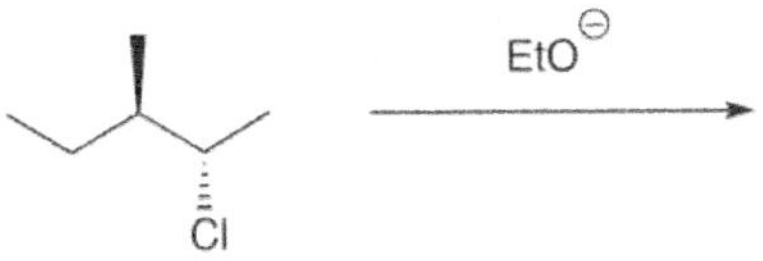

Answer   Let's first consider the expected regiochemical outcome of the reaction. The reaction does not employ a sterically hindered base, so we expect the major product to be the more-substituted alkene (the Zaitsev product):

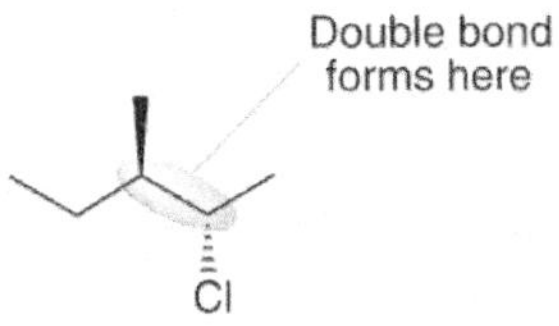

Now let's consider the stereochemical outcome. In this case, the beta position (where the reaction is taking place) has only one proton:

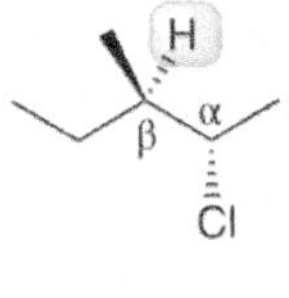

So, we expect this reaction to be stereospecific, rather than stereoselective. That is, we expect only one alkene, rather than a mixture of stereoisomeric alkenes. In order to determine which alkene is obtained, we begin by drawing the Newman projection:

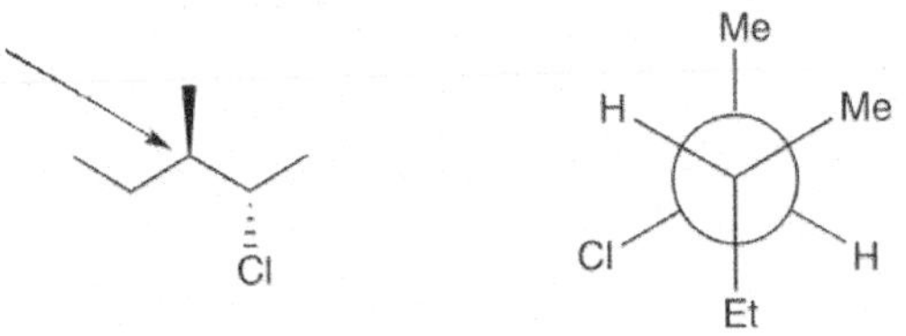

Next, we need to draw the conformation in which the H (on the front carbon) and the leaving group (Cl) are antiperiplanar:

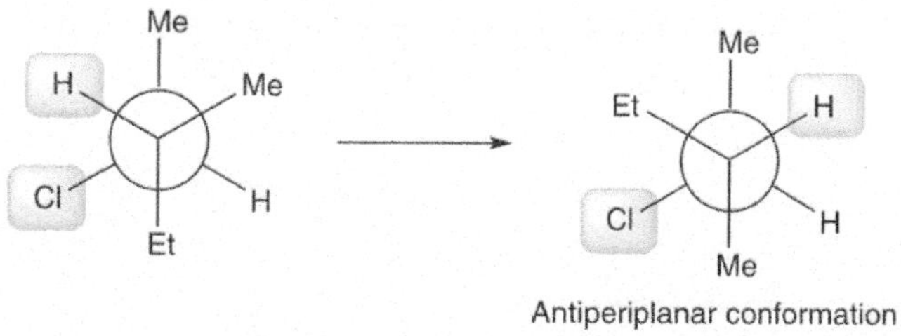

Antiperiplanar conformation

This is the conformation from which the reaction can take place. The double bond is being formed between the front carbon and the back carbon, and this Newman projection shows us the stereochemical outcome:

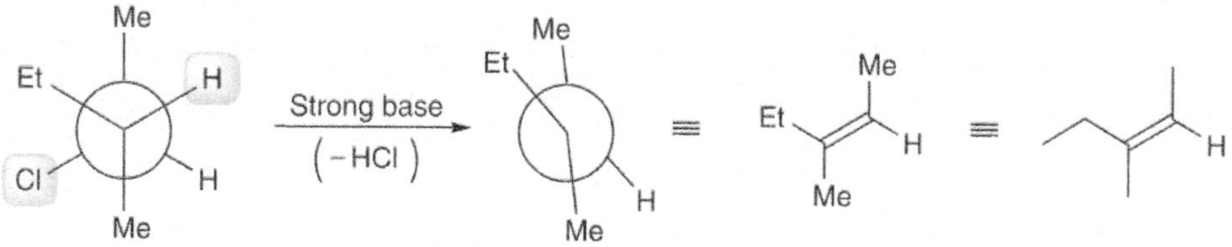

This is the major product that we expect. The stereoisomer of this alkene is not produced, because the E2 process is stereospecific:

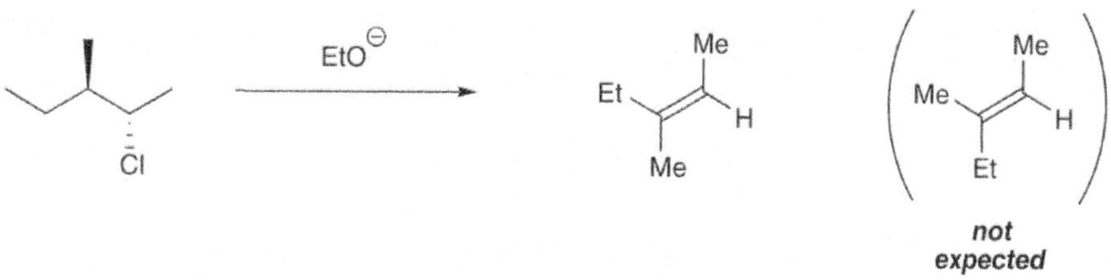

It is useful to draw Newman projections so that you can determine the stereoisomer that is expected from an E2 reaction. If you are rusty on Newman Projections, you should go back and review the first two sections in Chapter 6 in this book. Then come back to here, and try to use Newman projections to determine the stereochemical outcome of the following reactions.

PROBLEMS   Draw the major product that is expected when each of the following substrates is treated with ethoxide (a strong base that is not sterically hindered) to give an E2 reaction:

10.5

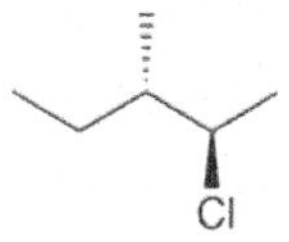

———————————————   ———————————————
Newman projection   Final answer

10.6

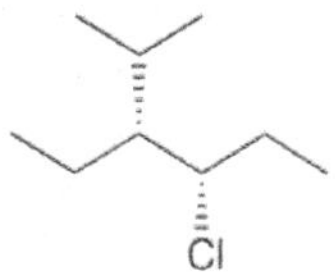

———————————————   ———————————————
Newman projection   Final answer

10.7

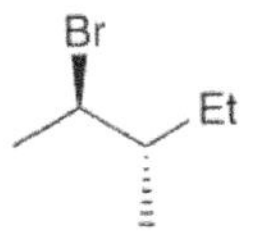

     Newman projection   Final answer

10.8

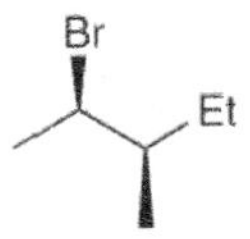

     Newman projection   Final answer

10.4 THE E1 MECHANISM

In an E1 process, there are two separate steps: the leaving group first leaves, generating a carbocation intermediate, which then loses a proton in a separate step:

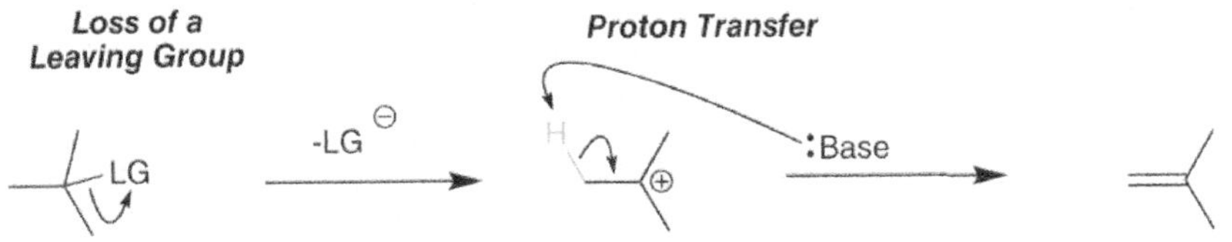

The first step (loss of the leaving group) is the rate-determining step, much like we saw for S_N1 processes. The base does not participate in this step, and therefore, the concentration of the base does not affect the rate. Because this step involves only one chemical entity, it is said to be unimolecular. Unimolecular elimination reactions are called E1 reactions, where the "1" stands for "unimolecular."

    Notice that the first step of an E1 process is identical to the first step of an S_N1 process. In each process, the first step involves loss of the leaving group to form a carbocation intermediate:

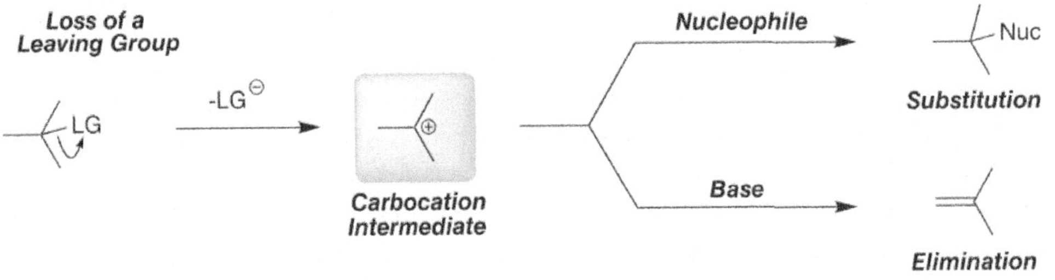

An E1 reaction is generally accompanied by a competing S_N1 reaction, and a mixture of products is generally obtained. At the end of this chapter, we will explore the main factors that affect the competition between substitution and elimination reactions.

    For now, let's consider the effect of the substrate on the rate of an E1 process. The rate is found to be very sensitive to the nature of the starting alkyl halide, with tertiary halides reacting more readily than secondary halides; and primary halides generally do not undergo E1 reactions. This trend is identical to the trend we saw for S_N1 reactions, and the reason for the trend is the same as well. Specifically, the rate-determining step of the mechanism involves formation of a carbocation intermediate, so the rate of the reaction will be dependent on the stability of the carbocation (recall that tertiary carbocations are more stable than secondary carbocations, which are more stable than primary carbocations).

In the previous chapter, we saw that an OH group is a terrible leaving group, and that an S_N1 reaction can only occur if the OH group is first protonated to give a better leaving group:

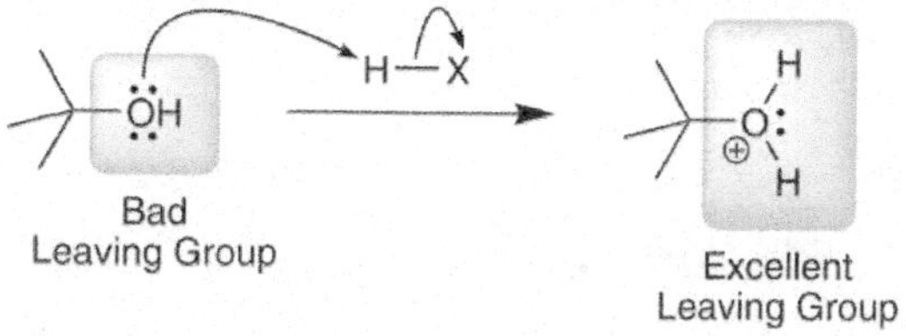

Bad
Leaving Group

Excellent
Leaving Group

The same is true with an E1 process. If the substrate is an alcohol, a strong acid (such as concentrated sulfuric acid) will be required in order to protonate the OH group:

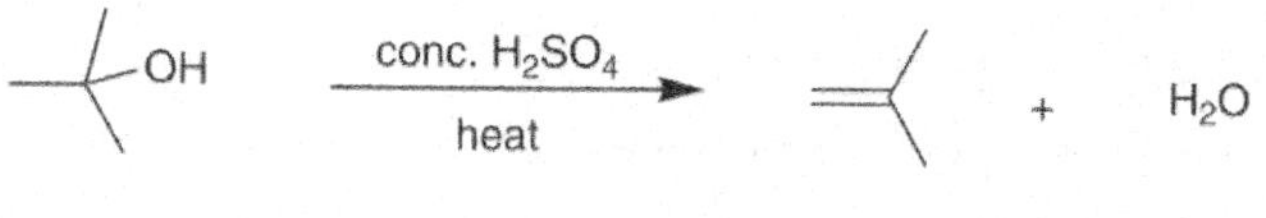

10.5 THE REGIOCHEMICAL OUTCOME OF AN E1 REACTION

E1 processes show a regiochemical preference for the Zaitsev product, just as we saw for most E2 reactions. For example:

Major   *Minor*

The more-substituted alkene (Zaitsev product) is the major product. However, there is one critical difference between the regiochemical outcomes of E1 and E2 reactions. Specifically, we have seen that the regiochemical outcome of an E2 reaction can often be controlled by carefully choosing the base (sterically hindered or not sterically hindered). In contrast, the regiochemical outcome of an E1 process *cannot* be controlled. The Zaitsev product will generally be obtained.

PROBLEMS   Draw the major and minor products that are expected when each of the following substrates is heated in the presence of concentrated sulfuric acid to give an E1 reaction:

10.9

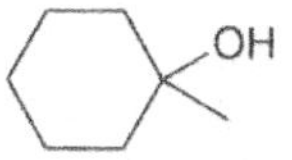

Major product   Minor product

10.10

Major product   Minor product

10.11

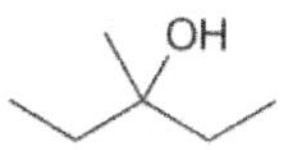

Major product   Minor product

10.6 THE STEREOCHEMICAL OUTCOME OF AN E1 REACTION

E1 reactions are not stereospecific. That is, they do not require antiperiplanarity in order for the reaction to occur. Nevertheless, E1 reactions are stereoselective. In other words, when *cis* and *trans* products are possible, we generally observe a preference for formation of the *trans* stereoisomer:

10.7 SUBSTITUTION VS. ELIMINATION

Substitution and elimination reactions are almost always in competition with each other. In order to predict the products of a reaction, you must determine which mechanism(s) win the competition. In some cases, there is one clear winner. For example, consider a case in which a tertiary alkyl halide is treated with a strong base, such as hydroxide:

In a case like this, E2 wins the competition, and no other mechanisms can successfully compete. Why not? An S_N2 process cannot occur at a reasonable rate because the substrate is tertiary (steric hindrance prevents an S_N2 from occurring). And unimolecular processes (E1 and S_N1) cannot compete because they are too slow. Recall that the rate-determining step for an E1 or S_N1 process is the loss of a leaving group to form a carbocation, which is a slow step. Therefore, E1 and S_N1 could only win the competition if the competing E2 process is extremely slow (when a weak base is used). However, when a strong base is used, E2 occurs rapidly, so E1 and S_N1 cannot compete.

Now let's consider a case where there is more than one winner, for example:

In this case, there are two winners! Don't fall into the trap of thinking that there must always be one clear winner. Sometimes there is, but sometimes, there are multiple products (perhaps even more than two). The goal is to predict *all* of the products, and to predict which products are major and which are minor. To accomplish this goal, you will need to perform the following three steps:

1. Determine the function of the reagent.
2. Analyze the substrate and determine the expected mechanism(s).
3. Consider regiochemical and stereochemical requirements.

The last three sections of this chapter are devoted to helping you become competent in performing all three steps. Let's begin with Step 1: Determining the function of the reagent.

10.8 DETERMINING THE FUNCTION OF THE REAGENT

We have seen earlier in this chapter that the main difference between substitution and elimination is the function of the reagent. A substitution reaction occurs when the reagent functions as a nucleophile, while an elimination reaction occurs when the reagent functions as a base. So the first step in any specific case is to determine whether the reagent is a strong or weak nucleophile, and whether it is a strong or weak base. Students generally assume that a strong base must also be a strong nucleophile, but this is not always true. It is possible for a reagent to be a weak nucleophile and a strong base. Similarly, it is possible for a reagent to be a strong nucleophile and a weak base. In other words, basicity and nucleophilicity do not always parallel each other. Let's begin by seeing when they *do* parallel each other.

When comparing atoms *in the same row* of the periodic table, basicity and nucleophilicity *do* parallel each other:

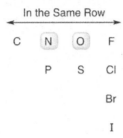

For example, let's compare H_2N^- and HO^-. The difference between these two reagents is the identity of the atom bearing the charge (O *vs.* N). We already saw in Chapter 3 (when we saw the factors determining charge stability) that oxygen, being more electronegative than nitrogen, can stabilize a charge better than nitrogen can. Therefore, HO^- will be more stable than H_2N^-, so H_2N^- will be a stronger base.

As it turns out, H_2N^- will also be a stronger nucleophile than HO^-, because basicity and nucleophilicity parallel each other when comparing atoms in the same row of the periodic table.

When comparing atoms *in the same column* of the periodic table, basicity and nucleophilicity *do not* parallel each other:

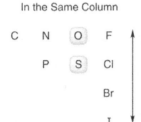

For example, let's compare HO^- and HS^-. Once again, the difference between these two reagents is the identity of the atom bearing the charge (O *vs.* S). We already saw in Chapter 3 that sulfur, being larger than oxygen, can stabilize a charge better than oxygen can (remember we saw that size is more important than electronegativity when comparing atoms in the same column). Therefore, HS^- will be more stable than HO^-, so HO^- will be a stronger base. Nevertheless, HS^- is a better nucleophile than HO^-. Why?

Recall that basicity and nucleophilicity are different concepts. Basicity measures stability of the charge (a thermodynamic argument), whereas nucleophilicity measures how fast a nucleophile attacks something (a kinetic argument). When you have a large atom, like sulfur, an interesting effect comes into play. As the sulfur atom approaches an electrophile (a compound with δ+), the electron density within the sulfur atom gets polarized, meaning that the electron density can move around. This property, called polarizability, increases the force of attraction between the nucleophile and the electrophile, so the rate of attack is very fast. Since *nucleophilicity* is a measure of *how fast* the nucleophile attacks, this effect renders the sulfur atom very nucleophilic. As a result, HS^- functions almost exclusively as a nucleophile and rarely functions as a base. The same is true for most of the halides (especially iodide), which function exclusively as nucleophiles. The halides are generally too weakly basic to function as bases. So, when you see one of these nucleophiles, you do not need to worry about elimination reactions—you should just expect substitution reactions. It is very common to see the halides being used as nucleophiles, so it is very helpful to know that you do not need to worry about elimination reactions when you see a halide as the reagent.

Armed with the understanding that nucleophilicity and basicity are not the same concepts, we can now categorize reagents into the following four groups:

NUCLEOPHILE (ONLY)		BASE (ONLY)	STRONG / STRONG NUC / BASE	WEAK / WEAK NUC / BASE

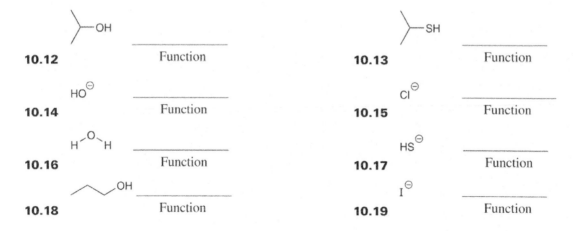

Let's quickly review each of these four categories. The first category contains reagents that function only as nucleophiles. They are strong nucleophiles because they are highly polarizable, but they are weak bases. When you see a reagent from this category, you should focus exclusively on substitution reactions (not elimination). Notice that sulfuric acid is NOT in this category (or any of the categories above). It is true that sulfuric acid contains sulfur, but the sulfur atom in sulfuric acid does not possess a lone pair, so it cannot function as a nucleophile. As its name implies, sulfuric acid functions only as an acid, so it is not listed in any of the four categories above.

The second category contains reagents that function only as bases; not as nucleophiles. The first reagent on this list is the hydride ion, usually shown as NaH, where Na^+ is the counter ion. The hydride ion of NaH is not a good nucleophile, despite the presence of a negative charge, because hydrogen is very small so it is not sufficiently polarizable. Nevertheless, the hydride ion is a very strong base. The use of a hydride ion as the reagent indicates that elimination will occur rather than substitution.

Notice that *tert*-butoxide appears in both the second and third categories. Technically, it is a strong nucleophile and a strong base, so it belongs in the third category. But practically, *tert*-butoxide is sterically hindered, which prevents it from functioning as a nucleophile in most cases. Therefore, it is often used as a base, to favor E2 over S_N2.

The third category contains reagents that are both strong nucleophiles and strong bases. These reagents include hydroxide (HO^-) and alkoxide ions (RO^-), and are generally used for bimolecular processes (S_N2 and E2).

The fourth and final category contains reagents that are weak nucleophiles and weak bases. These reagents include water (H_2O) and alcohols (ROH), and are generally used for unimolecular processes (S_N1 and E1).

In order to predict the products of a reaction, the first step is determining the identity and nature of the reagent. That is, you must analyze the reagent and determine the category to which it belongs. Let's get some practice with this critical skill.

PROBLEMS   Identify the function of each of the following reagents. In each case, the reagent will fall into one of the following four categories:

(a) strong nucleophile and weak base

(b) weak nucleophile and strong base

(c) strong nucleophile and strong base

(d) weak nucleophile and weak base

10.12 ─────── Function

10.13 ─────── Function

10.14 $HO^{\ominus}$ ─────── Function

10.15 $Cl^{\ominus}$ ─────── Function

10.16 ─────── Function

10.17 $HS^{\ominus}$ ─────── Function

10.18 ─────── Function

10.19 $I^{\ominus}$ ─────── Function

10.9   IDENTIFYING THE MECHANISM(S)

We mentioned that there are three main steps for predicting the products of substitution and elimination reactions. In the previous section, we explored the first step (determining the function of the reagent). In this section, we now explore the second step of the process in which we analyze the substrate and identify which mechanism(s) operates.

As described in the previous section, there are four categories of reagents. For each category, we must explore the expected outcome with a primary, secondary, or tertiary substrate. All of the relevant information is summarized in the following flow chart. It is important to know this flow chart extremely well, but be careful not to memorize it. It is more important to "understand" the reasons for all of these outcomes. A proper understanding will prove to be far more useful on an exam than simply memorizing a set of rules.

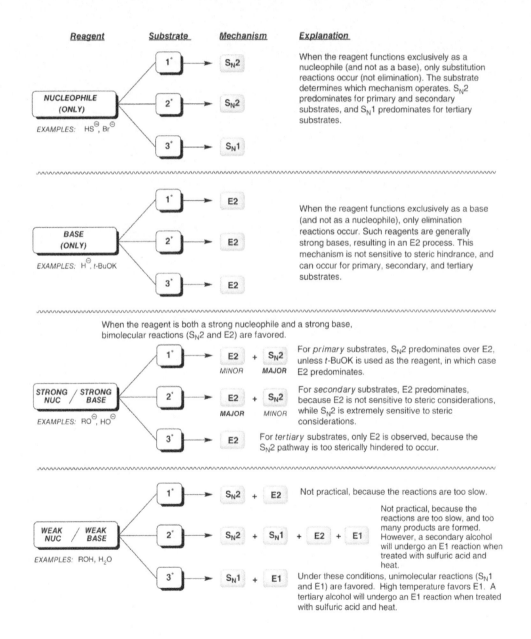

The flow chart above can be used to determine which mechanism(s) operate for a specific case. Let's get some practice.

EXERCISE 10.20 Identify the mechanism(s) expected to operate when 3-bromopentane is treated with sodium hydroxide:

$$\text{Br} \qquad \xrightarrow{\text{NaOH}}$$

Answer Our first step is to identify the function of the reagent. Using the skills developed in the previous section, we know that sodium hydroxide is both a strong nucleophile and a strong base:

$$\text{Na}^{\oplus} \qquad {}^{\ominus}\!\!:\!\ddot{\text{O}}\text{H}$$

Strong nucleophile/
strong base

Our next step is to identify the substrate. In this case, the substrate is 3-bromopentane, which is a secondary substrate, and therefore, we expect E2 and S_N2 mechanisms to operate:

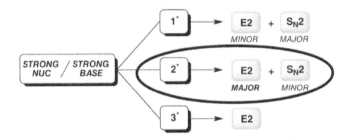

The E2 pathway is expected to provide the major product, because the S_N2 pathway is more sensitive to steric hindrance provided by secondary substrates.

PROBLEMS Identify the mechanism(s) expected to occur in each of the following cases. Do not worry about drawing the products yet. We will do that in the next section. For now, just identify which mechanisms are operating.

10.21 $\xrightarrow{\text{NaOEt}}$ (propyl bromide, Br)

10.22 (1-methylcyclopentanol, OH) $\xrightarrow[\text{heat}]{\text{conc. } H_2SO_4}$

10.23 (2-iodo-2-methylbutane, I) $\xrightarrow{\text{H}_2\text{O}}$

10.24 (1-chloropropane, Cl) $\xrightarrow{\text{NaBr}}$

10.25 (2-chloro-2-methylbutane, Cl) $\xrightarrow{\text{NaBr}}$

10.10   PREDICTING THE PRODUCTS

We mentioned that predicting the products of substitution and elimination reactions requires three discrete steps:

1. Determine the function of the reagent.
2. Analyze the substrate and determine the expected mechanism(s).
3. Consider any relevant regiochemical and stereochemical requirements.

In the previous two sections, we explored the first two steps of this process. In this section, we will explore the third and final step. After determining which mechanism(s) are expected to operate, the final step is to consider the regiochemical and stereochemical outcomes for each of the expected mechanisms. The following table provides a summary of guidelines that must be followed when drawing products.

	Regiochemical Outcome	**Stereochemical Outcome**
S_N2	The nucleophile attacks the α position, where the leaving group is connected.	The nucleophile replaces the leaving group with inversion of configuration.
S_N1	The nucleophile attacks the carbocation, which is generally where the leaving group was originally connected, unless a carbocation rearrangement took place (consult your textbook for a discussion of carbocation rearrangements during S_N1 reactions).	The nucleophile replaces the leaving group with racemization.
E2	The Zaitsev product is generally favored over the Hofmann product, unless a sterically hindered base is used, in which case the Hofmann product will be favored.	This process is both stereoselective and stereospecific. When applicable, a *trans* disubstituted alkene will be favored over a *cis* disubstituted alkene. When the β position of the substrate has only one proton, the stereoisomeric alkene resulting from antiperiplanar elimination will be obtained (exclusively, in most cases).
E1	The Zaitsev product is always favored over the Hofmann product.	The process is stereoselective. When applicable, a *trans* disubstituted alkene will be favored over a *cis* disubstituted alkene.

The table does not contain any new information. All of the information can be found in this chapter and in the previous chapter. The table is meant only as a summary of all of the relevant information, so that it is easily accessible in one location. Let's get some practice applying these guidelines.

EXERCISE 10.26   Predict the product(s) of the following reaction, and identify the major and minor products:

Answer   In order to draw the products, we must follow these three steps:

1. Determine the function of the reagent.
2. Analyze the substrate and determine the expected mechanism(s).
3. Consider any relevant regiochemical and stereochemical requirements.

We begin by analyzing the reagent. Methoxide is both a strong base and a strong nucleophile. Next, we move on to Step 2 and we analyze the substrate. In this case, the substrate is secondary, so we would expect E2 and S$_N$2 pathways to compete with each other:

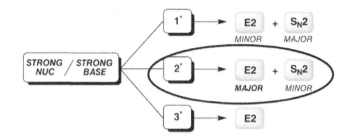

We expect the E2 pathway to predominate, because it is less sensitive to steric hindrance than the S$_N$2 pathway. Therefore, we would expect the major product(s) to be generated via an E2 process, and the minor product(s) to be generated via an S$_N$2 process. In order to draw the products, we must complete the third and final step. That is, we must consider the regiochemical and stereochemical outcomes for both the E2 and S$_N$2 processes. Let's begin with the E2 process.

For the regiochemical outcome, we expect the Zaitsev product to be the major product, because the reaction does not utilize a sterically hindered base:

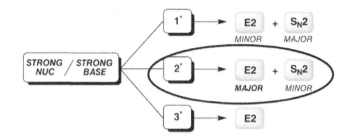

Next, look at the stereochemistry. The E2 process is stereoselective, so we expect *cis* and *trans* isomers, with a predominance of the *trans* isomer:

E2 processes are not only stereoselective, but are also stereospecific. However, in this case, the β position has more than one proton, so the stereospecificity of this reaction is not relevant.

Now consider the S$_N$2 product. This case involves a stereocenter, so we expect inversion of configuration:

In summary, we expect the following products:

PROBLEMS   Identify the major and minor product(s) that are expected for each of the following reactions:

10.27   NaOH

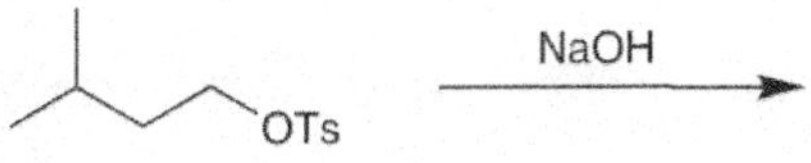

10.28   NaOMe

10.29   NaOEt

10.30   NaOH

10.31   conc. H_2SO_4 heat

10.32 H_2O

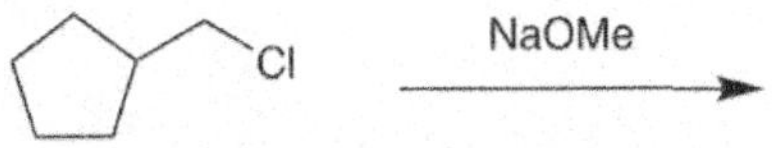

10.33   NaSH   DMSO

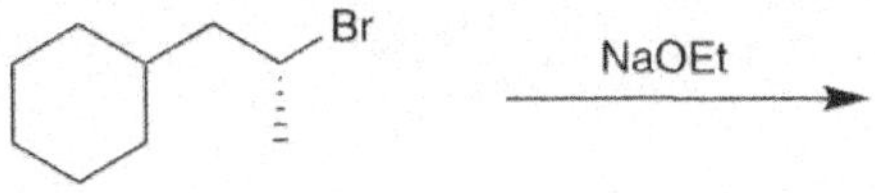

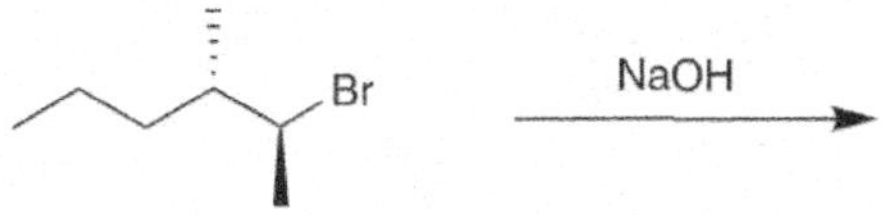

10.34   NaOEt

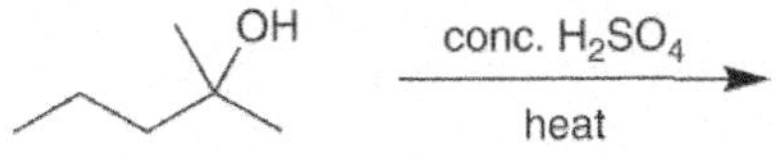

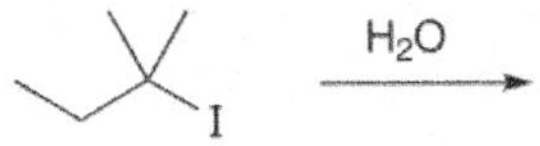

10.35 t-BuOK

10.36   NaOMe

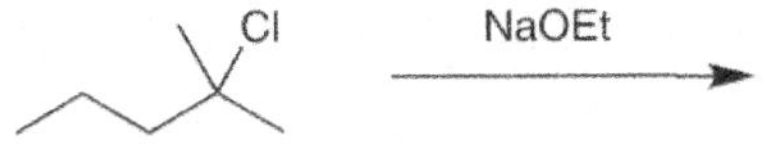

10.37   NaSH

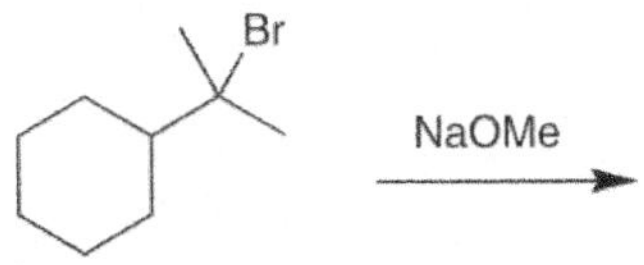

10.38   NaSMe

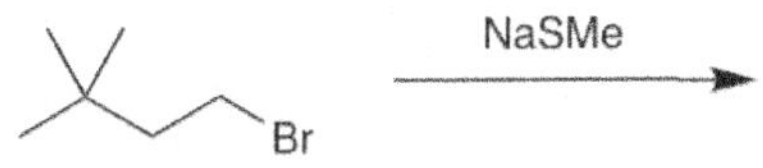

CHAPTER 11

ADDITION REACTIONS

Addition reactions are characterized by two groups adding across a double bond:

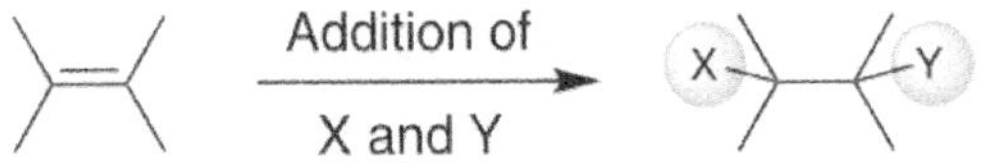

In the process, the double bond is destroyed, and we say that the two groups (X and Y) have "added" across the double bond. In this chapter, we will see many addition reactions, and we will focus on the following three types of problems: (1) predicting the products of a reaction, (2) drawing a mechanism, and (3) proposing a synthesis. In order to gain mastery over these types of problems, you must first become comfortable with some crucial terminology. We will focus on this terminology now, before learning any reactions.

11.1 TERMINOLOGY DESCRIBING REGIOCHEMISTRY

When adding two different groups across an unsymmetrical alkene, there is special terminology describing the *regiochemistry* of the addition. For example, suppose you are adding H and Br across an alkene. Regiochemistry refers to the positioning of the H and the Br in the product: where is H installed and where is Br installed?

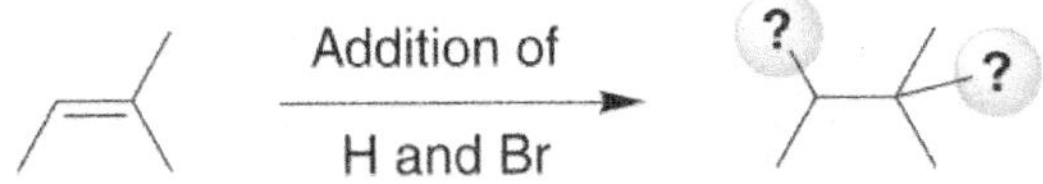

Regiochemistry is only relevant when adding two *different* groups (such as H and Br). However, when adding two of the same group (such as Br and Br), regiochemistry becomes irrelevant:

Similarly, when adding two different groups across a *symmetrical* alkene, regiochemistry is also irrelevant:

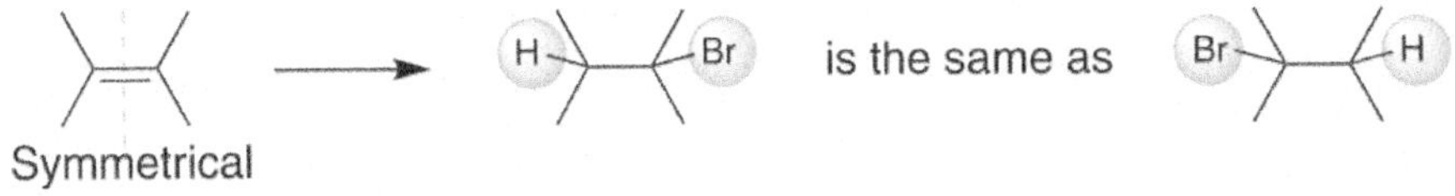

The bottom line is: regiochemistry is only relevant when adding two *different groups* across an *unsymmetrical* alkene.

As we learn addition reactions, we will be using two important terms to describe the regiochemistry: *Markovnikov* and *anti-Markovnikov*. To use these terms properly, we must be able to recognize which carbon is more substituted. Consider the following example:

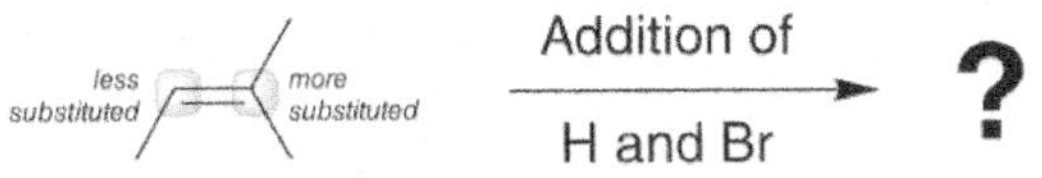

There are two vinylic positions, highlighted in grey above. The vinylic position on the right has more alkyl groups; it is more substituted. When Br ends up on the more substituted carbon, we call it a *Markovnikov addition:*

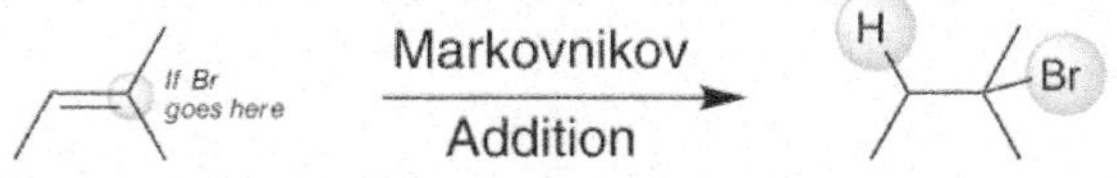

When Br ends up on the less substituted carbon, we call it an *anti-Markovnikov addition:*

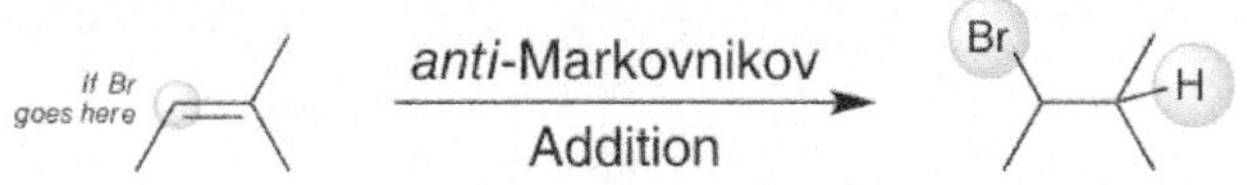

When we explore the mechanisms of addition reactions, we will see why some reactions proceed through a Markovnikov addition while others proceed through an *anti*-Markovnikov addition. For now, let's make sure that we are comfortable using the terms.

EXERCISE 11.1   Draw the product that you would expect from an *anti*-Markovnikov addition of H and Br across the following alkene:

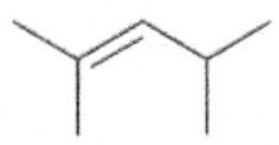

Answer   In an *anti*-Markovnikov addition, the Br ends up at the less substituted carbon, so we draw the following product:

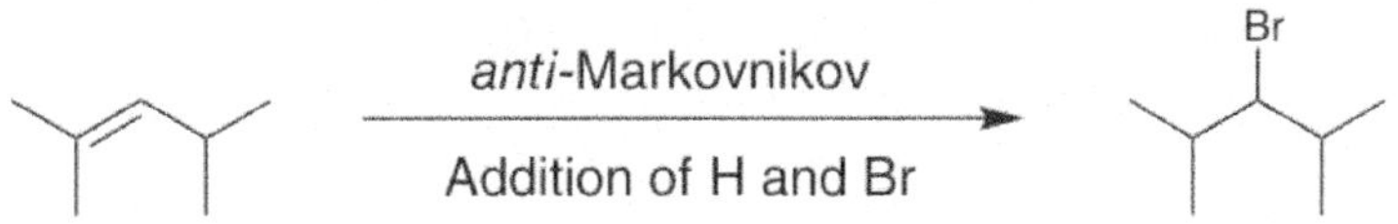

Remember that in bond-line drawings, it is not necessary to draw the H that was added.

PROBLEMS   In each of the following cases, use the information provided to draw the product that you expect.

11.2
anti-Markovnikov
Addition of H and Br

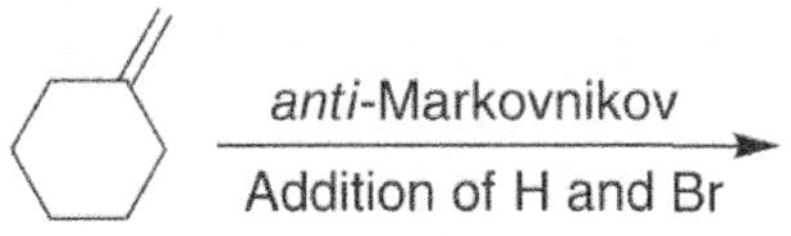

11.3
Markovnikov
Addition of H and Cl

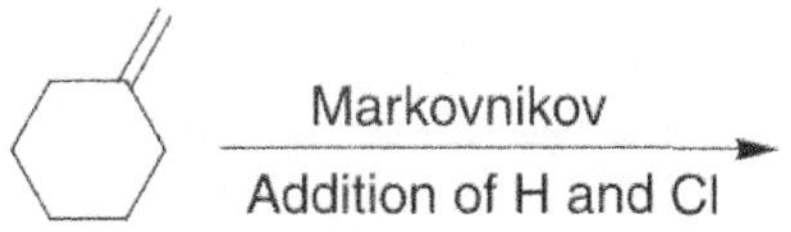

11.4
anti-Markovnikov
Addition of H and OH

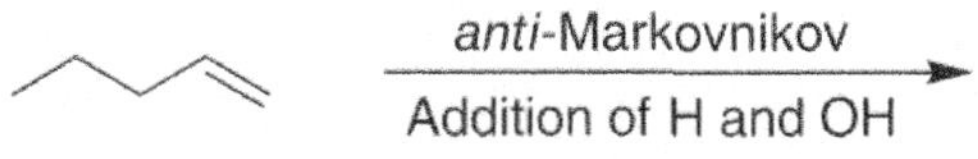

11.5
Markovnikov
Addition of H and OH

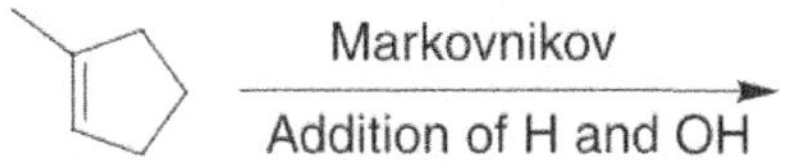

11.2 TERMINOLOGY DESCRIBING STEREOCHEMISTRY

In addition to the regiochemistry, there is also special terminology used to describe the stereochemistry of a reaction. As an example, consider the following simple alkene:

Suppose that we have an *anti*-Markovnikov addition of H and OH across this alkene:

OH goes here   *H goes here*

We know which two groups are adding to the double bond, and we know the regiochemistry of the addition. But in order to draw the products correctly, we also need to know the stereochemistry of the reaction. To better explain this, we will redraw the alkene in a different way.

The *vinylic* carbon atoms, highlighted above, are both sp^2 hybridized, and therefore trigonal planar. As a result, all four groups (connected to the vinylic positions) are in one plane. In order to discuss stereochemistry, we will rotate the molecule so that the plane is coming in and out of the page:

This is an unusual way to draw an alkene (where all bonds are shown as wedges and dashes, rather than straight lines), but this way of drawing the alkene will make it easier to explore stereochemistry.

We can imagine both groups being added on the same side of the plane (both groups can be installed from above the plane, or both groups can be installed from below the plane), which we call a *syn* addition:

and

Or, we can imagine both groups being added on opposite sides of the plane, which we call an *anti* addition:

and

Note: Do not confuse the term *"anti"* with the term *"anti*-Markovnikov." The term *"anti* addition" describes the stereochemistry, while the term *"anti*-Markovnikov addition" describes the regiochemistry. It is possible for an *anti*-Markovnikov reaction to be a *syn* addition. In fact, we will see such an example very soon.

We see that there are two products that arise from a *syn* addition, and two products that arise from an *anti* addition:

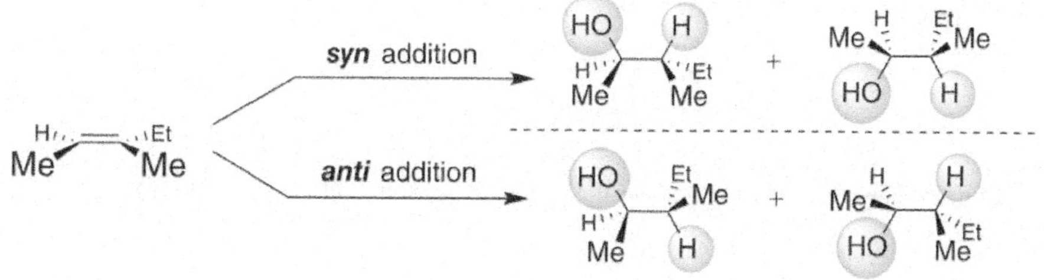

In total, there are four possible products (two pairs of enantiomers). The two products of a *syn* addition represent one pair of enantiomers. And the two products from an *anti* addition represent the other pair of enantiomers.

Some reactions are not stereospecific, and we might expect all four possible products (both pairs of enantiomers). Other reactions *are* stereospecific—we might only observe formation of the two enantiomers from a *syn* addition, or we might only observe formation of the two enantiomers from an *anti* addition. It is important to know which reactions occur through an *anti* addition, which reactions occur through a *syn* addition, and which reactions are not stereospecific at all. As we go through each reaction in this chapter, we will look closely at the accepted mechanism for each reaction, because the mechanism must always justify the observed stereochemistry of the reaction. For now, let's make sure that we know the terminology.

EXERCISE 11.6   Consider the following alkene:

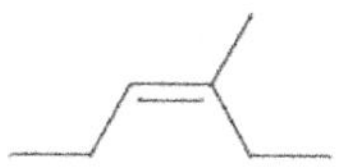

Draw the products that you would expect when adding H and OH in the following way:

   Regiochemistry = *anti*-Markovnikov

   Stereochemistry = *syn* addition

Answer   We begin by looking at the *regio*chemistry. The reaction is *anti*-Markovnikov, which means that the OH group will be positioned at the less substituted carbon:

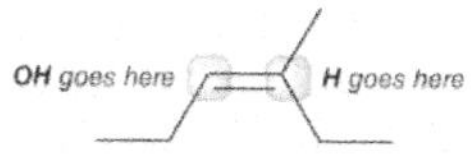

Next, we look at the *stereo*chemistry. The reaction is a *syn* addition, which means that the H and OH both add on the same side of the double bond. In order to see this more clearly, we rotate the molecule so that the plane of the double bond is coming out of the page, and we draw the pair of enantiomers that we expect from a *syn* addition:

Our products are a pair of enantiomers, so we can record our answer more quickly by drawing one enantiomer and then indicating the presence of the other, like this:

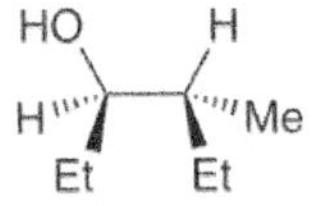

HO   H

H""⟍⟋""Me   +   Enantiomer

Et   Et

PROBLEMS   For each of the following problems, predict the products using the information provided:

11.7

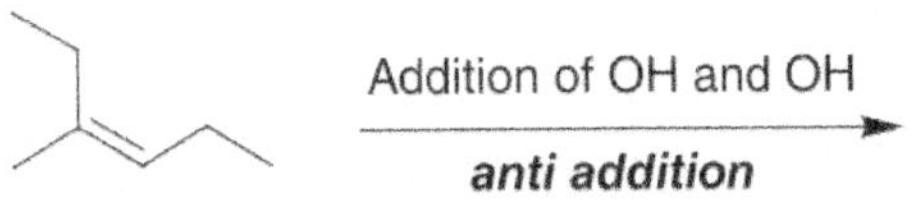

Addition of OH and OH
—————————————→
anti addition

11.8

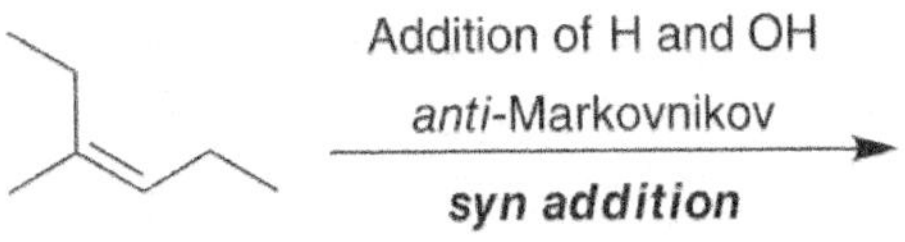

Addition of H and OH
anti-Markovnikov
—————————————→
syn addition

11.9

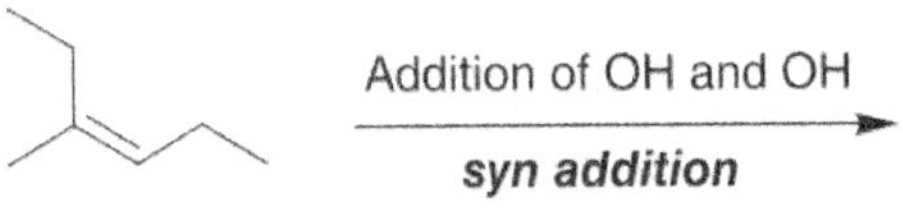

Addition of OH and OH
—————————————→
syn addition

11.10

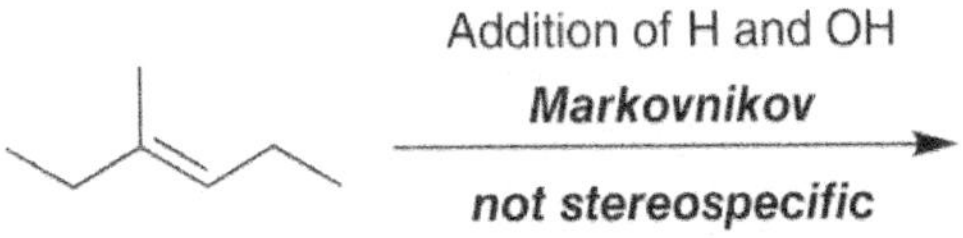

Addition of H and OH
Markovnikov
—————————————→
not stereospecific

The examples we have seen so far have been acyclic alkenes (not containing a ring). When we add across cyclic alkenes, the products are easier to draw because we don't have to rotate and redraw the alkene before starting. Let's see an example:

EXERCISE 11.11   Predict the products using the following information:

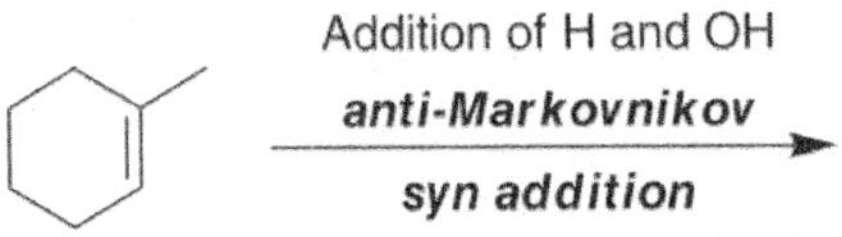

Addition of H and OH
anti-Markovnikov
—————————————→
syn addition

Answer   We begin by looking at the regiochemistry. The reaction is *anti*-Markovnikov, which means that the OH group will be positioned at the less substituted carbon:

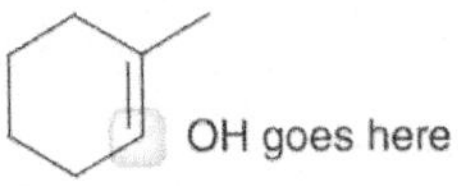

OH goes here

Next, we look at the stereochemistry. A *syn* addition means that the H and OH both add on the same side of the double bond. Since the alkene is a cyclic compound, the products are easy to draw (without having to rotate the alkene first). We simply place the groups on wedges and dashes, like this:

H   H

OH   +   OH

Our two products represent a pair of enantiomers, so we can record our answer more quickly in the following way:

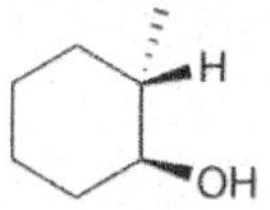

+   Enantiomer

PROBLEMS   For each of the following problems, draw the products using the information provided:

11.12

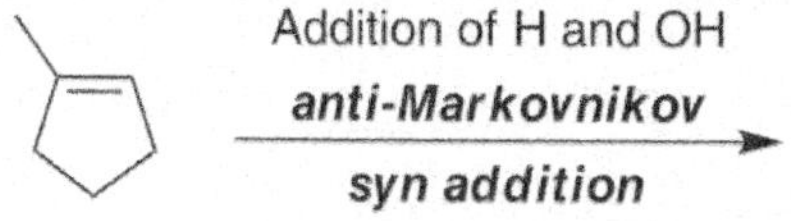

Addition of H and OH
anti-Markovnikov
syn addition

11.13

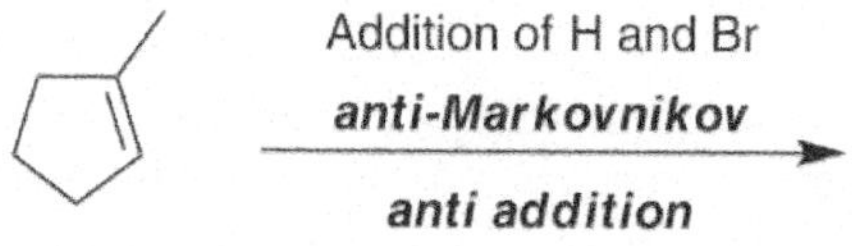

Addition of H and Br
anti-Markovnikov
anti addition

11.14

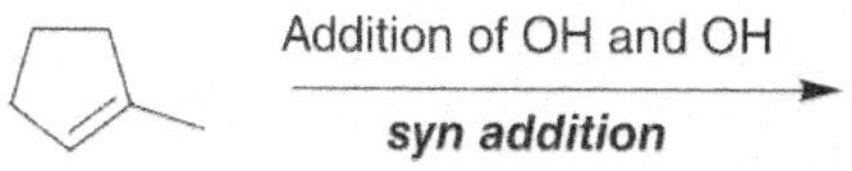

Addition of OH and OH
syn addition

11.15

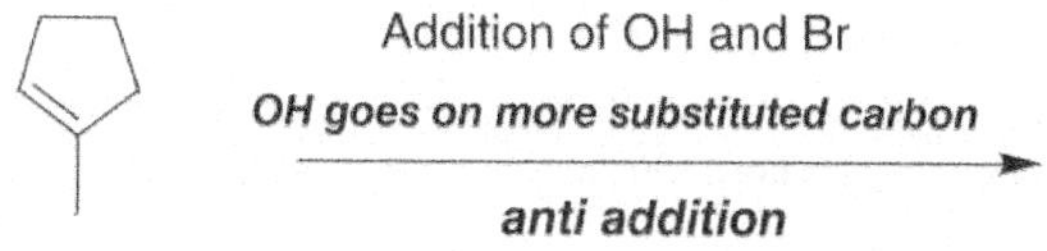

Addition of OH and Br
OH goes on more substituted carbon
anti addition

In most of the examples we have seen so far, we were creating *two* chiral centers:

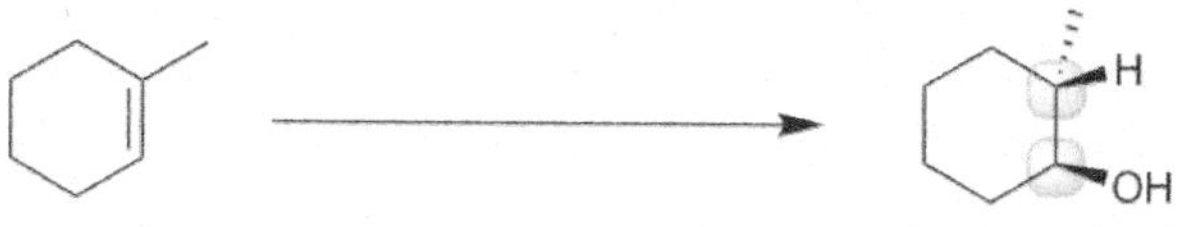

+   Enantiomer

two chiral centers

However, you may encounter examples where no chiral centers are being formed. For example,

no chiral centers

In situations like this, the stereochemistry is irrelevant. There is only one product (no stereoisomers).
    Similarly, you may encounter situations where only *one* chiral center is being formed. For example,

+   Enantiomer

only one chiral center

In cases like this, the stereochemistry is still irrelevant (as long as the compound does not possess any other chiral centers). Why? With only one chiral center, there will only be two possible products (not four). These two products will represent a pair of enantiomers (one will be *R* and the other will be *S*). Both of these products will be obtained, whether the reaction proceeds through a *syn* addition *or* through an *anti* addition. If the reaction is a *syn* addition, the OH group can come from above the plane or from below the plane of the double bond, giving both possible products. Similarly, if the reaction is an *anti* addition, the OH group can come from above the plane or from below the plane of the double bond, giving both possible products. Either way, a racemic mixture is obtained.

The bottom line is: the stereochemistry is most relevant when the addition reaction involves the creation of *two* new chiral centers.

EXERCISE 11.16   Given the following information, determine if the stereochemistry of the reaction is relevant, and draw the expected products:

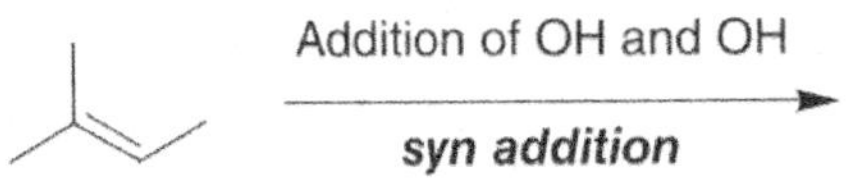

Answer   We begin by looking at the regiochemistry. We are not told what the regiochemistry is, because it is irrelevant (we are adding two groups of the same kind: OH and OH). Next, we look at the stereochemistry. The reaction is a *syn* addition. But in this case, we are only creating one chiral center:

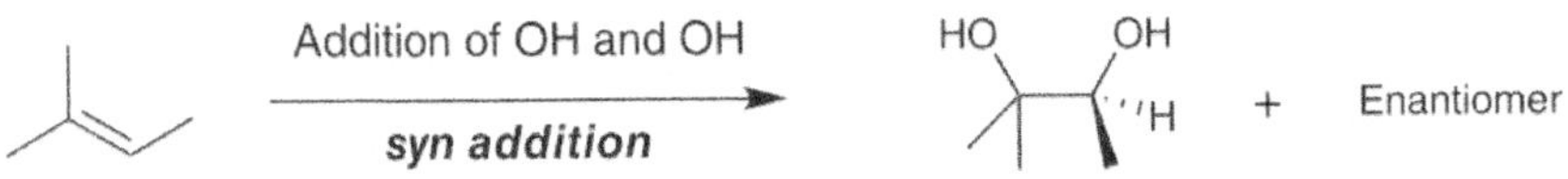

Therefore, the fact that the reaction proceeds through a *syn* addition is not important for predicting the products. If the reaction had been an *anti* addition, we would have obtained the same products. In fact, if the reaction had not been stereospecific at all, we still would have obtained the same two products (the pair of enantiomers above).

PROBLEMS   For each of the following problems, draw the products using the information provided:

11.17

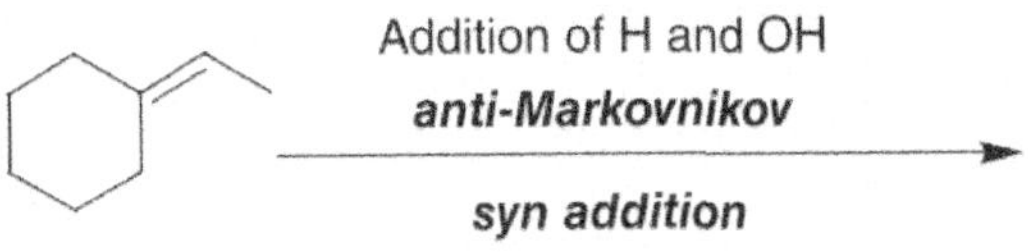

11.18

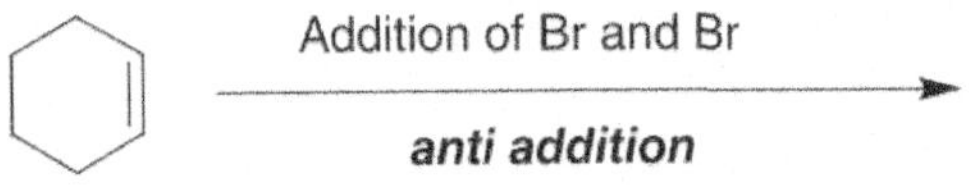

11.19

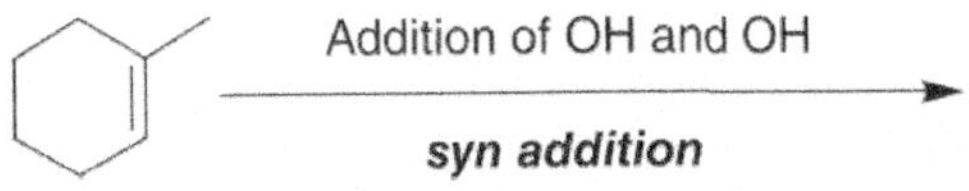

11.20

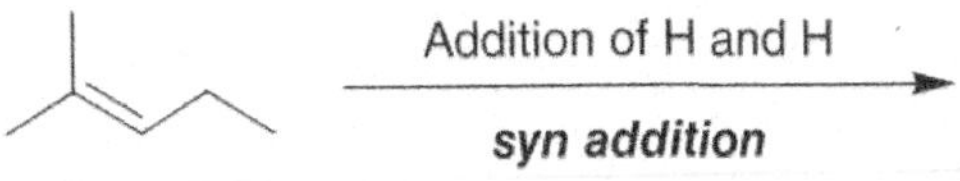

We are almost ready to begin learning the actual reactions. But first, we must explore one more subtlety associated with the stereochemistry of addition reactions. Consider the following example:

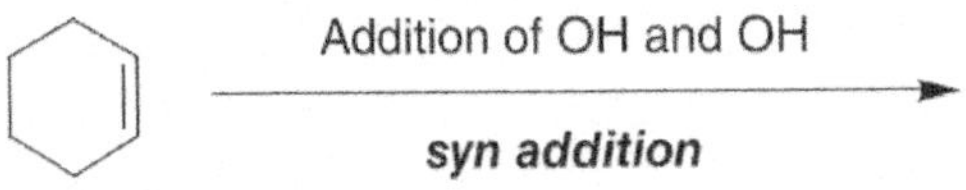

An analysis of the information given should lead us to draw the following products:

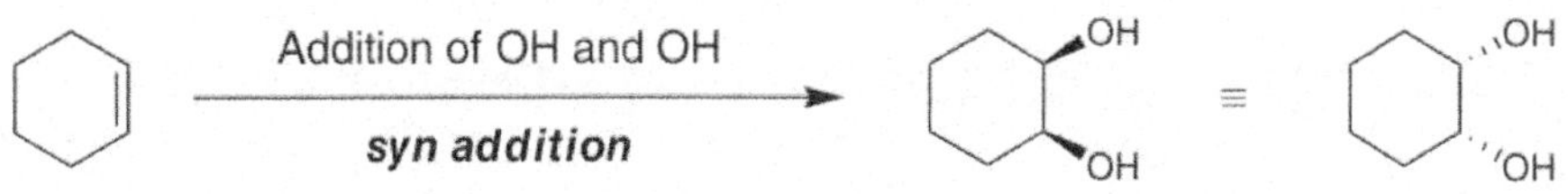

However, there are not two products here. Look closely and you will see that the two drawings above are actually the same compound. This compound is a *meso* compound, because it has an internal plane of symmetry. In a case like this, there is actually only one product. You can either draw the product with both OH groups on wedges, or you can draw it with both OH groups on dashes. Either way, you are drawing the same product. Just make sure not to draw *both* drawings, because that would imply that you don't recognize that it is a *meso* compound.

Here is another case where it is not so simple to see:

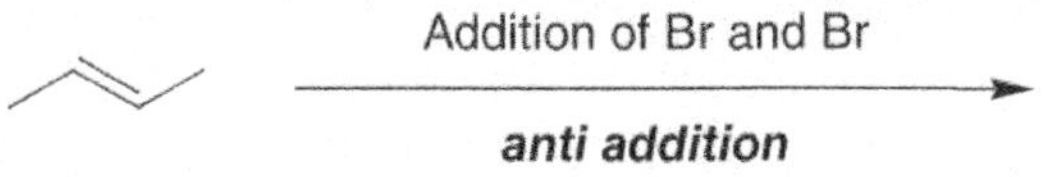

An analysis of the information should lead us to predict the following products:

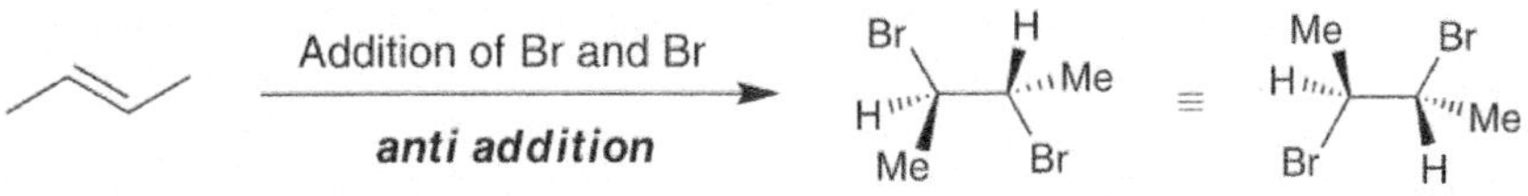

At first glance, it might be difficult to see that these two compounds are the same (they represent two drawings of the same *meso* compound). But if you rotate about the C—C single bond, you can see that it actually does have a plane of symmetry:

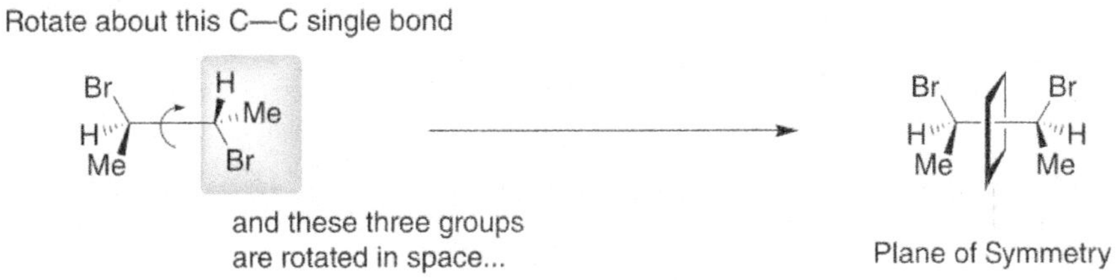

This is a subtle but important point. If you are not comfortable with identifying *meso* compounds, you should go back and review *meso* compounds in Chapter 7.

EXERCISE 11.21   Given the following information, draw the expected products:

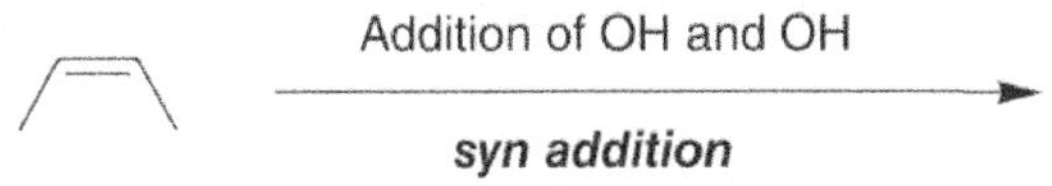

Answer   We begin by looking at the regiochemistry. In this case, we are adding two groups of the same kind (OH and OH), so the regiochemistry is irrelevant.

Next, we look at the stereochemistry. We are creating two new chiral centers in this case, so the requirement for *syn* addition determines the stereochemical outcome. That is, we expect only the pair of enantiomers that would result from a *syn* addition. To draw this pair of enantiomers, we do not have to rotate the alkene, as we have done in previous examples. In this example, it is simple enough to draw the products without rotating the alkene (sometimes, it will be simpler to do it this way):

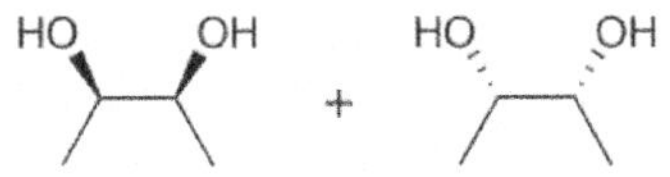

But wait! As a final step, we must determine whether the products are actually a pair of enantiomers or whether they are just two different ways to draw one *meso* compound. We look for a plane of symmetry, and in this case, we do have a plane of symmetry. Therefore, we don't draw both drawings above as our answer. We choose one drawing (either one). So our answer would look like this:

meso

PROBLEMS For each of the following, predict the products using the information provided. (Be on the lookout for *meso* compounds.)

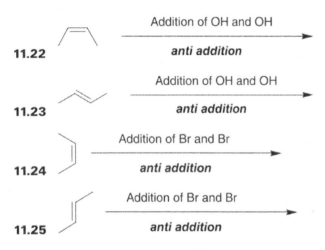

11.22 Addition of OH and OH — *anti addition*

11.23 Addition of OH and OH — *anti addition*

11.24 Addition of Br and Br — *anti addition*

11.25 Addition of Br and Br — *anti addition*

Until now, we have learned the basic terminology that you will need in order to predict products of addition reactions. To summarize, there are three pieces of information that you must have in order to predict products:

1. Which two groups are being added across the double bond (X and Y)?
2. What is the regiochemistry? (Markovnikov or *anti*-Markovnikov)
3. What is the stereochemistry? (*syn* or *anti*)

With these three pieces of information, you should be able to predict products with ease. Until now, you have been given all three pieces of information in each problem. However, as we progress through this chapter, this information will not be given to you. Rather, you will have to look at the reagents being used, and you will have to determine all three pieces of information on your own. That might sound like it involves a lot of memorization. Not so. We will soon see that the mechanism of each reaction contains all three pieces of information that you need. By understanding the mechanism for each reaction, you will "know" all three pieces of information about each reaction. We will focus on understanding, rather than memorization.

11.3 ADDING H AND H

It is possible to add H and H across an alkene. Here are two examples:

In this type of reaction, called *hydrogenation*, the regiochemistry will always be irrelevant, regardless of what alkene we use (we are adding two of the same group). However, we do need to explore the stereochemistry of hydrogenation reactions. In order to do this, let's take a close look at how the reaction takes place.

Notice the reagents that we use to accomplish a hydrogenation reaction (H_2 and a metal catalyst). A variety of metal catalysts can be used, such as Pt, Pd, or Ni. The hydrogen molecules (H_2) interact with the surface of the metal catalyst, effectively breaking the H—H bonds:

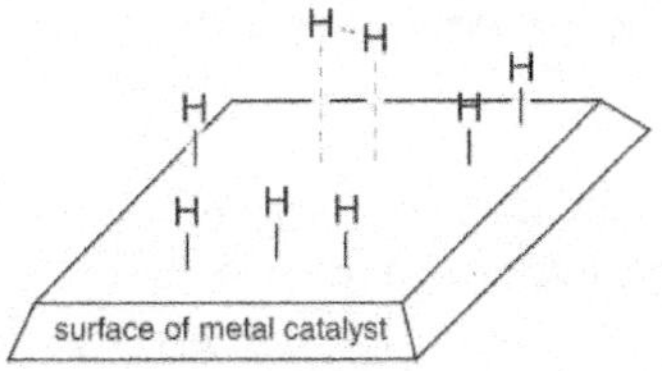

This forms individual hydrogen atoms adsorbed to the surface of the metal. These hydrogen atoms are now available for addition across the alkene. The addition reaction begins when the alkene coordinates with the metal surface:

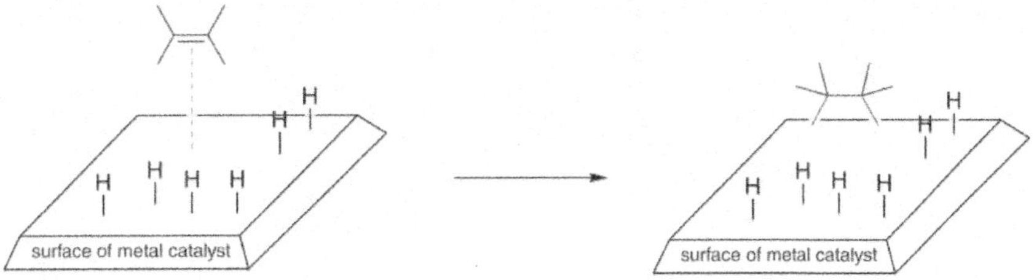

Surface chemistry then allows for the following two steps, effectively adding H and H across the alkene:

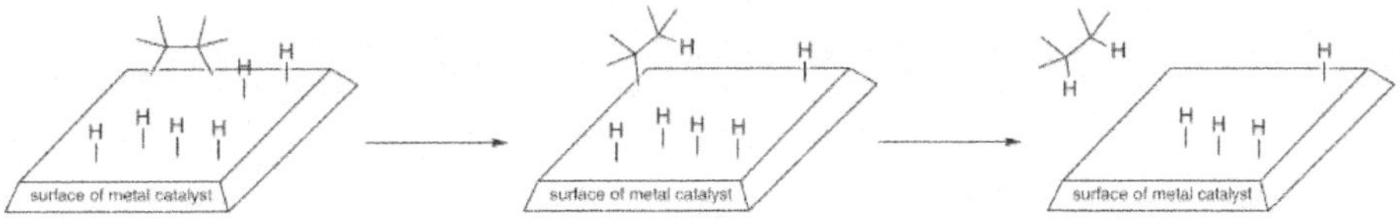

Notice that both hydrogen atoms are installed *on the same side of the alkene*, giving a *syn* addition.

The requirement for *syn* addition can be seen in the following example:

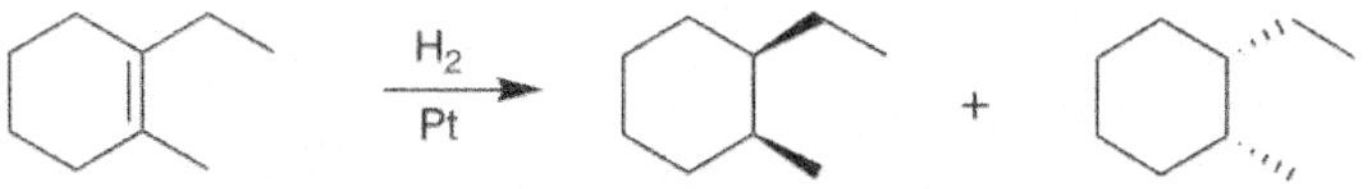

The hydrogen atoms that were added are not explicitly shown, but remember that H's don't need to be drawn in bond-line drawings. You should be able to see them, even though they are not drawn.

In the example above, we are creating two new chiral centers. So, theoretically, we could imagine four possible products (two pairs of enantiomers):

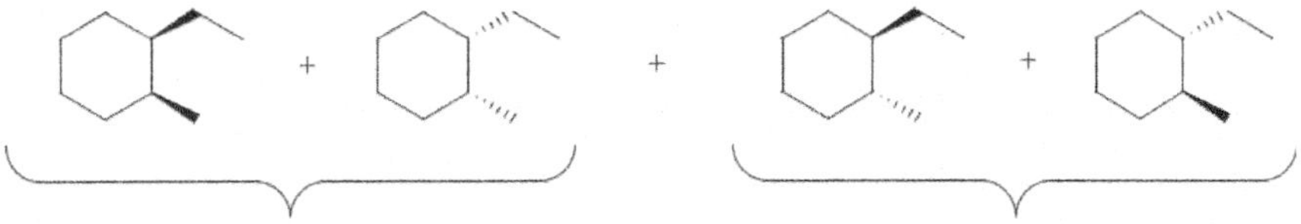

Pair of enantiomers   Pair of enantiomers

But we don't observe all four products. We only observe the pair of enantiomers that come from a *syn* addition (above left):

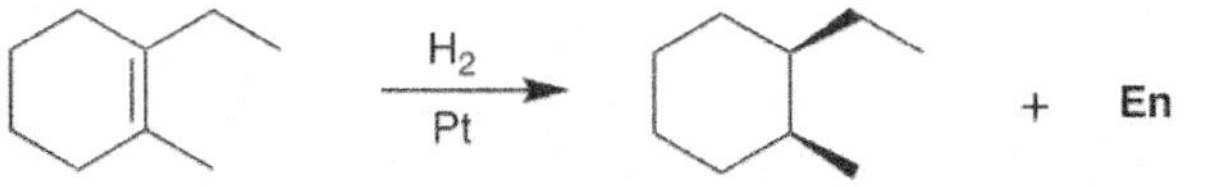

But be careful—make sure to be on the lookout for *meso* compounds. Consider this example:

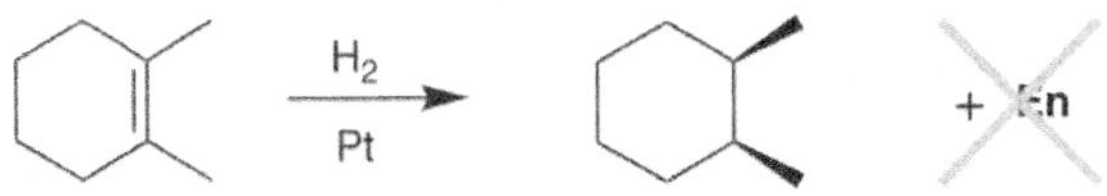

In this example, we would not write "+ Enantiomer," because the product is a *meso* compound.

Now we can summarize the reaction profile of a hydrogenation reaction:

$\xrightarrow[\text{Pt}]{\text{H}_2}$	**H and H**
*Regio*chem	not relevant
*Stereo*chem	*syn*

EXERCISE 11.26 Predict the products for each of the following reactions:

(a) 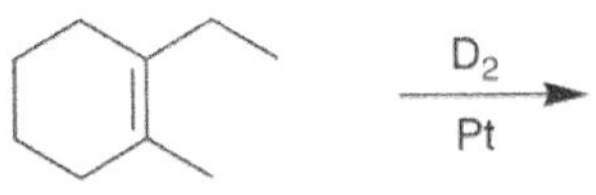　　　　　　　　　　　　　　　　　　　　(b)

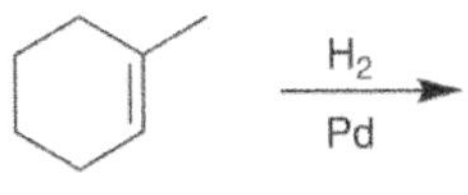

Answer

(a) Just as we can hydrogenate an alkene, we can also deuterate an alkene (deuterium is just an isotope of hydrogen). Therefore, we will be adding D and D across the alkene. We do not need to worry about regiochemistry, because we are adding two of the same group. However stereochemistry *is* relevant here, because we are creating two new chiral centers. Of the four possible products, we expect only the pair of enantiomers that would come from a *syn* addition:

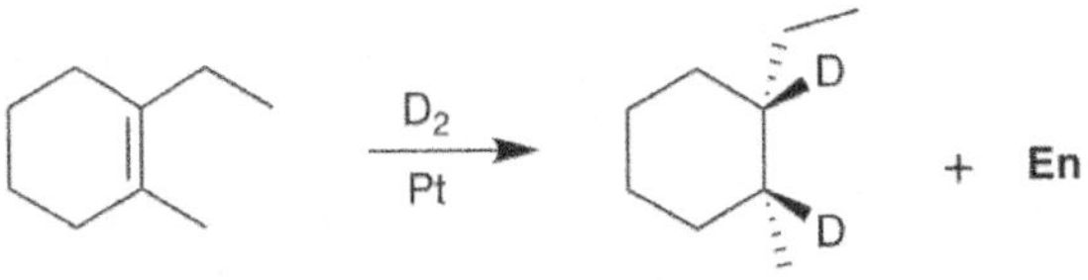

(b) These reagents will add H and H across a double bond. We do not need to worry about regiochemistry, because we are adding two of the same group. To determine whether stereochemistry is relevant, we must ask whether we are creating two new chiral centers. In this example, we are *not* creating two new chiral centers. In fact, we are not even creating one chiral center. Therefore, stereochemistry is irrelevant in this example. There will only be one product here:

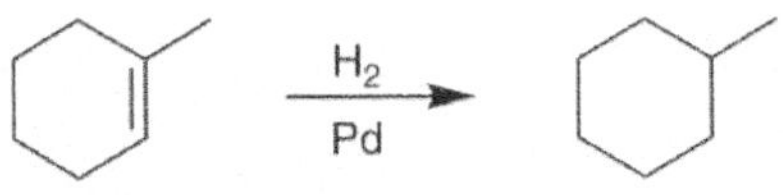

PROBLEMS Predict the products for each of the following reactions. In each example, make sure to determine whether or not you are forming any new chiral centers. If not, then the stereochemistry will be irrelevant.

11.27

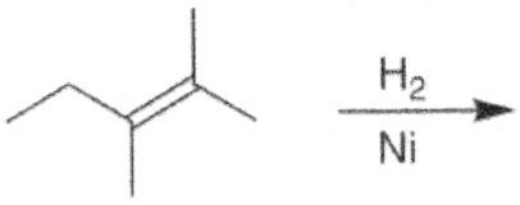

11.28

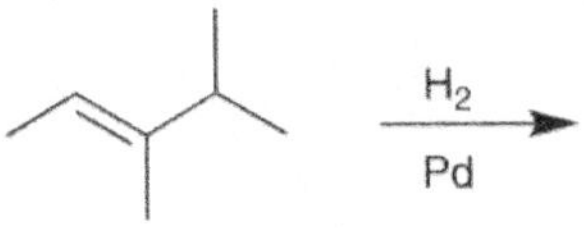

11.29

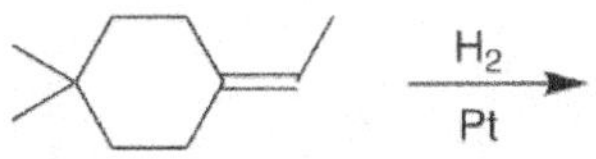

11.30

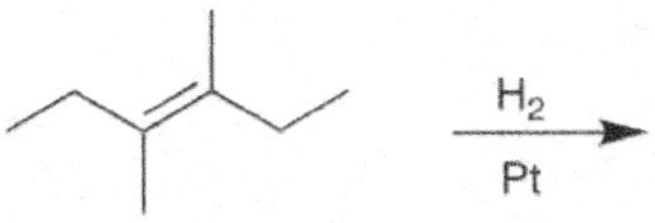

11.31

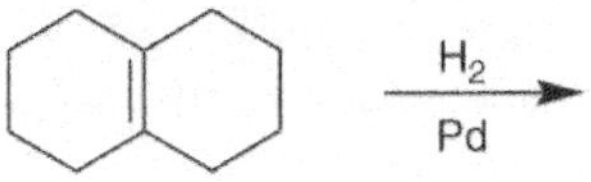

11.32

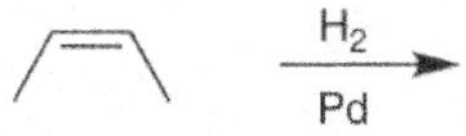

11.4 ADDING H AND X, MARKOVNIKOV

We will now explore the details of adding a hydrogen halide (HX) across a double bond. Here are two examples:

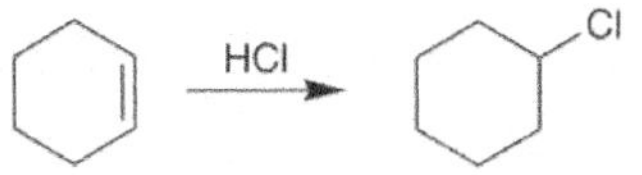

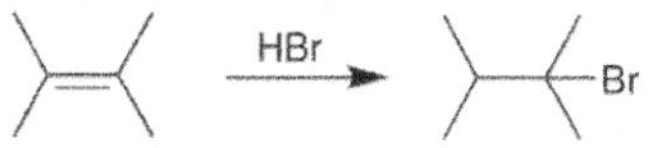

In order to understand the regiochemistry and stereochemistry of HX addition, we must analyze the accepted mechanism. When adding HX across a double bond, there are two key steps involved in the mechanism:

Step 1: Proton Transfer:

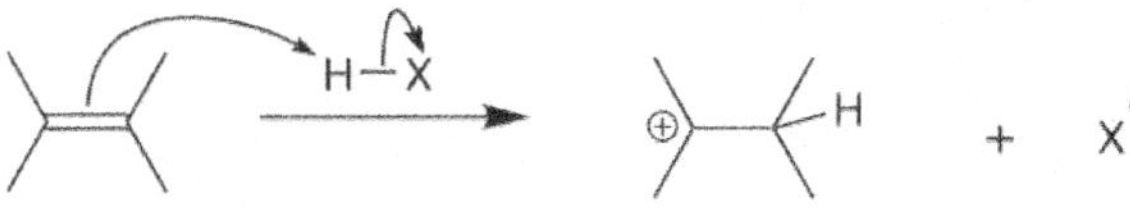

In this step, a proton is transferred to the alkene, which generates a carbocation. This carbocation is then attacked by the halide in step 2:

Step 2: Nucleophilic Attack:

The overall result is the addition of H and X across the double bond. We have specifically used a starting alkene that avoids issues of regiochemistry or stereochemistry; we will soon see other examples in which we must explore both of those issues. For now, focus on the curved arrows used in both steps. It is absolutely critical to master the art of drawing curved arrows properly. Let's quickly practice:

EXERCISE 11.33   Draw a mechanism for the following reaction:

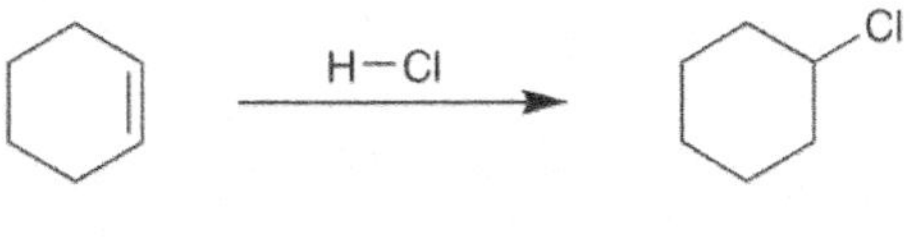

Answer   In the first step, there are two curved arrows:

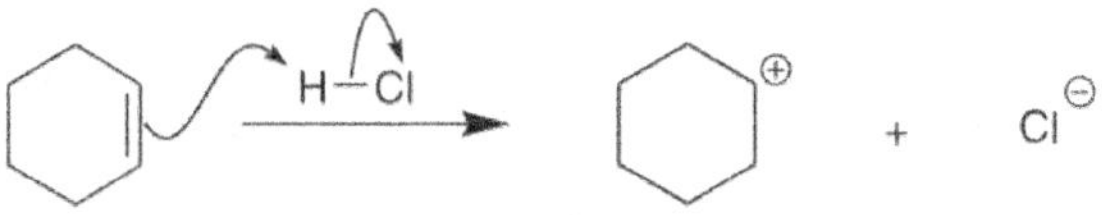

One curved arrow is drawn coming *from the alkene* and pointing *to the proton* (take special notice of this arrow, as it is a common mistake to draw this arrow in the wrong direction). The second curved arrow is drawn coming *from* the H—Cl bond and pointing *to* Cl.

In the second step, there is just one curved arrow. The chloride ion, formed in the previous step, now attacks the carbocation:

PROBLEMS   Draw a mechanism for each of the following reactions:

11.34

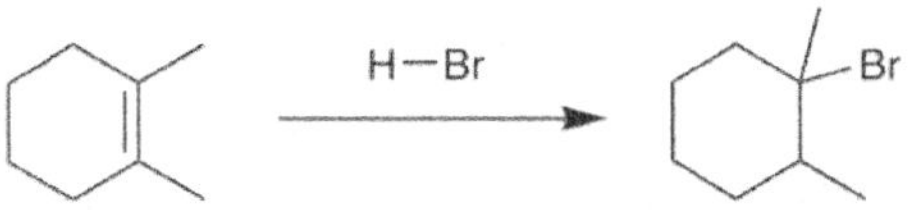

11.35

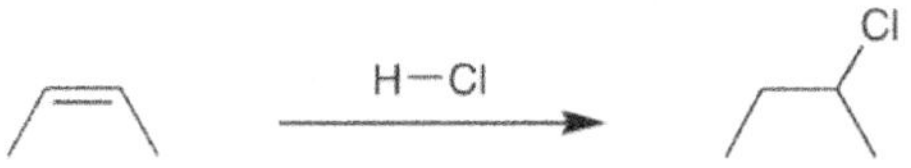

11.36

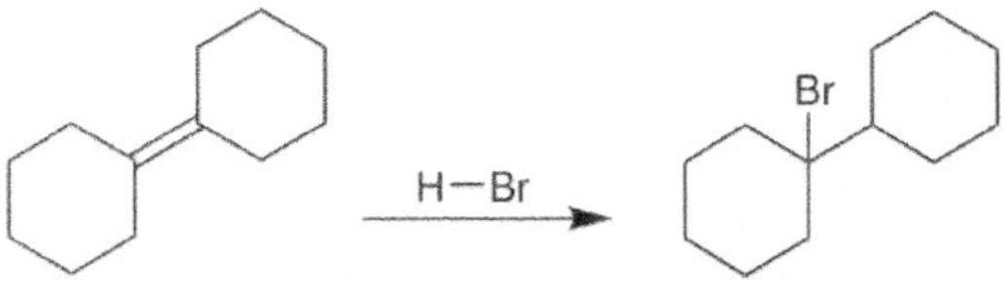

11.37

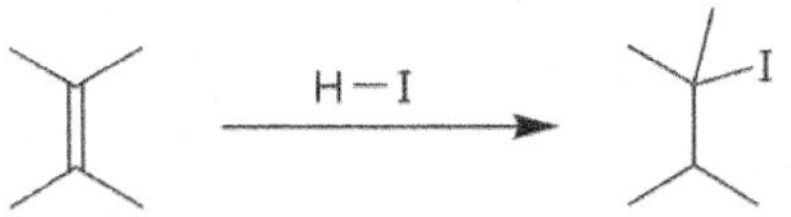

All of the examples above were symmetrical alkenes, so regiochemistry was not relevant. Now let's consider a case where regiochemistry is relevant. With an unsymmetrical alkene, we must decide where to put the H and where to put the X. For example,

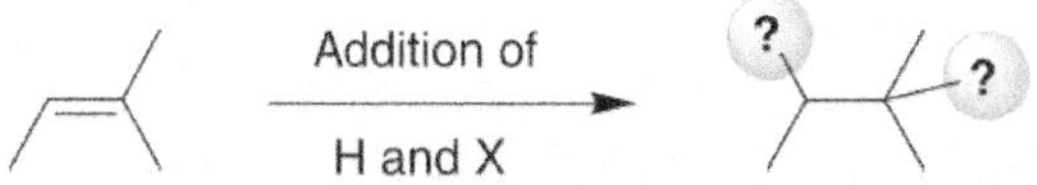

In other words, we must determine whether the reaction is a Markovnikov addition or an *anti*-Markovnikov addition. As promised, the answer to this question is contained in the mechanism. In the first step of the mechanism, a proton was transferred to the alkene, to form a carbocation. When starting with an unsymmetrical alkene, we are confronted with two possible carbocations that can form (depending on where the proton is installed):

Is the proton transferred to the more-substituted position, like this?

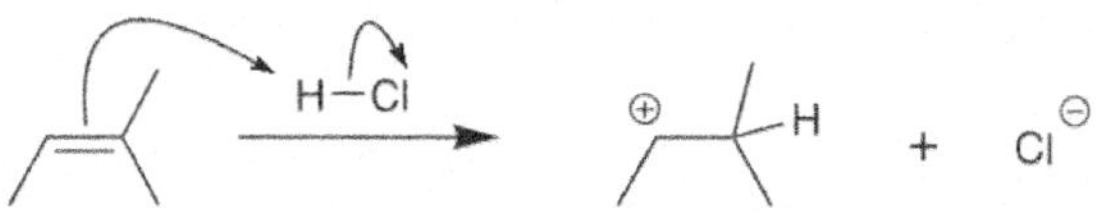

Or is the proton transferred to the less-substituted position, like this?

To answer this question, we compare the carbocations that would be formed in each scenario:

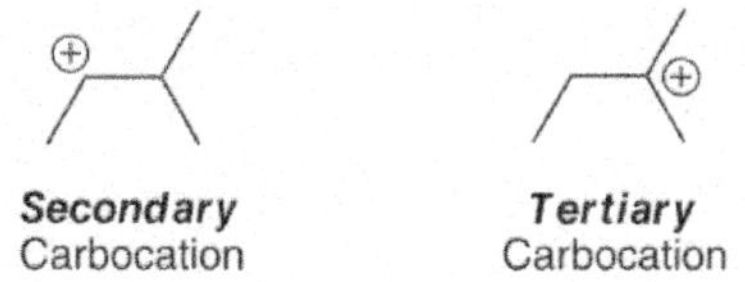

Secondary Carbocation   **Tertiary** Carbocation

Recall that tertiary carbocations are more stable than secondary carbocations. When given the choice, we expect the alkene to accept the proton in such a way as to form the more stable carbocation intermediate. In order to accomplish this, the proton must add to the less substituted carbon, generating the more substituted carbocation:

proton goes here   *carbocation ends up here*

The last step of the mechanism involves the halide attacking the carbocation. As a result, the halide will end up on the more substituted carbon (where the carbocation was). Therefore, this reaction is said to follow Markovnikov's rule:

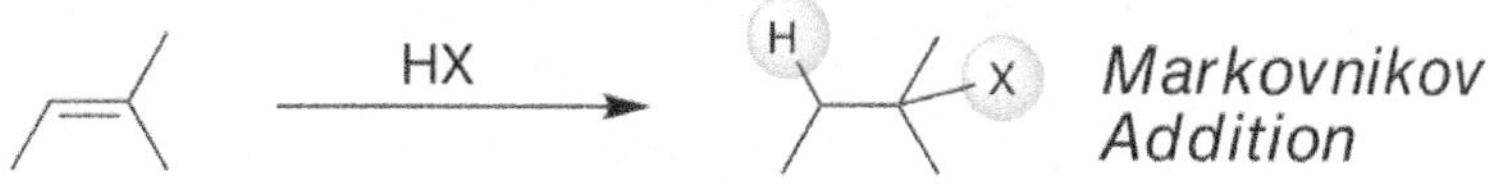

Markovnikov Addition

X ends up on more substituted carbon

As we saw in the previous section, Markovnikov's rule tells us to place the H on the less substituted carbon, and to place the X on the more substituted carbon. The rule is named after Vladimir Markovnikov, a Russian chemist, who first showed the regiochemical preference of HBr additions to alkenes. When Markovnikov recognized this pattern in the late nineteenth century, he stated the rule in terms of the placement *of the proton* (specifically, that the proton will end up on the less substituted carbon atom). Now that we understand the reason for the regiochemical preference (carbocation stability), we can state Markovnikov's rule in a way that more accurately reflects the underlying principle: *The regiochemistry will be determined by the preference for the reaction to proceed via the more stable carbocation intermediate.*

Notice that the regiochemistry of this reaction is explained by the mechanism. Don't try to memorize that the regiochemistry of this reaction is Markovnikov. Rather, try to "understand why" it must be that way.

In any reaction, the mechanism should explain not only the regiochemistry, but the stereochemistry as well. In this particular reaction (addition of H—X across alkenes), the stereochemistry is generally not relevant. Recall from the previous section that we need to consider stereochemistry (*syn* vs. *anti*) only in cases where the reaction generates *two* new chiral centers. If only one chiral center is formed, then we expect a pair of enantiomers (racemic mixture), regardless of whether the reaction was *anti* or *syn*. You will probably not see an example where two new chiral centers are formed, because the stereochemical outcome in such a case is complex and is beyond the scope of our conversation.

The details of this reaction can now be summarized with the following chart:

$\xrightarrow{\text{HX}}$	**H and X**
***Regio*chem**	Markovnikov
***Stereo*chem**	both *syn* and *anti*

EXERCISE 11.38 Predict the product of the following reaction:

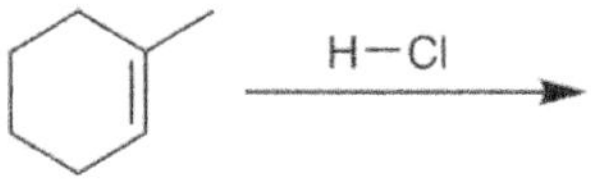

Answer We begin by focusing on the regiochemistry. This alkene is unsymmetrical, so we must decide where to place the H and where to place the Cl. To do this, we must identify which carbon is more substituted:

This reaction proceeds according to Markovnikov's rule, which tells us to place the H on the less substituted carbon, and to place the Cl on the more substituted carbon:

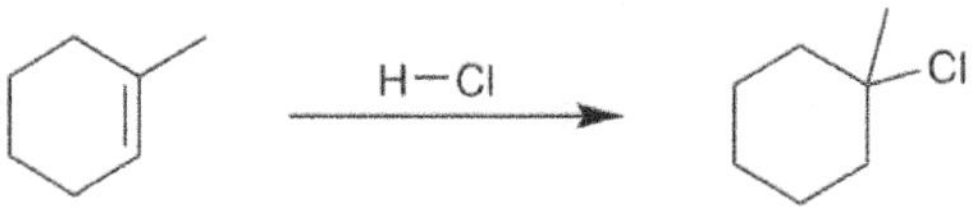

We do not need to think about stereochemistry here, because the product does not contain two chiral centers (in fact, it doesn't even have one chiral center). Therefore, stereochemistry is irrelevant in this example. As we have said, the stereochemistry of this reaction (H—X addition) will generally not be relevant in the problems that you will encounter.

PROBLEMS Predict the products for each of the following reactions. After finishing each problem, also try to draw the mechanism so that you can "see" exactly why the reaction proceeds through Markovnikov's rule.

11.39

11.40

11.41

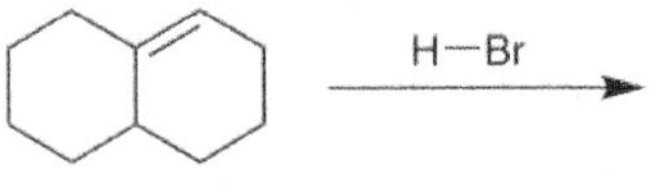

11.42

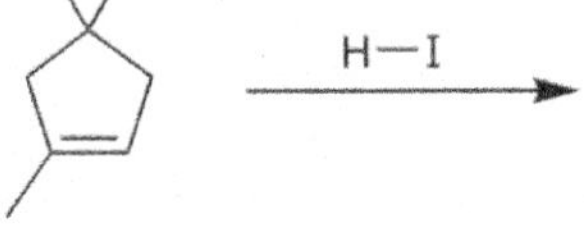

In order to understand the regiochemistry of HX additions to alkenes, we focused our attention on the intermediate carbocation. We argued that the reaction would proceed via the more stable carbocation. This all-important principle will also help explain why some reactions will involve a rearrangement. For example, consider the following reaction:

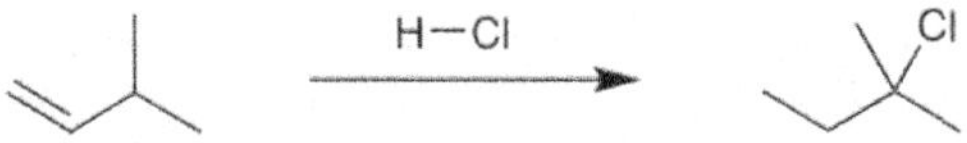

At first glance, the product is not what we might have expected. Once again, we turn to the mechanism for an explanation. The first step of the mechanism is identical to what we have seen so far—we protonate the double bond to produce the more stable carbocation (secondary, rather than primary):

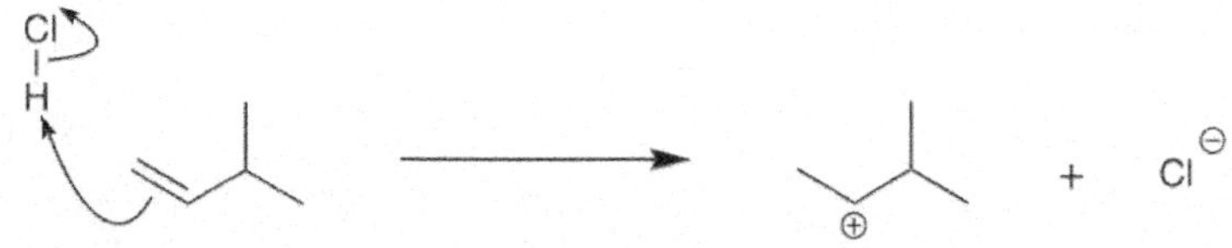

Now the carbocation is ready to be attacked by chloride. However, there is something else that can happen first (before chloride has a chance to attack): A hydride shift can produce a more stable carbocation:

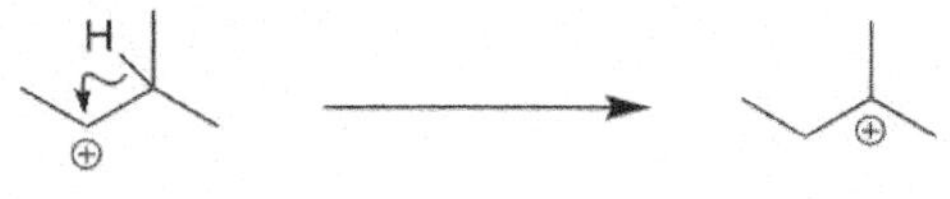

Secondary   *Tertiary*

This tertiary carbocation can now be attacked by chloride to give the product:

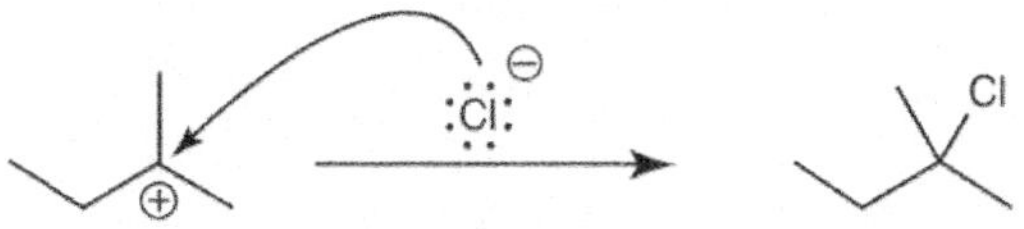

Clearly, we must be able to predict when to expect a carbocation rearrangement. There are two common ways for a carbocation to rearrange: either through a hydride shift or through a methyl shift. Your textbook will have examples of each. Carbocation rearrangements are possible for any reaction that involves an intermediate carbocation (not just for addition of HX across an alkene). Later in this chapter, we will see other addition reactions that also proceed through carbocation intermediates. In those cases, you will be expected to know that there will be a possibility for carbocation rearrangements.

Let's get some practice.

EXERCISE 11.43   Draw a mechanism for the following reaction:

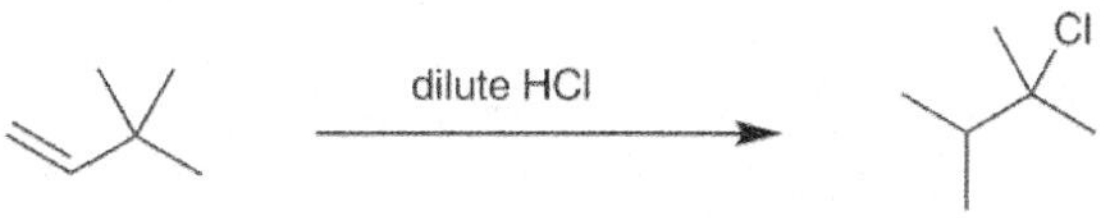

dilute HCl

Answer   In the first step, we protonate the alkene:

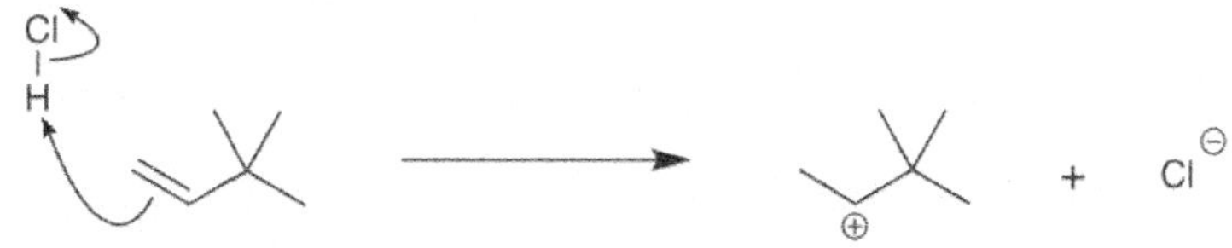

There were two ways that we could have protonated, and we chose the way that would produce the secondary carbocation (rather than producing a primary carbocation). Before we simply attack with the halide to end the reaction, we consider whether a rearrangement can take place. In this case, a methyl shift will produce a more stable, tertiary carbocation:

Secondary   *Tertiary*

Finally, the chloride ion now attacks the tertiary carbocation to give our product:

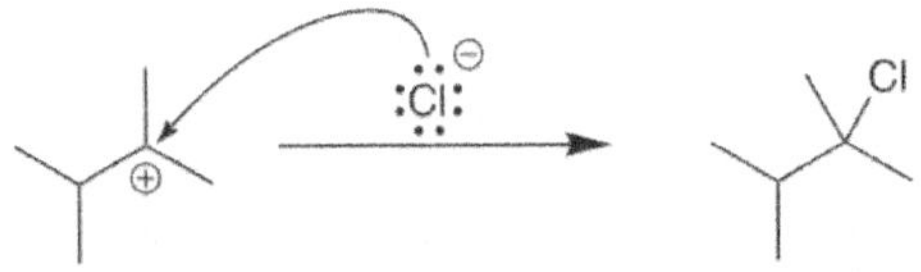

PROBLEMS Draw a mechanism for each of the following reactions:

11.44 dilute HBr

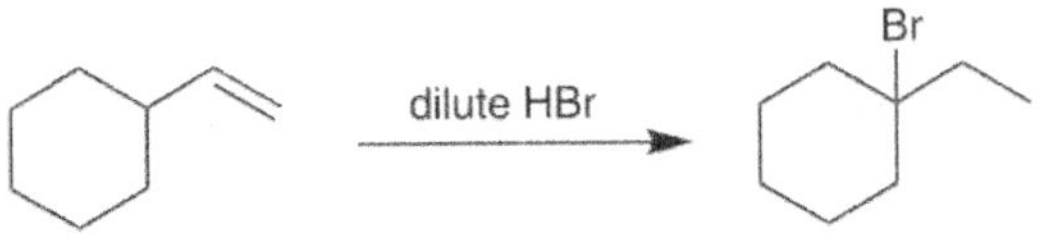

11.45 dilute HCl

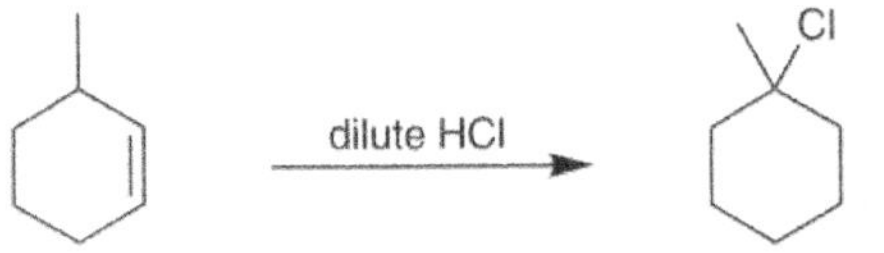

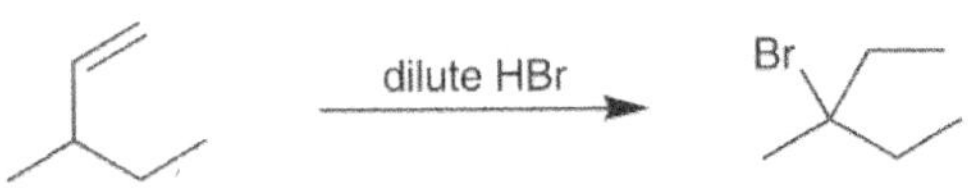

11.46 dilute HBr

11.47 dilute HCl

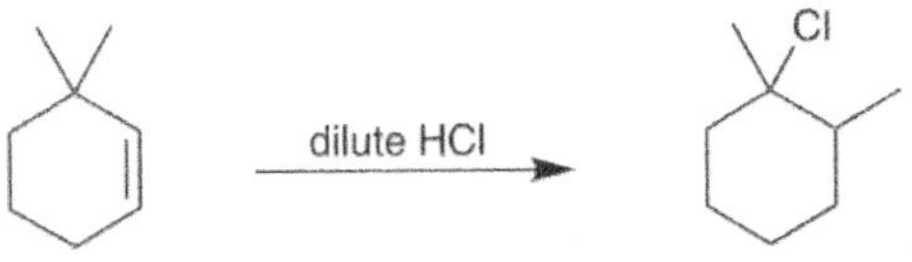

11.5 ADDING H AND Br, *ANTI*-MARKOVNIKOV

In the previous section, we saw how to add H and X, placing X at the more substituted carbon (Markovnikov addition). There is another reaction that will allow us to add H and X *anti*-Markovnikov, but it only works well with HBr (not any other H—X).

If we use HBr in the presence of peroxides (ROOR), Br ends up on the less substituted carbon:

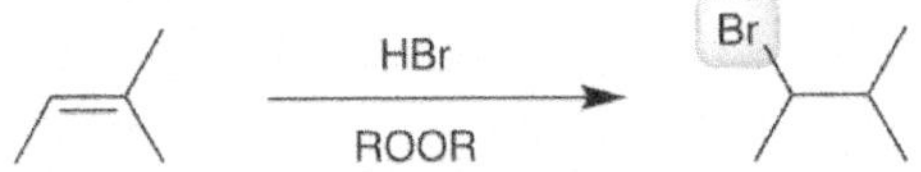

HBr / ROOR

Why does the presence of peroxides cause the addition to be *anti*-Markovnikov? In order to understand the answer to this question, we will need to explore the mechanism in detail. This reaction follows a mechanism that involves radical intermediates (such as Br•), rather than ionic intermediates (such as Br⁻). Peroxides are used to generate bromine radicals, in the following way:

The O—O bond of the peroxide is easily broken in the presence of light ($h\nu$) or heat. When this happens, the bond is broken homolytically, which means that two radicals are formed:

$$RO—OR \xrightarrow{h\nu} RO• \quad •OR$$

Each of these RO radicals can then abstract a hydrogen atom from HBr, to form the reactive intermediate (Br•):

In your mind, you can compare the step above to a proton transfer. But there is one important difference. In a proton transfer, we are transferring H⁺ (a proton is the nucleus of a hydrogen atom, *without* its corresponding electron) from one place to another, via an ionic process. But here, we are transferring an H• (an entire hydrogen atom: proton *and* electron), and therefore, we are dealing with a radical process.

Now that Br• has formed, it can attack the alkene, like this:

Notice that we have been using one-headed curved arrows exclusively:

rather than

These arrows (called fishhook arrows) are the hallmark of radical reactions. We use fishhook arrows in radical mechanisms, because they indicate the movement of only *one* electron, rather than *two* electrons (by contrast, two-headed curved arrows are used in ionic mechanisms to show the movement of two electrons).

In the step above, Br• attacked the alkene at the less substituted carbon, in order to form the more substituted carbon radical (C•). Tertiary radicals are more stable than secondary radicals, for the same reason that tertiary carbocations are more stable than secondary carbocations. Just as alkyl groups donate electron density to stabilize a neighboring, *empty p*-orbital, so too, alkyl groups can stabilize a neighboring, *partially filled* orbital. This preference for forming a tertiary radical (rather than a secondary radical) dictates that Br• will attack the *less substituted* carbon. This explains the observed *anti*-Markovnikov regiochemistry.

As a final step, the carbon radical then abstracts a hydrogen atom from HBr to give the product:

As a side product of this reaction, we regenerate another Br•, which can go and react with another alkene. We call this a chain reaction, and the reaction occurs very rapidly. In fact, when peroxides are present (to jump-start this chain process), the reaction occurs much more rapidly than the competing ionic addition of HBr that we saw in the previous section.

Compare the intermediate of this radical mechanism with the intermediate of an ionic mechanism:

Ionic Mechanism

Tertiary carbocation
intermediate

Radical Mechanism

Tertiary radical
intermediate

In both mechanisms, the regiochemical outcome is determined by a preference for forming the most stable intermediate possible. For example, in the ionic mechanism, H⁺ adds *to produce a tertiary carbocation,* rather than a secondary carbocation. Similarly, in the radical mechanism, Br• adds *to produce a tertiary radical,* rather than a secondary radical. In this respect, the two reactions are very similar. But take special notice of the fundamental difference. In the ionic mechanism, a *proton* is installed first. However, in the radical mechanism, *bromine* is installed first. This critical difference explains why an ionic mechanism gives a Markovnikov addition while a radical mechanism gives an *anti*-Markovnikov addition.

Now let's review the profile for the radical addition of HBr:

HBr ROOR	Br and H
***Regio*chem**	*anti*-Markovnikov
***Stereo*chem**	both *syn* and *anti*

Until now we have focused on the regiochemistry of this reaction. Let's now explore the stereochemistry. In situations where two chiral centers are formed, a mixture of all possible diastereomers is expected. Therefore, we will only present problems where no chiral centers are formed, or where only one chiral center is formed.

EXERCISE 11.48 Predict the product for the following reaction:

Answer HBr indicates that we will be adding H and Br across the double bond. The presence of peroxides indicates that the regiochemistry will be *anti*-Markovnikov. To determine whether stereochemistry is relevant in this particular case, we need to look at whether we are creating two new chiral centers. When we place the Br on the less substituted carbon (and the H on the more substituted carbon), we will only be creating one new chiral center. With only one chiral center, there are not four possible stereoisomers but just two possible products (a pair of enantiomers). And we expect this pair of enantiomers regardless of whether the reaction was *syn* or *anti:*

PROBLEMS Predict the products for each of the following reactions:

11.49

11.50

11.51

11.52

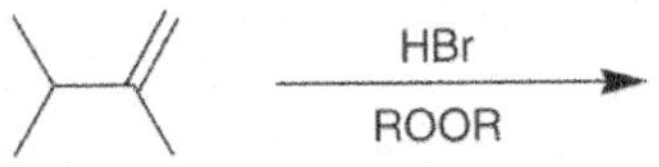

We have now seen two pathways for adding HBr across a double bond: the ionic pathway (which gives Markovnikov addition) and the radical pathway (which gives *anti*-Markovnikov addition). Both pathways are actually in competition with each other. However, the radical reaction is a much faster reaction. Therefore, we can control the regiochemistry of addition *by carefully choosing the conditions.* If we use a radical initiator, like ROOR, then the radical pathway will predominate, and we will see an *anti*-Markovnikov addition. If we do *not* use a radical initiator, then the ionic pathway will predominate, and we will see a Markovnikov addition:

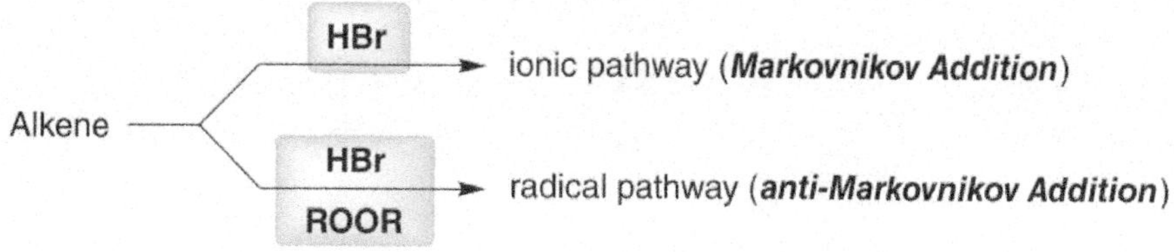

Let's get a bit of practice with choosing the appropriate conditions for addition of HBr.

EXERCISE 11.53   In the following hydrobromination reaction, determine whether or not you should use peroxides:

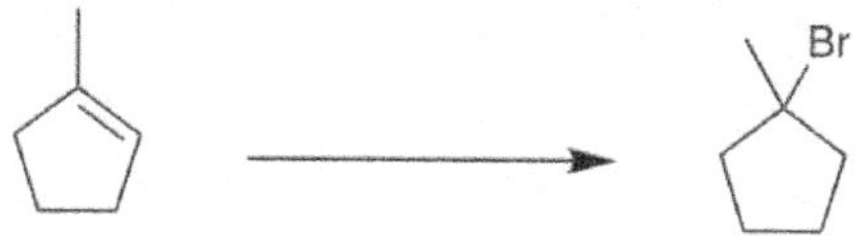

Answer   In order to determine whether or not to use peroxides, we must decide whether the desired transformation represents a Markovnikov addition or an *anti*-Markovnikov addition. When we compare the starting alkene above with the desired product, we see that we need to place the Br at the more substituted carbon (i.e., Markovnikov addition). Therefore, we need an ionic pathway to predominate, and we should *not* use peroxides. We just use HBr:

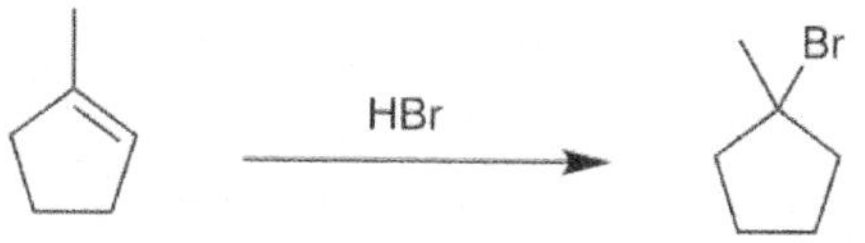

PROBLEMS   Identify what reagents you would use to carry out each of the following transformations:

11.54

11.55

11.56

11.57

11.6 ADDING H AND OH, MARKOVNIKOV

Over the next two sections, we will learn how to add H and OH across a double bond. The process of adding water across an alkene is called *hydration,* and it can be achieved through a Markovnikov addition or through an *anti*-Markovnikov addition. We just need to carefully choose our reagents. In this section, we will explore the reagents that give a Markovnikov addition of water. Then, in the next section, we will explore reagents that give an *anti*-Markovnikov addition of water.

Consider the following reaction:

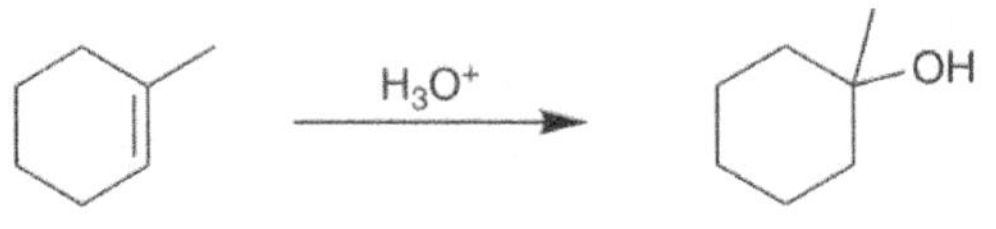

Careful comparison of the starting material and product will reveal that we have performed a hydration, with Markovnikov regiochemistry. Notice the reagent that we used (H_3O^+). Essentially, this is water (H_2O) and an acid source (such as sulfuric acid). There are many ways to show this reagent. Sometimes, it is written as H_3O^+ (as above), while at other times, it might be written like this: H_2O, H^+. You might even see it like this, with brackets around the acid:

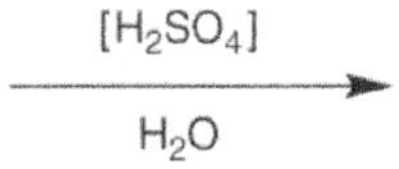

These brackets indicate that H^+ is not consumed in the reaction. In other words, H^+ is a catalyst, and therefore, we call this reaction an *acid-catalyzed hydration.* In order to understand why this reaction proceeds via a Markovnikov addition, we turn our attention to the mechanism. The accepted mechanism of an acid-catalyzed hydration is similar to the mechanism that we presented for addition of HX (the ionic pathway). Compare these two mechanisms:

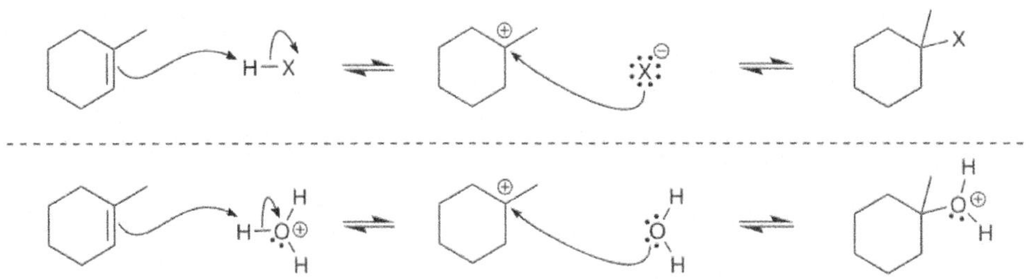

In each mechanism above, the first step involves protonation of the alkene to form a carbocation. Then, in both cases, a nucleophile (either X^- or H_2O) attacks the carbocation. The difference between these two reactions is in the nature of the product. The first reaction above (hydrohalogenation) gives a product that is neutral (no charge). However, the second reaction above (hydration) produced a charged species. Therefore, one more step must occur at the end of the process in order to remove the positive charge. Since the reaction occurs in the presence of water, it is likely that water serves as the base for this step:

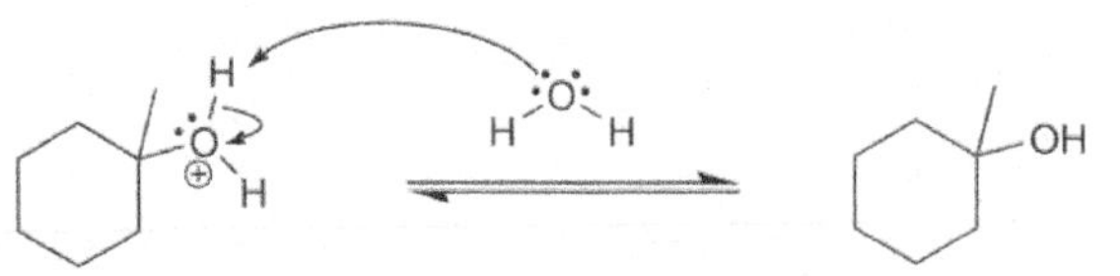

When drawing this final deprotonation step, notice that water is shown as the base, not hydroxide (HO^-). To understand why, remember that we are in acidic conditions; there really aren't many hydroxide ions floating around. But there is plenty of water, and a mechanism must always be consistent with the conditions that are present.

Now that we have seen all of the individual steps, let's look at the entire mechanism:

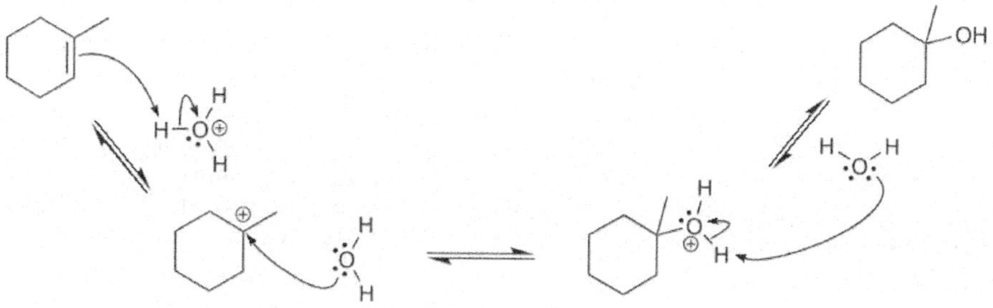

Notice that we are using equilibrium arrows here (⇌ rather than ⟶). These equilibrium arrows indicate that the reaction actually goes in both directions. In fact, the reverse path (starting from the alcohol and ending with the alkene) is a reaction that we have already studied. It is just an E1 reaction (follow the sequence above from the end to the beginning, and convince yourself that it is the E1 process). The truth is that most reactions represent equilibrium processes; however, organic chemists (generally) only draw equilibrium arrows in situations where the equilibrium can be easily manipulated (allowing us to control which products are favored). This reaction is one of those situations. By carefully controlling the amount of water present (using either concentrated acid or dilute acid), we can favor one side of the equilibrium or the other:

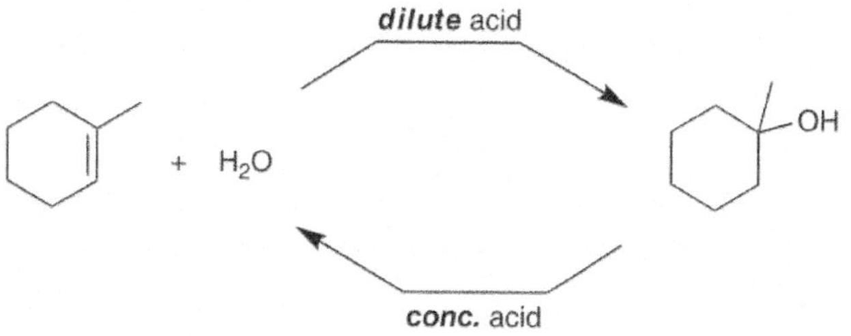

We are exploiting Le Chatelier's principle, which tells us that the equilibrium can be pushed toward one side or the other by removing or adding reagents. Imagine that you have the system shown above (alkene + water on the left side; alcohol on the right side), and this system has reached equilibrium. Then you add water. The concentrations would no longer be at equilibrium, and the system would have to adjust to re-establish new equilibrium concentrations. The end result: adding water would cause more alkene to turn into alcohol. Therefore, we would use dilute acid (which is mostly water) to favor the alcohol. If we wanted to favor the alkene, then we would want to remove water, which would push the equilibrium toward the left. Therefore, if we want to form the alkene, we would use concentrated acid (which is mostly acid and very little water).

Once again, we see that carefully choosing reaction conditions can greatly affect the outcome of a reaction.

We have already explained the regiochemistry of acid-catalyzed hydration; there is a strong preference for Markovnikov addition. But what about the stereochemistry?

The stereochemistry of acid-catalyzed hydration is very similar to the stereochemistry of ionic addition of HX (this should make sense, as we have already seen that the mechanisms for each of these reactions are similar). If only one chiral center is formed, then we expect a pair of enantiomers (racemic mixture), regardless of whether the reaction was *anti* or *syn*. You will probably not see an example where two new chiral centers are formed, because the stereochemical outcome in such a case is complex and is beyond the scope of our conversation.

Now we can summarize the profile for acid-catalyzed hydration:

$\xrightarrow{\text{H}_3\text{O}^+}$	**H** and **OH**
*Regio*chem	Markovnikov
*Stereo*chem	both *syn* and *anti*

EXERCISE 11.58 Predict the product for the following reaction, and then propose a mechanism for formation of that product:

Answer This reagent (H_3O^+) suggests that we have an acid-catalyzed hydration. Therefore, we are adding H and OH, and the regiochemistry will follow a Markovnikov addition. The stereochemistry of an acid-catalyzed hydration is only complex when two new chiral centers are formed. In this case, we are not forming two new chiral centers. In fact, we are not even forming one new chiral center. Without any chiral centers, we expect only one product:

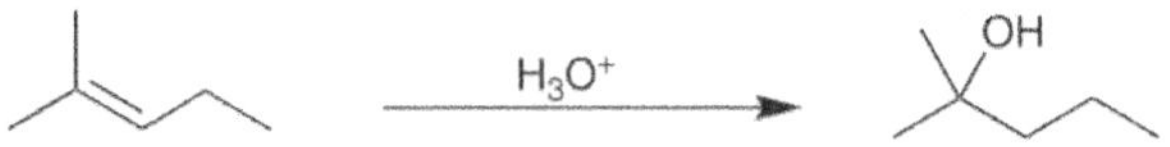

The mechanism of the reaction will have three steps: (1) protonate the alkene to form a carbocation, (2) water attacks the carbocation, and (3) deprotonate to form the product:

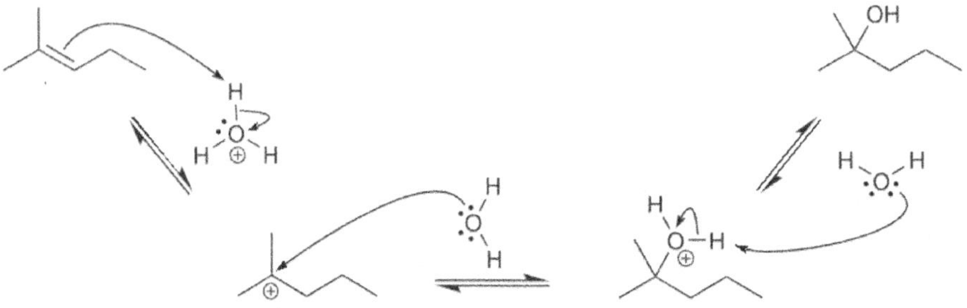

PROBLEMS For each of the following reactions, predict the expected product, and propose a plausible mechanism for formation of the product:

11.59

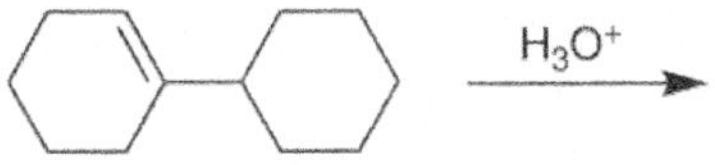

11.60

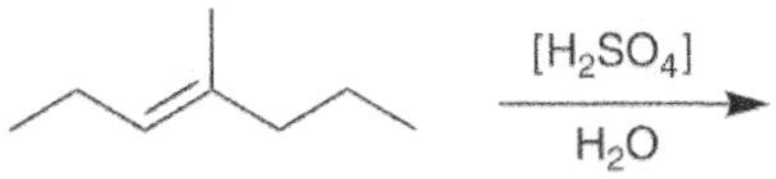

11.61

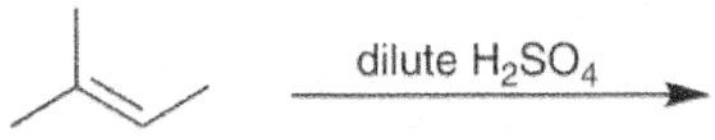

11.62

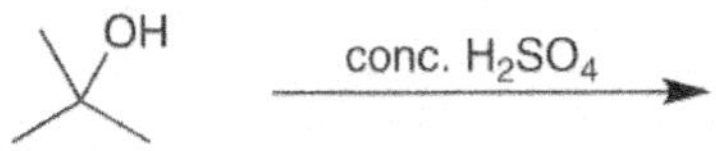

11.7 ADDING H AND OH, *ANTI*-MARKOVNIKOV

In the previous section, we saw how to perform a Markovnikov addition of H and OH across a double bond. In this section, we will learn how to perform an *anti-Markovnikov* addition of H and OH, for example:

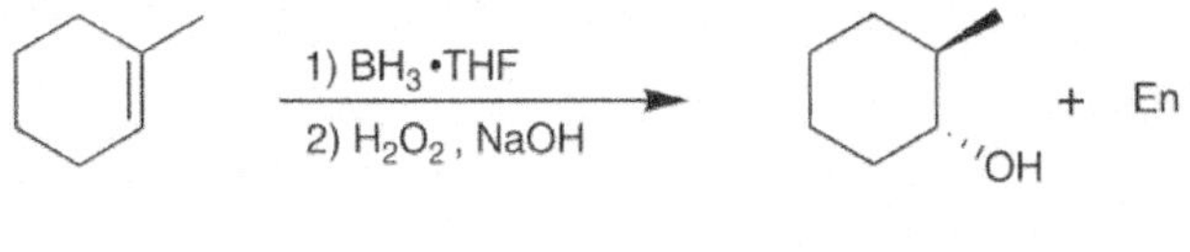

A quick glance at the products indicates that we are adding H and OH across the alkene. Let's take a closer look and carefully analyze the regiochemistry *and* stereochemistry of this reaction. The OH group is ultimately positioned at the less substituted carbon, and therefore, the regiochemistry represents an *anti*-Markovnikov addition. But what about the stereochemistry? Are we seeing a *syn* addition here, or is this *anti* addition?

Be careful. The example above represents somewhat of an optical illusion. The products seem to suggest an *anti* addition (the methyl and the OH are *trans* to each other). But think about what we added in this reaction: we did not add OH and a *methyl* group. The methyl group was already there. Rather, we added OH and H. The H that we added is not shown in the product above (because it does not have to be drawn in a bond-line drawing). If you draw that H, you will see that it is on a dash—therefore, this was a *syn* addition of H and OH.

To recap, the reaction above is an addition of H and OH, with *anti*-Markovnikov regiochemistry, and *syn* stereochemistry. Now we have three important questions to answer:

1. How do these reagents (BH_3, etc.) cause an addition of H and OH?

2. Why *anti*-Markovnikov?

3. Why *syn* addition?

The answers to all three of these questions are encapsulated in the mechanism as usual. In order to explore the accepted mechanism, we must first acquaint ourselves with the reagents. In the first step, the reagents are BH_3 and THF. The former is called borane. The element boron uses its three valence electrons to comfortably form three bonds:

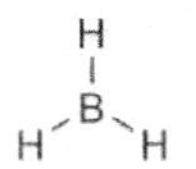

However, in this structure, boron does not have an octet. It has an empty *p* orbital, (very similar to a carbocation, except there is no positive charge here). Therefore, borane is very reactive. In fact, it reacts with itself to give dimeric structures, called diborane:

$$2\ BH_3 \rightleftharpoons B_2H_6$$

The empty *p* orbital in borane can be somewhat stabilized if we use a solvent (like THF) that can donate electron density into the empty *p* orbital of boron:

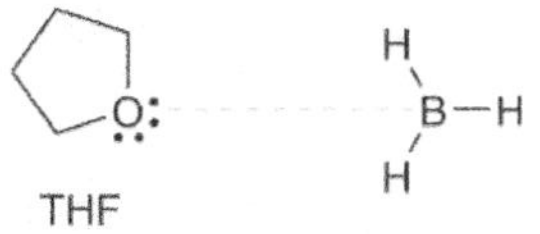

THF

This solvent is called tetrahydrofuran, or THF for short. Even though it somewhat stabilizes the empty *p* orbital on the boron atom in BH_3, nevertheless the boron atom is very eager to look for any other sources of electron density that it can find. It is an electrophile—it is scavenging for sites of high electron density to fill its empty orbital. A pi bond is a site of high electron density, and therefore, a pi bond can attack borane. In fact, this is the first step of our mechanism. A pi bond attacks the empty *p* orbital of boron, which triggers a *simultaneous* hydride shift:

Notice that it all happens in one concerted process (via a four-membered transition state). Let's take a close look at this first step, and consider the regiochemistry and stereochemistry.

For the regiochemistry, we notice that boron ends up on the less substituted carbon (and that is where the OH group will ultimately end up). Now we can understand one of the sources of this regiochemical preference. H and BH_2 are being added across the double bond. BH_2 is bigger and bulkier than H, so it will have an easier time fitting over the less substituted carbon (the less sterically hindered position). This results in an *anti*-Markovnikov addition.

The stereochemical preference (for *syn* addition) can now also be understood. The step above represents a concerted process. Both BH_2 and H are adding simultaneously, so they must end up on the same face of the alkene. In other words, the reaction must be a *syn* addition.

The structure shown above still has two remaining B—H bonds (look at the BH$_2$ group), and so the reaction can occur again with those B—H bonds. In other words, one molecule of BH$_3$ can react with three molecules of alkene to give a trialkylborane:

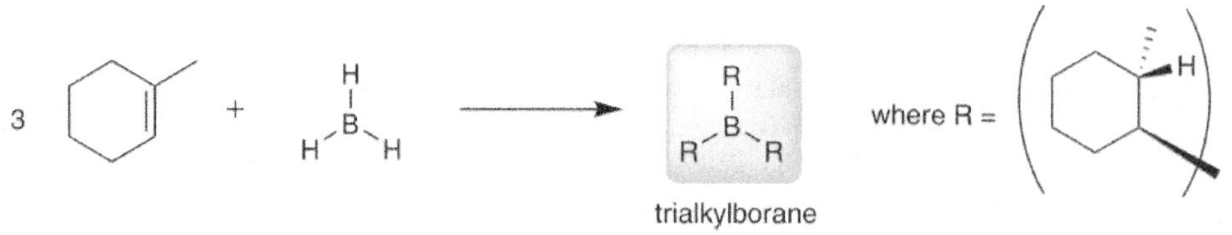

trialkylborane

What we have seen until now (formation of the trialkylborane above) is called hydroboration, which occurs when you mix an alkene with BH$_3$ in THF. Now, we move on to the next set of reagents, which accomplish an oxidation reaction: H$_2$O$_2$ and hydroxide. These reagents give us an oxidation reaction:

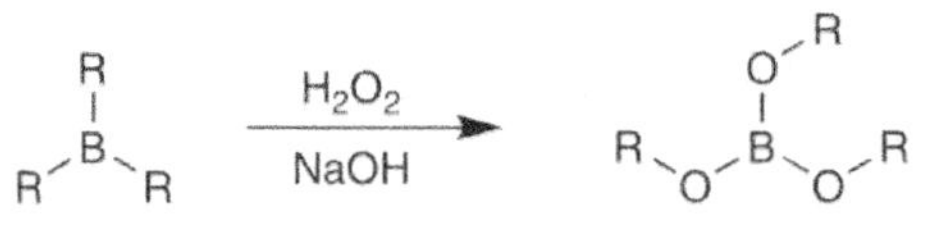

How does oxygen insert in between the B—R bonds? Let's take a closer look at the reagents—a hydroxide ion can deprotonate hydrogen peroxide to form a hydroperoxide anion:

This hydroperoxide anion can attack the trialkylborane (remember that the boron atom still has an empty p orbital, and therefore, it is still scavenging for electron density):

At this point, something fairly unique happens (we have not seen anything like this until now). One of the alkyl groups migrates over (an alkyl shift) to expel hydroxide:

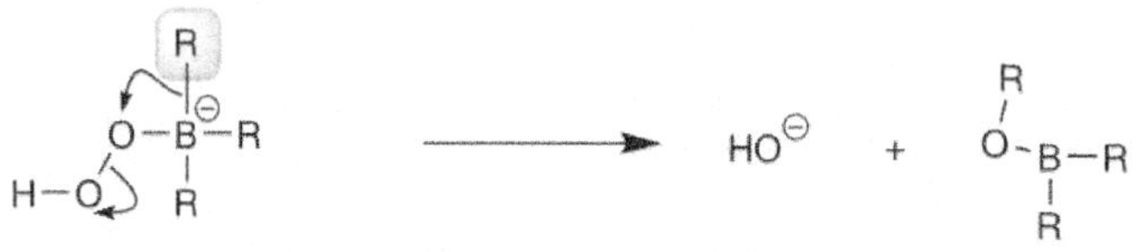

Notice the overall result. When R migrates over, the net result is to place oxygen in between B and R. Focus your attention on the chiral center of the alkyl group (R). As R migrates, the configuration of the chiral center is unaffected by the migration. In other words, the configuration of the chiral center is preserved. This happens to all three B—R bonds:

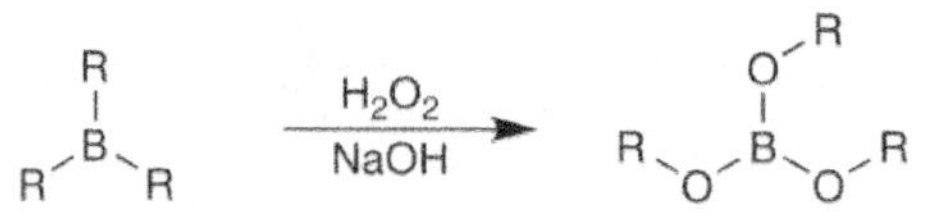

The final step involves removing the OR groups from B, which happens like this:

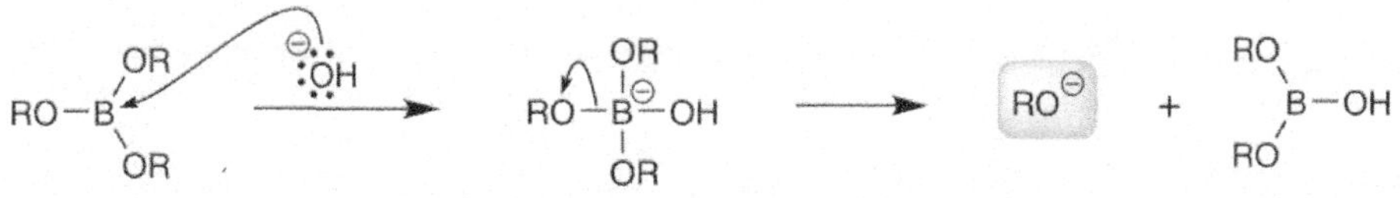

RO⁻ then removes a proton from water, and the final product is an alcohol. Overall, we have a two-step process for converting an alkene into an alcohol. This two-step process is called *hydroboration–oxidation*. Let's now summarize the profile of this process:

1) BH₃•THF	**H and OH**	
2) H₂O₂, NaOH		
*Regio*chem	*anti*-Markovnikov	
*Stereo*chem	*syn*	

EXERCISE 11.63   Predict the products for each of the following reactions:

(a)

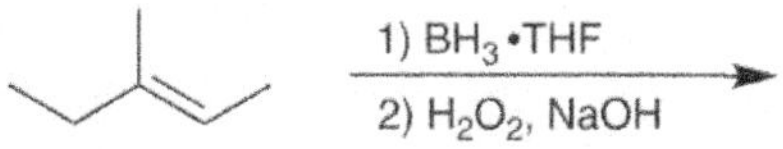

1) BH₃•THF
2) H₂O₂, NaOH

(b)

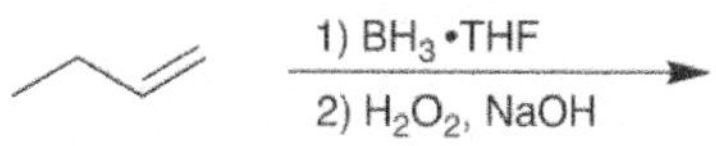

1) BH₃•THF
2) H₂O₂, NaOH

Answer

(a) These reagents will accomplish an *anti*-Markovnikov addition of OH and H. The stereochemical outcome will be a *syn* addition. But we must first decide whether stereochemistry will even be a relevant factor in how we draw our products. To do that, remember that we must ask if we are creating *two* new chiral centers in this reaction. In this example, we *are* creating two new chiral centers. So, stereochemistry *is* relevant. With two chiral centers, there theoretically could be four possible products, but we expect only two of them; we expect only the pair of enantiomers that result from a *syn* addition. In order to get it right, let's redraw the alkene (as we have done many times earlier), and add OH and H like this:

(b) These reagents will accomplish an *anti*-Markovnikov addition of OH and H. The stereochemical outcome will be a *syn* addition. But we must first decide whether stereochemistry will even be a relevant factor in how we draw our products. To do this, we ask if *two* new chiral centers will be forming. In this example, we are not creating two new chiral centers. In fact, we are not even creating one chiral center. So, stereochemistry is not relevant for us in this problem:

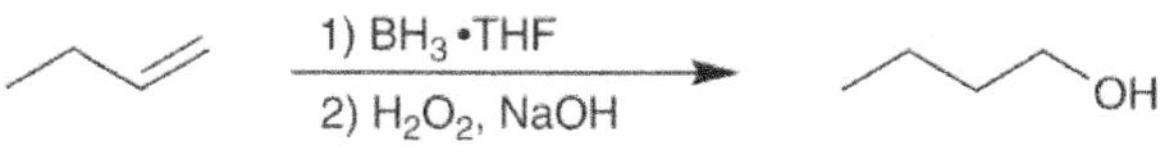

1) BH₃•THF
2) H₂O₂, NaOH

Whenever stereochemistry is irrelevant (i.e., whenever we are not creating two new chiral centers) the problem becomes a bit easier to solve.

PROBLEMS Predict the products of the following reactions:

11.64

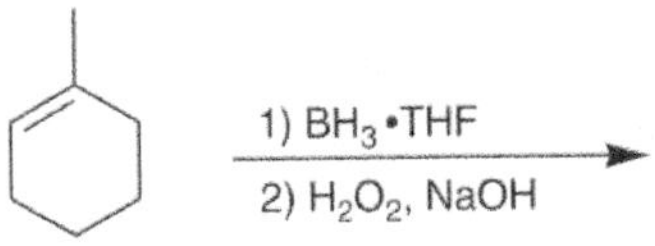

1) BH$_3$•THF
2) H$_2$O$_2$, NaOH

11.65

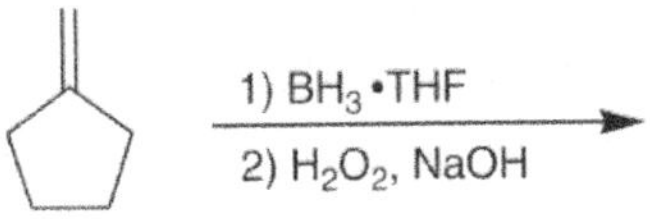

1) BH$_3$•THF
2) H$_2$O$_2$, NaOH

11.66

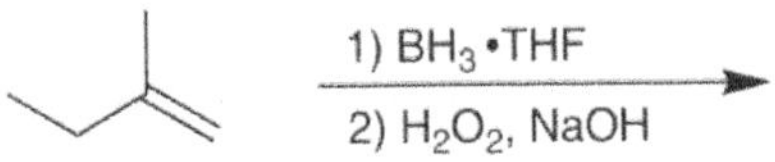

1) BH$_3$•THF
2) H$_2$O$_2$, NaOH

11.67

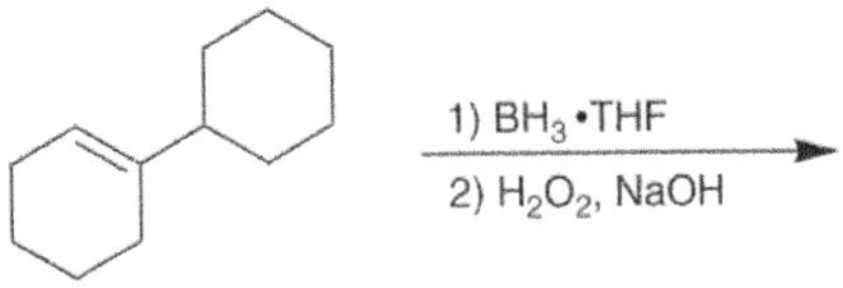

1) BH$_3$•THF
2) H$_2$O$_2$, NaOH

11.68

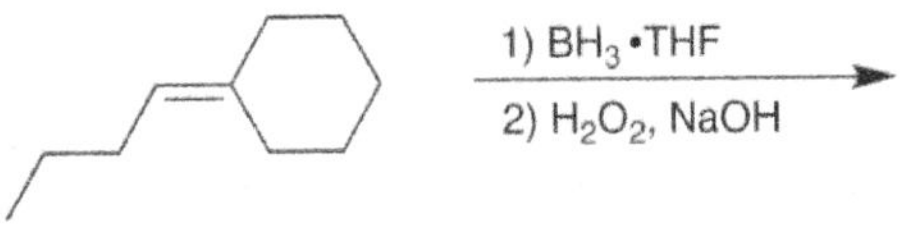

1) BH$_3$•THF
2) H$_2$O$_2$, NaOH

11.69

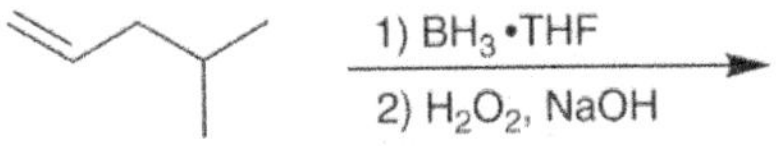

1) BH$_3$•THF
2) H$_2$O$_2$, NaOH

11.8 SYNTHESIS TECHNIQUES

11.8A One-Step Syntheses

In order to begin practicing synthesis problems, it is absolutely essential that you master all of the individual reactions that we have seen so far. You must learn how to walk before you can start to run. Therefore, we will first focus on one-step synthesis problems. Once you feel comfortable with the individual reactions, then we can start stringing them together in various sequences to form synthesis problems.

Until now, we have seen substitution reactions (S$_N$1 and S$_N$2), elimination reactions (E1 and E2), and five addition reactions. Let's quickly review what these reactions can accomplish. *Substitution* allows us to interconvert groups:

Elimination allows us to form alkenes:

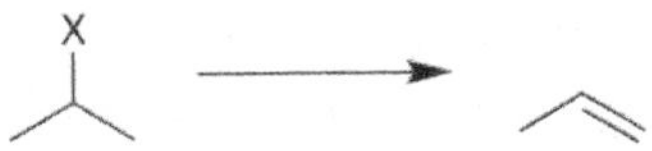

Addition reactions allow us to add two groups across a double bond. So far, we have seen the following five addition reactions:

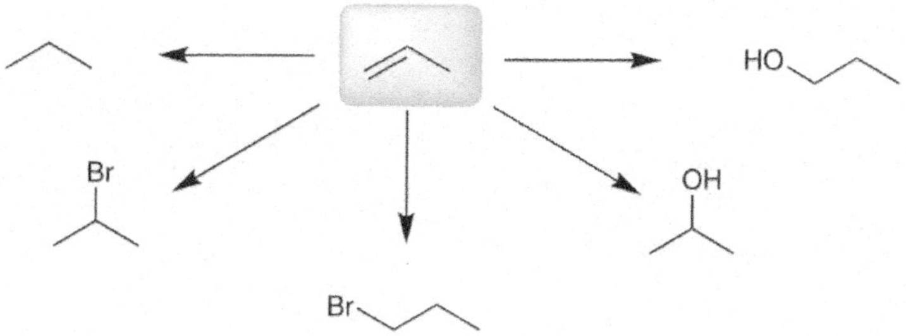

Can you fill in the reagents necessary to accomplish each of these five transformations? Try it ....

EXERCISE 11.70 What reagents would you use to accomplish the following transformation?

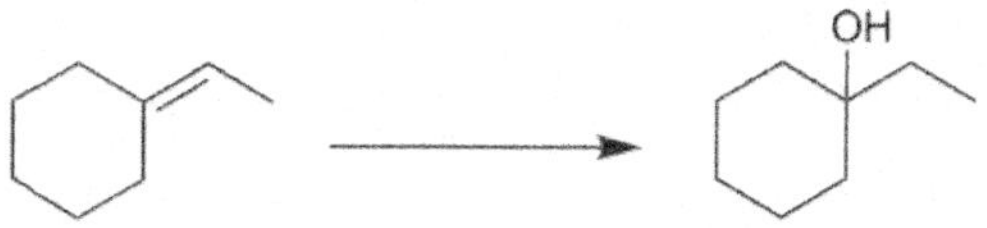

Answer If we compare the starting material and product, we see that we must add H and OH. We look at the regiochemistry, and we see that OH is ending up at the more substituted carbon—so we need a Markovnikov addition. Then, we look at the stereochemistry and we see that we are *not* creating two chiral centers in this reaction (in fact, we are not even creating one chiral center). Therefore, the stereochemistry of the reaction will be irrelevant. So we need to choose reagents that will give a Markovnikov addition of H and OH. We can accomplish this with an acid-catalyzed hydration:

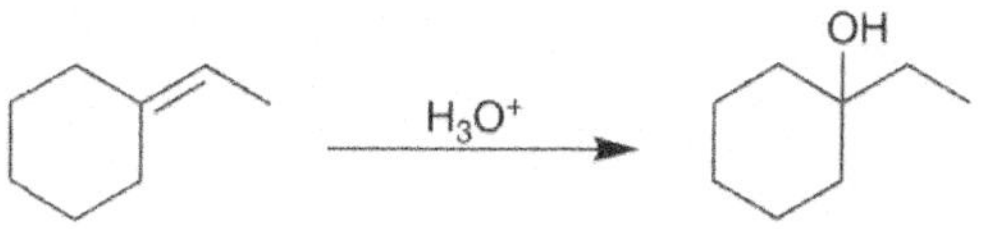

PROBLEMS What reagents would you use to accomplish each of the following transformations?

11.71

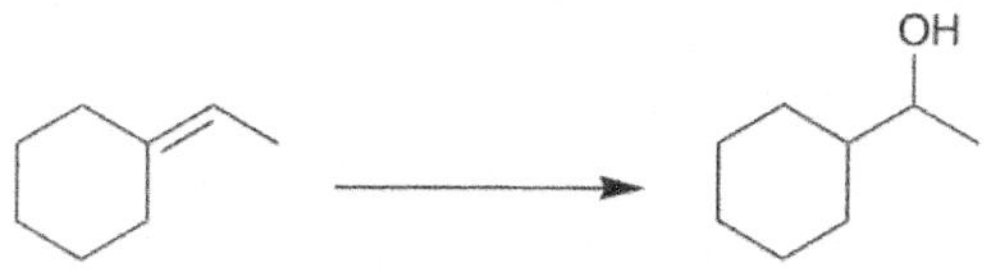

11.72

11.73

11.74

11.75

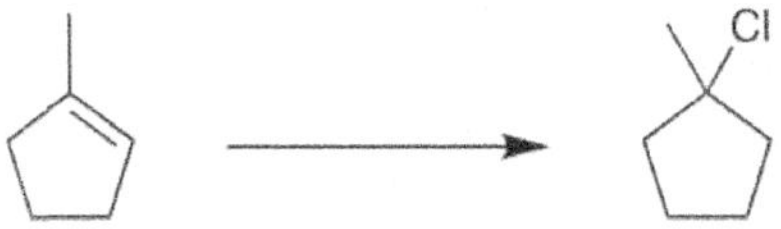

11.76 + En

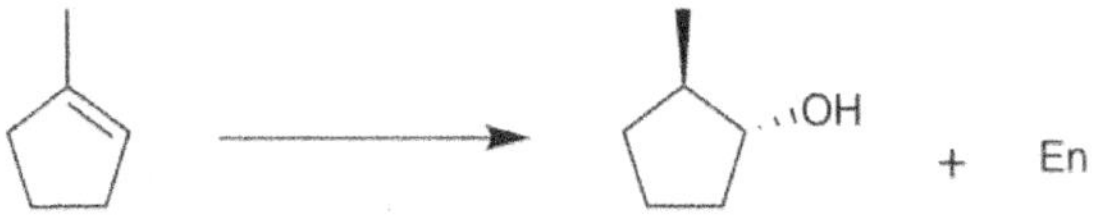

11.77

11.78

11.8B   Changing the Position of a Bromine Atom

Now let's get some practice *combining* our reactions and proposing syntheses.
    Consider the following transformation:

The net result is the change in position of the Br atom. It has "moved" over. How can we accomplish this type of transformation? We don't have any one-step method for doing this. But we can achieve the desired outcome if we do it in two steps: We eliminate, and then we add:

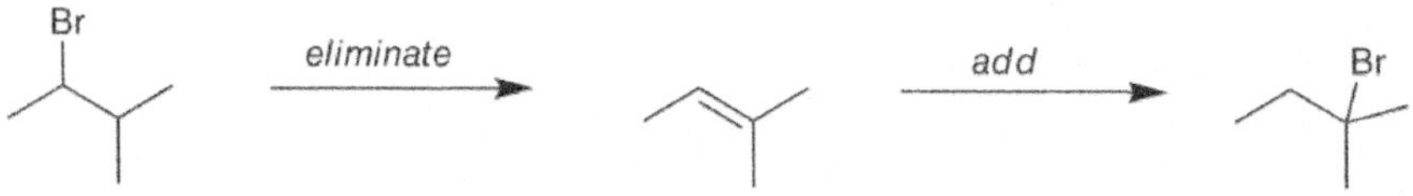

eliminate   *add*

When doing this type of sequence, there are a few important things to keep in mind. In the first step (elimination), we have a choice regarding which way to eliminate: do we form the more substituted alkene (Zaitsev product) or do we form the less substituted alkene (Hofmann product)?

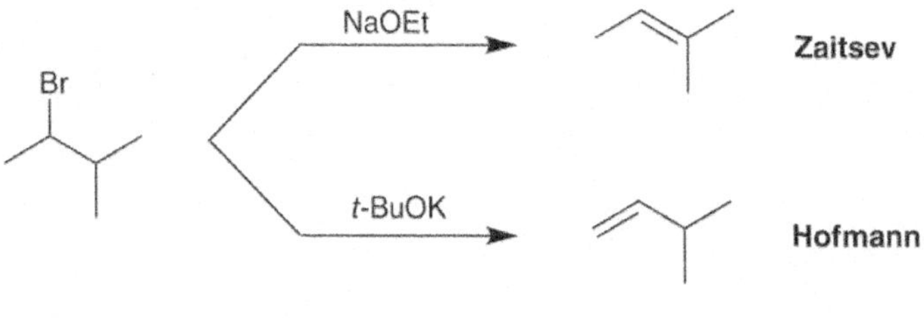

NaOEt   **Zaitsev**

t-BuOK   **Hofmann**

We can control which product is favored by carefully choosing our base. If we use a strong base (like methoxide or ethoxide), then the more substituted alkene will be favored. However, if we use a strong, *sterically hindered* base, such as *tert*-butoxide, then the less substituted alkene will be favored.

After forming the double bond, we must also carefully consider the regiochemistry of how we add HBr across the double bond. Once again, by carefully choosing our reagents, we can control the regiochemical outcome. We can either use HBr to achieve a Markovnikov addition, or we can use HBr with peroxides to achieve an *anti*-Markovnikov addition. Let's see an example:

EXERCISE 11.79   What reagents would you use to accomplish the following transformation?

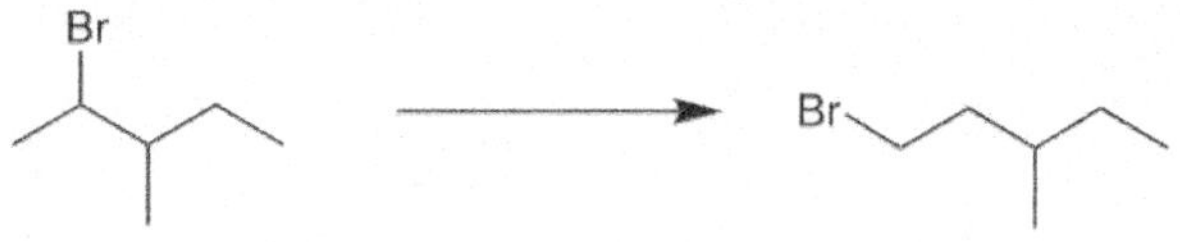

Answer   This problem requires that we move the Br over to the left. We can accomplish this by eliminating first (to form a double bond), and then adding across that double bond:

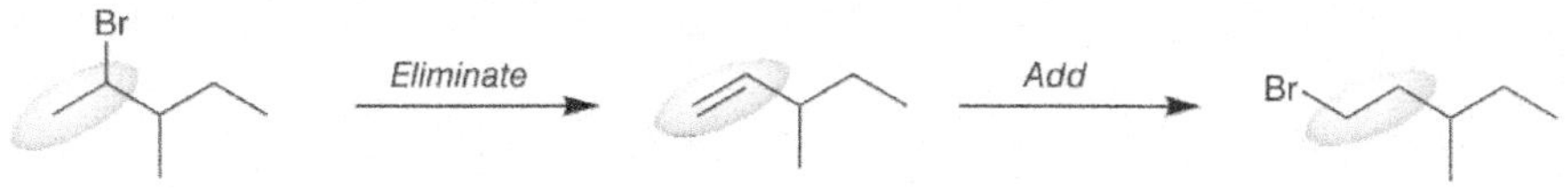

However, we must be careful to control the regiochemistry properly in each of these two steps. In the elimination step, we need to form the less substituted double bond (i.e., the Hofmann product), and therefore, we must use a sterically hindered base. Then, in the addition step, we need to place the Br on the less substituted carbon (*anti*-Markovnikov addition), so we must use HBr *with peroxides*. This gives us the following overall synthesis:

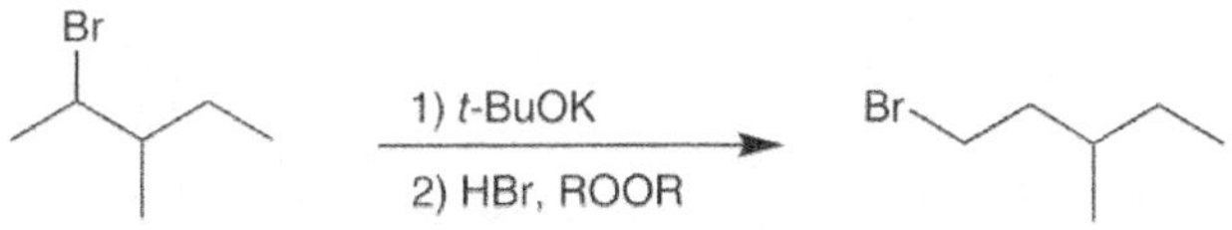

1) *t*-BuOK
2) HBr, ROOR

There is one more thing to keep in mind when using this type of technique: *hydroxide is a terrible leaving group*. Let's see what to do when dealing with an OH group. As an example, suppose we wanted to do the following transformation:

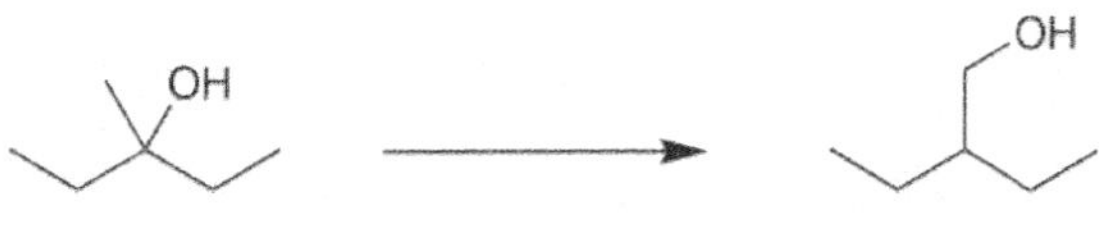

Our technique would suggest the following steps:

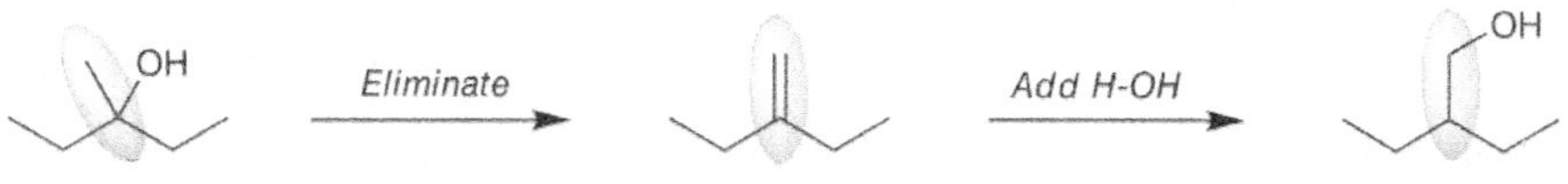

Eliminate   Add H-OH

The first step of our technique requires an elimination reaction to give the Hofmann product, so we must use an E2 reaction, employing a sterically hindered base. In order to do this, we must first convert the OH group into a good leaving group. This can be accomplished by converting the OH group into a tosylate, which is a much better leaving group than OH:

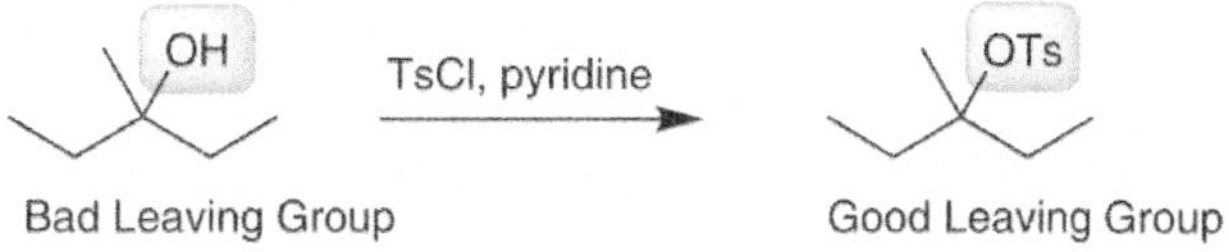

TsCl, pyridine

Bad Leaving Group   Good Leaving Group

If you have not yet learned about tosylates in your lecture course, you might want to consult your textbook for more information on tosylates. Note that you cannot use an acid to protonate the OH group (thereby converting it into a better leaving group) and then treat the protonated alcohol with a strong base to give the desired alkene, because a strong acid cannot be used at the same time as a strong base (they would neutralize each other).

After we have converted the OH into a tosylate, then we can do our technique (using a strong, sterically hindered base to eliminate, followed by *anti*-Markovnikov addition of H and OH):

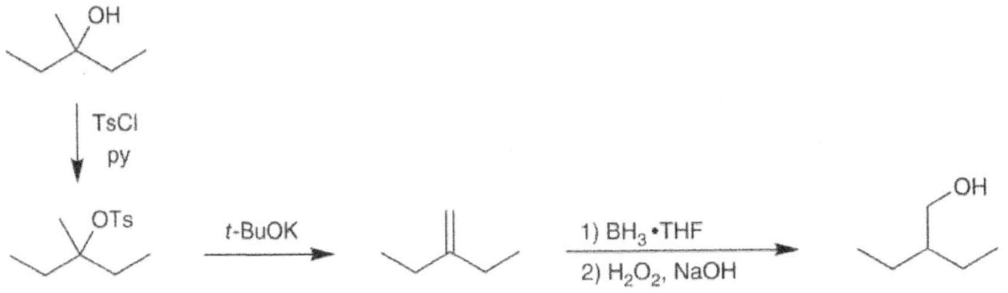

PROBLEMS What reagents would you use to accomplish each of the following transformations?

11.80

11.81

11.82

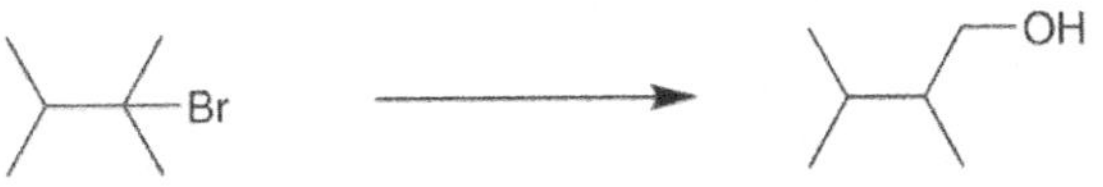

11.83

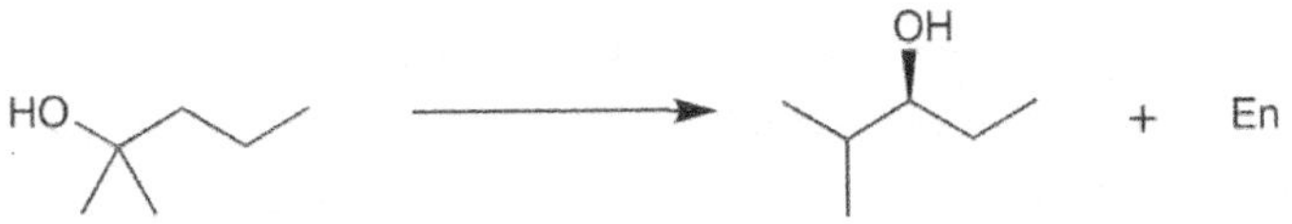

11.84

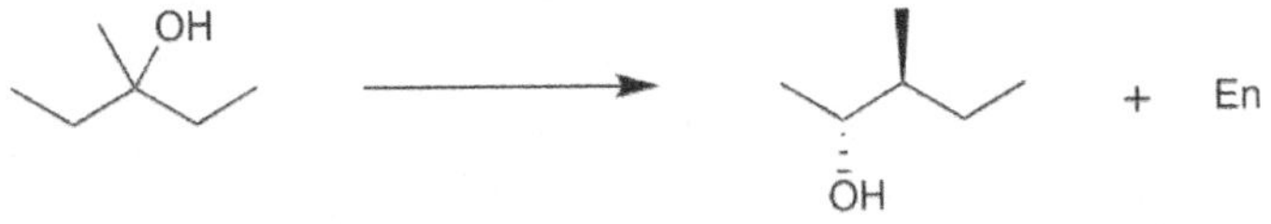

11.85

11.8C Changing the Position of a π Bond

Until now, we have seen how to combine two reactions into one synthetic technique: eliminate and then add. Now, let's focus on another type of technique: add and then eliminate. Let's see an example:

EXERCISE 11.86 What reagents would you use to accomplish the following transformation?

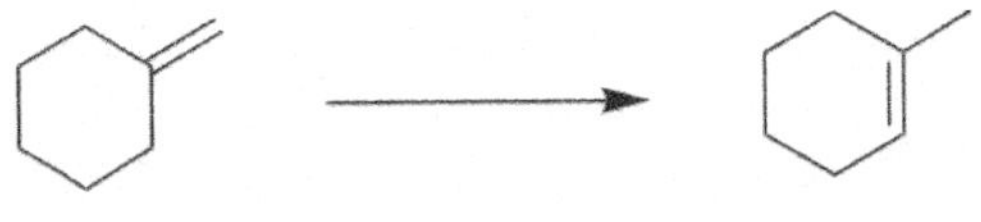

Answer In this example, we are tasked with moving the position of a double bond. We have not seen a way to do the transformation above in one step. However, we can easily accomplish it with two steps—add and then eliminate:

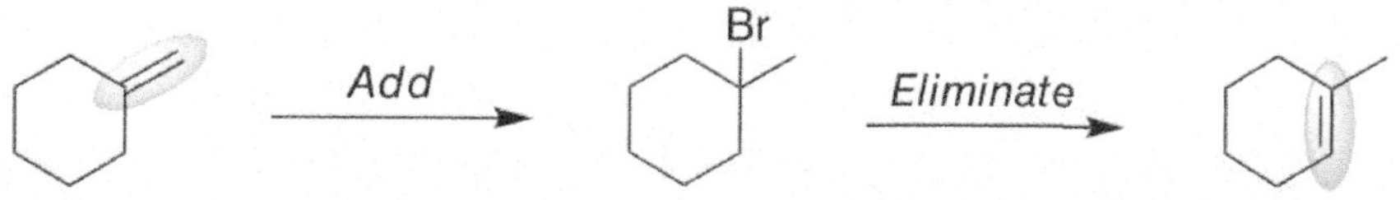

In the first step above, we need a Markovnikov addition (Br needs to end up at the more substituted carbon), which we can easily accomplish by using HBr. The second step requires an elimination to give the Zaitsev product, which we can accomplish if we choose our base carefully (we will need to use a base that is NOT sterically hindered). Therefore, the overall synthesis is:

Take special notice of what we can accomplish when we use this technique: it gives us the power *to move the position of a double bond*. When using this technique, we must carefully consider the regiochemistry of each step. In the first step (addition), we must decide whether we want a Markovnikov addition (HBr) or an *anti*-Markovnikov addition (HBr with peroxides). Also, in the second step (elimination), we must decide whether we want to form the Zaitsev product or the Hofmann product (which we can control by carefully choosing our base, ethoxide or *tert*-butoxide). Get some practice with this technique in the following problems.

PROBLEMS What reagents would you use to accomplish each of the following transformations?

11.87

11.88

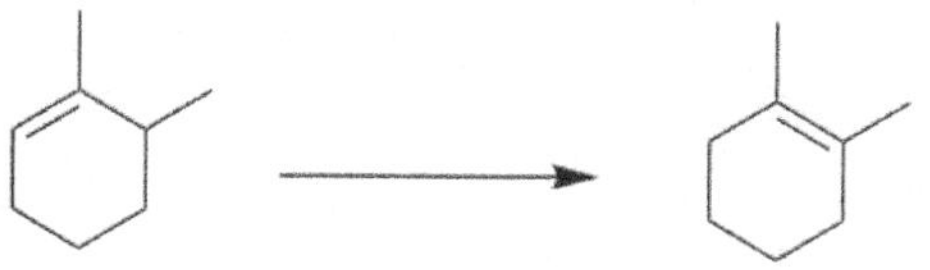

11.89

11.90

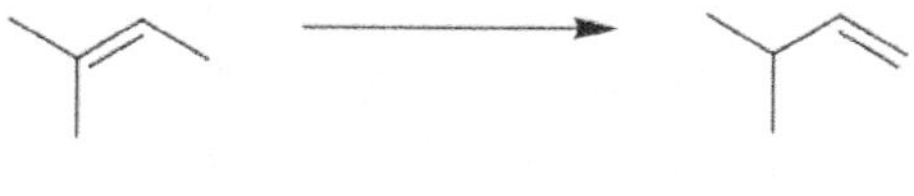

11.8D Introducing Functionality

In the techniques we have seen so far, the starting compound always had some functional group that we could manipulate. Either the starting material had a leaving group or it had a double bond. But what about situations in which the starting material has no functional groups—no leaving group, and no double bond? What can be done? In situations like this, there is only one possible course of action to take: *radical bromination*. You should consult your textbook and/or lecture notes for a discussion of radical bromination of alkanes. Once you have read that section, the following example will illustrate how radical bromination can be used in a synthesis.

EXERCISE 11.91 What reagents would you use to accomplish the following transformation?

Answer   In this problem, we are starting with an alk*ane*. There are no leaving groups, so we cannot do a substitution or an elimination reaction. There are also no double bonds, so we cannot do an addition. It seems that we are stuck, with nothing to do. Clearly, our only way out of this situation is to introduce a functional group into the compound, via radical bromination. Radical bromination will place a Br at the most substituted position (the tertiary position), and then we can eliminate:

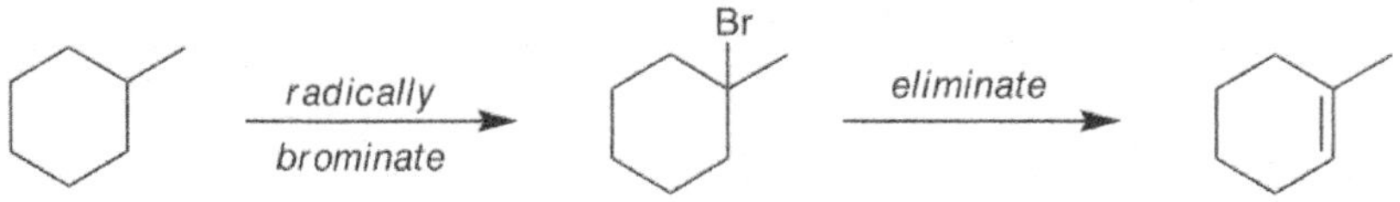

So, our overall synthesis is:

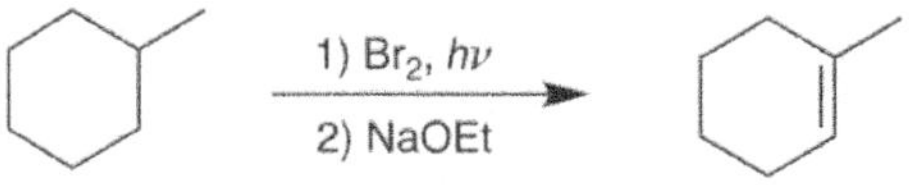

There are a few things to keep in mind when using this technique. First of all, radical bromination will selectively place a Br on the most substituted position. Therefore, you should always look for the tertiary position to see where the Br will go. Then, when doing the elimination, make sure to choose the base carefully, in order to achieve the desired regiochemistry. Let's get some practice with this.

PROBLEMS   What reagents would you use to accomplish each of the following transformations?

11.92

11.93

11.94

11.95

11.96

11.97

11.9   ADDING Br AND Br; ADDING Br AND OH

We now turn our attention to the next addition reaction: adding Br and Br across a double bond:

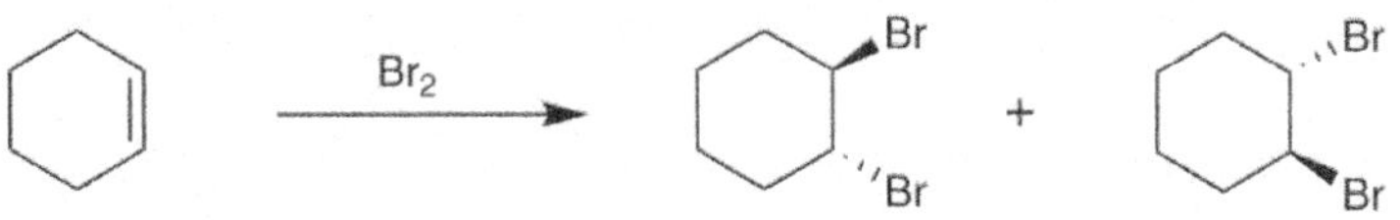

We will begin by analyzing the regiochemistry and stereochemistry of the above reaction. The regiochemistry will be irrelevant, regardless of what alkene we start with, because we are adding two of the same group (Br and Br). However, the stereochemistry *is* relevant, because we are creating two new chiral centers in the example above. If we carefully examine the products, we will notice that we have a pair of enantiomers. (*Caution:* Many students mistakenly believe that these two products are the same, but they are not—they are in fact enantiomers. If you have trouble seeing this, I recommend building molecular models of both compounds.) Our products represent the pair of enantiomers that would result from an *anti* addition. So, we must try to understand *why* this reaction proceeds through an *anti* addition. To do this, we turn once again to the accepted mechanism.

In the first step, we have an alkene reacting with Br_2. To understand this step of the mechanism, we must determine which reagent is the nucleophile, and which reagent is the electrophile. The alkene possesses a pi bond, which represents a region in space of electron density. Therefore, the alkene functions as the nucleophile.

This implies that Br_2 is the electrophile. But how does Br_2 function as an electrophile? The bond between the two bromine atoms is a covalent bond, and we therefore expect the electron density to be equally distributed over both Br atoms. However, an interesting thing happens when a Br_2 molecule approaches an alkene. The electron density of the pi bond *repels* the electron density in the Br_2 molecule, creating a temporary dipole moment in Br_2.

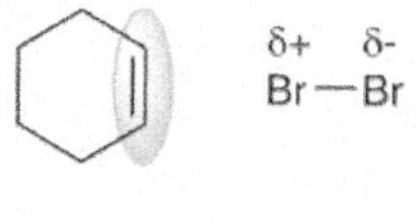

As the Br_2 molecule gets closer to the alkene, this temporary effect becomes more pronounced. Now we can understand why Br_2 functions as an electrophile in this reaction: there is a temporary $\delta+$ on the bromine atom that is closer to the pi bond of the alkene. The electron-rich alkene attacks the electron-poor bromine, giving the following first step of our mechanism:

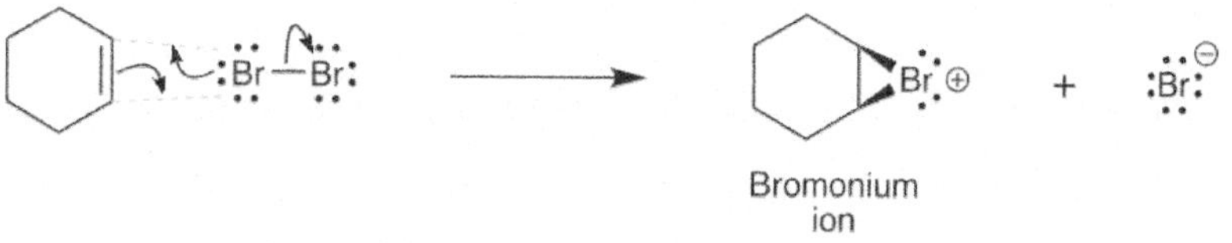

Notice that there are three curved arrows here. For some reason, students drawing this mechanism commonly forget to draw the third curved arrow (the one that shows the expulsion of Br^-). The product of this first step is a bridged, positively charged intermediate, called a *bromonium* ion ("onium" because there is a positive charge). In the second step of our mechanism, the bromonium ion is attacked by Br^- (formed in the first step):

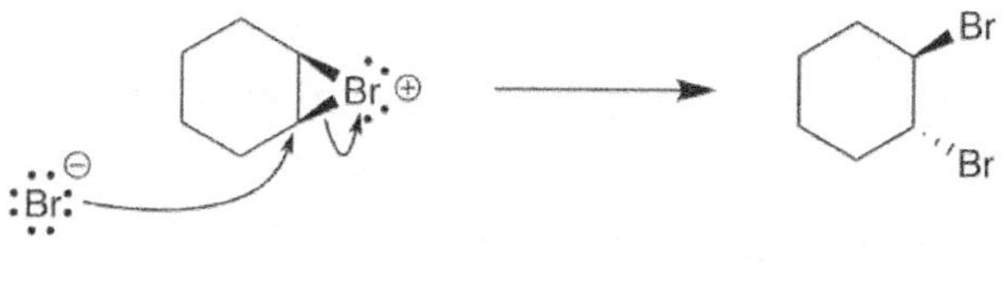

This step is an S_N2-type process, and therefore, must be a back-side attack. In other words, the attacking bromide ion must come *from behind* (from behind the bridge), and therefore, we observe an *anti* addition. There are some alkenes for which a *syn* addition predominates. Clearly, a different mechanism is operating in those cases. For the alkenes that you will encounter in this course, this reaction will always be an *anti* addition, proceeding via the mechanism that we showed.

Let's now summarize the profile for this reaction:

Br_2 $\longrightarrow$	Br and Br
*Regio*chem	not relevant
*Stereo*chem	*anti*

The outcome is more interesting when we use water as the solvent for the reaction. For example:

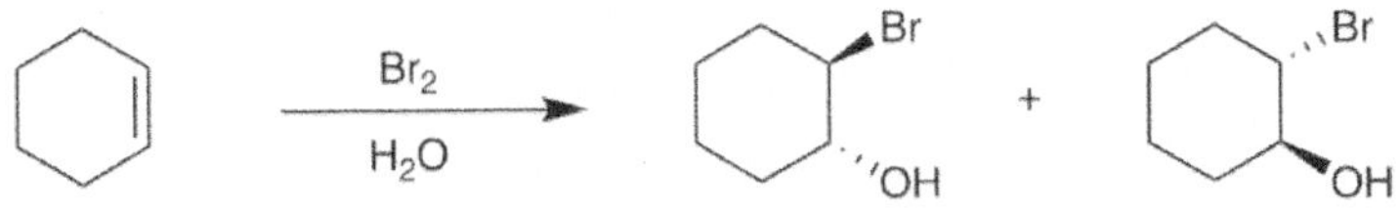

If we look at the products, we will see that we are now adding Br and OH, instead of Br and Br. In order to understand what is going on, we refer back to the accepted mechanism.

The first step is identical to what we saw just a few moments ago: the alkene attacks Br_2 to form a bridged, bromonium intermediate. But now we have a new possibility, because there are now two nucleophiles present: bromide and water. Rather than Br⁻ attacking the bromonium ion, a water molecule can attack instead, which will ultimately give the products shown above. The obvious question is: Why should H_2O attack instead of bromide? Isn't bromide a better nucleophile? Yes, it is true that bromide is a better nucleophile than H_2O. However, think about it from the point of view of a bromonium ion. It is a very high-energy intermediate (it has ring strain from the three-membered ring, and it has a positive charge on a bromine atom). Therefore, it is very eager to react with any nucleophile. It will not be very picky. It will react with the first nucleophile that it encounters. Since we are using water as the solvent here, the bromonium ion will most likely encounter a water molecule before it has a chance to get attacked by a bromide ion.

In the absence of water, we did not need to think about regiochemistry, because we were adding Br and Br. But now, in the presence of water, we are adding Br and OH (two different groups). As a result, the regiochemistry will be relevant if we start with an unsymmetrical alkene—for example,

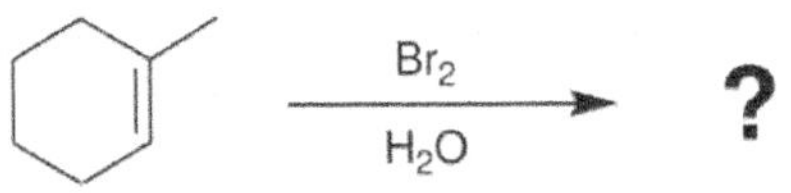

Which group will end up on the more substituted carbon? Br or OH? In other words, does water attack the more substituted carbon or the less substituted carbon? To answer this question, we need to look at the structure of the bromonium ion more carefully than we did earlier. When we drew the bromonium ion earlier, the positively charged bromine atom was drawn to form a perfect three-membered ring (a nice triangle). But bromine does not have to be perfectly in the center; it can lean to one side or the other:

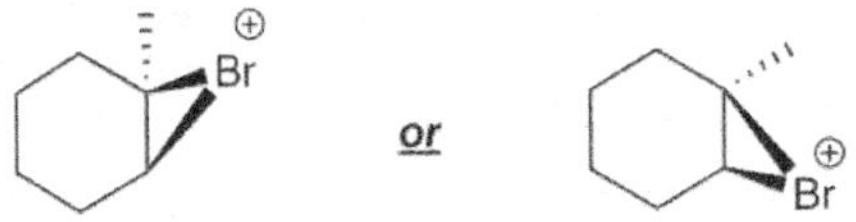

or

Imagine if the bromine atom were leaning *all the way* to one side. That would just give us a carbocation:

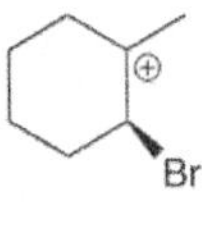

It does not actually lean completely to one side, but it does lean a little toward one side—which gives more "carbocationic character" to the tertiary carbon:

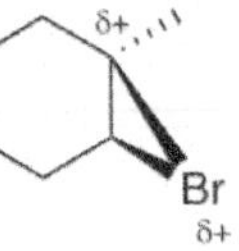

The more substituted carbon (the tertiary carbon) can best handle this carbocationic character. Therefore, the more substituted carbon will possess more δ+ than the less substituted carbon. This means that the tertiary carbon atom will be close in character to an sp^2 hybridized carbon atom. It is not a full-fledged carbocation, so it is not fully sp^2 hybridized. But it is also not a regular sp^3 hybridized carbon. Rather, it is somewhere in between these two extremes (somewhere in between sp^2 and sp^3). Therefore, the geometry will not really be trigonal planar, nor will it be tetrahedral either. The geometry of the tertiary carbon atom above will be somewhere in between trigonal planar and tetrahedral. This helps explain how we can have an S_N2 attack at a tertiary center in the second step of the mechanism. Normally, we can't have an S_N2 at a tertiary center. But here it's OK, because the geometry is closer to being trigonal planar, which allows water to attack:

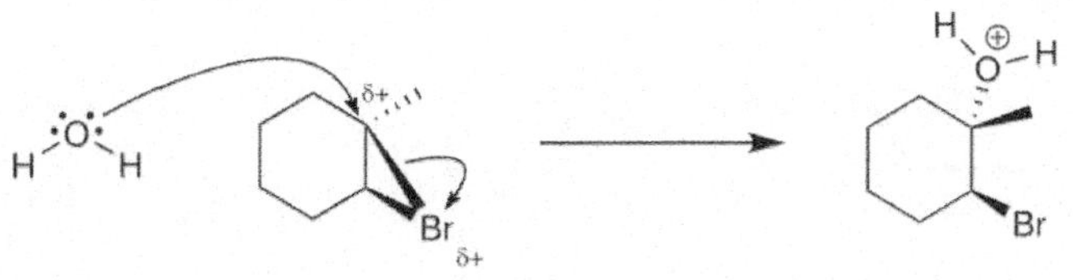

The resulting cation (called an oxonium ion) is then deprotonated by water to give the final product:

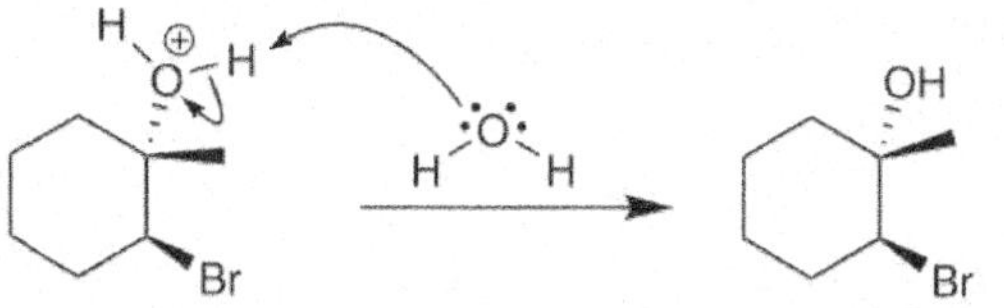

Notice that we show water as the base that removes the proton—we do not show hydroxide (HO^-) as the base, because there isn't much hydroxide around. It is always important to stay consistent with the conditions that are stated. In this case, the reagents are Br_2 and H_2O (not hydroxide). Therefore, we must draw the last step of the mechanism with water functioning as the base that removes the proton.

The final product is called a *halohydrin* (indicating that we have a halogen—Br—and an OH group in the same compound). This reaction is commonly called *halohydrin formation*.

The profile of halohydrin formation can be summarized in the following chart:

Br_2 / H_2O →	Br and OH
Regiochem	OH goes on more substituted C
Stereochem	*anti*

EXERCISE 11.98   Predict the products for each of the reactions below, and then propose a mechanism for formation of those products:

(a)    (b)

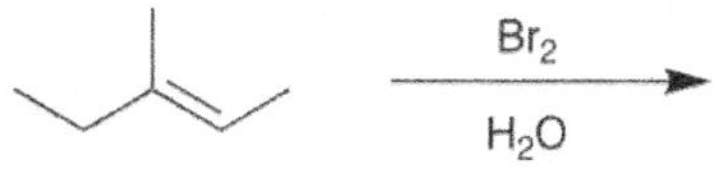

Answer
(a) We are adding Br and Br, so the regiochemistry is irrelevant. But what about the stereochemistry? We look to see whether we are creating two new chiral centers. In this case, we are. So, the stereochemistry is relevant. We have explored the mechanism and justified why the reaction must be an *anti* addition. So, we must draw the pair of enantiomers that would result from an *anti* addition. To do this properly, it will be helpful to redraw the alkene, as we have done many times before:

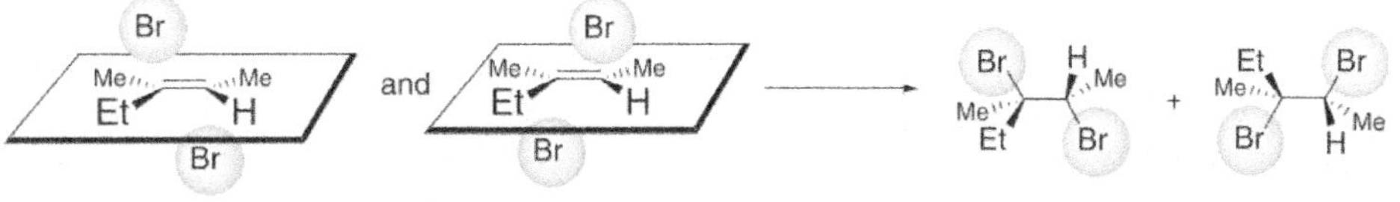

The accepted mechanism of this reaction involves formation of a bromonium ion, followed by attack to open it up:

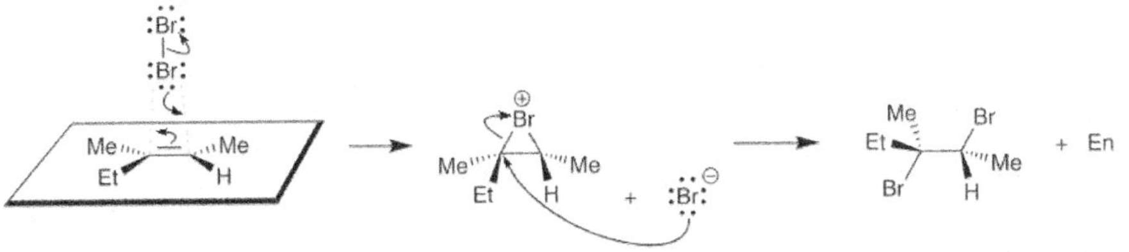

(b) In this problem, we are adding two different groups (Br and OH). So, the regiochemistry *is* relevant. We will need to draw our products with the OH on the more substituted carbon. What about the stereochemistry? We look to see whether we are creating two new chiral centers. In this case, we are. So, the stereochemistry *is* relevant. We have already explained why this reaction must be an *anti* addition. So, we must draw the pair of enantiomers that we expect from an *anti* addition. To do this properly, it will be helpful to redraw the alkene, like this:

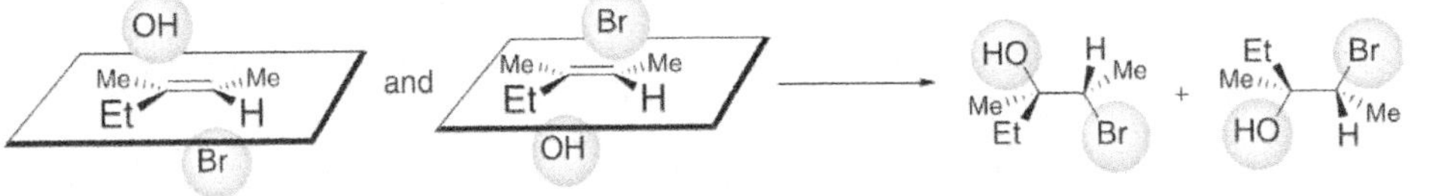

The mechanism of this reaction will have three steps: (1) formation of bromonium ion, (2) attack by water, and (3) deprotonation:

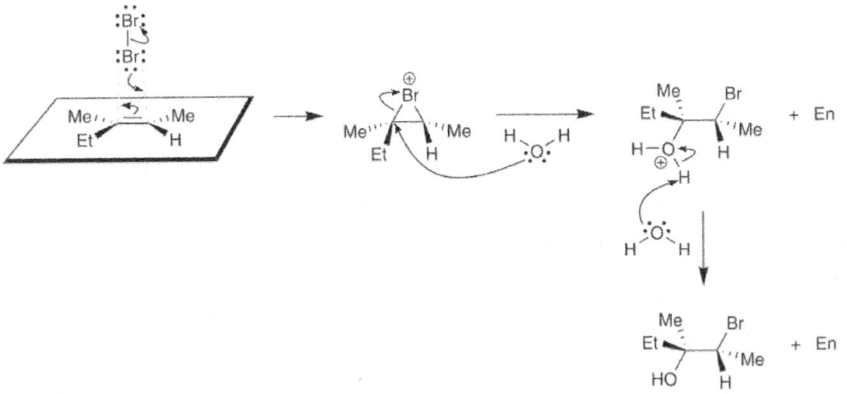

PROBLEMS Predict the products for each of the following reactions:

11.99

Br_2

11.100

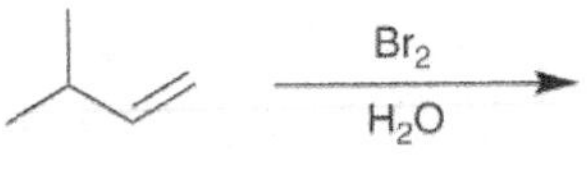

Br_2 / H_2O

11.101

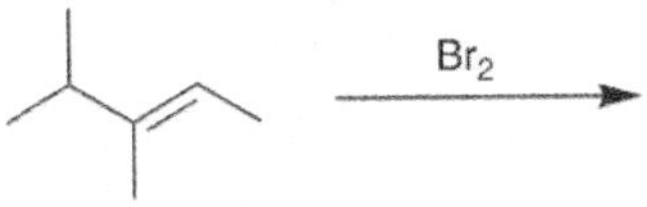

Br_2

11.102

$$\xrightarrow[\text{H}_2\text{O}]{\text{Br}_2}$$

11.103

$$\xrightarrow{\text{Br}_2}$$

11.104

$$\xrightarrow[\text{H}_2\text{O}]{\text{Br}_2}$$

11.10 ADDING OH AND OH, *ANTI*

Now we will see how to add two OH groups across an alkene. It is possible to control whether the OH groups add via a *syn* addition, or via an *anti* addition (by choosing which reagents we will use to add the OH groups). In this section, we will learn how to add two OH groups in an *anti* addition. In the next section, we will learn how to add two OH groups in a *syn* addition.

To add two OH groups in an *anti* addition, we will employ a two-step process: we will first make an epoxide, and then we will open the epoxide with water under conditions of acid-catalysis:

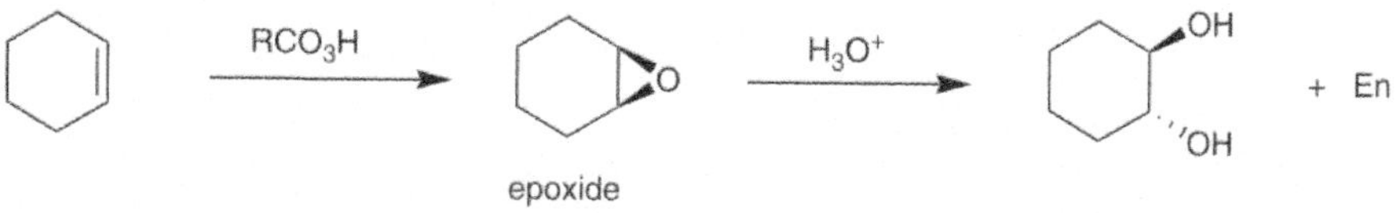

epoxide

We will now explore each aspect of this two-step process. In the first step, a peroxyacid (RCO_3H), sometimes called a per-acid, reacts with the alkene. Compare the structure of a carboxylic acid with the structure of a peroxy acid:

carboxylic acid   peroxy acid

Peroxy acids resemble carboxylic acids in structure, possessing just one additional oxygen atom. Peroxy acids are not very acidic, but they are used as strong oxidizing agents. Common examples of peroxy acids are:

The second example above is called *meta*-chloroperbenzoic acid (MCPBA). It is perhaps the most common example in this course. Whenever you see MCPBA, you should immediately recognize it as an example of a peroxy acid. Similarly, whenever you see RCO_3H, you should also see it as the generic formula for a peroxy acid.

Peroxy acids will react with an alkene to form an epoxide. The mechanism is somewhat complicated, and may or may not be in your textbook, depending on which textbook you are using.

You will not see many mechanisms that are quite as complicated as this one, so we will not spend time on this mechanism. Let's turn our attention to the product of this reaction. We call the product an *epoxide*, which is the term for a three-membered, cyclic ether.

We then take this epoxide and open it with water under conditions of acid-catalysis. Let's explore the mechanism of how this occurs. First, the epoxide is protonated:

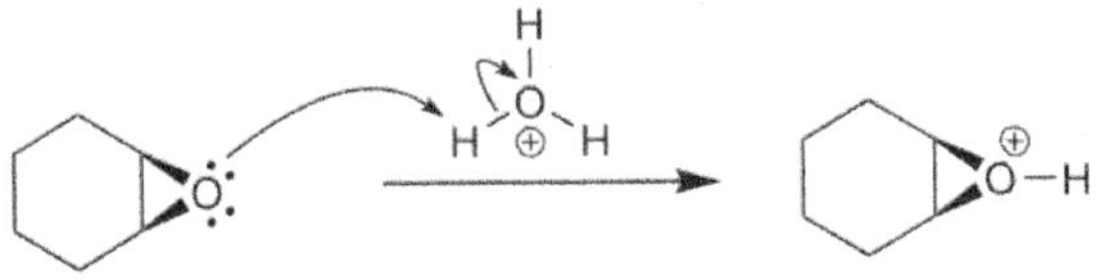

This proton transfer step produces an intermediate that is very similar to a bromonium ion (a three-membered ring with a positive charge on an electronegative atom). Just as a bromonium ion can be attacked by water, similarly, a protonated epoxide can also be attacked by water:

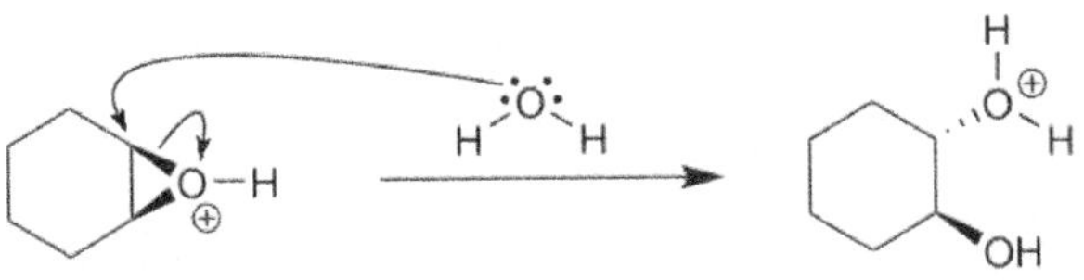

Once again (just like the attack of the bromonium ion in the previous section), water must attack from the back side, which explains the observed stereochemical preference for *anti* addition.

As a final step, water serves as a base to remove a proton and give the product:

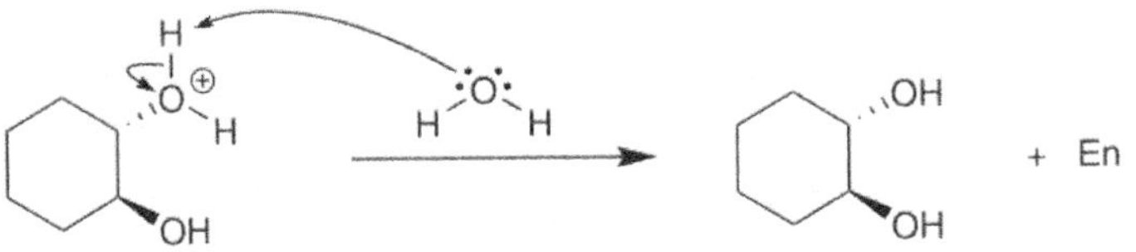

Once again, notice that we show water as the base that deprotonates the oxonium ion to give the product (rather than showing hydroxide as the base) in order to stay consistent with the conditions. In acidic conditions, we most certainly *cannot* draw hydroxide ions in our mechanism.

Now we can summarize the profile for the two-step synthesis that we saw in this section:

1) MCPBA 2) H$_3$O$^+$	OH and OH
*Regio*chem	not relevant
*Stereo*chem	*anti*

EXERCISE 11.105   Predict the products for each of the reactions below:

(a) 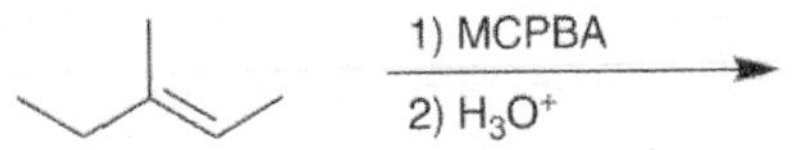   1) MCPBA   2) H$_3$O$^+$

(b) 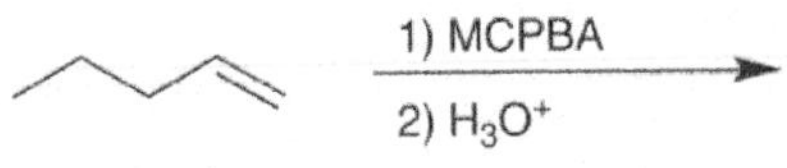   1) MCPBA   2) H$_3$O$^+$

Answer
(a) We are adding OH and OH, and therefore, regiochemistry will be irrelevant. What about stereochemistry? In this case, we are creating two new chiral centers, so we must carefully consider the stereochemistry of this reaction in order to draw the correct pair of enantiomers. This two-step synthesis gives an *anti* addition of OH and OH. Therefore,

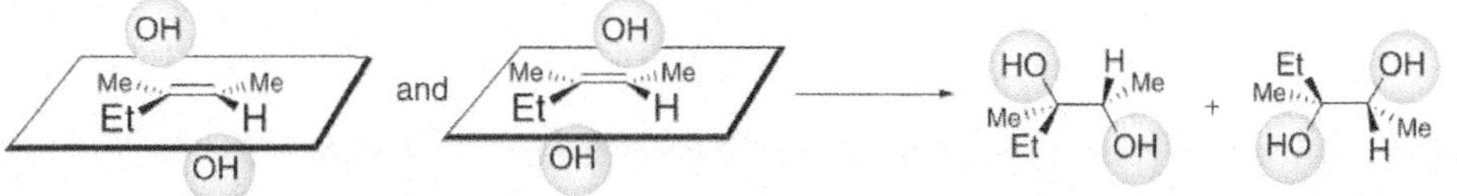
OH
Me
Me
Et
H
OH
and
OH
Me
Me
Et
H
OH
HO
H
Me
Me
Et
OH
+
Et
OH
Me
HO
H
Me

(b) In this example, we are *not* creating two new chiral centers. We are only creating one new chiral center. So, the stereochemistry in this case becomes irrelevant. It is true that the reaction proceeds through an *anti* addition of OH and OH. However, with only one chiral center in the product, the preference for *anti* addition becomes irrelevant. We expect the following products, which are a pair of enantiomers:

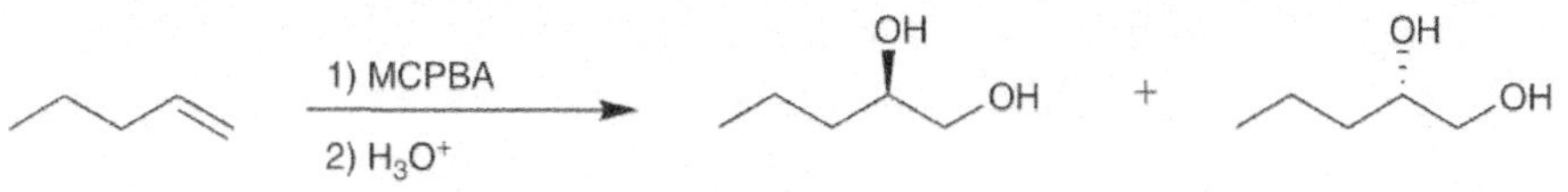
1) MCPBA
2) H_3O^+
OH
OH
+
OH
OH

PROBLEMS Predict the products for each of the following reactions:

11.106

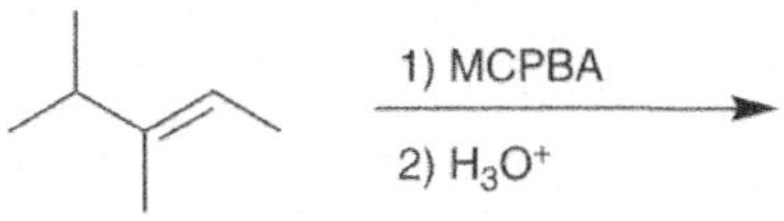
1) MCPBA
2) H_3O^+

11.107

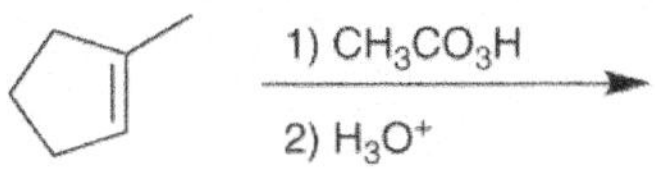
1) CH_3CO_3H
2) H_3O^+

11.108

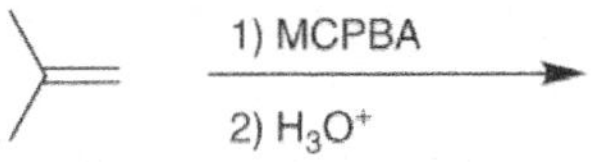
1) MCPBA
2) H_3O^+

11.109

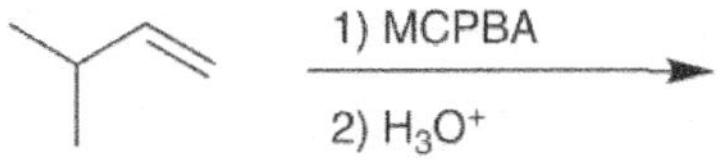
1) MCPBA
2) H_3O^+

11.11 ADDING OH AND OH, *SYN*

In the previous section, we saw how to perform an *anti* addition of OH and OH across a double bond. In this section, we will explore the conditions that will allow us to perform a *syn* addition of OH and OH across a double bond. This reaction is often called a *syn dihydroxylation*.

Consider the following example:

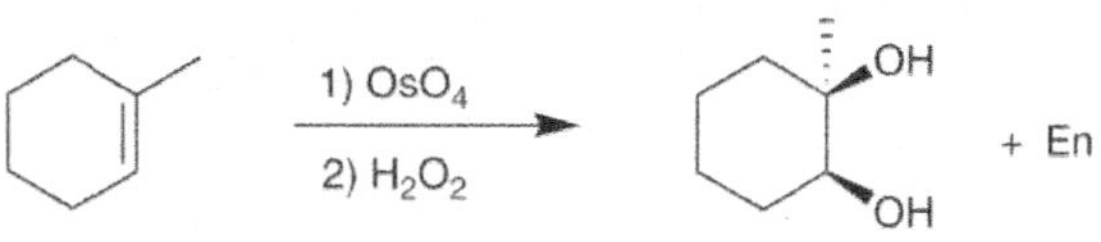
1) OsO_4
2) H_2O_2
OH
OH
+ En

In this example, the OH groups clearly added in a *syn* addition. In order to explain why this reaction proceeds via a *syn* addition, we must look at the first step of the mechanism:

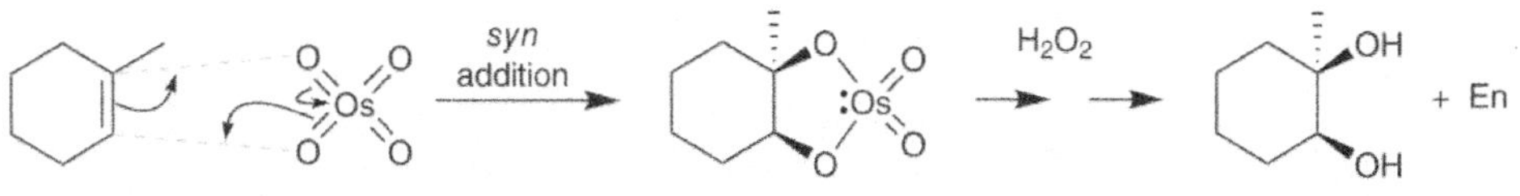
syn addition
H_2O_2
OH
OH
+ En

It is this first step that allows us to understand why the reaction follows a *syn* addition. In this step, osmium tetroxide (OsO$_4$) adds across the alkene *in a concerted process*. In other words, both oxygen atoms attach to the alkene simultaneously. This effectively adds two groups *across the same face* of the alkene.

The pertinent details of this reaction can be summarized in the following chart:

1) OsO$_4$ 2) H$_2$O$_2$	**OH and OH**
*Regio*chem	not relevant
*Stereo*chem	*syn*

The same transformation (*syn* addition of OH and OH) can also be accomplished with cold KMnO$_4$ and hydroxide. Again, we will only look at the first step of the mechanism:

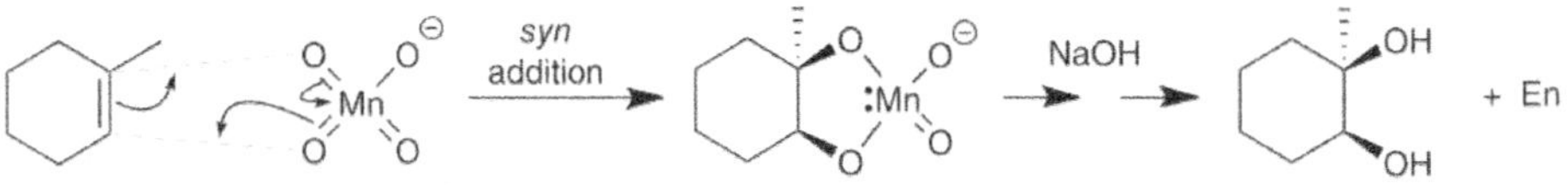

Once again, we have a concerted process that adds both oxygen atoms simultaneously across the double bond. Notice the similarity between the mechanisms of these two methods (OsO$_4$ vs. KMnO$_4$).

EXERCISE 11.110 Predict the products for each of the reactions below:

$$\xrightarrow[\text{cold}]{\text{KMnO}_4 \text{, NaOH}}$$

Answer In this reaction, we are adding two OH groups, so we don't need to think about regiochemistry. As always, stereochemistry will only be relevant if we are forming two new chiral centers. In this example, we are creating two new chiral centers, and therefore, we must be careful to draw only the pair of enantiomers that represent the products of a *syn* addition:

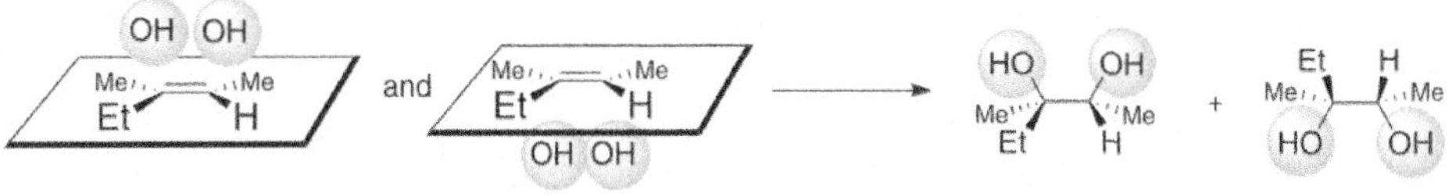

PROBLEMS Predict the products for each of the following reactions:

11.111

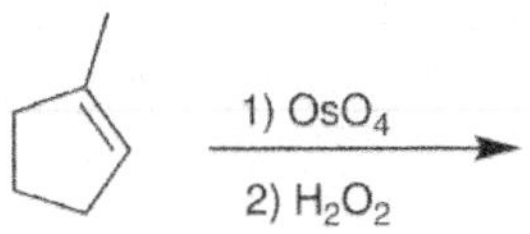

$$\xrightarrow{\begin{array}{c}\text{1) OsO}_4\\\text{2) H}_2\text{O}_2\end{array}}$$

11.112

$$\xrightarrow{\begin{array}{c}\text{1) OsO}_4\\\text{2) H}_2\text{O}_2\end{array}}$$

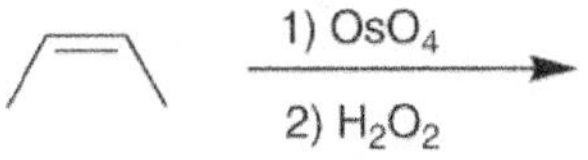

11.113 $\xrightarrow[\text{cold}]{\text{KMnO}_4,\ \text{NaOH}}$

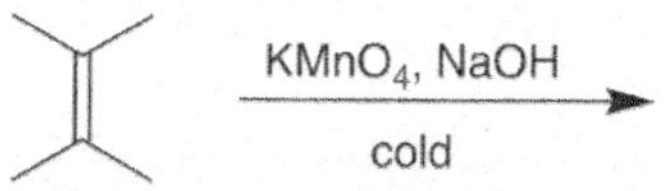

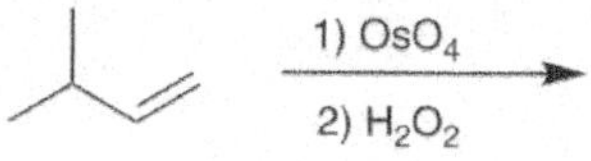

11.114 $\xrightarrow[\text{cold}]{\text{KMnO}_4,\ \text{NaOH}}$

11.115 $\xrightarrow[\text{2) H}_2\text{O}_2]{\text{1) OsO}_4}$

11.116 $\xrightarrow[\text{2) H}_2\text{O}_2]{\text{1) OsO}_4}$

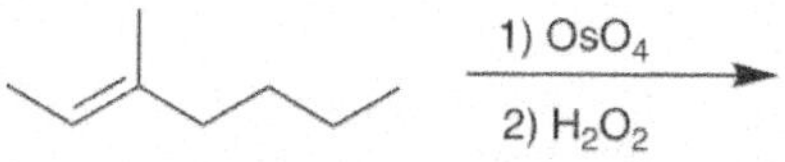

11.12 OXIDATIVE CLEAVAGE OF AN ALKENE

There are many reagents that will add across an alkene and completely cleave the C=C bond. In this section, we will learn about one such reaction, called *ozonolysis*. Consider the following example:

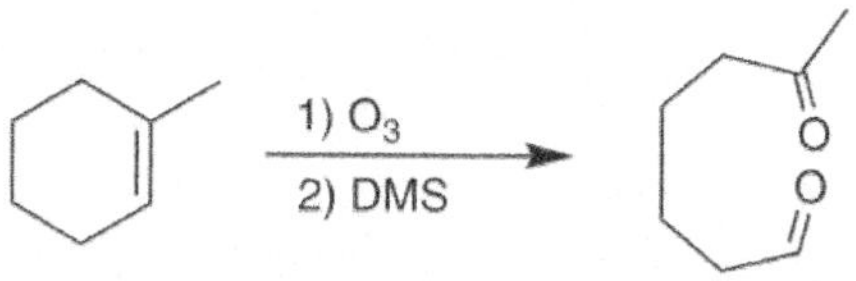

Notice that the C=C bond is completely split apart to form two C=O double bonds. Therefore, issues of stereochemistry or regiochemistry become irrelevant. In order to understand how this reaction occurs, we must first explore the reagents.

Ozone is a compound with the following resonance structures:

Ozone is formed primarily in the upper atmosphere, where oxygen gas (O$_2$) is bombarded with ultraviolet light, although ozone can also be produced under laboratory conditions.

As we did in the previous sections, we will only explore the first step of the mechanism:

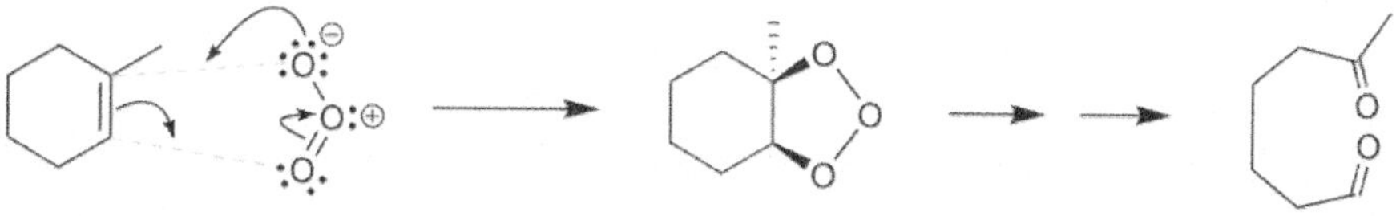

Compare this step to the mechanisms we saw in the previous section (*syn* dihydroxylation), and you should see striking similarities. The initial product (shown above) is called a molozonide, and it subsequently undergoes further rearrangements, before ultimately giving the product upon treatment with dimethyl sulfide (DMS). The structure of DMS is:

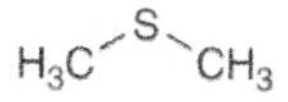

DMS is a mild reducing agent. There are many other reducing agents that can be used in the final step of an ozonolysis, but DMS is common.

There is a simple technique for drawing the products of an ozonolysis: just split up each C=C bond into two C=O bonds. Let's see an example of how this works:

EXERCISE 11.117 Predict the products of the following reaction:

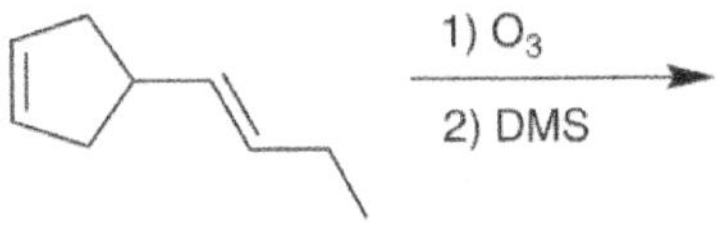

Answer There are two C=C double bonds in this compound. For each C=C double bond, we simply erase the C=C double bond, and place two C=O bonds in its place. Our answer looks like this:

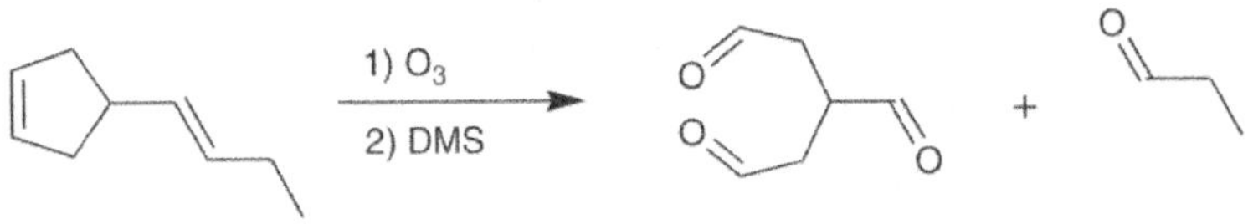

PROBLEMS Predict the products for each of the following reactions:

11.118

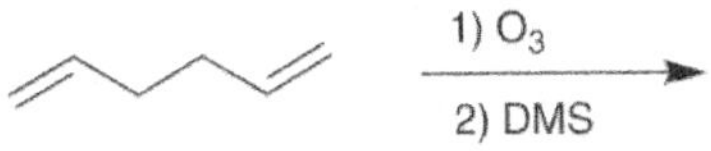

11.119

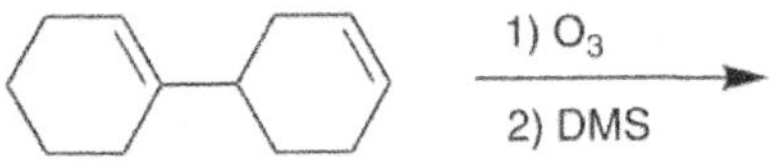

11.120

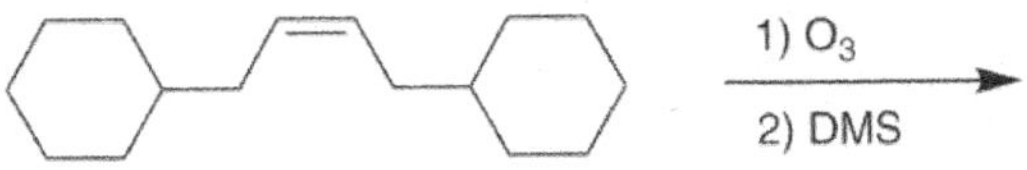

11.121

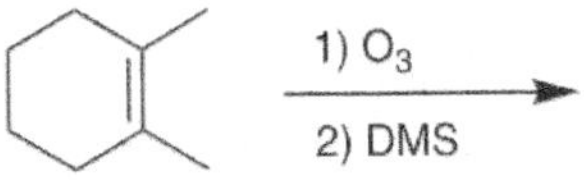

11.122

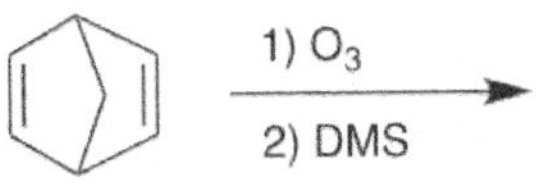

11.123

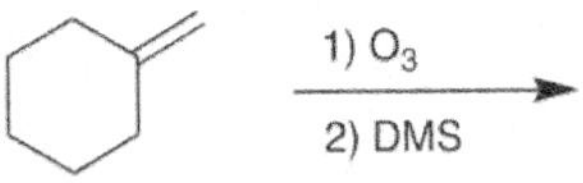

SUMMARY OF REACTIONS

Below is a diagram that shows the key reactions in this chapter. If you have completed all of the problems in this chapter up to this point, you should be able to fill in the reagents necessary for every transformation shown. Try doing that now. If you have trouble remembering the reagents for a particular reaction, then just flip back to the appropriate section, and find the reagents:

After you have filled in the necessary reagents above, you should study this diagram carefully. Look it over, and make sure that every part of it makes sense to you. Go over it ten times in your head. Review it until you are able to reconstruct the whole diagram on a blank piece of paper (with nothing else in front of you).

CHAPTER 12

ALKYNES

12.1 STRUCTURE AND PROPERTIES OF ALKYNES

A compound containing a C≡C triple bond is called an alkyne. In an alkyne, each of the two carbon atoms bearing the triple bond is *sp* hybridized:

$$R - C \equiv C - R$$

sp hybridized   *sp* hybridized

Recall (from Chapter 4) that *sp* hybridized carbon atoms have linear geometry. As such, alkynes should be drawn in a linear fashion. Make sure to draw them correctly:

R—≡—R   R⟋R

CORRECT   INCORRECT

We already saw how to name alkynes in Chapter 5, when we covered rules of nomenclature, but there is additional terminology (not covered in Chapter 5) that we will use frequently in this chapter. Specifically, a *terminal* alkyne has a proton connected directly to the C≡C triple bond, while an *internal* alkyne does not have such a proton:

R—≡—H   R—≡—R

Terminal   Internal

The highlighted proton (in a terminal alkyne) is weakly acidic, and can be removed upon treatment with a strong base, giving a conjugate base called an alkynide ion:

$$R - \equiv - H \quad \overset{\ominus}{:}Base \quad \longrightarrow \quad R - \equiv : \ominus$$

An alkynide ion

How strong does the base need to be? In order to deprotonate a terminal alkyne, we must use a base that is less stable (a stronger base) than the resulting alkynide ion. That way, the deprotonation process is downhill in energy (thermodynamically favorable). Recall from Chapter 3 that there are four factors to consider (**ARIO**) when comparing the stability of bases. In our case, the first factor (**A**tom) and the fourth factor (**O**rbital) are the most relevant. Let's explore each of them now.

The first factor (**A**tom) tells us that O⁻ is more stable than N⁻, which is more stable than C⁻ (because of electronegativity):

Stability

And factor #4 (**O**rbital) tells us that C⁻ is more stable when it is *sp* hybridized:

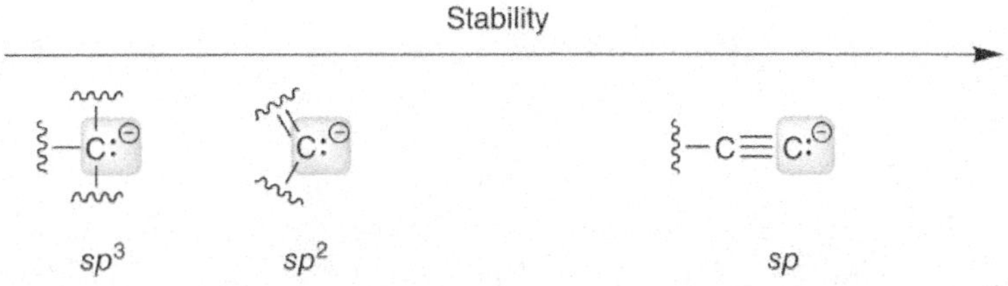

Stability

sp^3 sp^2 sp

In fact, an *sp* hybridized C⁻ is even more stable than N⁻:

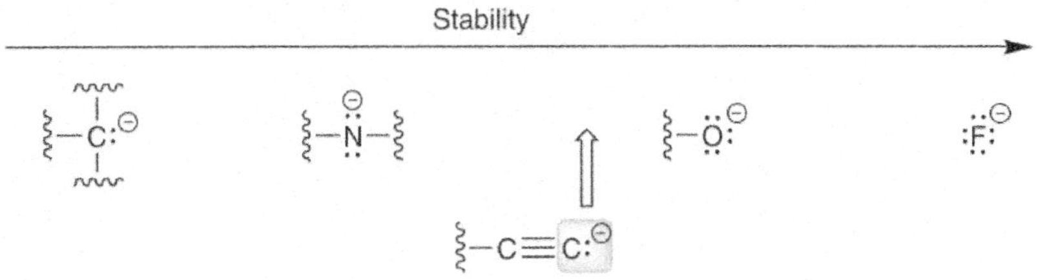

Stability

Note that in this case, factor #4 actually trumps factor #1 (C⁻ is more stable than N⁻, even though factor #1 would predict otherwise). Recall that we first encountered this exception in Section 3.5.

While an alkynide ion is more stable than N⁻, it is not more stable than O⁻. Therefore, if you want to deprotonate a terminal alkyne, you cannot use an alkoxide ion (RO⁻), such as NaOEt or NaOMe. These bases are more stable than an alkynide ion, so they can't be used to produce one. Such a process would be uphill in energy (thermodynamically unfavorable). In order to produce an alkynide ion, we must use a base that is stronger (less stable) than an alkynide ion. For example, we can use a carbanion (C⁻) or an amide ion (N⁻). Sodium amide (NaNH$_2$) is commonly used:

Similarly, sodium hydride (NaH) can also be used to deprotonate a terminal alkyne, because the hydride ion is a sufficiently strong base:

When NaH is used, hydrogen gas evolves as a byproduct. The evolution of a gas forces the reaction to completion.

EXERCISE 12.1   Consider the following alkyne and the following base:

(a) Determine whether the base shown is sufficiently strong to deprotonate the alkyne. If not, then suggest a base that IS sufficiently strong.

(b) Draw the alkynide ion that would be obtained if/when the alkyne above is deprotonated.

Answer
(a) This base (sodium phenolate) has a negative charge on an oxygen atom (O⁻), and it is therefore not sufficiently strong to deprotonate a terminal alkyne. Indeed, a phenolate ion is even more stable than an alkoxide ion (RO⁻), such as NaOMe or NaOEt, because the phenolate ion is resonance-stabilized. In order to deprotonate the alkyne, we will need a stronger base, such as NaNH$_2$ or NaH.

(b) If the alkyne is treated with a stronger base, such as $NaNH_2$ or NaH, deprotonation would result, giving the following alkynide ion:

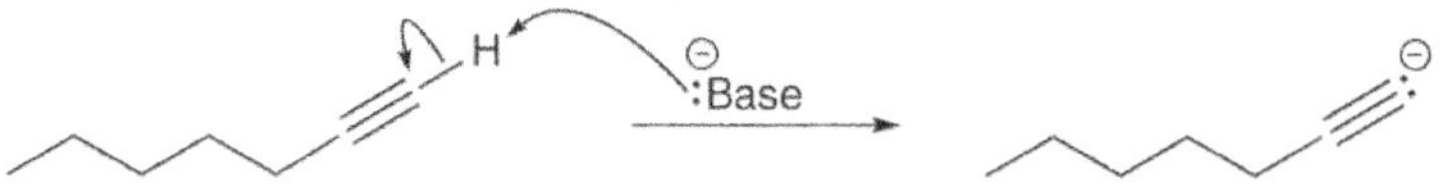

PROBLEMS In each of the following cases, determine whether the base shown is sufficiently strong to deprotonate the alkyne, and draw the alkynide ion that would be obtained if/when the alkyne is deprotonated.

12.2

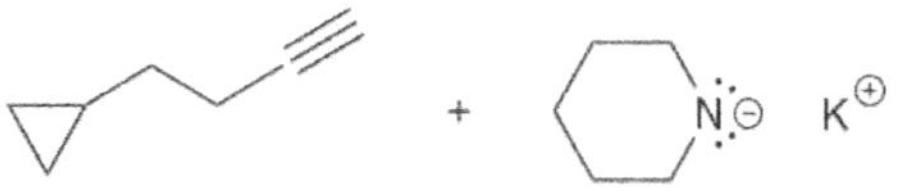

12.3

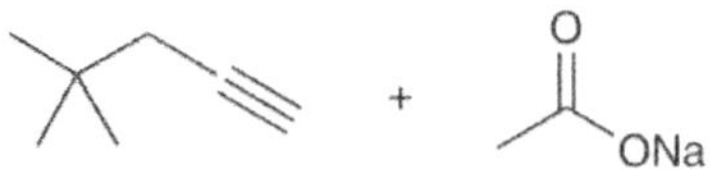

12.4

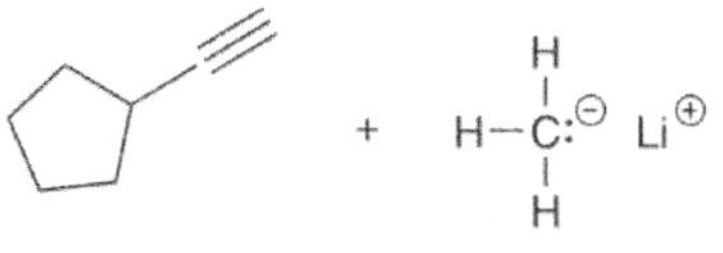

12.5

12.2 PREPARATION OF ALKYNES

Alkynes can be prepared by treating a dihalide (either geminal or vicinal) with a very strong base:

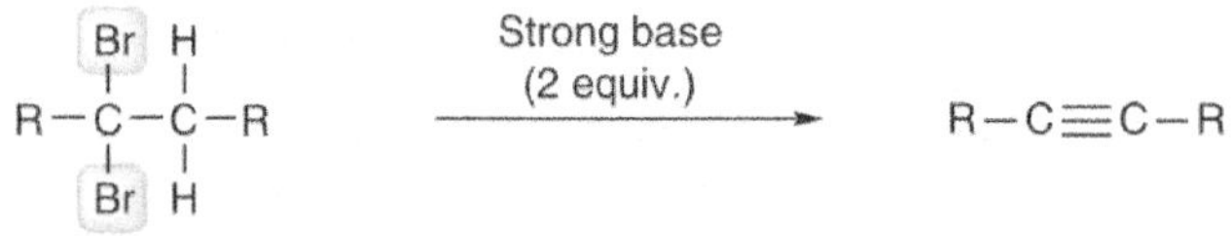

geminal dihalide

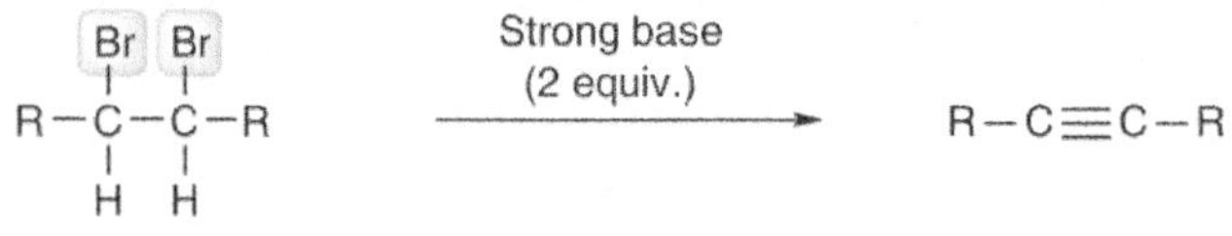

vicinal dihalide

In each case, two equivalents of the base are required (two moles of base for every mole of dihalide) because the transformation occurs via two successive elimination (E2) reactions, as seen here:

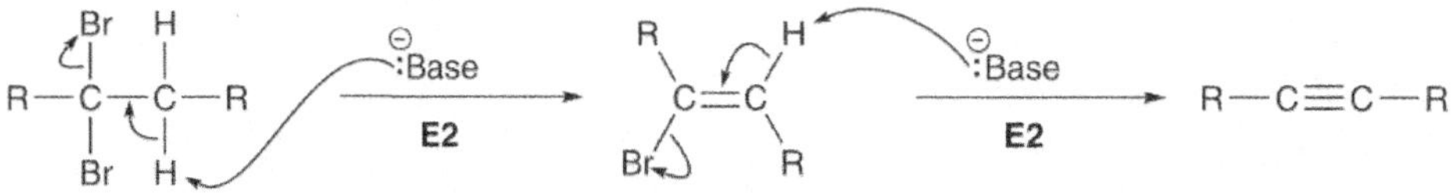

In order to achieve both elimination reactions, we must use a very strong base, such as $NaNH_2$.

When this process is used to prepare terminal alkynes, we use three equivalents of the base, giving an alkynide salt as the product:

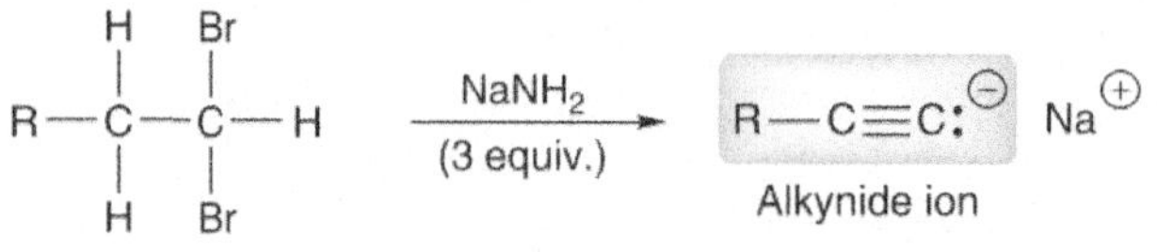

Alkynide ion

Once the reaction is complete, and the alkynide is obtained, a proton source is introduced into the reaction flask, thereby protonating the alkynide and giving the corresponding alkyne. This protonation process can be achieved with a hydronium ion (H_3O^+), or even just with water (H_2O):

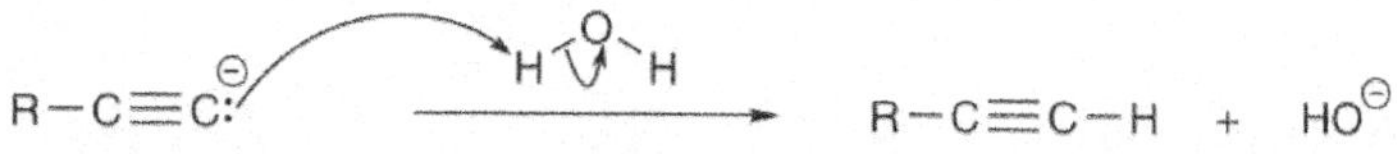

Water is a suitable proton source because the resulting hydroxide ion is more stable than the starting alkynide ion, so this protonation step is thermodynamically favorable (the process favors formation of the more stable base).

In summary, the conversion of a dihalide into a terminal alkyne requires the use of three equivalents of $NaNH_2$, followed by a proton source, which can be shown like this:

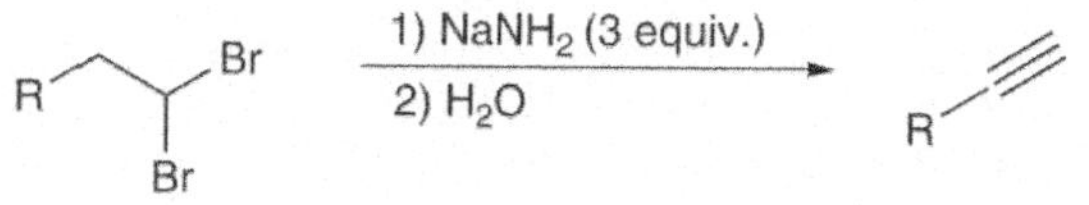

Alternatively, we can simply indicate the use of excess $NaNH_2$, which can be shown like this:

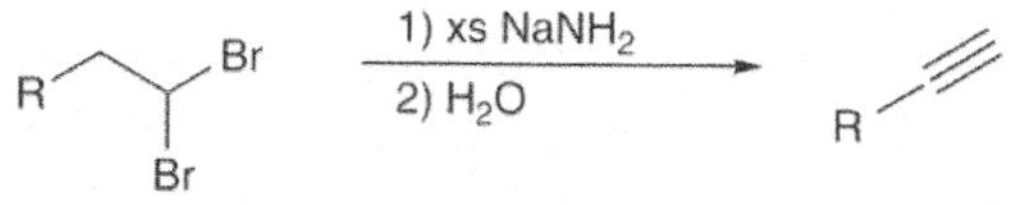

PROBLEMS   Predict the major product for each of the following reaction sequences:

12.6

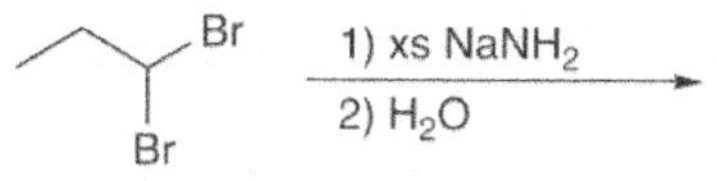

12.7

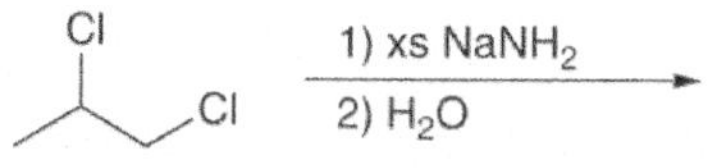

12.8

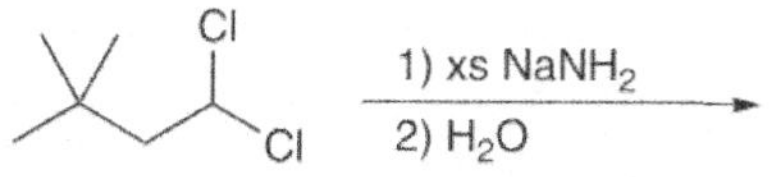

12.9

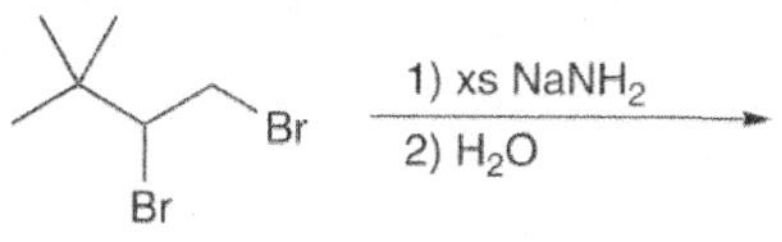

12.3   ALKYLATION OF TERMINAL ALKYNES

In Section 12.1, we saw that a terminal alkyne can be deprotonated upon treatment with a strong base (such as sodium amide) to give an alkynide ion:

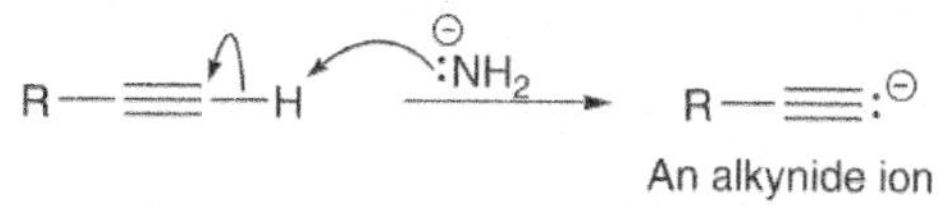

An alkynide ion

Alkynide ions are very strong nucleophiles. Indeed, an alkynide ion can serve as a nucleophile in an S_N2 reaction, upon treatment with a suitable electrophile:

R—≡:⊖ + R—X $\xrightarrow{\text{S}_\text{N}2}$ R—≡—R

Nucleophile   Electrophile

The process is only efficient when the electrophile is a methyl halide, CH_3X, or a primary halide, RCH_2X (for secondary or tertiary halides, elimination products are favored over substitution).

This S_N2 process achieves the installation of an alkyl group, and is therefore called an *alkylation* process. As an example, consider the following alkylation process in which ethyl iodide is used as the electrophile:

$\xrightarrow[\text{2) EtI}]{\text{1) NaNH}_2}$

Notice that the process requires two steps. In the first step, the terminal alkyne is deprotonated to give an alkynide ion. Then, in the second step, the alkynide ion functions as a nucleophile and attacks ethyl iodide in an S_N2 reaction, giving the product (an internal alkyne).

If the starting material is acetylene (H—C≡C—H), each side can be alkylated separately, for example:

$\xrightarrow[\text{2) EtI}]{\text{1) NaNH}_2}$ $\xrightarrow[\text{2) MeI}]{\text{1) NaNH}_2}$

In this case, an ethyl group has been installed on one side, and a methyl group has been installed on the other side. The order of events (whether we first install the ethyl group or whether we first install the methyl group) is not important. That is, either group can be installed first.

In cases where we are installing two of the same R group (for example, two ethyl groups), we still need to install each group separately. That is, the following four steps are required:

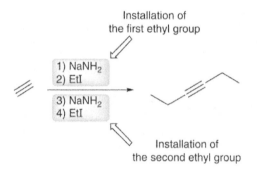

Installation of
the first ethyl group

1) NaNH₂
2) EtI

3) NaNH₂
4) EtI

Installation of
the second ethyl group

The first two steps are used to install one ethyl group, and the next two steps are used to install another ethyl group. Do NOT show the reagents like this:

$\xrightarrow[\text{2) EtI (2 eq.)}]{\text{1) NaNH}_2 \text{ (2 eq.)}}$ ✗

This won't work. Why not? Because acetylene is not deprotonated twice (when treated with two equivalents of $NaNH_2$) to give a dianion:

⊖:≡:⊖

Dianion
(too high in energy)

This dianion is too high in energy and does not form. Therefore, if we treat one mole of acetylene with two moles of sodium amide, we will find that one mole of amide ions remains in the mixture. Then, when ethyl iodide is introduced into the reaction flask, these excess amide ions would react with ethyl iodide (either via substitution or elimination). Therefore, if we want to install an alkyl group on each side of acetylene, we must install each alkyl group separately.

PROBLEMS   Identify the reagents necessary to achieve each of the following transformations:

12.10

12.11

12.12

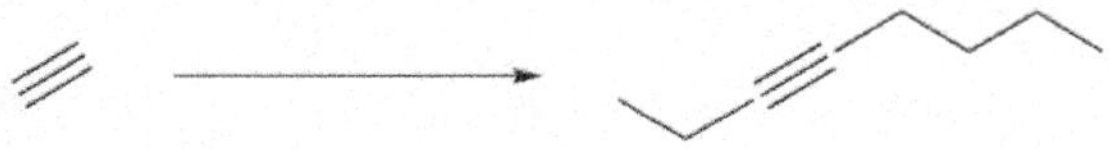

12.13

12.4   REDUCTION OF ALKYNES

Recall from Section 11.3 that the π bond of an alkene will react with molecular hydrogen (H_2) in the presence of a suitable catalyst (such as Pt, Pd, or Ni). The product is an alkane:

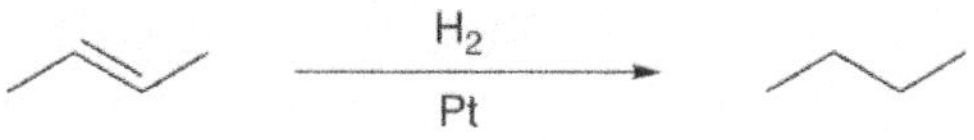

$$\xrightarrow[\text{Pt}]{H_2}$$

During this process, called hydrogenation, the alkene is said to be *reduced* to an alkane (see Section 13.5 for a definition of the term *reduction*). Hydrogenation is also observed for alkynes. Upon treatment with H_2, in the presence of a catalyst such as Pt, an alkyne is reduced to give an alkene. As soon as the alkene is formed, it is further reduced (under these conditions) to give an alkane:

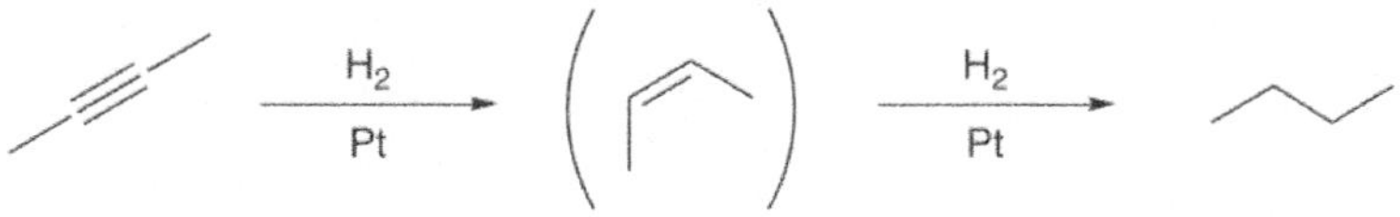

$$\xrightarrow[\text{Pt}]{H_2} \quad \left(\right) \quad \xrightarrow[\text{Pt}]{H_2}$$

The parentheses indicate that the alkene is difficult to isolate under these conditions, because it is even more reactive toward hydrogenation than the starting alkyne. If we want the alkene as our product, we must use a partially deactivated catalyst, also called a poisoned catalyst. One example is Lindlar's catalyst, which is a Pd catalyst prepared with $CaCO_3$ and traces of PbO_2. When an alkyne is treated with H_2 in the presence of Lindlar's catalyst, a *cis* alkene is obtained as the product:

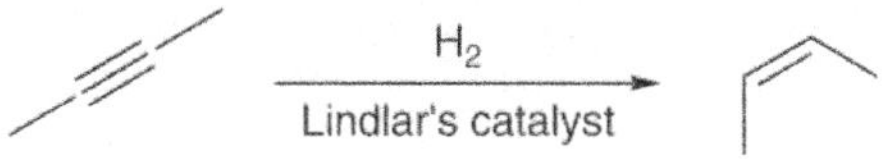

$$\xrightarrow[\text{Lindlar's catalyst}]{H_2}$$

Notice that hydrogenation occurs via a *syn* addition (just as we saw in Section 11.3). Therefore, this process cannot be used to make a *trans* alkene. If we want to make a *trans* alkene, we must use an entirely different process, called a dissolving metal reduction, shown here:

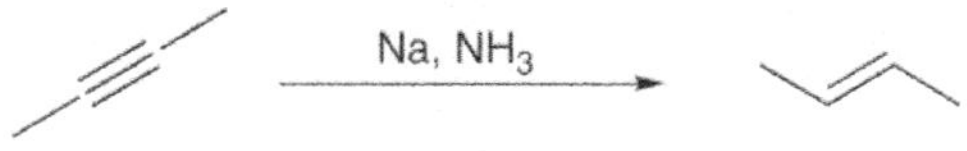

$$\xrightarrow{\text{Na, NH}_3}$$

Notice that the reagents for this reaction are elemental sodium (Na) and liquid ammonia (NH_3). Be careful not to confuse these reagents with sodium amide ($NaNH_2$). We have seen that $NaNH_2$ is a strong base that can be used to deprotonate a terminal alkyne. In contrast, elemental sodium (Na) is a source of electrons, and NH_3 is a solvent and a source of protons:

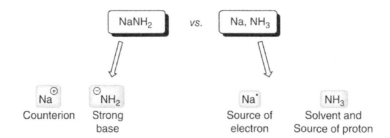

When elemental sodium is dissolved in liquid ammonia (Na, NH$_3$), the resulting mixture is both a source of electrons and a source of protons. This can be seen in the steps of the accepted mechanism for dissolving metal reductions. The first two steps of the mechanism are shown here:

In the first step, an electron is transferred to the alkyne. Then, the second step of the mechanism is a proton transfer. Together, these two steps install a hydrogen atom (one electron + one proton = one hydrogen atom). Before we move on to the last two steps of the mechanism, let's explore some important features of the first two steps above. Notice that the first step uses curved arrows that are single-barbed. Each of these arrows is called a fishhook arrow (because it resembles a fishhook), and it represents the motion of only one electron. For example, the sodium atom transfers only one electron in the first step, so a fishhook arrow is used to show the motion of that one electron. The other two arrows in the first step are also fishhook arrows, as they also each represent one electron. However, in the second step of the mechanism, the more familiar double-barbed arrows are used. Each of these curved arrows represents the motion of two electrons. This mechanism is relatively unique because it employs both types of curved arrows (single-barbed and double-barbed).

There is one other feature of the first step that is worthy of our attention. Focus on the structure of the intermediate produced by the first step of the mechanism:

Radical anion

This intermediate is called a radical anion because it has both an unpaired electron (thus, a radical) and a negative charge (thus, an anion). Notice that the unpaired electron and the lone pair are positioned as far apart as possible, in order to minimize their repulsion. This causes the R groups to adopt a *trans* configuration, where they remain for the rest of the reaction. This explains the stereochemical outcome (*anti* addition).

The last two steps of the mechanism are (once again) the transfer of an electron, followed by the transfer of a proton:

Together, these last two steps install a hydrogen atom (one electron + one proton = one hydrogen atom). In total, the mechanism has four steps. The first two steps install one hydrogen atom, and the last two steps install another hydrogen atom. The net result is the installation of two hydrogen atoms in an *anti* addition.

In summary, we have seen three different ways to reduce an alkyne:

PROBLEMS   Draw the expected product for each of the following reactions:

12.14 $\dfrac{H_2}{Pt}$

12.15 $\dfrac{H_2}{\text{Lindlar's catalyst}}$

12.16 $\xrightarrow{\text{Na, NH}_3}$

12.17 $\dfrac{H_2}{\text{Lindlar's catalyst}}$

12.18 $\xrightarrow{\text{Na, NH}_3}$

12.19   Complete the following reaction sequence by drawing the expected product in each box:

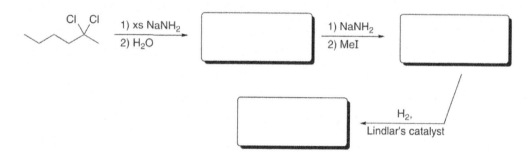

12.20   Identify reagents that can be used to convert 1-pentyne into each of the following compounds:

(a) Pentane

(b) 2-Hexyne

(c) 3-Heptyne

(d) *cis*-4-Octene

(e) *trans*-4-Octene

12.5   HYDRATION OF ALKYNES

Recall from Section 11.6 that an alkene will react with aqueous acid to give an alcohol:

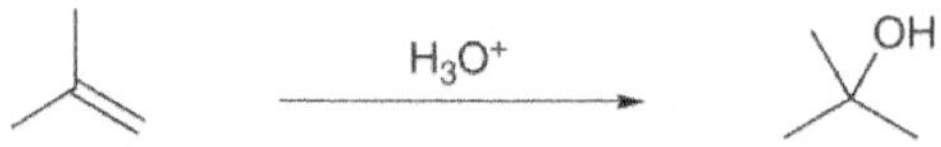

This process, called *acid-catalyzed hydration*, involves the addition of H and OH across the π bond. The OH group is observed to be installed at the more substituted position (Markovnikov addition). Alkynes are also observed to undergo acid-catalyzed hydration via Markovnikov addition, albeit at a slower rate. The rate of reaction is significantly enhanced when mercuric sulfate ($HgSO_4$) is added to the reaction mixture:

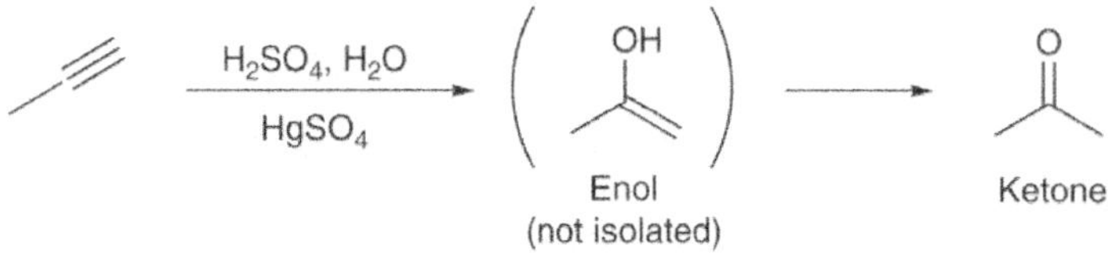

Enol
(not isolated)   Ketone

Don't be confused by the reagents here (H_2SO_4 and H_2O). These reagents are the same as those shown for hydration of an alkene (H_2SO_4, H_2O is the same as H_3O^+). The only real difference here (in reagents) is the use of mercuric sulfate ($HgSO_4$) as a catalyst.

Notice that the initial product has both a π bond (alk**ene**) and an OH group (alcoh**ol**), and is therefore called an *enol*. In the reaction scheme above, the parentheses indicate that the enol cannot be isolated, as it rapidly converts into a ketone. To understand why this happens, let's first consider the relationship between the enol and the ketone. They have the same molecular formula, but they differ in their constitution (connectivity of atoms), so they are constitutional isomers:

C_3H_6O C_3H_6O

More specifically, they are called *tautomers* (a subcategory of constitutional isomers) because they can rapidly interconvert via the migration of a proton, highlighted below:

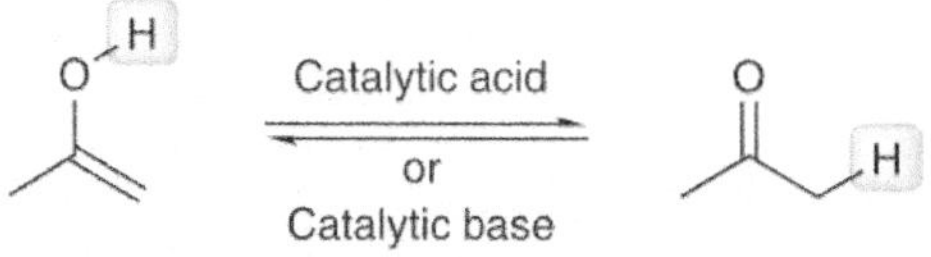

Catalytic acid
or
Catalytic base

The migration of the proton (highlighted) is accompanied by a change in location of the π bond. This equilibrium process, called *tautomerization*, is catalyzed by the presence of an acid or base. The following is a mechanism for tautomerization under acidic conditions:

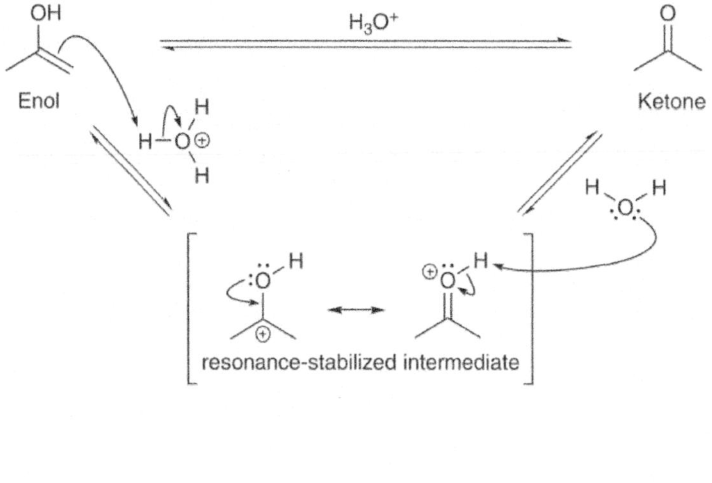

Enol H_3O^+ Ketone

resonance-stabilized intermediate

Notice that there are only two steps. The first step involves protonation to give a resonance-stabilized intermediate, while the second step involves deprotonation to give the ketone. The process is governed by an equilibrium that greatly favors the ketone. Indeed, it is often difficult to detect the presence of the enol (although some enol is always present, albeit in small quantities).

In summary, an alkyne will undergo acid-catalyzed hydration to give a ketone:

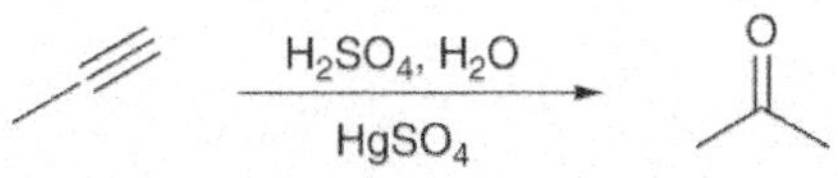
H2SO4, H2O
HgSO4
O

Notice that we do NOT draw the enol as the product. Avoid that mistake. The enol is not isolated, so it should not be drawn as the product. The tautomerization process generally favors formation of the ketone, rather than the enol (we will see an exception in the next section).

This process is most useful for terminal alkynes, which are converted into methyl ketones:

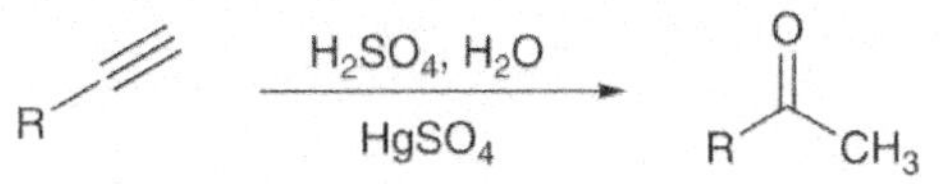
R
H2SO4, H2O
HgSO4
O
R
CH3

If the starting material is an internal alkyne (rather than a terminal alkyne), then a mixture of products is obtained, rendering the process inefficient as a synthetic method:

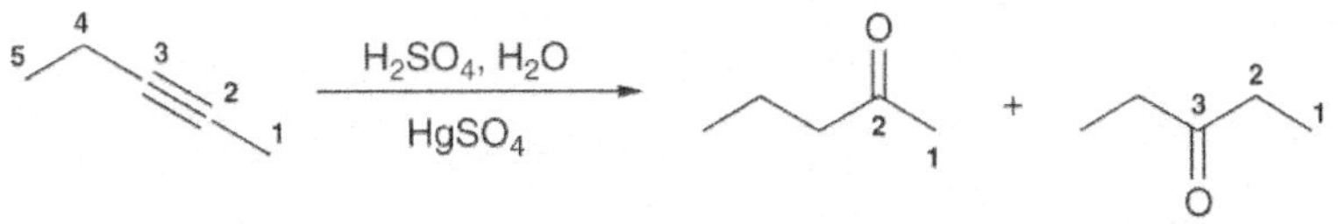
5
4
3
2
1
H2SO4, H2O
HgSO4
O
2
1
+
2
3
1
O

Hydration of an internal alkyne will only be efficient if the alkyne is symmetrical. For a symmetrical alkyne, we expect only one product, as seen in the following example:

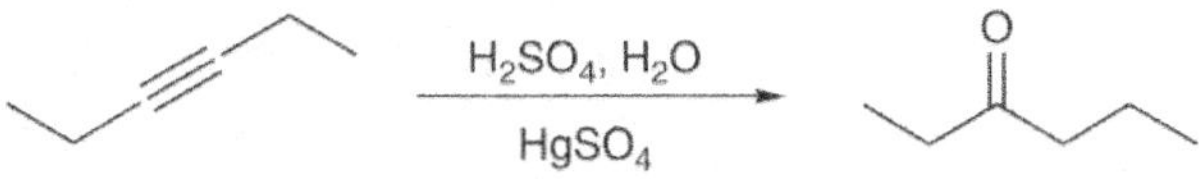
H2SO4, H2O
HgSO4
O

In this example, there is only one possible regiochemical outcome: 3-hexanone is the only product of hydration.

PROBLEMS Draw the expected product for each of the following reactions:

12.21 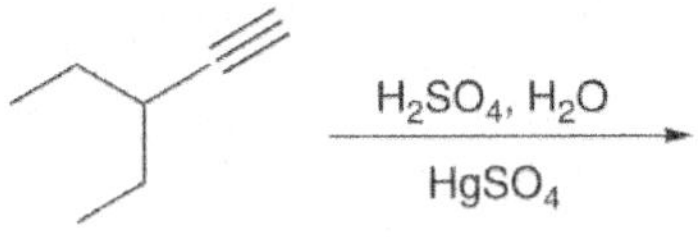
H2SO4, H2O
HgSO4

12.22 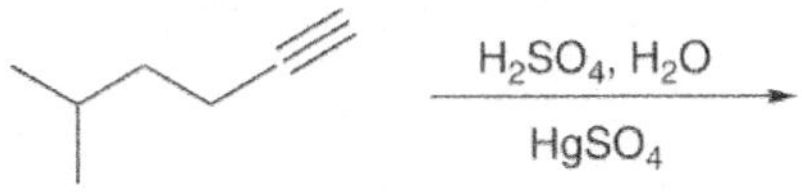
H2SO4, H2O
HgSO4

12.23 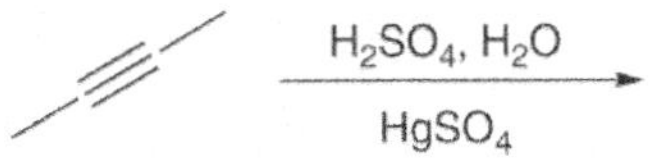
H2SO4, H2O
HgSO4

In Section 11.7, we saw yet another method for adding H and OH across a π bond. This other method, called hydroboration–oxidation, converts an alkene into an alcohol, with the OH group being installed at the less substituted position (*anti*-Markovnikov addition):

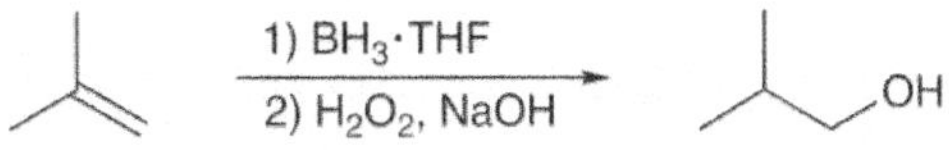
1) BH3·THF
2) H2O2, NaOH
OH

Alkynes are also observed to undergo hydroboration–oxidation:

Once again, parentheses are used to indicate that the resulting enol is not isolated. Under the conditions of its formation, the enol undergoes rapid tautomerization to give an aldehyde.

The following is a mechanism for tautomerization under basic conditions:

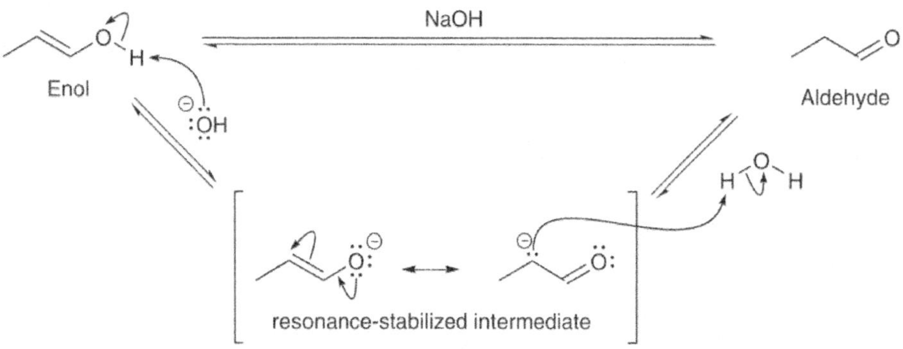

Notice that there are only two steps. The first step involves deprotonation to give a resonance-stabilized intermediate, while the second step involves protonation to give the aldehyde. These two steps resemble the two steps for tautomerization under acid-catalyzed conditions, but in reverse order. Under acid-catalyzed conditions, the first step is protonation to give a cation. Under base-catalyzed conditions, the first step is deprotonation to give an anion.

In summary, a terminal alkyne will undergo hydroboration–oxidation to give an aldehyde:

Notice the reagent in the hydroboration step. Rather than using BH_3, we use a dialkyl borane (R_2BH) for the hydroboration of alkynes. The alkyl groups provide steric bulk that prevent the starting alkyne from reacting with two molecules of the borane consecutively (the alkyne has two π bonds, each of which could conceivably react with a molecule of borane). You should consult your lecture notes and textbook to see if you must be familiar with the structures of any specific dialkyl boranes. Two examples are shown here:

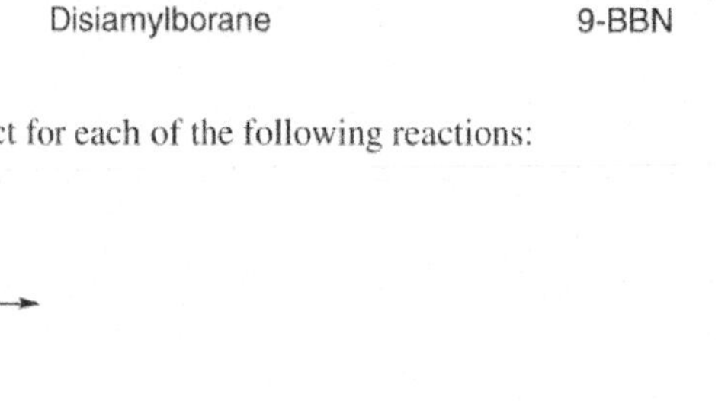

Disiamylborane   9-BBN

PROBLEMS   Draw the expected product for each of the following reactions:

12.24

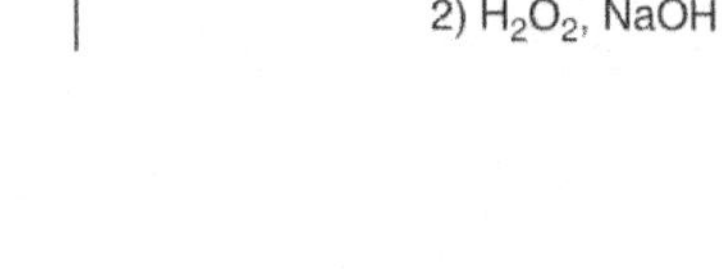

12.25

12.26

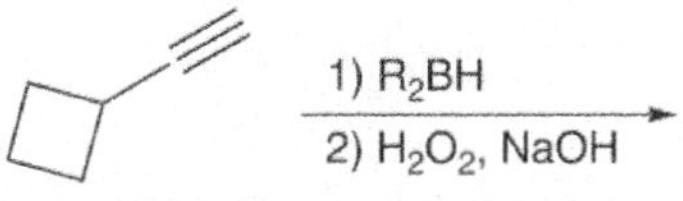

1) R$_2$BH
2) H$_2$O$_2$, NaOH

In this section, we have seen two methods for the hydration of a terminal alkyne. Acid-catalyzed hydration converts a terminal alkyne into a methyl ketone, while hydroboration–oxidation produces an aldehyde:

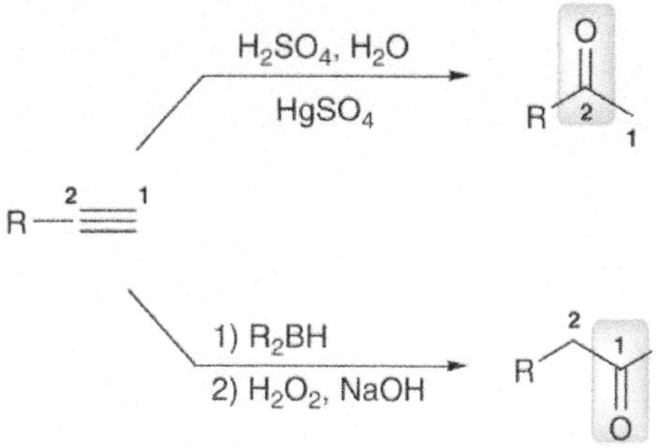

Notice that acid-catalyzed hydration installs a carbonyl (C=O) group at C2, while hydroboration–oxidation installs the carbonyl group at C1.

PROBLEMS   Identify the reagents you would use to achieve each of the following transformations:

12.27

12.28

12.29

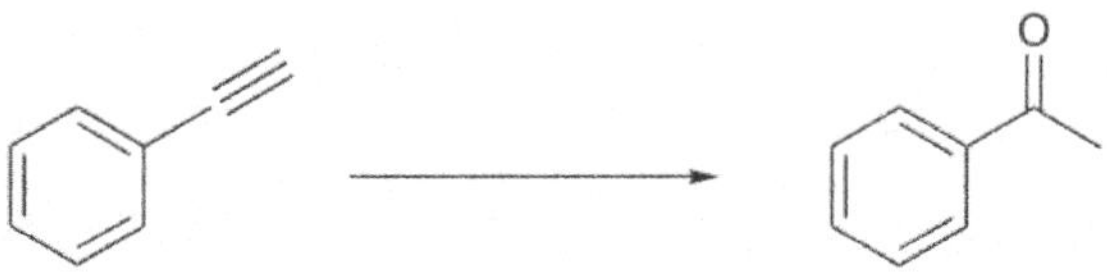

12.30

12.6   KETO-ENOL TAUTOMERIZATION

In the previous section, we saw that enols undergo tautomerization in the presence of catalytic acid or base:

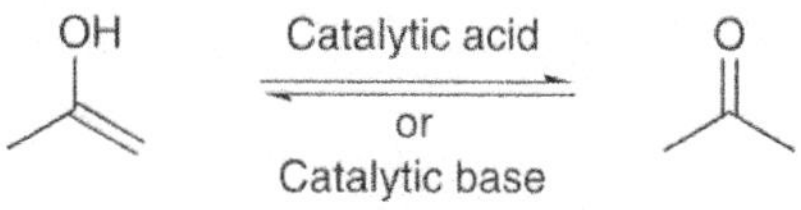

Catalytic acid
or
Catalytic base

In practice, it is exceedingly difficult to remove ALL traces of acid or base from a reaction vessel, and therefore, tautomerization is generally unavoidable. The amounts of enol and ketone are determined by the equilibrium, with the ketone generally being favored greatly.

Since you will encounter keto-enol tautomerization several times as you study organic chemistry, we will spend just a few moments on this important topic.

First and foremost, make sure that you don't confuse tautomers and resonance structures. To illustrate the difference, let's review the mechanism for acid-catalyzed tautomerization that was presented in the previous section:

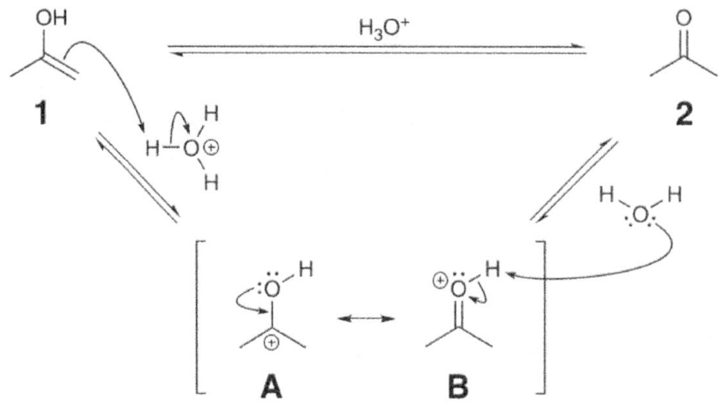

Compounds **1** and **2** (the enol and the ketone) are tautomers. That is, they are two different compounds, in equilibrium with each other, and both are present at equilibrium. In contrast, structures **A** and **B** do NOT represent two different compounds. They (together) represent one entity (an intermediate). This reaction has only one intermediate (not two), and it is resonance-stabilized (structures **A** and **B**). The relationship between **1** and **2** (tautomers) is very different than the relationship between **A** and **B** (resonance structures).

The same concept can be illustrated if we consider keto-enol tautomerization under basic conditions:

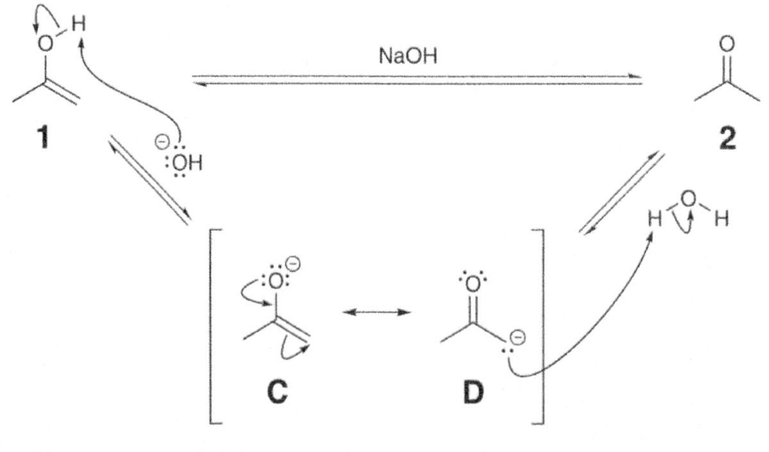

Once again, compounds **1** and **2** are tautomers of each other, while structures **C** and **D** are resonance structures of a single intermediate.

As we have mentioned, keto-enol tautomerization generally favors the ketone, which can be illustrated with unsymmetrical equilibrium arrows, like this:

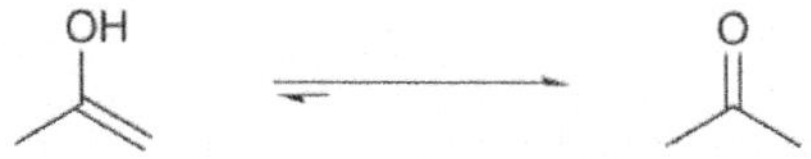

Of course, there are a few exceptions where the enol is actually favored, and you will certainly encounter at least one of these exceptions (shown below) as you progress through the course. Consider the structure of phenol and its tautomer:

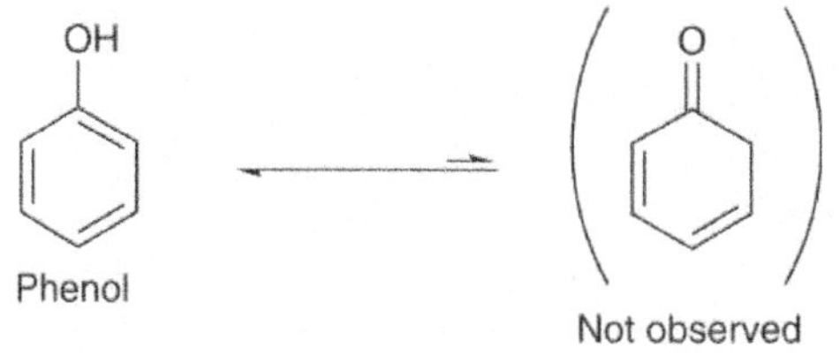

These two compounds are indeed tautomers. However, in this case, the ketone is not favored. In fact, the enol is so highly favored that the concentration of ketone is negligible at equilibrium. Why? Because the enol (in this case) has extra stability that the ketone lacks. Specifically, the enol has an aromatic ring, similar to the ring present in benzene:

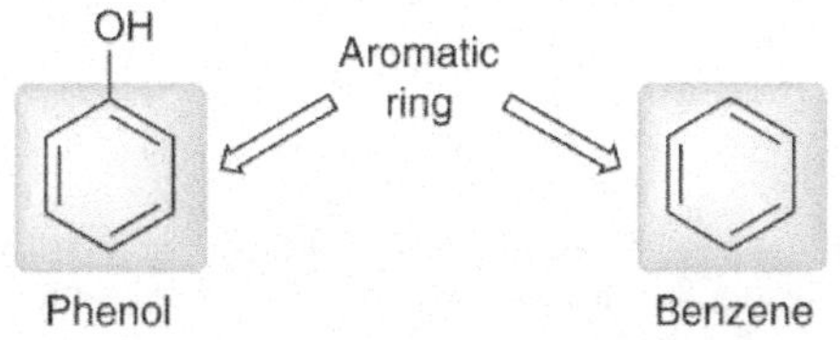

Phenol   Benzene

You may have already covered the concept of aromaticity in your organic chemistry course, or you may be covering that topic soon. Either way, it is important for you to know that the stability associated with an aromatic ring is significant, and in this case, it causes the tautomerization process to favor the aromatic enol, rather than the ketone.

Now let's focus on drawing mechanisms. When drawing the mechanism of a tautomerization process, we must decide whether to begin with protonation (to give a cation) or whether to begin with deprotonation (to give an anion). The choice should be consistent with the conditions that are employed. For example, during acid-catalyzed hydration of an alkyne (Section 12.5), an enol is formed in acidic conditions. Since the conditions are acidic, the tautomerization process begins with protonation (acid-catalyzed tautomerization) to give a cationic intermediate. It would be a mistake to draw deprotonation occurring first, because the resulting intermediate (an anion, which happens to be a very strong base) would be inconsistent with acidic conditions. Similarly, during hydroboration–oxidation of an alkyne, the tautomerization process begins with deprotonation (base-catalyzed tautomerization) to give an anionic intermediate. It would be a mistake to draw protonation occurring first, because the resulting intermediate (a cation, which happens to be a very strong acid) would be inconsistent with basic conditions.

Now let's focus on something else that you should consider when drawing mechanisms. For a tautomerization process under acid-catalyzed conditions, students often make an error that causes the entire mechanism to be incorrect. Specifically, during the first protonation step, students are tempted to protonate the OH group, rather than the π bond. This is incorrect, because the resulting intermediate is not resonance-stabilized:

NOT
resonance-stabilized

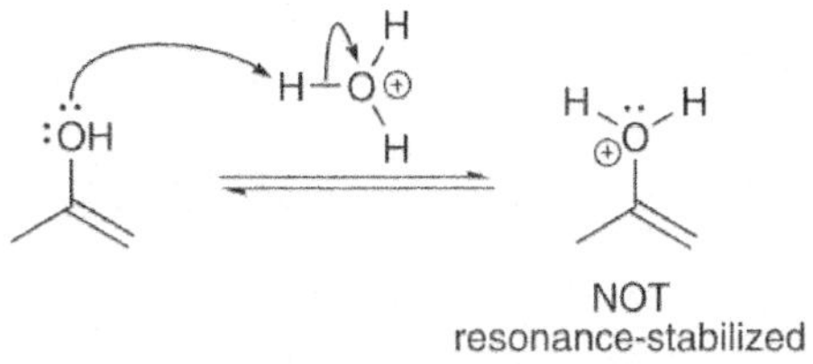

In contrast, protonation of the π bond gives a more stable, resonance-stabilized intermediate:

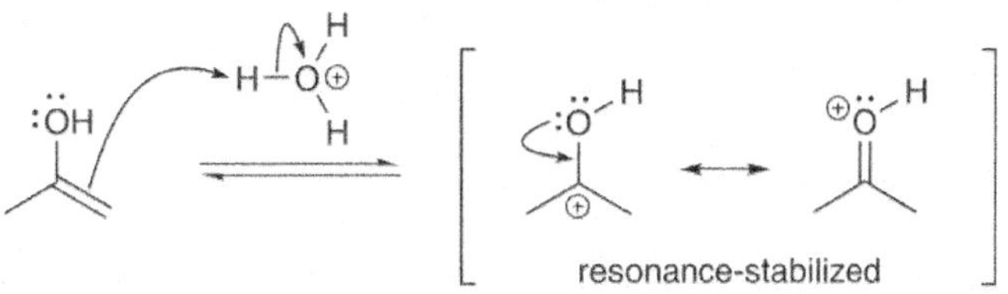

resonance-stabilized

And once you draw the resonance structure with a C=O bond, you should be able to see that you are on the right path, because deprotonation gives the ketone:

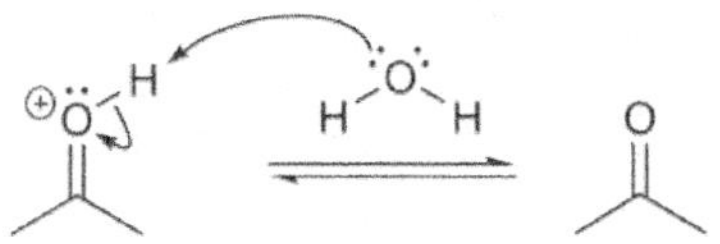

Notice that water (not hydroxide) is used as the base for this deprotonation step. Why? Once again, we must remain consistent with the acidic conditions employed. We cannot show hydroxide functioning as the base here because hydroxide is a strong base, which would be inconsistent with acidic conditions. Under acidic conditions, water (a weak base) is much more likely to function as the base for the deprotonation step.

For a similar reason, when tautomerization occurs under basic conditions, the final protonation step should be shown with water as the proton source (not with H_3O^+):

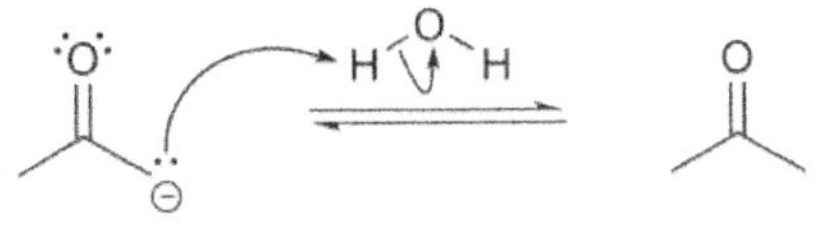

Why can't we show H_3O^+? Because we must remain consistent with the basic conditions employed. H_3O^+ is a strong acid, which would be inconsistent with basic conditions. Under basic conditions, water is much more likely to function as the proton source for the protonation step.

EXERCISE 12.31   Consider the structure of the enol shown below. If this compound were prepared, we would not be able to isolate or store it, because it would undergo rapid tautomerization.

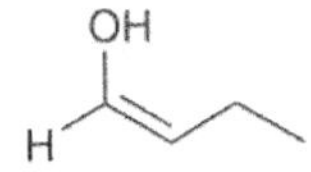

(a) Draw the structure of the resulting tautomer, and show a mechanism of its formation under acidic conditions.

(b) Draw a mechanism for this tautomerization process if it were to occur under basic conditions.

Answer
(a) In acidic conditions, the first step is protonation, using H_3O^+ as the proton source. Make sure to protonate the π bond, rather than the OH group. The resulting intermediate should be resonance-stabilized (if it isn't, then you have done something wrong!):

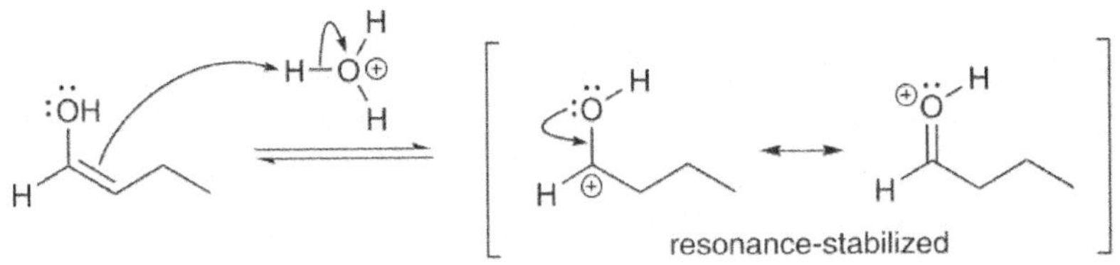

Finally, deprotonation gives the tautomer (an aldehyde, in this case):

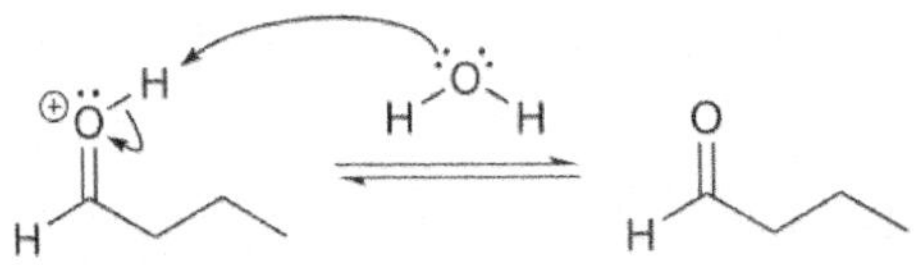

Notice that water functions as the base in this deprotonation step. It would be wrong to show hydroxide in your mechanism, because the process occurs under acidic conditions, and hydroxide (a strong base) is inconsistent with acidic conditions.

(b) In basic conditions, the first step is deprotonation. Hydroxide functions as the base, giving a resonance-stabilized anion:

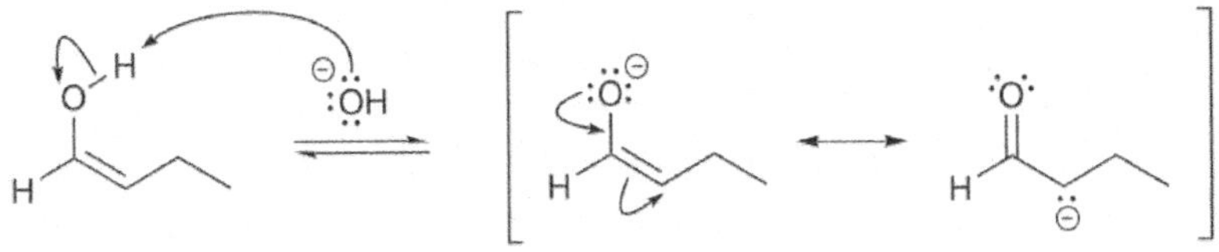

Finally, protonation gives the aldehyde:

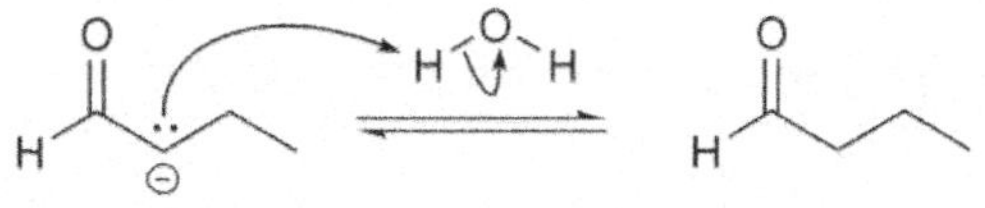

Notice that water functions as the proton source in this protonation step. It would be wrong to show H_3O^+ in your mechanism, because the process occurs under basic conditions, and H_3O^+ (a strong acid) is inconsistent with basic conditions.

PROBLEMS 12.32   The following compound cannot be isolated or stored, because it would undergo rapid tautomerization.

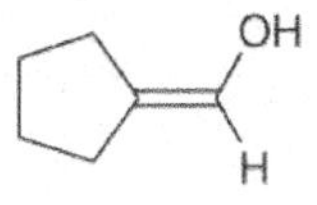

(a) In the space provided below, draw a mechanism for the tautomerization process if it were to occur under acidic conditions.

(b) In the space provided below, draw a mechanism for the tautomerization process if it were to occur under basic conditions.

12.33   The following compound cannot be isolated or stored, because it would undergo rapid tautomerization.

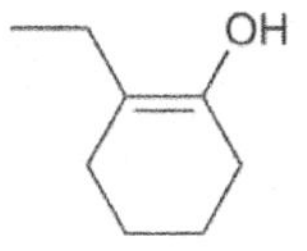

(a) In the space provided below, draw a mechanism for the tautomerization process if it were to occur under acidic conditions.

(b) In the space provided below, draw a mechanism for the tautomerization process if it were to occur under basic conditions.

12.34 The following compound cannot be isolated or stored, because it would undergo rapid tautomerization.

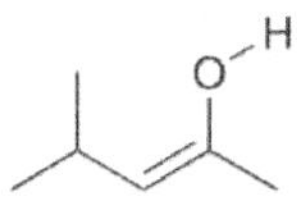

(a) In the space provided below, draw a mechanism for the tautomerization process if it were to occur under acidic conditions.

(b) In the space provided below, draw a mechanism for the tautomerization process if it were to occur under basic conditions.

12.7 OZONOLYSIS OF ALKYNES

Recall from Section 11.12 that alkenes undergo oxidative cleavage via a process called ozonolysis:

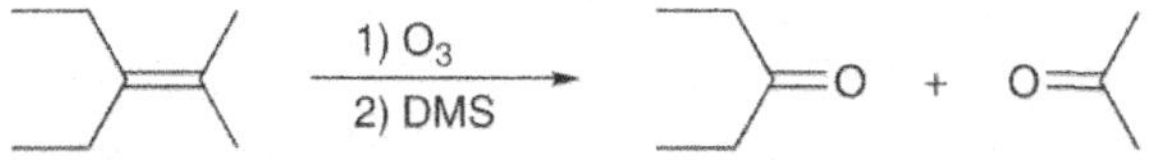

Alkynes are also observed to undergo oxidative cleavage when treated with ozone, followed by water. However, the products are carboxylic acids:

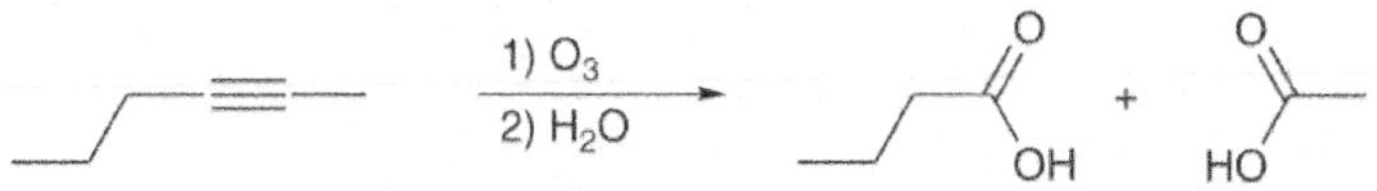

When this process is performed on a terminal alkyne, the terminal side is converted to carbon dioxide, and the other side is converted into a carboxylic acid:

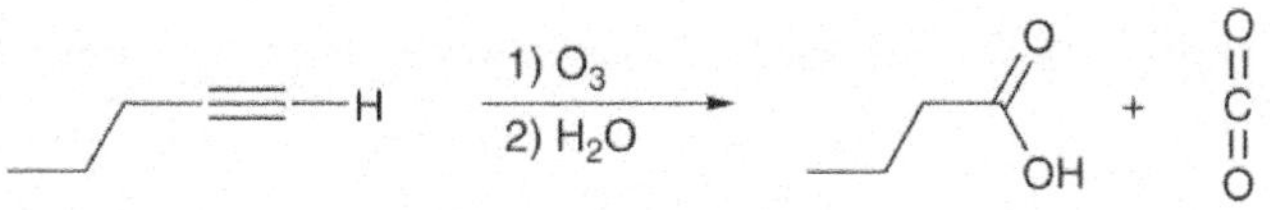

PROBLEMS   Draw the expected products for each of the following reactions:

12.35

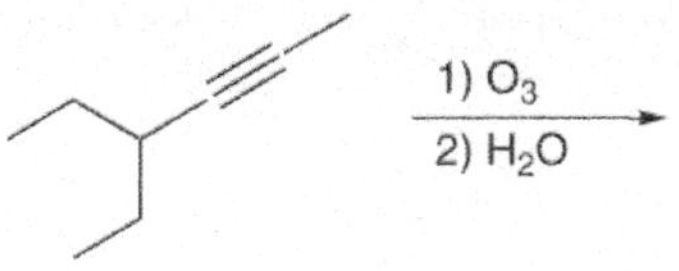

12.36

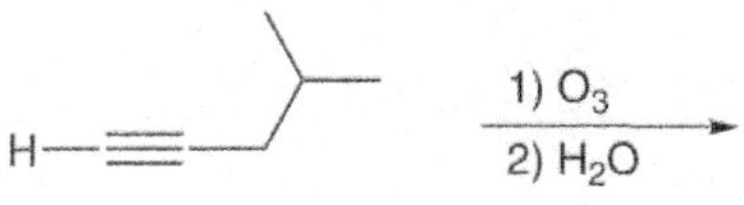

12.37

12.38

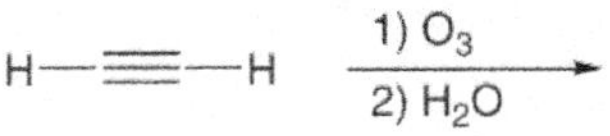

ALCOHOLS

Alcohols are compounds containing an OH group. In this chapter, we will learn how to make alcohols, and we will learn how to convert alcohols into a variety of other compounds.

13.1 NAMING AND DESIGNATING ALCOHOLS

We saw how to name alcohols in Chapter 5, when we learned the basic rules of nomenclature. In addition to their names, alcohols are also often classified into the following categories:

Primary   *Secondary*   *Tertiary*

This designation scheme (primary, secondary, or tertiary) indicates the degree of substitution of the carbon atom bearing the OH, called the alpha (α) carbon. This will become important later in this chapter when we explore reactions of alcohols. Specifically, we will encounter reactions where the designation (primary, secondary, or tertiary) will affect the reaction outcome.

EXERCISE 13.1   Identify whether the following alcohol is primary, secondary, or tertiary:

Answer   We begin by identifying the carbon atom connected directly to the OH group (the α carbon):

Now, we count the number of alkyl groups attached directly to the α carbon. In this case, there are two alkyl groups:

Therefore, this compound is a secondary alcohol.

234

PROBLEMS   Identify whether each of the following alcohols is primary, secondary, or tertiary:

13.2 _____   **13.3** _____

13.4 _____   **13.5** _____

13.2   PREDICTING SOLUBILITY OF ALCOHOLS

An alcohol, *by definition*, will contain an OH group in its structure. Therefore, we expect hydrogen bonding to occur:

Hydrogen bonding does NOT refer to a type of bond (even though it is called hydrogen *bonding*). This terminology is somewhat misleading. H-bonding actually refers to a force of attraction between molecules (an intermolecular force). This attraction is temporary, or fleeting, in the liquid and gaseous states. As two alcohol molecules approach each other, they are momentarily attracted to one another. At low enough temperatures (if the alcohol is a solid), then these forces will actually hold together the alcohol molecules in a crystalline state. But in the liquid or gaseous state, the molecules are bouncing off each other, and they are NOT permanently held together.

Why the misleading terminology (H-*bonding*)? At some point during your high school studies, you were likely exposed to a model of the DNA helix, which looks like a twisted ladder. That twisted ladder is actually two very large molecules (each of which is like a piece of cooked spaghetti), twisted around each other. Each rung of the ladder is a hydrogen bonding interaction between the two molecules. Each individual H-bond, by itself, would not be strong enough to hold the two large molecules together. However, the cumulative effect (of millions of H-bonds) holds the two molecules together in a twisted helix. Perhaps this can help us understand why we call these interactions (the rungs of the ladder) hydrogen *bonding*. This also explains why it is so easy to "unzip" the helix.

You might learn about the structure of DNA at the end of your organic chemistry course. For right now, we will be focused on problems that deal primarily with small molecules; and therefore, for our purposes, we should think of H-bonding as an interaction; a type of intermolecular force.

For any given alcohol, the strength of the H-bonding interactions will be greatly dependent on concentration. For a concentrated alcohol, there will be a fairly decent amount of H-bonding interactions at any given moment in time. However, if we dilute the alcohol in a solvent that cannot form hydrogen bonding interactions with the alcohol, then the H-bonding effect will be minimal:

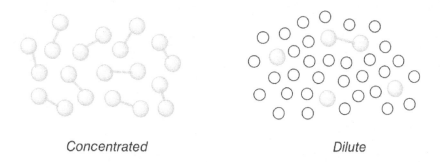

Concentrated   Dilute

Now consider diluting an alcohol in a solvent that *can* form hydrogen bonding interactions with the alcohol (for example, a solvent such as water). In such a case, the interaction is indeed very strong (it is similar to the effect of having a concentrated alcohol). This explains why methanol is miscible with water. The term *miscible* means that methanol can be mixed with water (they will dissolve in each other) *in all proportions*. However, not all alcohols are miscible with water. In fact, just the very small alcohols (methanol, ethanol, propanol, and *tert*-butanol) are miscible with water. To understand why, we must realize that every alcohol has two regions: the *hydrophobic* region, which does *not* interact well with water, and the *hydrophilic* region, which *does* interact well with water:

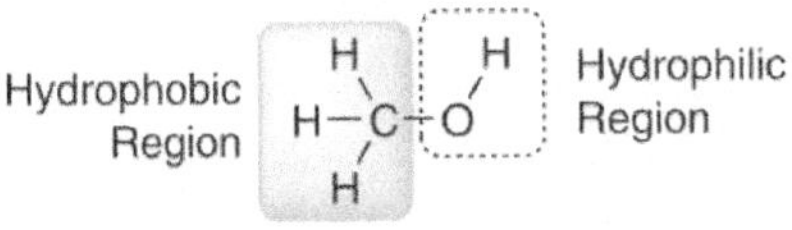

In the case of methanol, the hydrophobic end of the molecule is fairly small. This is true even of ethanol and propanol. But an interesting thing happens when we look at 1-butanol. The hydrophobic end of the molecule is now large enough to prevent miscibility:

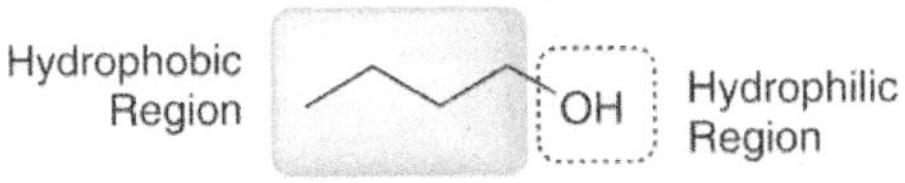

Water and 1-butanol will mix, but *not in all proportions*. That is, 1-butanol is considered to be soluble in water, rather than miscible. The term *soluble* means that only a certain volume of 1-butanol will dissolve in a specified amount of water at room temperature.

As we consider an even larger hydrophobic region, solubility decreases. For example, 1-octanol has a very low solubility in water at room temperature:

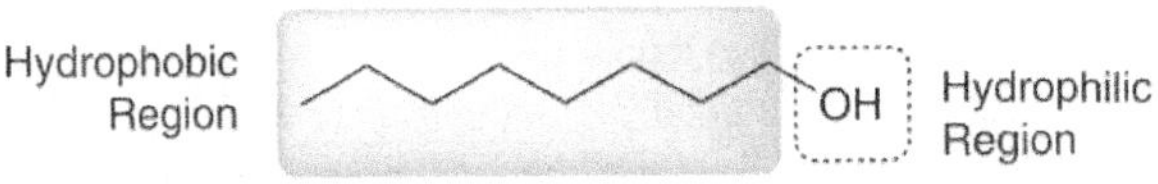

Here is a rule of thumb that can help us predict solubility: in order to have a high water-solubility, there should be no more than five carbon atoms *per OH group*. A compound with two OH groups and seven carbon atoms will be soluble in water. A compound with one OH group and seven carbon atoms will NOT be soluble in water. There are, of course, many exceptions to this simplified rule of thumb, but the rule can be helpful whenever we need to make a quick prediction about the solubility of an alcohol.

EXERCISE 13.6   Predict whether the following alcohol is soluble or insoluble in water:

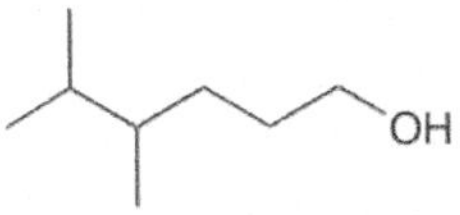

Answer   This compound has eight carbon atoms, and only one OH group. The hydrophobic region of the molecule is too large, and we expect the molecule to be insoluble in water.

PROBLEMS   Predict whether each of the following alcohols is soluble or insoluble in water:

13.7

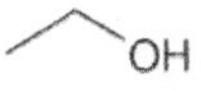

13.8

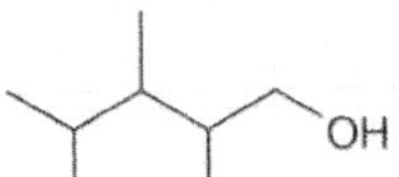

13.9

13.10

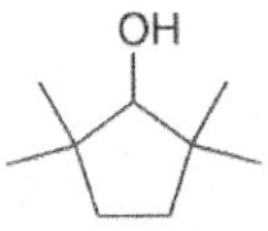

13.3 PREDICTING RELATIVE ACIDITY OF ALCOHOLS

As we learned in Chapter 3, in order to determine the relative acidity of a compound, we must carefully assess the stability of its conjugate base. For example:

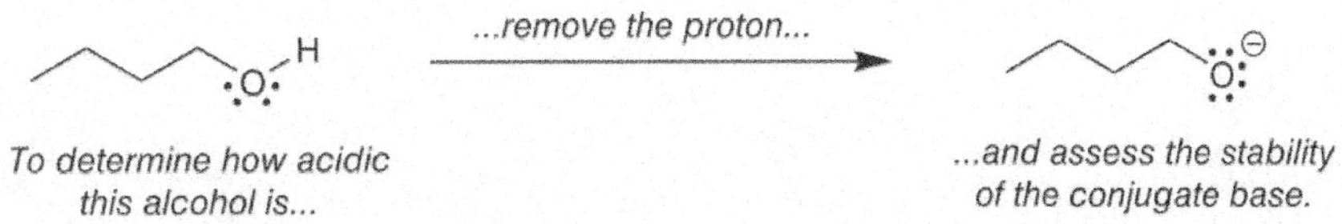

...remove the proton...

To determine how acidic
this alcohol is...

...and assess the stability
of the conjugate base.

When we draw the conjugate base of an alcohol, we see that the negative charge is on an oxygen atom. This is much more stable than a negative charge on a nitrogen atom, but it is not as stable as a negative charge on a halogen:

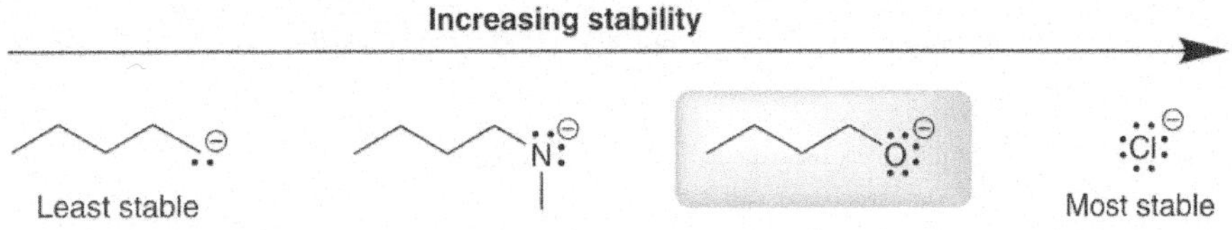

Increasing stability

Least stable   Most stable

Therefore, in terms of acidity, alcohols will lie somewhere between amines and hydrogen halides:

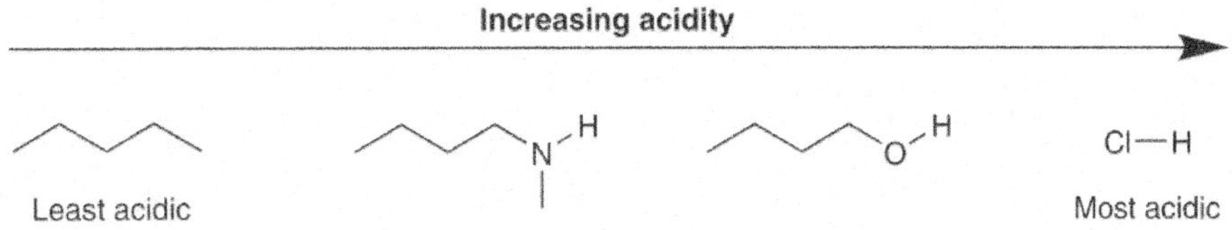

Increasing acidity

Least acidic   Most acidic

The pK_a for most alcohols will fall in the range 15–18. Remember, a low pK_a value means that a compound is fairly acidic, and a high pK_a value means that it is not very acidic. Compare the following pK_a values:

Increasing acidity

	R—H	R—NH$_2$	R—OH	X—H
pK_a values	Between 45 and 50	Between 35 and 40	Between 15 and 18	Between −10 and 3

In Chapter 3, we learned how to analyze the four factors that stabilize a negative charge (ARIO). The "I" in ARIO stood for "induction." We saw that neighboring halogens can withdraw electron density and stabilize the negative charge:

is more stable than

Therefore, the presence of halogens can lower the pK_a value of an alcohol to around 14. In other words, the presence of a halogen will make the compound more acidic.

The following two compounds provide an interesting comparison:

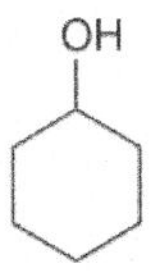

Cyclohexanol
$pK_a = 18$

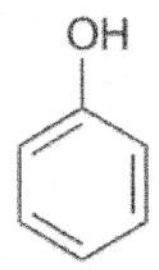

Phenol
$pK_a = 10$

The pK_a of cyclohexanol is 18, whereas the pK_a of phenol is 10. This means that phenol is eight orders of magnitude more acidic than cyclohexanol. In other words, phenol is 100 million times more acidic than cyclohexanol. Why is there such a huge difference? We can understand why cyclohexanol has a pK_a of 18, because that is within the expected range of an alcohol (15–18). So the real question is: Why is the pK_a of phenol so low? Why is phenol so much more acidic than a regular alcohol? To answer this question, we will use the "R" of ARIO (*Resonance*) to explain the acidity of phenol:

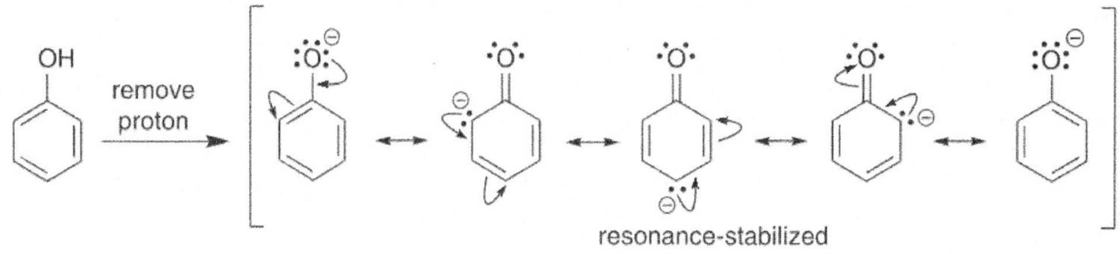

resonance-stabilized

The conjugate base of phenol is stabilized by resonance. This explains why phenolic protons are more acidic than typical alcoholic protons.

EXERCISE 13.11 Identify the most acidic proton in the following compound:

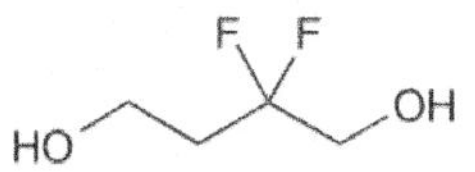

Answer The two OH groups will certainly be the two acidic protons in the compound, but we must choose which proton is more acidic. We must begin by drawing the conjugate base in each case, and then we will compare those conjugate bases, using our four factors (ARIO):

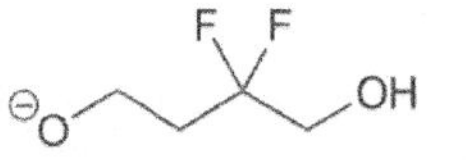

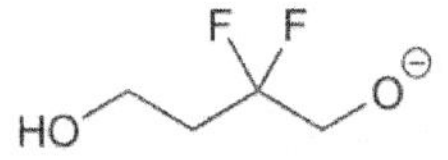

In both cases, the negative charge is on an oxygen atom, so our first factor (**A**RIO) does not help. Neither structure is stabilized by resonance, so our second factor (A**R**IO) also does not help. In this case, it is the third factor (induction) that indicates which structure is more stable. Both conjugate bases are stabilized by the electron-withdrawing effects of the two fluorine atoms. But the more stable conjugate base is the one where the two fluorine atoms are closer to the negatively charged oxygen atom:

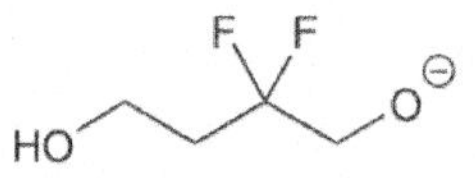

Therefore, we expect the following highlighted proton to be the most acidic proton:

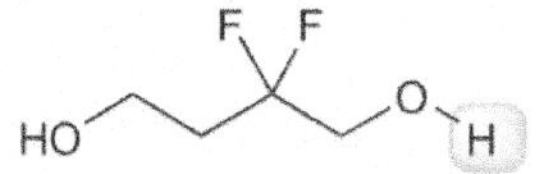

PROBLEMS For each pair of compounds, identify the compound that is more acidic. Make sure that you can explain your choice in each case.

13.12

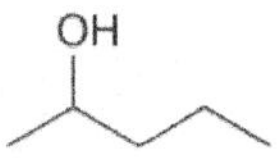

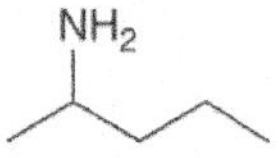

13.13

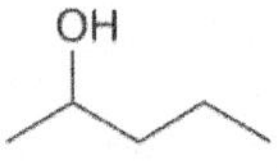

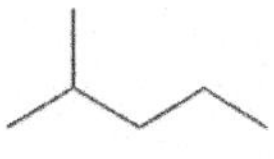

13.14

OH
⟍⟋⟍

OH
⟍⟋⟍F

13.15

OH
⟍⟋⟍

OH
⟍⟋⟍
       ‖
       O

13.16

OH
(phenol)

OH
(cyclohexanol)

13.17

OH   O
⟍   ‖
   ⟍⟋⟍

HO⟍⟋⟍
        ‖
        O

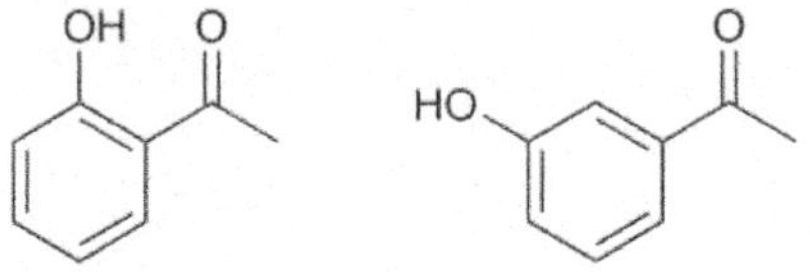

13.4 PREPARING ALCOHOLS: A REVIEW

We will learn many ways of making alcohols in this chapter. First, let's review the various reactions that we have already learned in previous chapters. All of these reactions can be used to produce an alcohol:

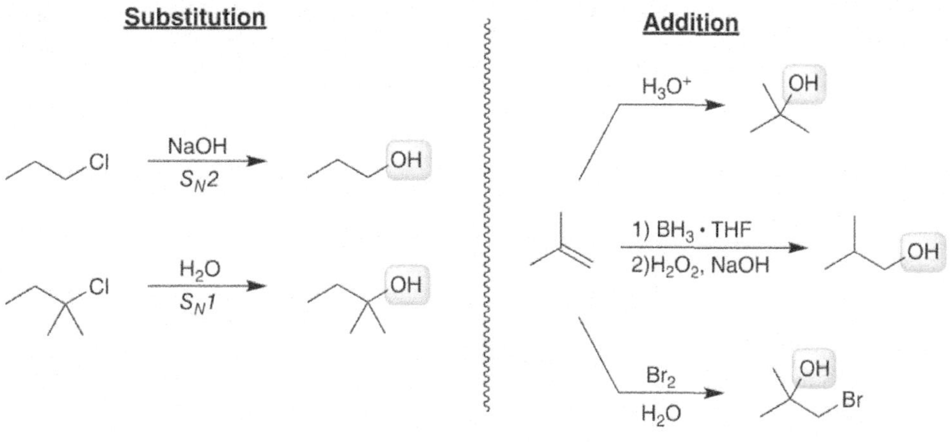

Substitution

⟍⟋⟍Cl $\xrightarrow[S_N2]{\text{NaOH}}$ ⟍⟋⟍OH

⟍⟋⟍Cl $\xrightarrow[S_N1]{\text{H}_2\text{O}}$ ⟍⟋⟍OH

Addition

$\xrightarrow{\text{H}_3\text{O}^+}$ OH

$\xrightarrow[\text{2)H}_2\text{O}_2\text{, NaOH}]{\text{1) BH}_3 \cdot \text{THF}}$ ⟍⟋⟍OH

$\xrightarrow[\text{H}_2\text{O}]{\text{Br}_2}$ OH, Br

Before we can learn *new* methods for making alcohols, you must make sure that you remember the methods that you have already learned (above). Let's get some practice:

EXERCISE 13.18 What reagents would you use to accomplish the following transformation:

⟍⟋⟍⟋ $\longrightarrow$ HO⟍⟋⟍⟋

Answer In this case, H and OH are added across the pi bond in an *anti*-Markovnikov addition. No chiral centers were formed, so the stereochemical outcome is not relevant. We have only seen one way to achieve an *anti*-Markovnikov addition of water across a pi bond: hydroboration–oxidation. Therefore, our answer is:

⟍⟋⟍⟋ $\xrightarrow[\text{2) H}_2\text{O}_2\text{, NaOH}]{\text{1) BH}_3 \cdot \text{THF}}$ HO⟍⟋⟍⟋

PROBLEMS   What reagents would you use to accomplish each of the following transformations:

13.19

13.20

13.21

13.22

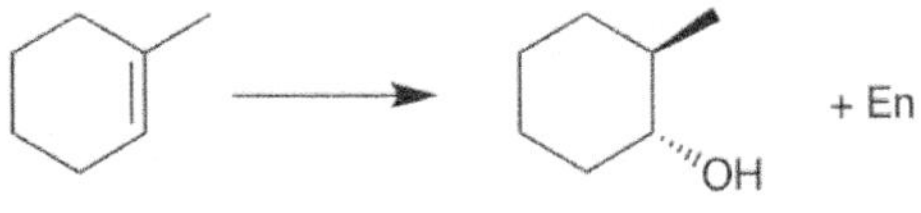

13.5   PREPARING ALCOHOLS VIA REDUCTION

In this section, we will learn how to prepare alcohols through a *reduction* process. In order to understand what the word *reduction* means, we must go back and review what oxidation states are. We will do that now:

There are two methods for counting electrons: *formal charge* and *oxidation state*. These two counting methods actually represent two flipsides of the same coin. To calculate formal charge, we treat all bonds as covalent, regardless of whether they are or not:

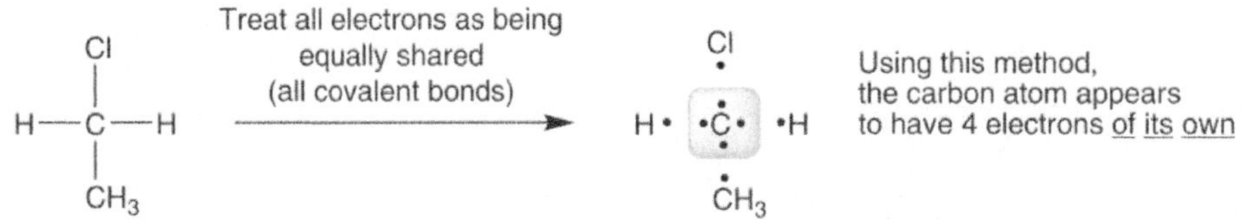

When we treat all bonds as covalent, the carbon atom appears to have four electrons of its own. Carbon is *supposed to have* four valence electrons. When we compare how many electrons carbon actually has with the number of electrons it is supposed to have, we see that everything is just right in this case. It is supposed to have four valence electrons, and it is clearly using four valence electrons. Therefore, there is no formal charge.

On the flipside, we will now calculate the *oxidation state* of the same carbon atom. To calculate the oxidation state of an atom, we treat all bonds as ionic, regardless of whether they are or not:

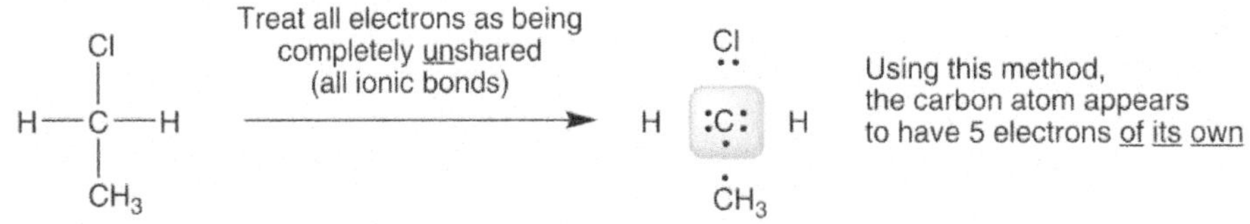

For each bond, we give both electrons to the more electronegative atom. For the C—Cl bond, chlorine is more electronegative, so chlorine gets both electrons. For each of the C—H bonds, carbon is *very slightly* more electronegative than hydrogen, so carbon will get both electrons in each case. But for the C—C bond, neither atom is more electronegative than the other, so we must place one electron on each carbon atom. Now when we count the electrons, it appears as though the carbon atom is using five electrons of its own. But we know that carbon is only supposed to have four valence electrons. This carbon atom is using one extra electron, and therefore, this carbon atom will have an oxidation state of −1.

Formal charges and *oxidation states* represent two different ways of estimating electron density. Neither method is perfectly accurate. Each method assumes an extreme that is not true. Formal charges are calculated based on the assumption that all bonds are *covalent* (generally an incorrect assumption), and oxidation states are calculated based on the assumption that all bonds are *ionic* (generally an incorrect assumption). Earlier in this book, we focused our attention on formal charges exclusively. For purposes of this section, we will now focus our attention exclusively on oxidation states.

Carbon can range in oxidation state from −4 to +4:

H
|
H—C—H
|
H

−4

O
‖
C
‖
O

+4

The oxidation state of the carbon atom in an alcohol will be dependent on the identities of the atoms that are attached to the carbon atom. Here are some examples (make sure that you can calculate and verify that the oxidation states shown here are correct):

OH
|
H⦂C⦂H
|
H

−2

OH
|
H⦂C—CH₃
|
H

−1

OH
|
H⦂C—CH₃
|
CH₃

0

OH
|
H₃C—C—CH₃
|
CH₃

+1

Let's get practice:

EXERCISE 13.23   Calculate the oxidation state of the carbon atom highlighted below:

OH

Answer   For each bond, we give both electrons to the more electronegative atom. For the C—O bond, oxygen is more electronegative, so oxygen gets both electrons. For each of the C—C bonds, we treat the electrons as being shared equally (one electron on one carbon atom, and the other electron on the other carbon atom). For the C—H bond, carbon is *very slightly* more electronegative than hydrogen, so carbon will get both electrons.

OH
|
H₃C—C—CH₃
|
H

Treat all electrons as being
completely <u>unshared</u>
(all ionic bonds)

⟶

OH
•
H₃C •   •C•   •CH₃
•
H

Using this method,
the carbon atom appears
to have 4 electrons <u>of its own</u>

When we count the electrons that carbon is using now, it appears as though the carbon atom is using four electrons of its own. Next, we compare that number to the number of valence electrons that carbon is supposed to have (four valence electrons). This carbon atom is using exactly the right number of electrons, and therefore, this carbon atom will have an oxidation state of 0.

PROBLEMS   Calculate the oxidation state of the carbon atom that is highlighted in each of the following compounds:

13.24   O
             ‖
             —H   _____

13.25   O
             ‖
             —OH   _____

13.26 _____

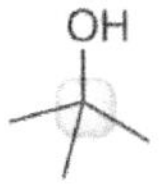

13.27 _____

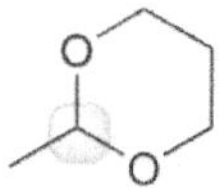

13.28 _____

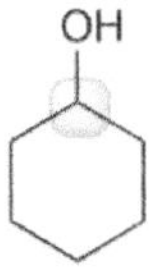

13.29 _____

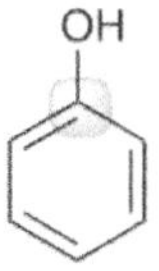

Now let's compare the oxidation states of the central carbon atom in each of the following compounds:

Alkane	Alcohol	Aldehyde	Carboxylic acid	Carbon dioxide

| -4 | -2 | 0 | $+2$ | $+4$ |

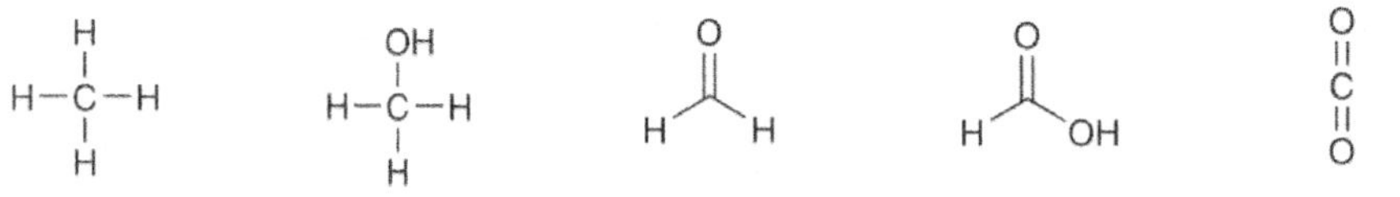

Notice the trend. Let's ignore the two extremes above (alkane and carbon dioxide), and let's focus on the middle three compounds: alcohols, aldehydes, and carboxylic acids. Carboxylic acids are at a higher oxidation state than aldehydes, which in turn are at a higher oxidation state than alcohols. Now imagine that we are running a reaction that converts an alcohol into an aldehyde or a carboxylic acid. This reaction would constitute an increase in oxidation state. Whenever we run a reaction that increases the oxidation state, we say that an *oxidation* has occurred. Therefore, converting a primary alcohol into an aldehyde or a carboxylic acid is called an oxidation:

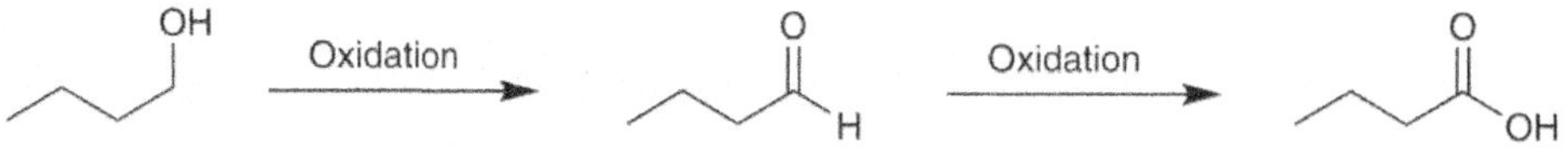

Similarly, it is also called an oxidation when we convert a secondary alcohol into a ketone:

In order to accomplish an oxidation, there is always some other compound that must itself get reduced (that compound is called the oxidizing agent, because it causes the desired oxidation).

Whenever we run a reaction that involves a *decrease* in oxidation state, we say that a *reduction* has occurred. For example, converting a ketone or aldehyde into an alcohol:

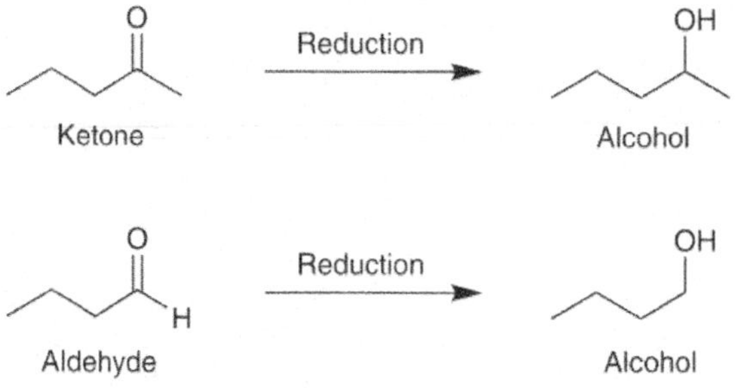

Ketone   Alcohol

Aldehyde   Alcohol

Notice that the product of reducing a C=O bond is simply an alcohol. Therefore, we can make alcohols by reducing ketones or aldehydes. To accomplish this reduction, we will need a reducing agent. There are two common reagents for doing this. In order to understand the structures of these reagents, let's quickly review something from the periodic table. The elements boron and aluminum are in the same column of the periodic table (just to the left of carbon):

Each of these elements (boron and aluminum) has three valence electrons. Therefore, each of these elements can form three bonds:

In the compounds shown above, boron and aluminum are using their valence electrons to form bonds, but notice that neither one has an octet. Each element is capable of forming a fourth bond in order to obtain an octet, but then each element will bear a formal charge of −1.

Sodium Borohydride
(NaBH₄)

Lithium Aluminum Hydride
(LiAlH₄)

In both of these compounds, the central atom (B or Al) has four bonds and a negative charge. In the first compound ($NaBH_4$), Na^+ is used as the counter-ion. In the second compound ($LiAlH_4$), Li^+ is used as the counter-ion. The choice of counter-ion is usually not relevant for our discussion, and we will choose to ignore it from now on.

Either one of these reagents ($NaBH_4$ or $LiAlH_4$) can attack a ketone or aldehyde to give an alcohol:

1) LiAlH₄

2) H₂O

NaBH₄

MeOH

The mechanisms are somewhat complex and beyond the scope of our course, but the following simplified mechanism will be useful:

As we carefully analyze the first step of this mechanism, we see that the reducing agent ($LiAlH_4$) is simply functioning as a source of H^-. The mechanism of this first step is the same whether we use $LiAlH_4$ or $NaBH_4$. Then, in the second step, a proton source is used to generate the alcohol.

We have now seen the way that LiAlH$_4$ and NaBH$_4$ work. Essentially, they are both just sources of nucleophilic H$^-$. You might wonder why we can't just use sodium hydride (NaH), like this:

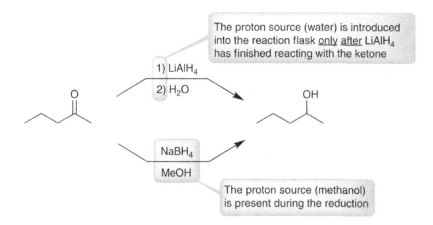

This actually doesn't work well. Recall (Chapter 10) that nucleophilicity is dependent on polarizability (the size of the atom). Large atoms (like sulfur or iodine) are very polarizable, and therefore, they are excellent nucleophiles. Small atoms are not polarizable, and they make poor nucleophiles. H$^-$ is as small as they come, and therefore H$^-$ is not a great nucleophile. H$^-$ is a good base (in fact, it is an excellent base), but it is not a good nucleophile. So, we use LiAlH$_4$ or NaBH$_4$ as a "source" of nucleophilic H$^-$. We can think of LiAlH$_4$ and NaBH$_4$ as *delivery agents of nucleophilic H$^-$*.

Although we can use either reagent to reduce a ketone or aldehyde, nevertheless, there are striking differences between LiAlH$_4$ and NaBH$_4$. LiAlH$_4$ is much more reactive than NaBH$_4$. If we look at the structure of LiAlH$_4$, we see that the central atom (bearing the negative charge) is aluminum. By contrast, in NaBH$_4$, the negative charge is on boron. Aluminum is much larger (more polarizable) than boron, and therefore, LiAlH$_4$ is much more reactive. The higher reactivity of LiAlH$_4$ is a topic that will be important in the second semester of organic chemistry. For now, the difference in reactivity is less apparent, because both LiAlH$_4$ and NaBH$_4$ will react with ketones and aldehydes to give alcohols.

For purposes of our discussion here, there is one noticeable difference between LiAlH$_4$ and NaBH$_4$. Whenever we use LiAlH$_4$, we must provide a proton source *after* the reaction (*after* LiAlH$_4$ has had a chance to attack the ketone or aldehyde), because LiAlH$_4$ is so reactive that it would violently react with water. However, when we use NaBH$_4$ (the milder reducing agent), we do not have to worry about NaBH$_4$ reacting with the proton source. The proton source (methanol is often used) can be present in the reaction flask at the same time as NaBH$_4$. This is shown in the following way:

Notice that with LiAlH$_4$, two separate steps are required.

So far, we have seen two sources of nucleophilic H$^-$ (LiAlH$_4$ and NaBH$_4$). There are many other hydride reagents that have been prepared (some that are even more reactive than LiAlH$_4$, and others that are even milder than NaBH$_4$), for example:

where R can be almost anything

By carefully choosing the R groups above (to be either electron-donating groups or electron-withdrawing groups), we can very carefully control the reactivity of the hydride reagent. For now, we will simply focus our attention on the two commonly used hydride reagents: $LiAlH_4$ and $NaBH_4$.

Let's do some problems where we use $LiAlH_4$ or $NaBH_4$ to reduce ketones or aldehydes into alcohols.

EXERCISE 13.30 Identify a ketone and any other reagents you would use to prepare the following alcohol:

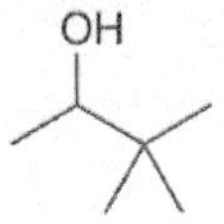

Answer This alcohol could be prepared from the following ketone:

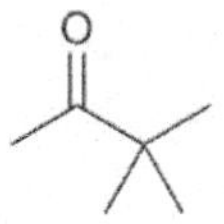

We can convert this ketone into our product (the desired alcohol) using either $LiAlH_4$ followed by water, or $NaBH_4$ together with methanol:

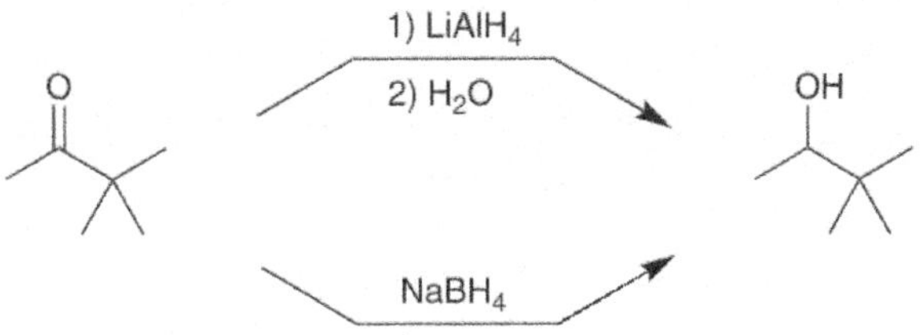

PROBLEMS Identify the starting ketone or aldehyde you would use to prepare each of the following alcohols through reduction reactions:

13.31

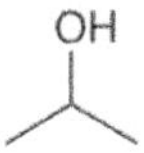

13.32

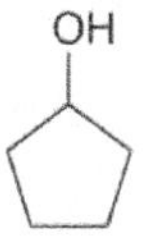

13.33

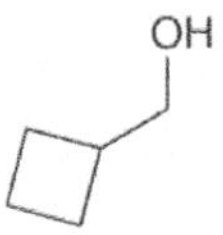

13.34

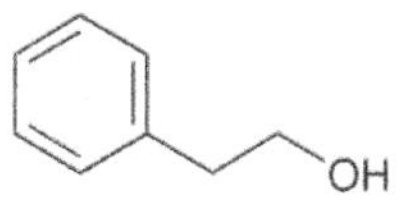

13.35

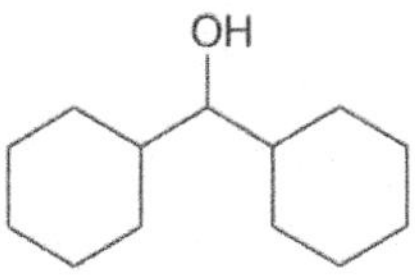

13.36

13.6 PREPARING ALCOHOLS VIA GRIGNARD REACTIONS

In the previous section, we saw that ketones and aldehydes can be attacked by a suitable source of H⁻. In a similar way, ketones and aldehydes can also be attacked by a suitable source of R⁻. Compare these two reactions:

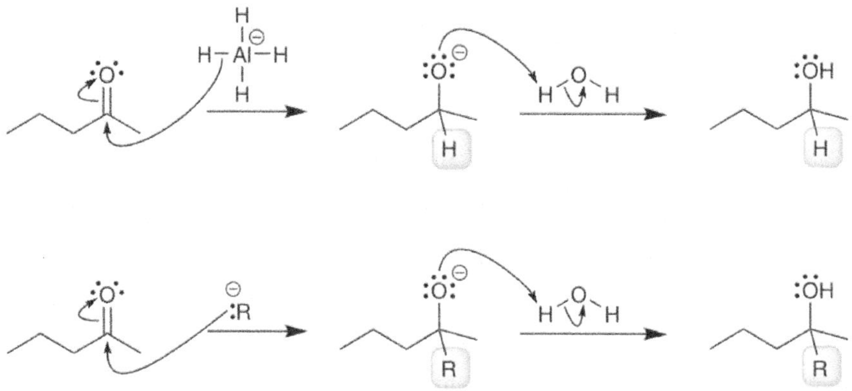

Both H⁻ and R⁻ can attack a ketone or aldehyde to give an alcohol. The main difference is the effect on the carbon skeleton. With H⁻, the carbon skeleton does not change at all. But with R⁻, the carbon skeleton gets larger. We are forming a C—C bond. We will soon see that this is very important for synthesis problems. For now, let's focus on how we can make R⁻ in the first place. After all, a negative charge on a carbon atom is not very stable (and therefore not trivial to make).

There are many ways to get a negative charge on a carbon atom. Later in this course, you will spend a lot of time learning about special C⁻ compounds. For now, we will just learn about one such compound, called a Grignard reagent:

$$R-MgX$$

where R is an alkyl group. In a Grignard reagent, there is a carbon atom that is directly connected to a magnesium atom. If we compare the electronegativity values of C and Mg, we will see that C is much more electronegative than Mg. Therefore, carbon pulls more strongly on the electron density and will develop a negative charge.

Grignard reagents are formed by inserting magnesium in between a C—X bond (where X is a halogen):

$$R-X \xrightarrow{\text{Mg}} R-Mg-X$$

Grignard reagent

The mechanism for this insertion of magnesium is beyond the scope of this course, and therefore, we will not go into it. For now, we should just know that we can insert Mg into a C—X bond (where X is Cl, Br, or I). Here are some examples:

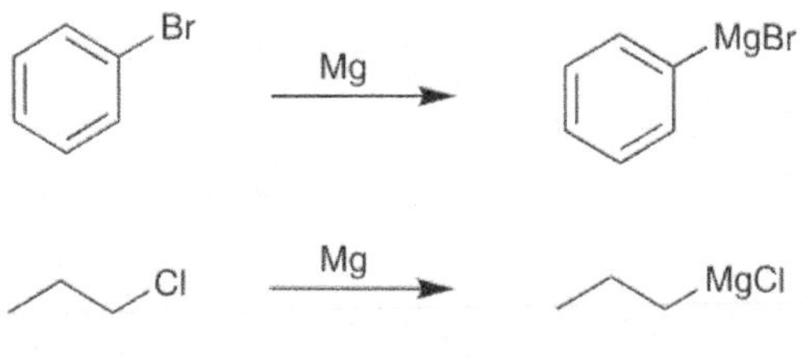

Once a magnesium atom is inserted between a C—X bond, the newly formed C—Mg bond is somewhat ionic in character (because carbon is so much more electronegative than magnesium). Therefore, there are two acceptable ways to draw a Grignard reagent:

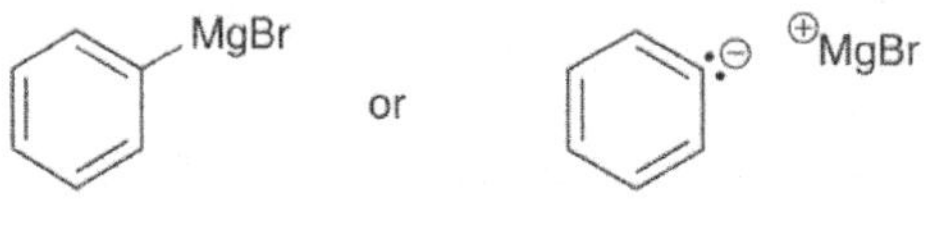

Neither drawing is absolutely correct. The first drawing assumes a perfectly *covalent* bond, which is certainly not accurate. The second drawing assumes an *ionic* bond. The reality actually lies somewhere in between these two extremes, although it is a lot closer to being ionic.

When a Grignard reagent attacks a ketone or aldehyde, it forms an alcohol, with a newly installed R group:

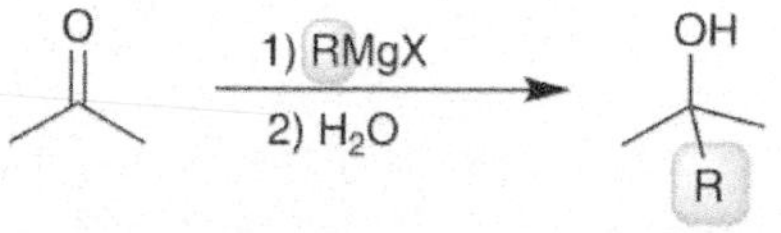

Here are two specific examples:

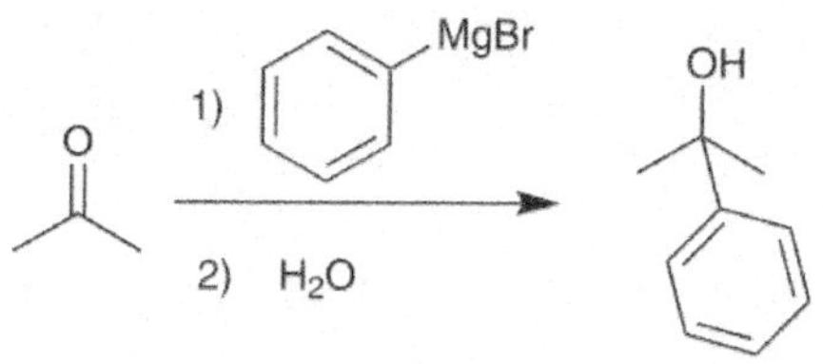

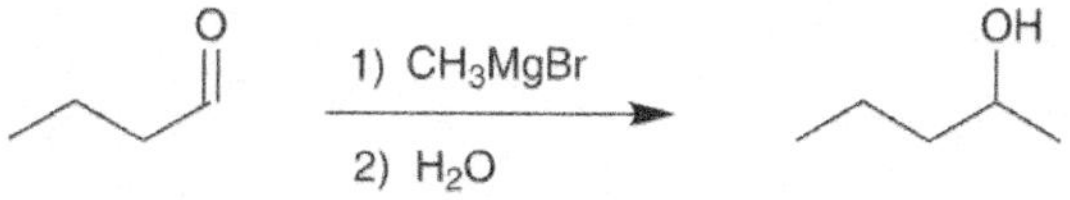

Grignard reagents also attack other compounds possessing a C=O bond (such as esters) to produce alcohols. For now, let's just focus on Grignard reagents attacking ketones and aldehydes.

EXERCISE 13.37   Show how you would use a Grignard reaction to prepare the following alcohol:

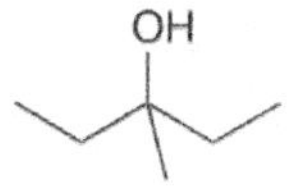

Answer   Working backwards, we could have formed the following bond (see the squiggly line), using a Grignard reaction:

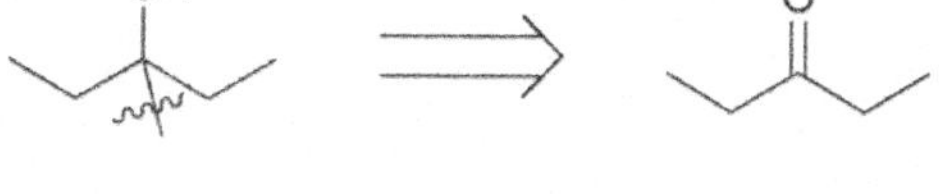

Notice the retrosynthetic arrow above. That arrow indicates that we can make the desired alcohol from the ketone above, like this:

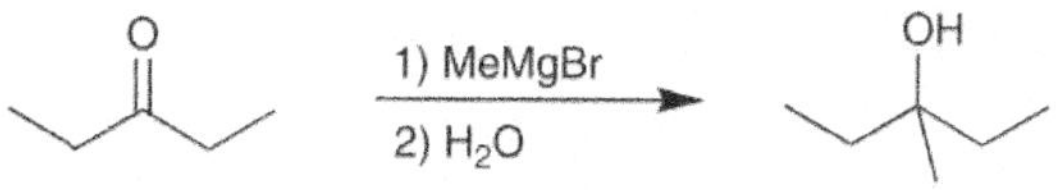

Alternatively, we could have formed this bond:

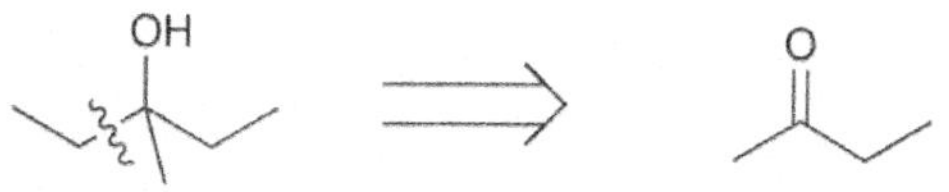

Once again, notice the retrosynthetic arrow above, which indicates that we can make the desired alcohol from the ketone above, like this:

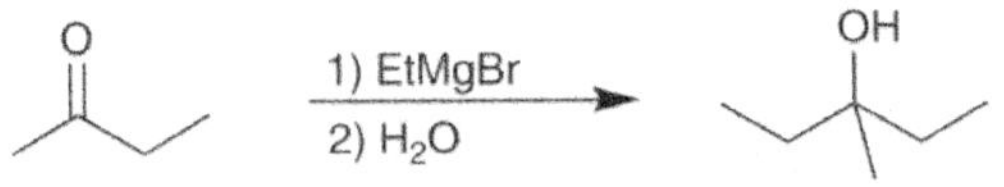

This problem illustrates an important point: we have seen two perfectly correct answers to this problem. In fact, from now on, we will rarely encounter synthesis problems that have only one solution. More often, we will find synthesis problems that have more than one acceptable answer.

PROBLEMS Show how you would use a Grignard reaction to prepare each of the following alcohols:

13.38

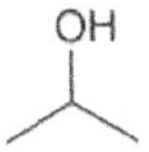

13.39

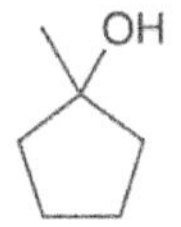

13.40

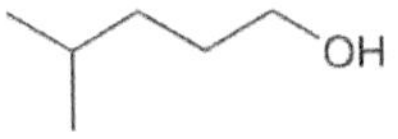

13.41

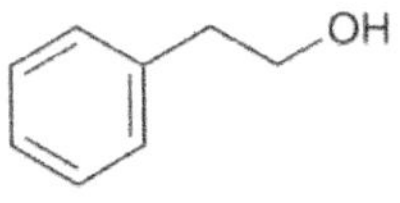

13.42

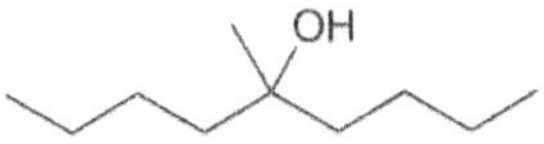

13.43

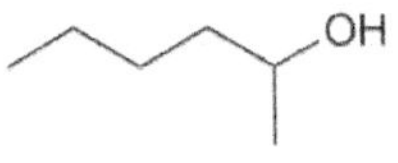

In this section, we learned that a Grignard reagent can attack a ketone or aldehyde to give an alcohol. This reaction is so incredibly important because we are not only converting a C=O bond into an alcohol, but we are also introducing an R group into the compound:

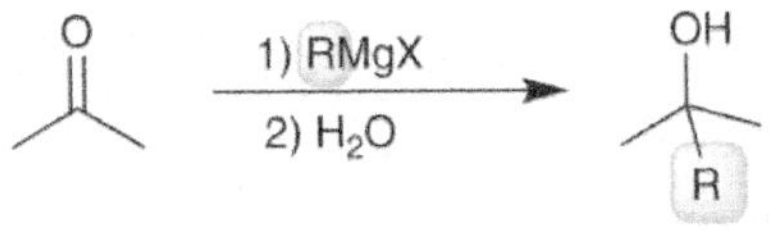

This is a *C—C bond-forming reaction*. Until now, we have only seen one kind of C—C bond-forming reaction (the alkylation of terminal alkynes). Now we have a second approach for creating a C—C bond. We must put this reaction into our synthetic "toolbag." Whenever we have a synthesis problem, we must always ask two questions: Is there a change in the carbon skeleton, and is there change in the functional group (as we will see in the next chapter). If we have a synthesis problem where the carbon skeleton is getting larger, then we know we will have to create a C—C bond. And so far in this book, we have seen two ways to do that. We can alkylate a terminal alkyne, or we can use a Grignard reagent to attack a ketone or aldehyde.

It is important to be familiar with Grignard reactions, because they will appear several times throughout the rest of your organic chemistry course. Let's practice a bit. We will begin with a few problems that are just one-step syntheses (Grignard) to make sure you got it. Then, the last two problems in this problem set will be multi-step syntheses (using a Grignard reaction *and* other reactions we have seen).

PROBLEMS Propose an efficient synthesis for each of the following transformations:

13.44

13.45

13.46

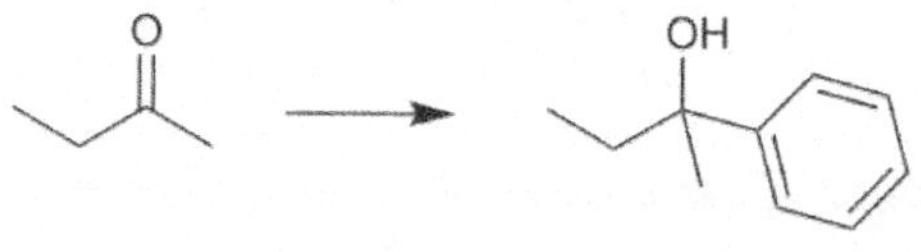

13.47

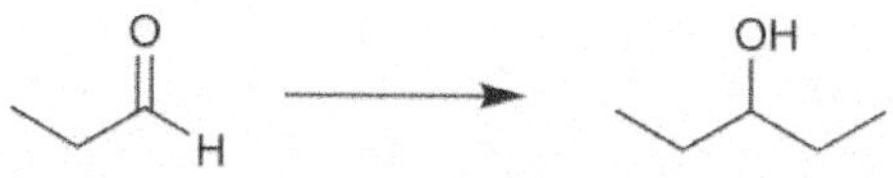

13.48

13.49

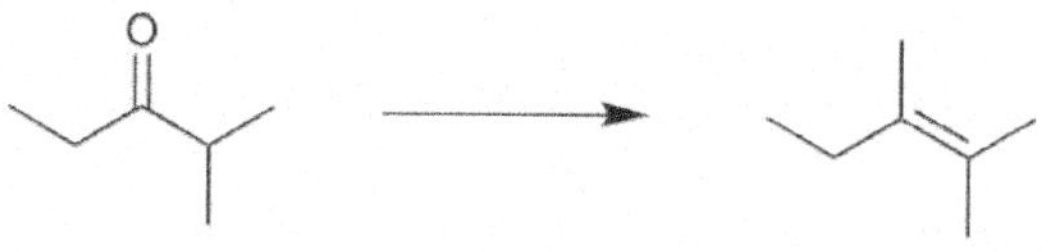

13.7 SUMMARY OF METHODS FOR PREPARING ALCOHOLS

So far, we have seen many ways to make alcohols. We can make primary, secondary, or tertiary alcohols through a variety of methods. As a review, identify the reagents necessary to accomplish each of the following transformations:

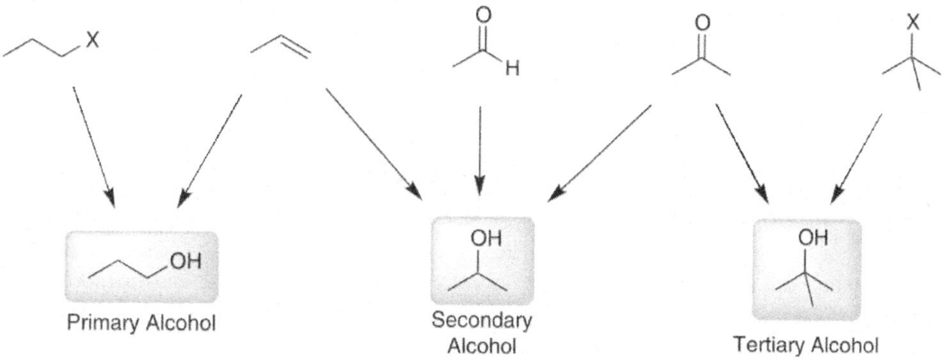

Primary Alcohol

Secondary
Alcohol

Tertiary Alcohol

EXERCISE 13.50   Using any reagents of your choice, show three different ways of preparing the following alcohol:

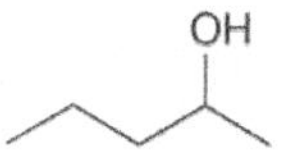

OH

Answer   This is a secondary alcohol. We have seen many ways to make secondary alcohols. We can start with a ketone, and reduce with $LiAlH_4$ or $NaBH_4$:

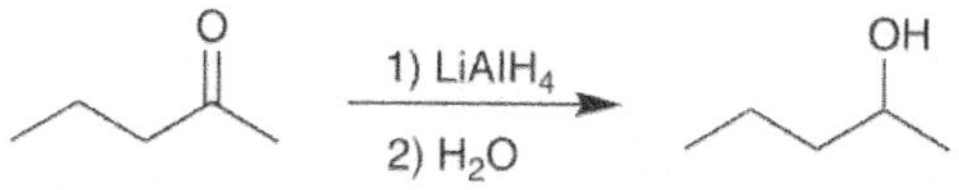

1) $LiAlH_4$

2) H_2O

OH

Or, we can start with an aldehyde, and perform a Grignard reaction:

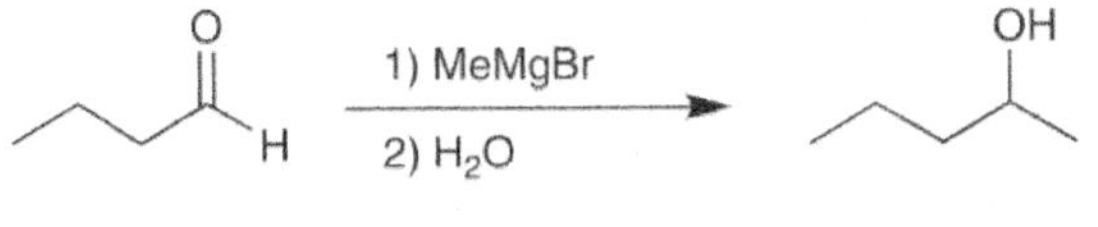

$$\text{(propanal)} \xrightarrow[\text{2) H}_2\text{O}]{\text{1) MeMgBr}} \text{(2-pentanol)}$$

Or we can start with an alkene and perform an addition (acid-catalyzed hydration):

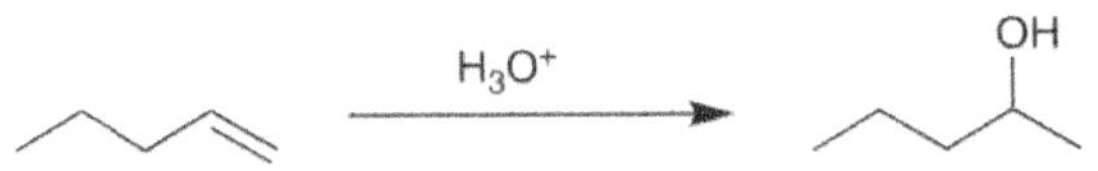

$$\text{(1-pentene)} \xrightarrow{\text{H}_3\text{O}^+} \text{(2-pentanol)}$$

PROBLEMS   Using any reagents of your choice, show at least two ways to make each of the following alcohols:

13.51

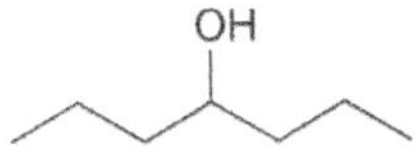

13.52

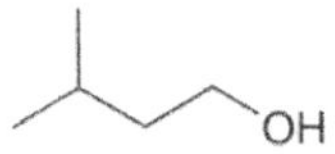

13.53

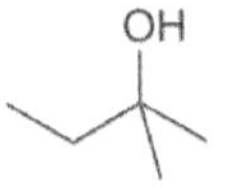

13.54

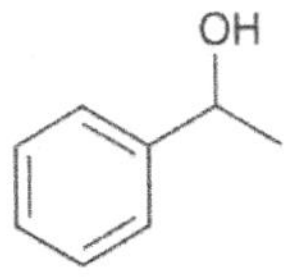

13.55

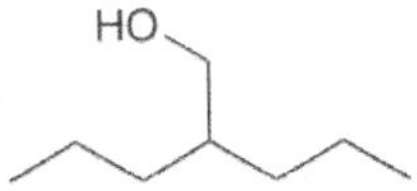

13.56

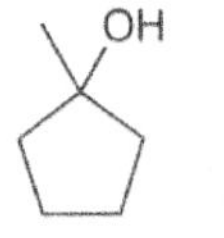

13.8   REACTIONS OF ALCOHOLS: SUBSTITUTION AND ELIMINATION

Now that we have seen how to make alcohols, we will focus our attention on reactions of alcohols. Let's start by reviewing reactions that we have already seen: substitution and elimination. Let's begin our review with elimination reactions first. We have seen two types of elimination reactions: E1 an E2.

To eliminate an OH group under E1 conditions, we need acidic conditions:

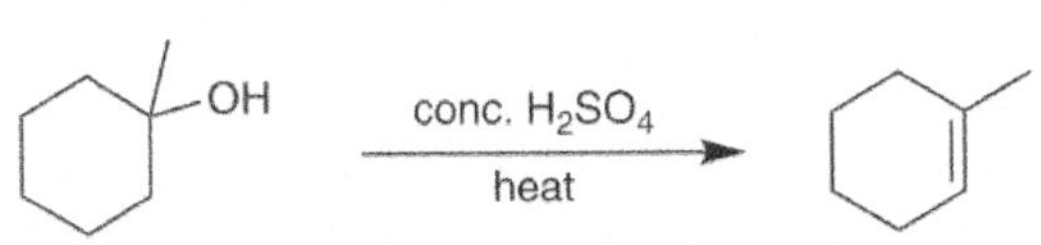

$$\text{(1-methylcyclohexanol)} \xrightarrow[\text{heat}]{\text{conc. H}_2\text{SO}_4} \text{(1-methylcyclohexene)}$$

In this reaction, acidic conditions are used in order to protonate the OH group, effectively converting the OH group into an excellent leaving group. Then, the leaving group leaves to give an intermediate carbocation, which then loses a proton to give the alkene (for a review of this mechanism, see Chapter 10). Because the E1 mechanism involves an intermediate carbocation, this reaction works well with a tertiary alcohol, although it can also be achieved with secondary alcohols.

For primary alcohols, we will *not* be able to form a carbocation. Therefore, we generally cannot use an E1 reaction on primary alcohols. Instead, we use an E2 reaction. But we still have the same problem—OH is still a terrible leaving group. That problem was resolved in an E1 reaction, because the OH group is protonated under strongly acidic conditions, thereby converting a bad leaving group into a good leaving group. In an E2 reaction, we will also need to convert the OH group into a better leaving group. But we cannot use a strong acid, because that would be incompatible with the strongly basic conditions that are required in order to perform an E2 process. Therefore, if we want to use E2 conditions, we will first need to convert the OH group into a tosylate group:

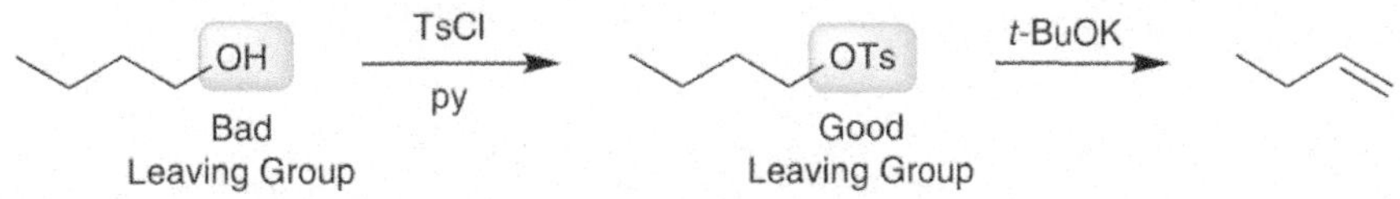

In the last step, *tert*-butoxide was used to favor elimination over substitution (see Section 10.10). In summary, we have seen that we can use either an E1 process or an E2 process to convert an alcohol into an alkene.

Now let's consider substitution reactions involving alcohols:

$$R-OH \longrightarrow R-X$$

If we want to perform a substitution reaction with an alcohol, we have the same issue that we had when we explored elimination reactions a few moments ago—the OH group is not a good leaving group. So, we must convert the OH into a better leaving group. There are several ways to do that for substitution reactions. We will look at four different ways:

1. *Via an S_N1 process performed under strongly acidic conditions.* With a tertiary alcohol, we can perform an S_N1 process by treating the alcohol with HX (where X is a halogen, such as Br or Cl). Under these strongly acidic conditions, the OH group is protonated to form a good leaving group, in the presence of a strong nucleophile (a halide ion, such as bromide or chloride). The first two steps below resemble the first two steps of the E1 process that we just saw. But in the last step, instead of forming an alkene, bromide attacks the carbocation in an S_N1 process, giving a tertiary alkyl halide:

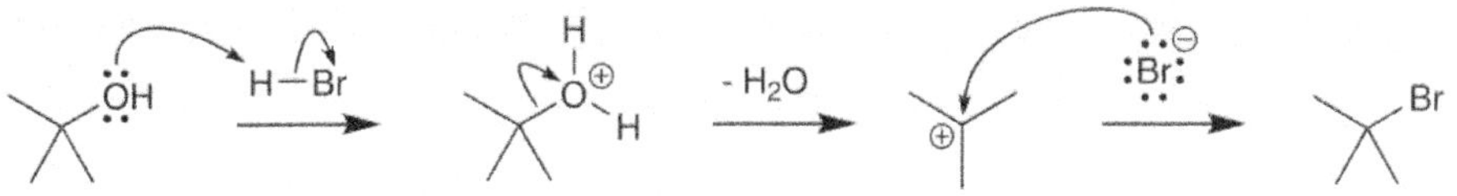

2. *Via an S_N2 process performed under strongly acidic conditions.* With a primary or secondary alcohol, we can perform an S_N2 process by treating the alcohol with HX (where X is a halogen, such as Br or Cl). Under these strongly acidic conditions, the OH group is protonated to form a good leaving group, in the presence of a strong nucleophile (a halide ion, such as bromide or chloride). The nucleophile then attacks the protonated alcohol (called an oxonium ion) in an S_N2 process, to give the corresponding alkyl halide:

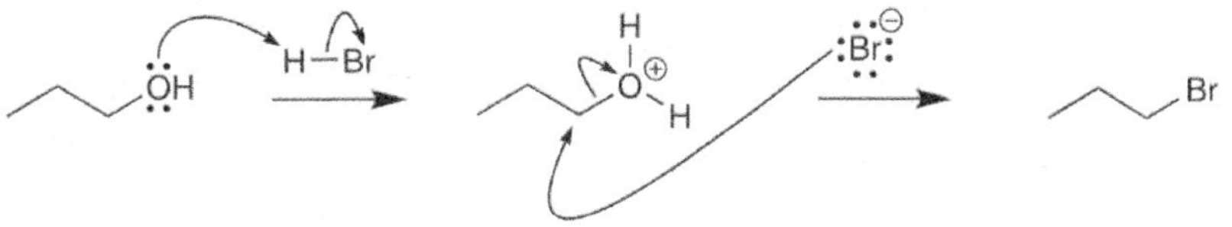

In this reaction, a carbocation would be too unstable to form. After protonating the OH group (converting it into a better leaving group), the leaving group is expelled when the nucleophile attacks, in an S_N2 process.

This reaction works well with HBr, but it does not work so well with HCl. Chloride is smaller than bromide, and therefore, chloride is less polarizable than bromide. This means that chloride is less nucleophilic than bromide. Although chloride is still fairly nucleophilic, the reaction is slow, and we can help it along with a catalyst, such as $ZnCl_2$:

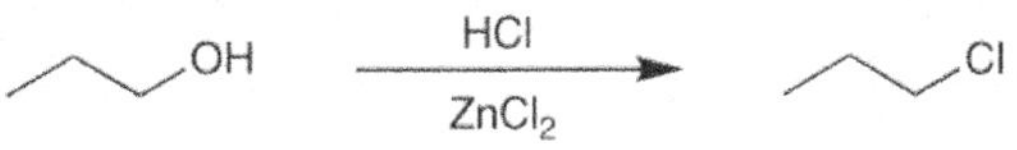

The effect of ZnCl$_2$ is similar to the effect of protonating an OH group, but ZnCl$_2$ just does a better job than H$^+$:

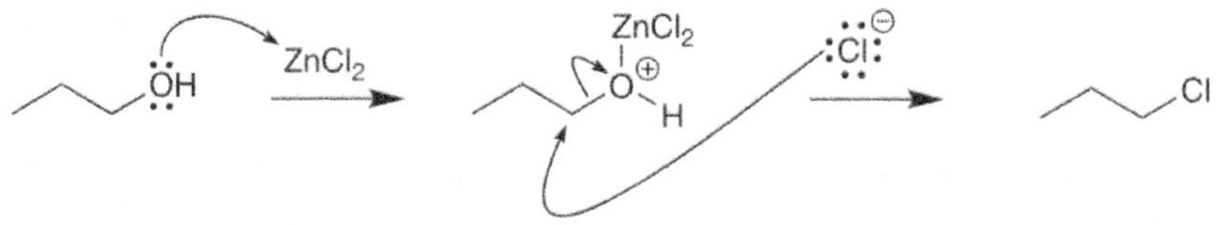

3. A third way to perform a substitution reaction is to convert the OH group into a tosylate, and then do an S$_N$2 reaction:

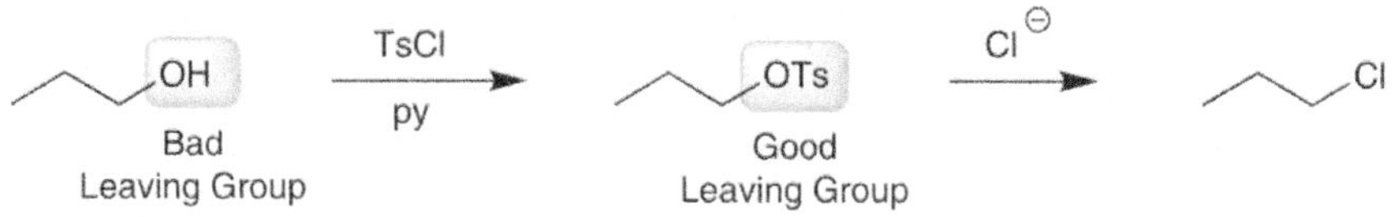

Bad Leaving Group   Good Leaving Group

4. There is another way to convert an OH group into a better leaving group for the purpose of doing a substitution reaction. We can use thionyl chloride (SOCl$_2$) to convert an alcohol into an alkyl chloride:

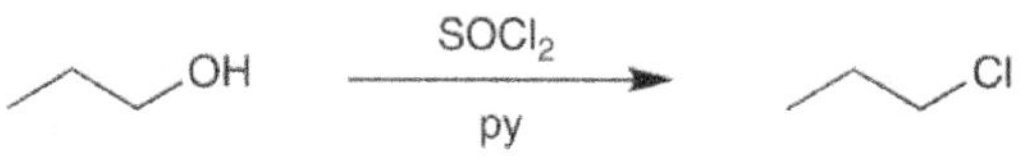

A mechanism for this process is shown below:

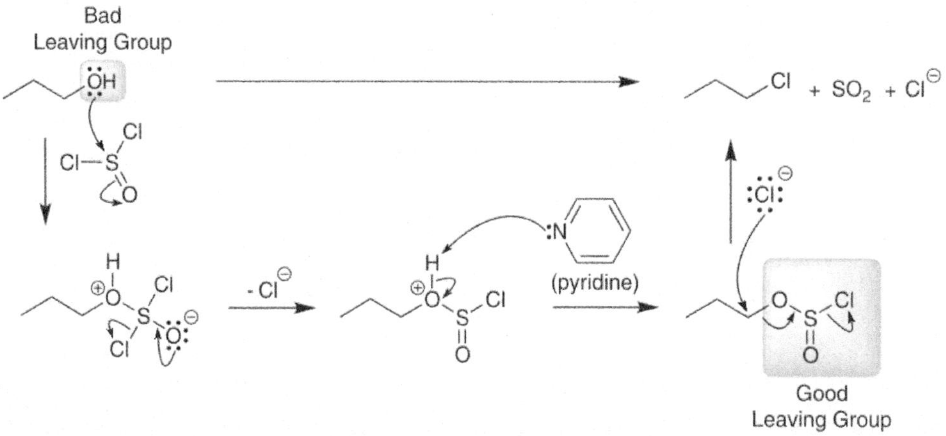

Bad Leaving Group

Good Leaving Group

Notice that the first three steps simply convert a bad leaving group into a good leaving group. Also notice that SO$_2$ (a gas) is produced as a side product. This gas can escape from the reaction flask, thereby pushing the equilibrium toward formation of products. In fact, if the gas is free to leave as it is formed, the reaction will be pushed to completion.

This last reaction might seem like a new reaction. But it really is not so new. It is just an S$_N$2 reaction where the OH group is first converted into a better leaving group, and then chloride functions as the nucleophile.

In this section, we have started looking at reactions of alcohols. So far in this section, we have focused on the details of familiar reactions (substitution and elimination). Before we learn some new reactions, let's practice the ones that we just reviewed:

EXERCISE 13.57   Identify the reagents you would use to achieve the following transformation:

Answer   We are forming an alkene, so we must perform an elimination reaction. This problem requires an elimination with specific regiochemistry. We must form the less substituted alkene (the Hofmann product). Therefore, we will need to use a sterically hindered base (*tert*-butoxide). The obstacle is that an OH group is a terrible leaving group. So, we must first convert the OH group into a better leaving group. We cannot use a strong acid in this case, because we want to treat the substrate with a strong base, so the use of an acid would be incompatible (we cannot use a strong acid and a strong base at the same time, as they will neutralize each other). Instead, we can convert the alcohol into a tosylate, which will then undergo elimination when treated with *tert*-butoxide:

1) TsCl, py
2) *t*-BuOK

PROBLEMS   Identify the reagents you would use to achieve each of the following transformations:

13.58

13.59

13.60

13.61

13.9   REACTIONS OF ALCOHOLS: OXIDATION

Earlier in this chapter, we learned definitions for the terms oxidation and reduction. We saw that oxidation involves an increase in oxidation state. For example, oxidation of a secondary alcohol will produce a ketone:

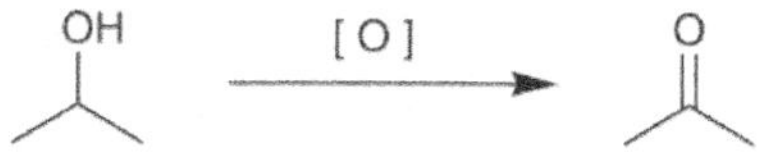

[O]

Notice the term used over the arrow, [O]. This term means that we are performing an oxidation. There are many oxidizing agents that can be used to accomplish this transformation. Chromic acid (H_2CrO_4) is a good example. Chromic acid can be formed from mixing sodium dichromate ($Na_2Cr_2O_7$) and sulfuric acid:

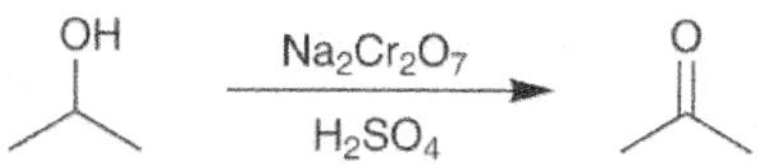

$Na_2Cr_2O_7$
H_2SO_4

In the example above, a *secondary* alcohol was oxidized to give a ketone. However, when we start with a *primary* alcohol, we will have two choices: 1) we can oxidize to give an aldehyde, or 2) we can oxidize even further to the carboxylic acid:

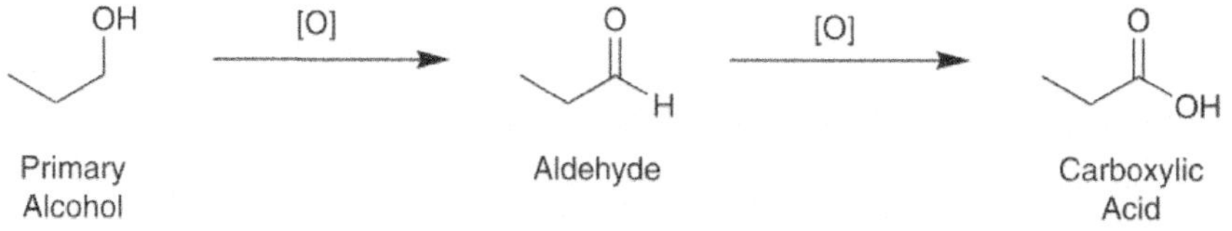

We can control how far we oxidize by carefully choosing our reagents. If we want to go all the way up to a carboxylic acid, then we just use chromic acid:

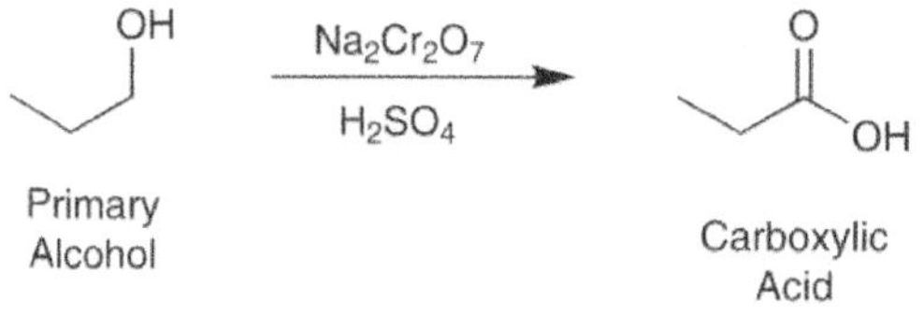

However, if we want to stop at the aldehyde, we will have to use a milder oxidizing agent. There are many such reagents that will oxidize a primary alcohol to give an aldehyde (and will not oxidize the aldehyde further to give a carboxylic acid):

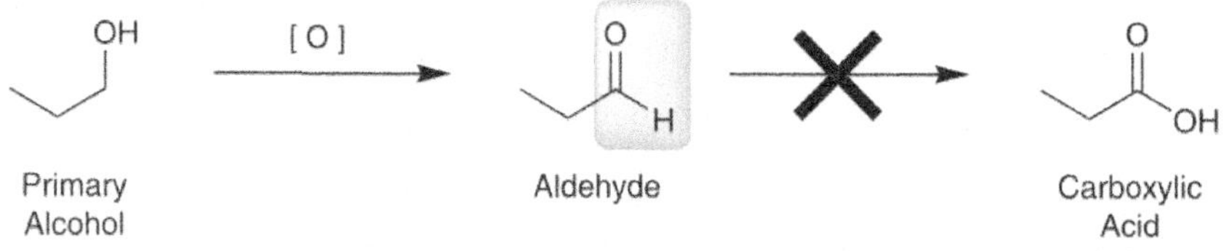

One such example is pyridinium chlorochromate, or PCC for short, which is prepared in the following way:

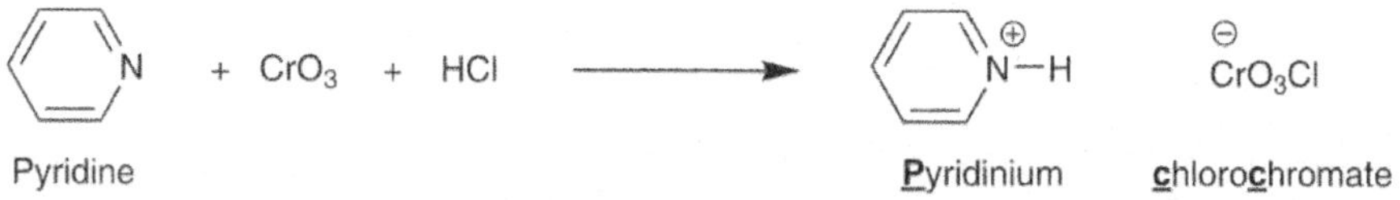

PCC is a mild oxidizing agent, and it will oxidize a primary alcohol to give an aldehyde. For example:

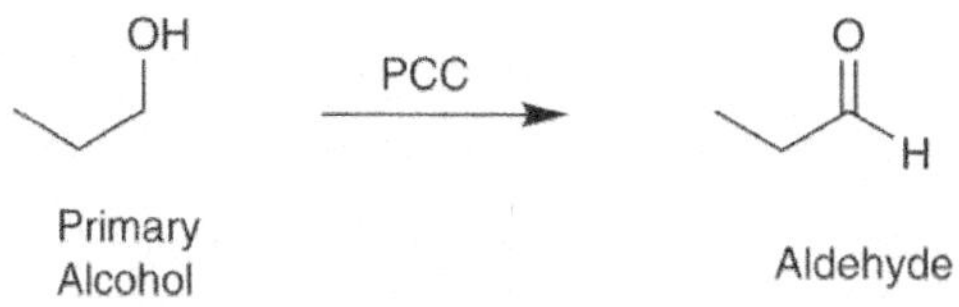

EXERCISE 13.62 Predict the major product of the following reaction:

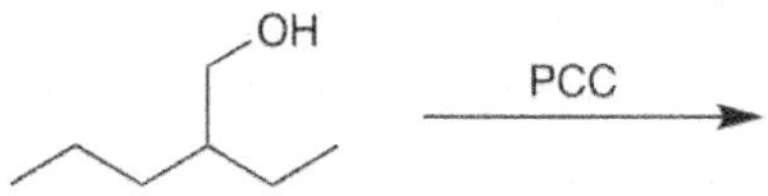

Answer   We are starting with a primary alcohol, and we are oxidizing with PCC. This will give an aldehyde as the product (rather than a carboxylic acid):

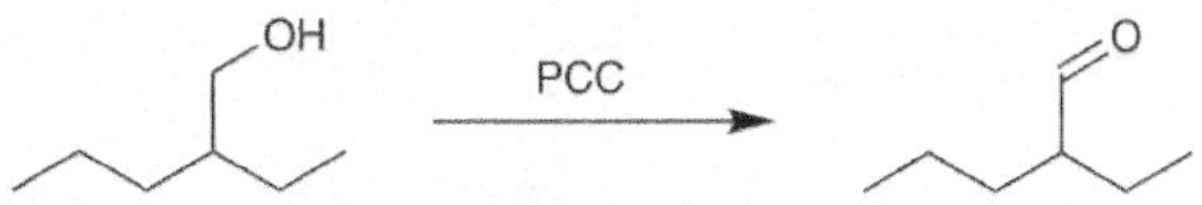

PROBLEMS   Identify what reagents you would use to accomplish each of the following transformations:

13.63

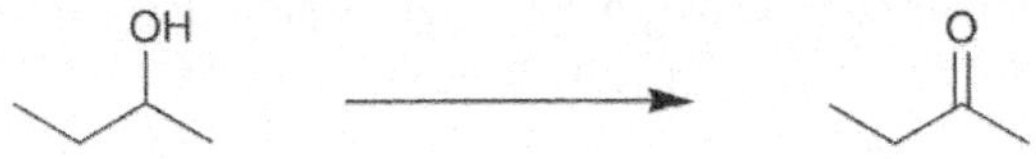

13.64

13.65

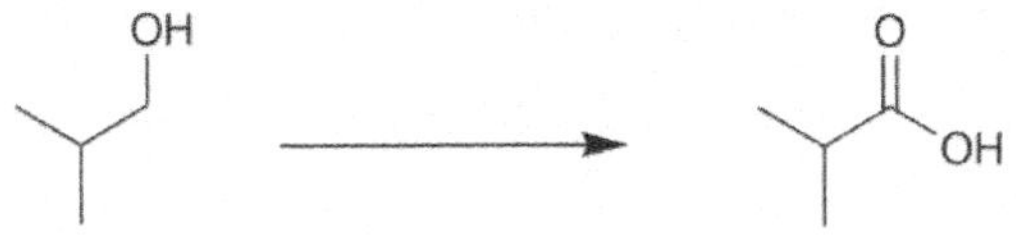

13.66

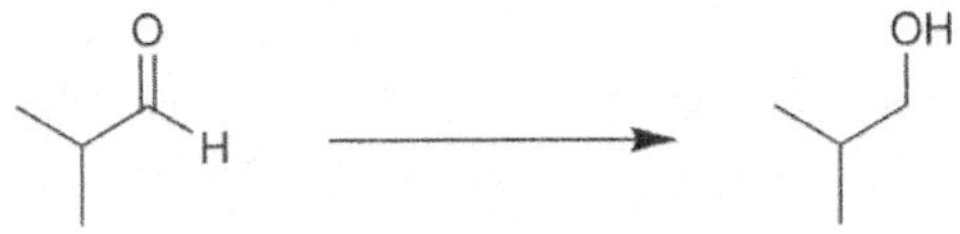

13.67

13.68

13.10   CONVERTING AN ALCOHOL INTO AN ETHER

Alcohols can be deprotonated with a strong base:

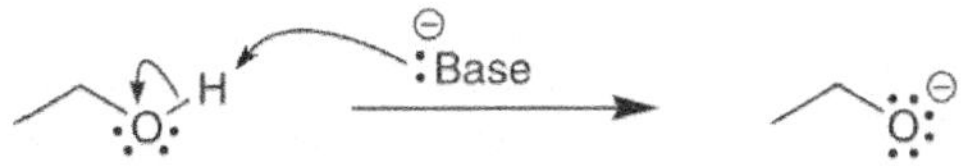

A fairly strong base is required. By removing the proton, we are forming a negative charge on an oxygen atom (an alkoxide ion). Therefore, in order to deprotonate, we will need a base that is even stronger than an alkoxide ion. As an example, we can use a base with a negative charge on a nitrogen atom, such as sodium amide ($NaNH_2$):

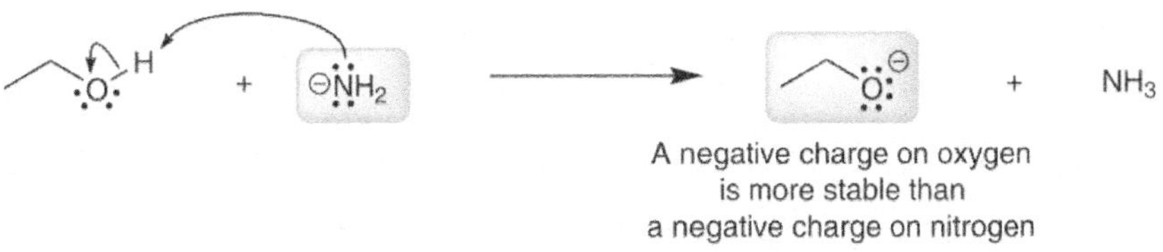

A negative charge on oxygen
is more stable than
a negative charge on nitrogen

However, there is a simpler way to deprotonate an alcohol. We can just use elemental sodium (Na), like this:

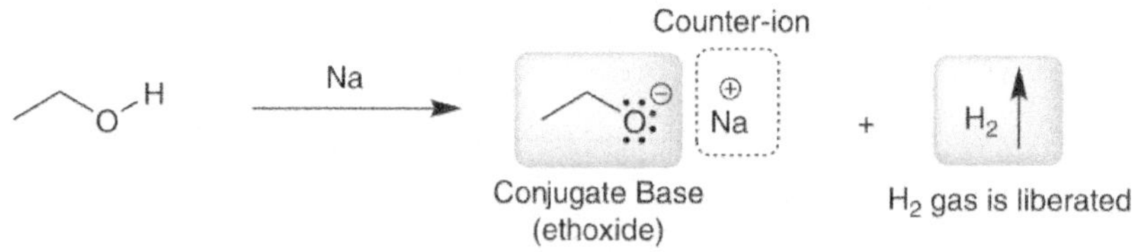

Elemental sodium (Na) has one electron which it can give up to form Na⁺. This electron can combine with the alcoholic proton to form a hydrogen atom (remember that a hydrogen atom is a proton *and* an electron). Two hydrogen atoms give hydrogen gas (H_2), which can escape from the reaction flask, pushing the reaction to completion. This process effectively deprotonates the alcohol (turns it into an alkoxide), forms Na⁺ as the counter-ion, and liberates hydrogen gas.

Now that we have seen how to deprotonate an alcohol (to form an alkoxide), the obvious question is: What can we do with alkoxide ions? We have already seen that alkoxide ions can be used as strong bases. But alkoxides can also function as strong nucleophiles. For example, consider the following S_N2 reaction:

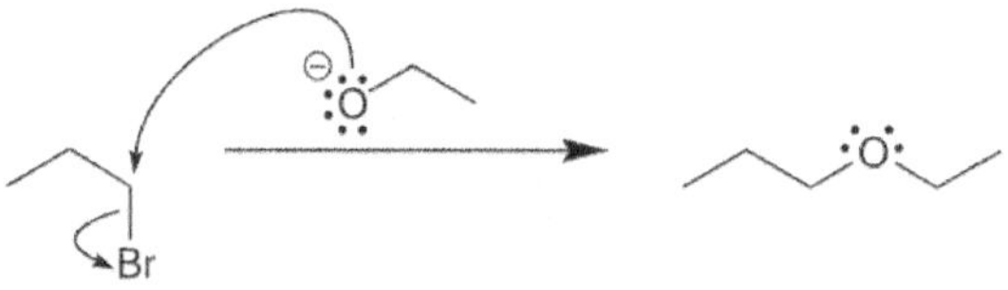

This is not a new reaction. This is just an S_N2 reaction. We are simply using the alkoxide ion (ethoxide in this case) to function as the attacking nucleophile. But notice the net result of this reaction: we have combined an alcohol and an alkyl halide to form an ether. This process has a special name. It is called the *Williamson ether synthesis*. This process relies on an S_N2 reaction as the main step, and therefore, we must be careful to obey the restrictions of S_N2 reactions. It is best to use a primary alkyl halide. Secondary alkyl halides are less effective substrates because elimination will predominate over substitution, and tertiary alkyl halides certainly cannot be used.

EXERCISE 13.69 Show how you could use a Williamson ether synthesis to make the following compound:

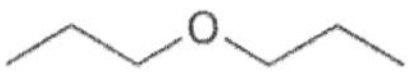

Answer We are using a Williamson ether synthesis, so we will need an alkoxide ion and an alkyl halide to form the ether linkage. Working backwards (retrosynthetic analysis), we get the following:

This retrosynthetic analysis indicates that we can make our product from sodium propoxide and propyl bromide. We can prepare sodium propoxide by treating propanol with Na. In summary, our answer is:

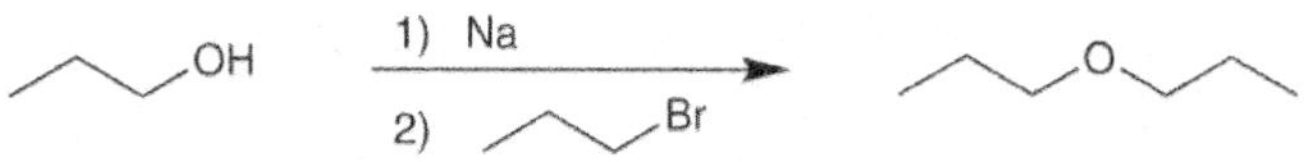

PROBLEMS Starting with 1-propanol, and using any other reagents of your choice, show how you could use a Williamson ether synthesis to make each of the following compounds:

13.70

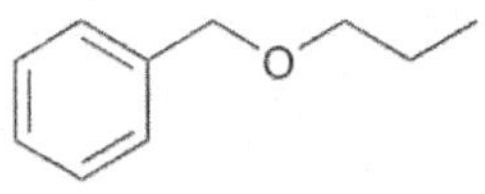

13.71

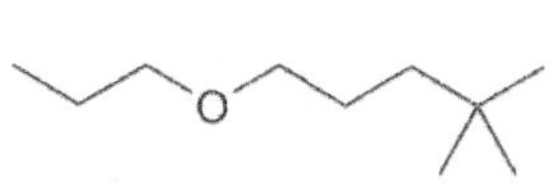

CHAPTER *14*

ETHERS AND EPOXIDES

14.1 INTRODUCTION TO ETHERS

An ether is a compound that has an oxygen atom sandwiched between two R groups, where the R groups can be alkyl, aryl or vinyl, as seen in the following examples:

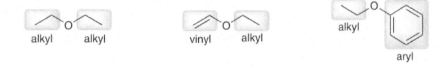

We have already covered many of the rules of nomenclature in Chapter 5, but we did not encounter many examples of ethers, so we will spend just a few moments on naming ethers.

Common names are constructed by identifying each R group and then adding the suffix "ether," as seen in the following example:

methyl O ethyl

Ethyl methyl ether

Notice that the alkyl groups are arranged in alphabetical order, so <u>e</u>thyl precedes <u>m</u>ethyl in the name. If the two R groups are identical, then the compound is named as a dialkyl ether, as seen in the following example:

Diethyl ether

Systematic names are assigned by choosing the larger group to serve as the parent alkane, while the smaller group is listed as an alkoxy substituent:

2-Ethoxyhexane

ethoxy group

In this example, the larger, six-carbon chain is chosen as the parent (hexane), and the smaller ethyl group is listed (together with the oxygen atom) as an ethoxy substituent. Notice that the ethoxy substituent is located at C2, so the number 2 is used to identify its position (2-ethoxy), just as we would do for any other substituent. For a review of naming parents and substituents, see Chapter 5.

Ethers make excellent solvents for organic reactions, for a number of reasons: 1) they are fairly unreactive, 2) they dissolve a wide variety of organic compounds, and 3) they have low boiling points (making it easy to evaporate the solvent after the reaction is complete). Below are two examples of ethers that are commonly used as solvents for organic reactions. You might recall the second example (THF) as a reagent used in hydroboration–oxidation (Section 11.7).

Diethyl ether

Tetrahydrofuran (THF)

In THF, the oxygen atom is incorporated into a ring, so THF is a *cyclic ether*. If a compound has a ring with multiple ether groups, then the compound is called a *cyclic polyether*. Several examples are shown below:

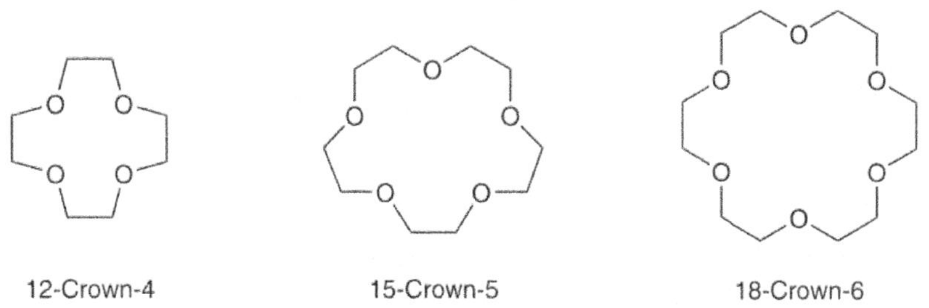

12-Crown-4   15-Crown-5   18-Crown-6

These compounds are often called *crown ethers*, and they have common names that indicate the size of the ring and the number of oxygen atoms incorporated in the ring. For example, in 18-crown-6, the ring is comprised of 18 atoms, 6 of which are oxygen atoms.

Crown ethers are useful in their ability to interact with (and solvate) metal ions. For example, the central cavity in 18-crown-6 is just the right size to accommodate a potassium ion (K^+). As a result, potassium fluoride (KF) will dissolve in benzene, if 18-crown-6 is present:

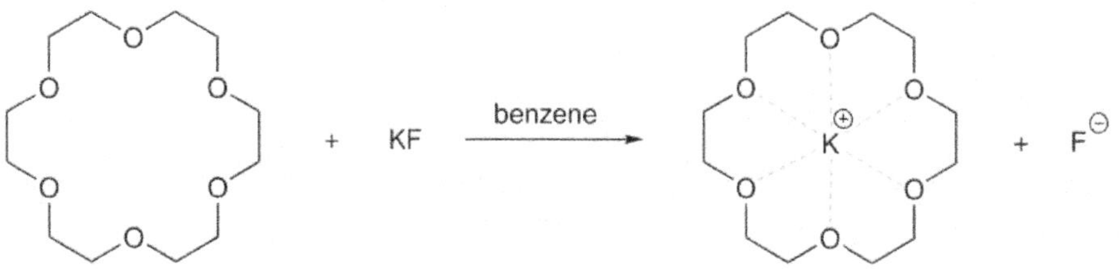

Without the crown ether, KF would not dissolve in a nonpolar solvent like benzene. The presence of 18-crown-6 generates a complex (shown above, right) that dissolves in benzene. This mixture can serve as an excellent source of fluoride ions for use as nucleophiles in substitution reactions. This is useful because nucleophilic fluoride ions are difficult to prepare (when KF is dissolved in polar solvents, the fluoride ions interact strongly with the solvent, rendering them unavailable to serve as nucleophiles in a substitution reaction). There are indeed many other useful applications of crown ethers as well, and you should consult your lecture notes and/or textbook to see if you are responsible for any specific ways in which crown ethers can be used.

PROBLEMS   Provide a common name for each of the following compounds:

14.1 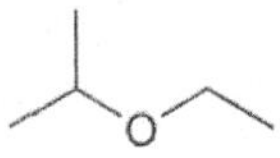   Name: _____

14.2 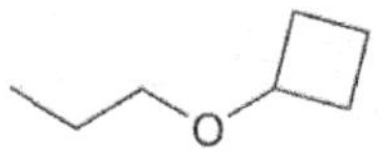   Name: _____

14.3 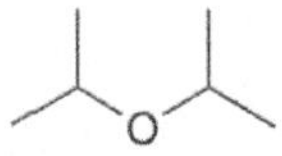   Name: _____

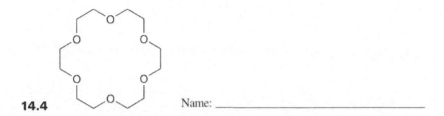

14.4   Name: _____

PROBLEMS   Provide a systematic name for each of the following compounds:

14.5   Name: _____

14.6   Name: _____

14.7   Name: _____

14.2 PREPARATION OF ETHERS

We have already seen (in Section 13.10) that ethers can be prepared from alcohols:

$$R-OH \xrightarrow[\text{2) R'X}]{\text{1) Na or NaH}} R-O-R'$$

An alcohol   An ether

This process has two steps, and is called the *Williamson ether synthesis*. In the first step, the alcohol is deprotonated to give an alkoxide ion. This deprotonation step can be accomplished by treating the alcohol with elemental sodium, Na (as described in Section 13.10). Alternatively, the deprotonation step can also be accomplished by treating the alcohol with a strong base, such as sodium hydride (NaH):

$$R-O-H \quad :H^{\ominus} \longrightarrow R-\overset{..}{\underset{..}{O}}:^{\ominus} \quad + \quad H_2$$

An alkoxide ion

Then, in the second step of the Williamson ether synthesis, this alkoxide ion is used as a nucleophile in an S_N2 reaction, giving an ether:

$$R-\overset{..}{\underset{..}{O}}:^{\ominus} \quad + \quad R'-X \xrightarrow{\;S_N2\;} R-O-R'$$

Nucleophile   Electrophile
(alkoxide ion)

Since the reaction occurs via an S_N2 pathway, it is subject to the restrictions of S_N2 reactions. Specifically, the substrate (electrophile) must be a methyl or primary halide. Secondary halides are less efficient because elimination is favored over substitution, and tertiary halides cannot be used. Also recall that S_N2 reactions only occur at sp^3 hybridized centers, not at sp^2 hybridized centers. So, the following compound cannot be made via a Williamson ether synthesis:

The C—O bond on the left cannot be made because formation of that bond would require an S_N2 reaction with a tertiary halide, which won't work. And the C—O bond on the right cannot be made because formation of that bond would require an S_N2 reaction at an sp^2 hybridized center, which also won't work.

Now let's get some practice planning a Williamson ether synthesis.

EXERCISE 14.8   Show how you would use a Williamson ether synthesis to make the following compound:

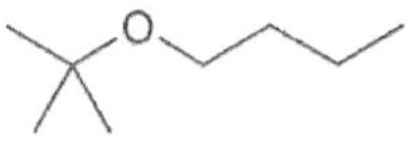

Answer   We begin by inspecting both C—O bonds:

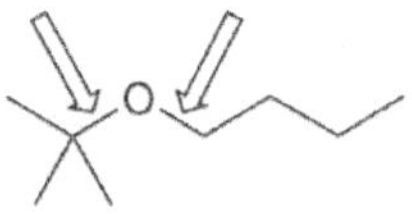

The C—O bond on the left cannot be made via a Williamson ether synthesis because formation of that bond would require an S_N2 reaction with a tertiary halide, which won't work. However, we can make the other C—O bond (on the right) with a Williamson ether synthesis because formation of that bond requires an S_N2 reaction with a primary halide, which will work:

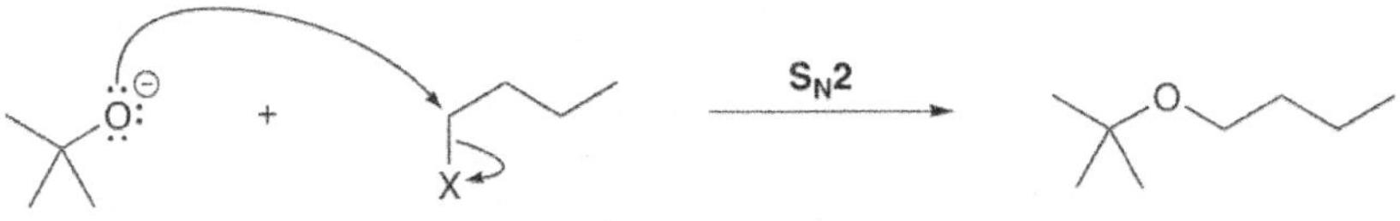

Primary halide

We would therefore use the following reaction sequence to convert an alcohol (*t*-butanol) into the desired ether.

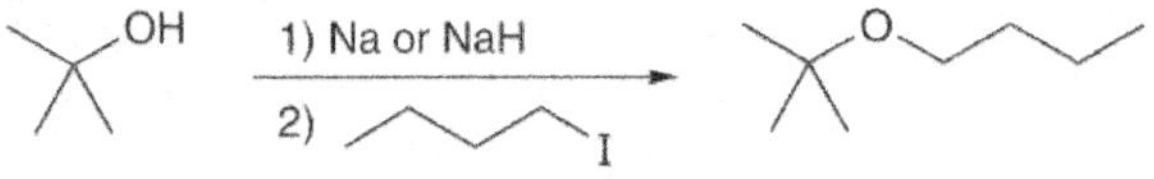

Notice that we chose to use an alkyl iodide, because alkyl iodides are more reactive than alkyl bromides or chlorides (but the process would also work with an alkyl bromide or an alkyl chloride).

PROBLEMS   Show how you would use a Williamson ether synthesis to make each of the following compounds. In each case, draw the structure of the starting alcohol, and show the reagents necessary to convert the alcohol into the desired ether:

14.9

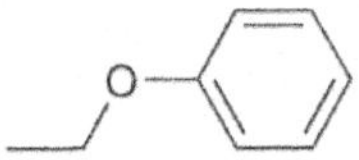

14.10

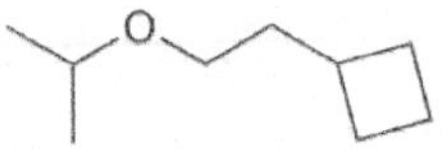

14.11

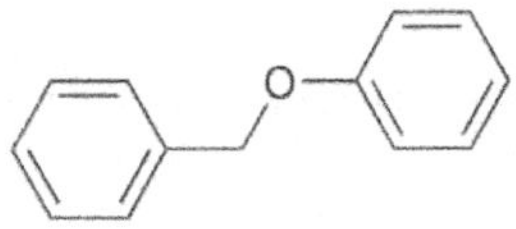

14.12   Show two different ways to make the following compound via a Williamson ether synthesis.

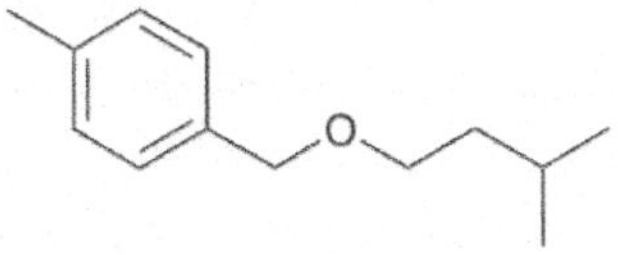

14.13   Identify a reagent that can be used to achieve the following intramolecular Williamson ether synthesis:

14.3 REACTIONS OF ETHERS

Ethers are generally unreactive under basic or mildly acidic conditions. But under strongly acidic conditions, the C—O bonds undergo bond cleavage (they are broken). For example, upon treatment with HX (where X = Br or I), the C—O bonds undergo acidic cleavage, generating two alkyl halides:

$$R—O—R \xrightarrow[\text{heat}]{\text{excess HX}} R—X \ + \ X—R \ + \ H_2O$$

Each of the R groups is released as an alkyl halide (RX), and the oxygen atom is released as water (H_2O), which is a byproduct of the reaction. The mechanism by which each C—O bond is cleaved depends strongly on the identity of each R group. Primary alkyl groups are cleaved via an S_N2 pathway, while tertiary alkyl groups are cleaved via an S_N1 pathway. As an illustration of this point, consider the following example:

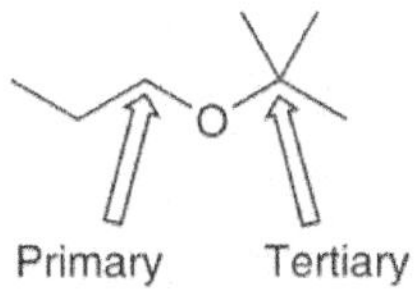

Primary   Tertiary

Upon treatment with HX, both C—O bonds undergo acidic cleavage, yet each C—O bond is cleaved via a different mechanism. For the C—O bond on the left, the R group is primary, so an S_N2 pathway predominates:

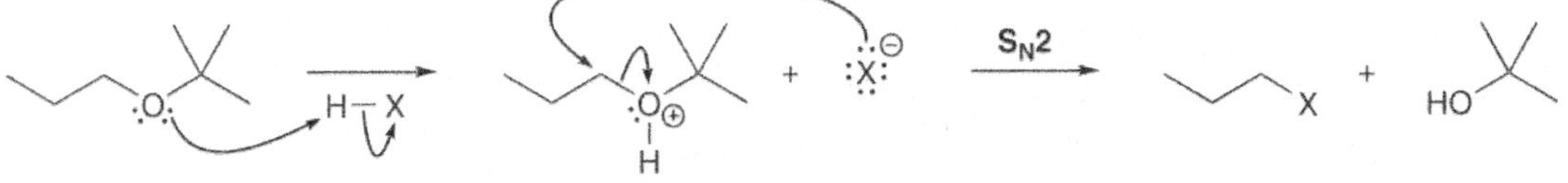

As shown, the ether is protonated under acidic conditions, thereby converting a bad leaving group (*t*-butoxide) into an excellent leaving group (*t*-butanol). The resulting oxonium ion (intermediate with a positive charge on oxygen) is then attacked by a halide in an S_N2 process.
    The other C—O bond is cleaved via an S_N1 pathway because the R group is tertiary:

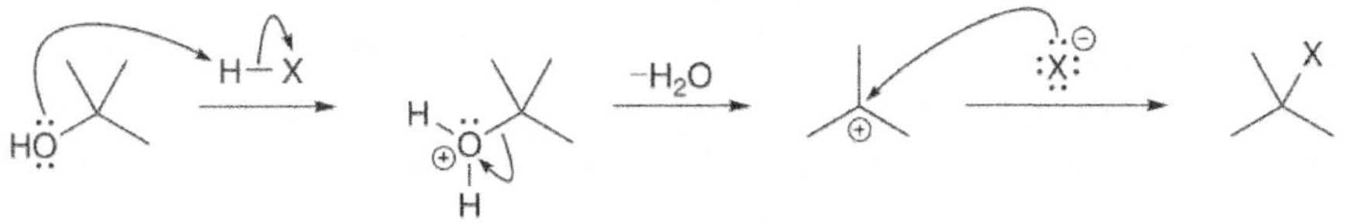

Protonation gives another oxonium ion, which then undergoes an S_N1 process to give a tertiary alkyl halide.

Some R groups cannot be cleaved via either mechanistic pathway (S_N2 or S_N1). For example, consider the highlighted bond in the following compound:

R—O—⟨benzene ring⟩

⇧

Not cleaved in HX

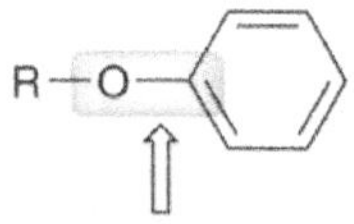

This C—O bond does not undergo acidic cleavage, because S_N2 and S_N1 processes are not efficient at sp^2 hybridized centers. Treatment with strong acid therefore gives the following products:

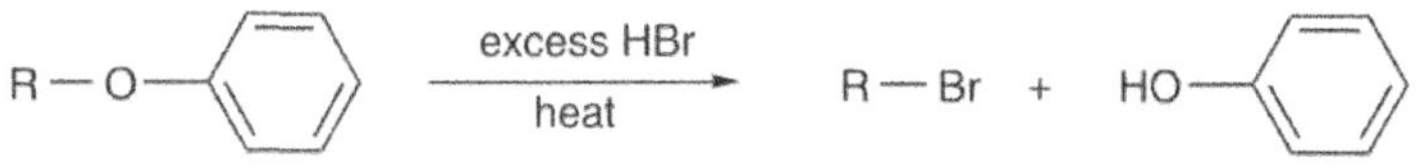

R—O—⟨benzene ring⟩ $\xrightarrow[\text{heat}]{\text{excess HBr}}$ R—Br + HO—⟨benzene ring⟩

One C—O bond is cleaved to give an alkyl halide, and the other C—O bond is not cleaved.

PROBLEMS Draw the products that are expected when each of the following compounds is heated with excess HBr.

14.14

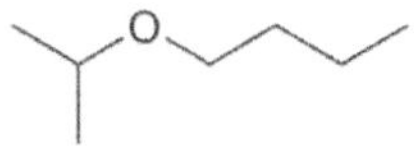

14.15

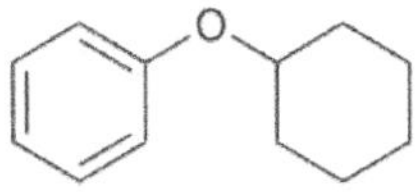

14.16

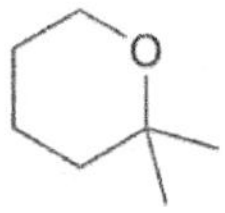

14.17

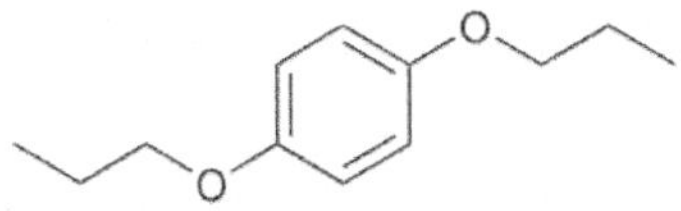

14.4 PREPARATION OF EPOXIDES

The following compounds are cyclic ethers, because each has an oxygen atom incorporated into a ring:

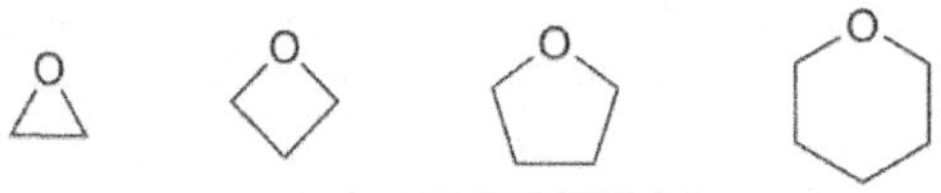

The first example has a three-membered ring, and is called an oxirane. The three-membered ring of an oxirane has significant ring strain, and is therefore more reactive than typical ethers. Oxiranes, often called *epoxides*, can have up to four substituents, highlighted here:

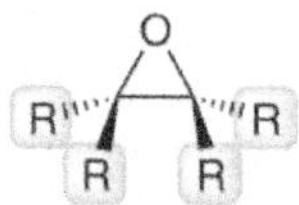

R''''—⟨epoxide⟩—''''R
R   R

Recall from Section 11.10 that epoxides can be made from alkenes, upon treatment with a peroxy acid such as MCPBA:

This transformation can be achieved with a variety of peroxy acids (RCO_3H). Two commonly used peroxy acids are MCPBA and peroxyacetic acid, shown here:

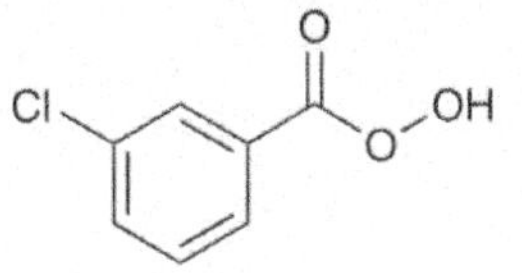

meta-Chloroperoxybenzoic acid
(MCPBA)

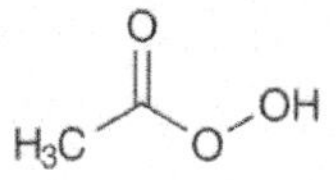

Peroxyacetic acid

Epoxide formation is observed to be a stereospecific process. A *cis* alkene is converted into a *cis* epoxide:

And a *trans* alkene is converted into a *trans* epoxide:

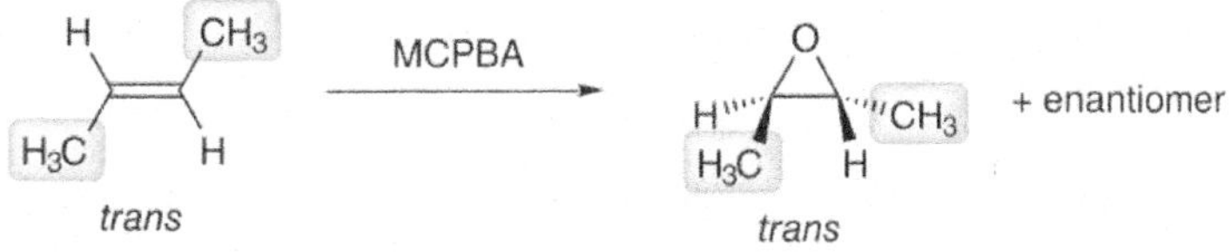

In this case, the product is a racemic mixture, because epoxide formation can occur on either face of the alkene, giving rise to a pair of enantiomers.

There are many ways to make epoxides. Indeed, there are even ways to form just one enantiomer (rather than a racemic mixture), called enantioselective epoxidation. Consult your textbook and lecture notes to see if you are responsible for any additional methods for making epoxides.

PROBLEMS   Draw the product that is expected when each of the following compounds is treated with a peroxy acid (RCO_3H) such as MCPBA.

14.18

14.19

14.20

14.21

14.22

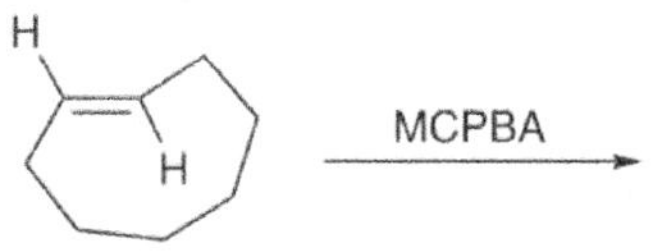

PROBLEMS Show what reagents you would use to make each of the following epoxides.

14.23

+ Enantiomer

14.24

14.25

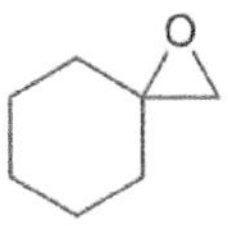

14.5 RING-OPENING REACTIONS OF EPOXIDES

Epoxides are susceptible to reactions in which the ring is opened, thereby alleviating the ring strain associated with the three-membered ring. For example, when an epoxide is treated with a strong nucleophile, the nucleophile will attack the epoxide in an S_N2-type reaction, thereby opening the ring:

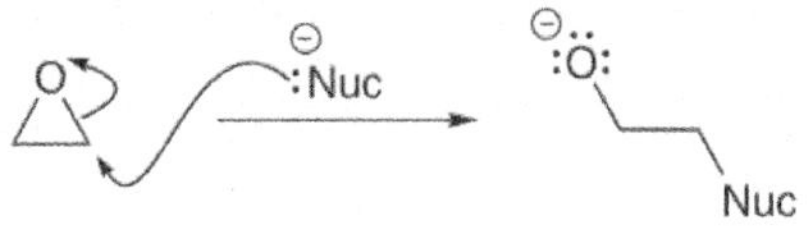

In this substitution process, the epoxide serves as the substrate (electrophile), and one of the C—O bonds is broken. A variety of nucleophiles can be used to open the ring. For example, consider the following case, in which a hydroxide ion (HO^-) is used as the nucleophile:

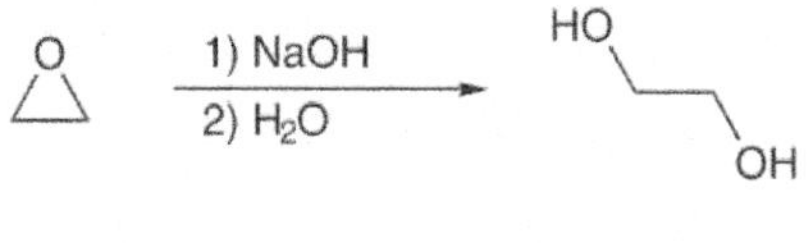

Notice that this process has two steps. In the first step, the hydroxide ion functions as a nucleophile and attacks the epoxide, thereby opening the ring and giving an alkoxide ion (an intermediate with a negative charge on oxygen). Then, in step 2, the alkoxide ion is treated with water to give the product:

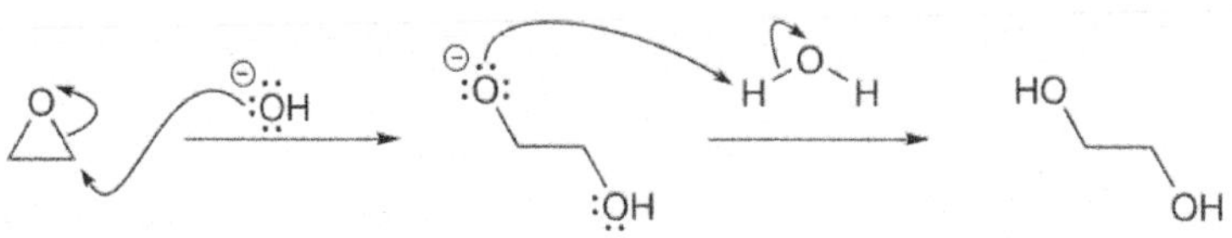

The second step (treatment with H_2O) is necessary in order to protonate the alkoxide ion and obtain the product.

Hydroxide is certainly not the only nucleophile that will open an epoxide. Indeed, a variety of nucleophiles will react with epoxides to give ring-opening reactions. For example, consider the following reaction, in which an alkoxide ion (RO⁻) is used as the nucleophile:

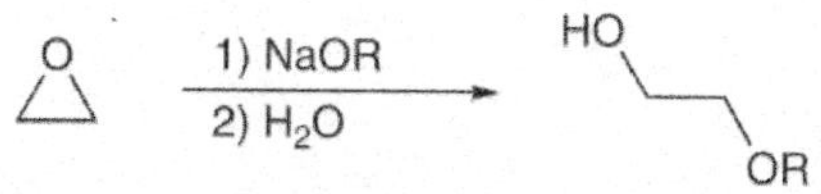

Once again, water is used in the second step, to serve as a source of protons. Notice that, in this case, an alkoxy group (OR) is installed, because the nucleophile was an alkoxide ion. The following are several other examples in which an epoxide ring is opened upon treatment with a strong nucleophile (followed by water work-up):

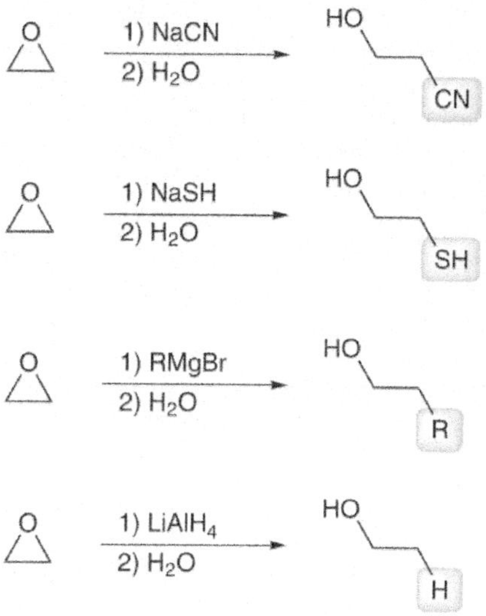

In all of these cases, the starting epoxide is symmetrical, so we don't have to consider the regiochemical outcome. That is, we don't have to consider which side of the epoxide is attacked by the nucleophile (attacking either side leads to the same outcome). But now let's consider an unsymmetrical epoxide, where one side is more substituted than the other:

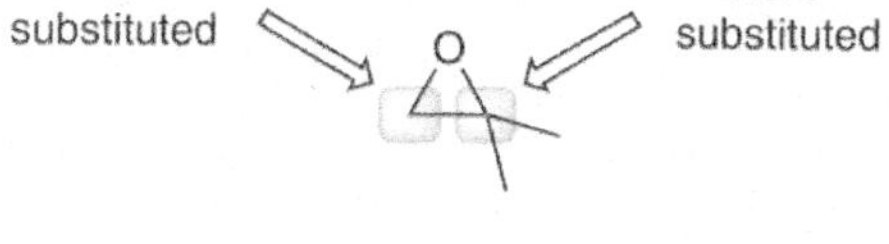

When an unsymmetrical epoxide is treated with a strong nucleophile, the nucleophile is observed to attack the less-substituted side of the epoxide. This is a logical outcome because the S_N2 backside attack is faster where there is less steric hindrance.

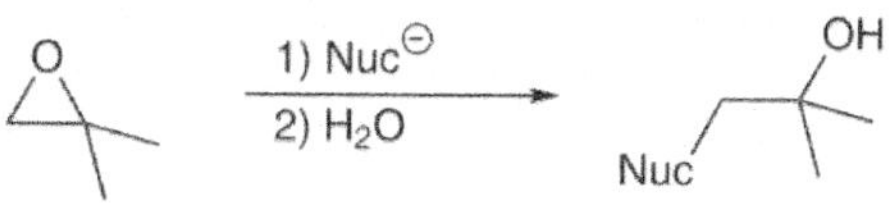

The nucleophile is installed at the less-substituted position, and the oxygen atom of the epoxide is converted into an OH group at the more-substituted position. Below are two specific examples:

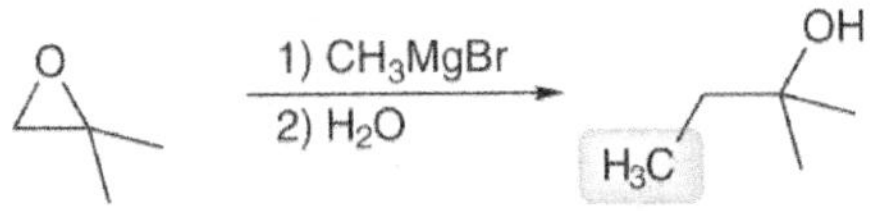

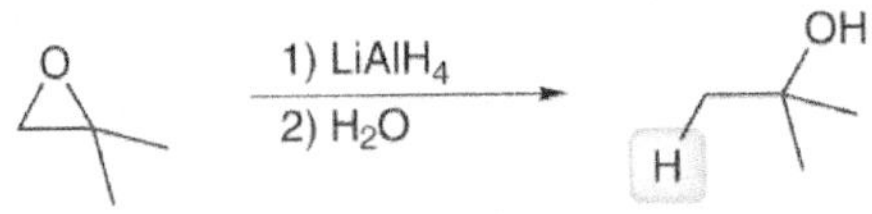

In the first example, a Grignard reagent (RMgX) is used as a nucleophile (for a review of Grignard reagents, see Section 13.6). In the second example, lithium aluminum hydride (LiAlH$_4$) is used as a source of nucleophilic H$^-$ (for a review of LiAlH$_4$ as a nucleophile, see Section 13.5). These nucleophiles are also very strong bases, and they are therefore incompatible with the presence of an acid, even a weak acid (these reagents would be destroyed in the presence of an acid). This is why protonation (treatment with H$_2$O) must be performed as a separate step after the reaction is complete, as shown in each case above.

Not all nucleophiles are strong bases. Indeed, there are some strong nucleophiles, such as halide ions (X$^-$), that are very weak bases, and therefore, they CAN exist in acidic conditions. For example, HX provides both a strong nucleophile (X$^-$) and acidic conditions (H$^+$) simultaneously. If an epoxide is treated with HX, the epoxide undergoes a ring-opening reaction, just as we might expect. But the regiochemical outcome is the opposite of what we have seen so far. Specifically, the halide is installed at the *more*-substituted position, and the oxygen atom of the epoxide is converted into an OH group at the less-substituted position:

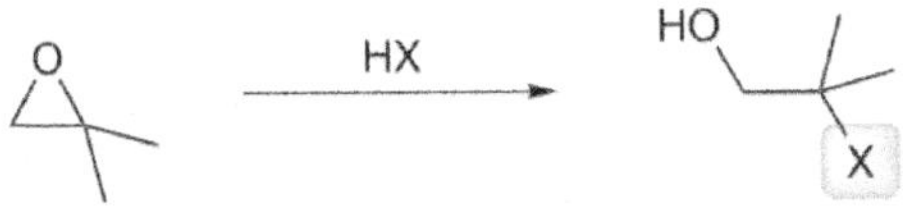

Why? To understand the reason for this reversed regiochemical outcome, let's consider the following mechanism, which is believed to operate under acidic conditions:

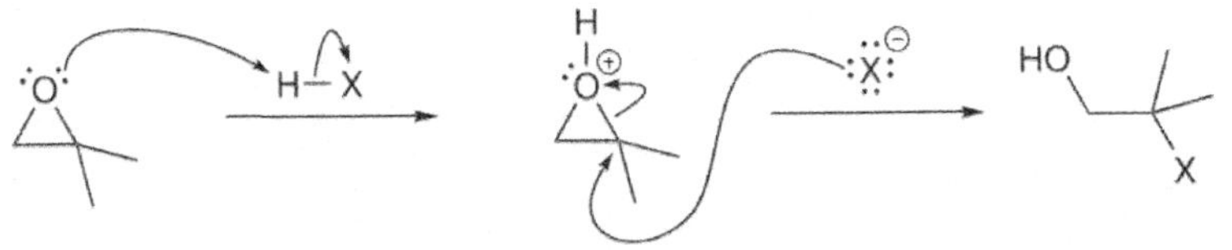

Under acidic conditions, the epoxide is first protonated, followed by an S$_N$2-type process. The intermediate in this process (a protonated epoxide) bears a positive charge and is therefore much more electrophilic (electron deficient) than the starting epoxide:

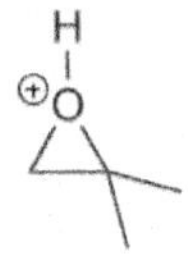

The electron-deficient oxygen atom withdraws electron density (via induction) from the neighboring carbon atoms. As a result, those carbon atoms (highlighted) are significantly electron deficient ($\delta+$):

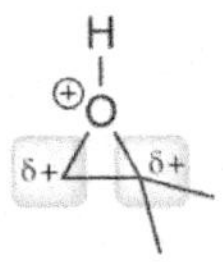

While both highlighted positions are electron deficient, one of them is better suited to tolerate the electron deficiency. Specifically, the more-substituted (tertiary) position is significantly better at supporting a partial positive charge than the less-substituted (primary) position. As a result, the tertiary position has significantly more carbocationic character than the primary position. And this explains why the nucleophile attacks the more-substituted position (in acidic conditions). You may be wondering how an S_N2-type process can occur at a tertiary substrate, but this situation is unique, because the tertiary position has significant carbocationic character. This means that its geometry is not purely tetrahedral, but rather, it is somewhere between tetrahedral and trigonal planar. This allows the nucleophile to attack at the more-substituted position even though it is tertiary.

The following reaction is another example of a ring-opening reaction under acidic conditions (in the presence of an acid catalyst, such as H_2SO_4):

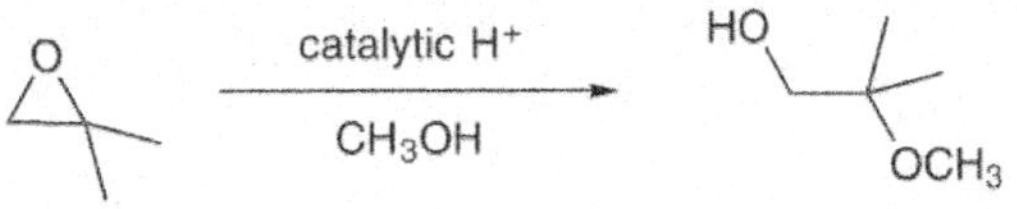

In this reaction, the nucleophile is an alcohol (CH_3OH), which is a weak nucleophile. A mechanism for this process is shown below. Notice that the nucleophile attacks the more-substituted position, because the conditions are acidic:

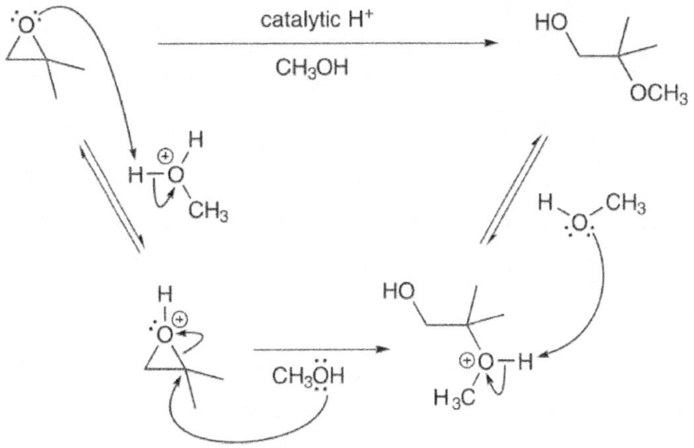

Thus far, we have focused exclusively on the *regiochemistry* of ring-opening reactions, and we have seen that the presence or absence of acidic conditions can have a significant impact on the regiochemical outcome. Now we must also consider the *stereochemistry* of ring-opening reactions. We have seen that ring-opening reactions of epoxides occur via an S_N2-type process (whether the reaction occurs under acidic conditions or not). Therefore, if the nucleophile is attacking a chiral center, then we expect inversion of configuration. Consider the following example:

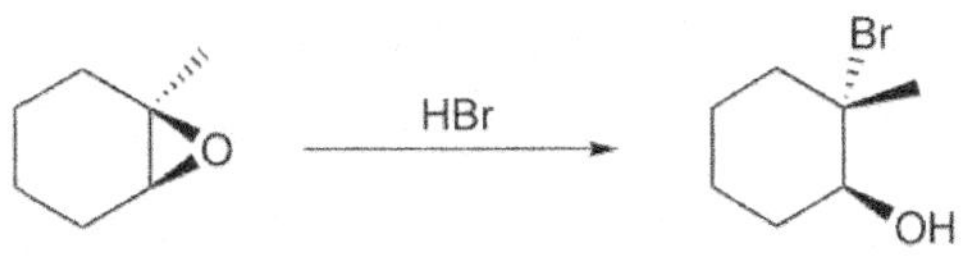

Notice that under acidic conditions, the nucleophile (a bromide ion) attacks the more-substituted position, because the nucleophile is attacking a protonated epoxide (rather than an epoxide):

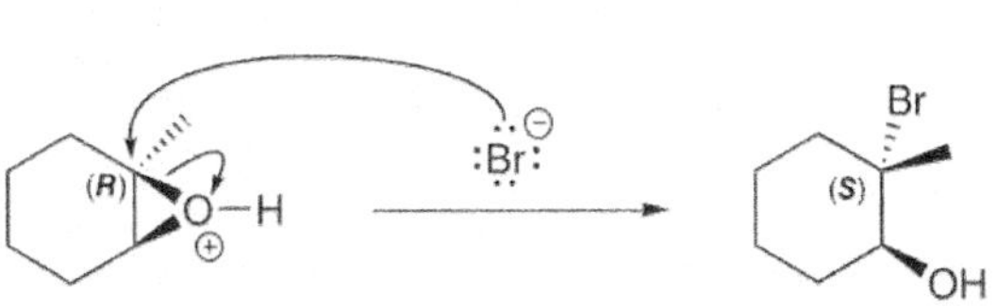

In this case, the more-substituted position is a chiral center, so inversion of configuration is observed at that center (*R* becomes *S*). The bromide attacks from the backside of the epoxide (S$_N$2 is characterized by back-side attack), so the Br ends up on a dash (because the leaving group was on a wedge).

Be careful, though. Inversion of configuration only occurs at the position being attacked by the nucleophile. In the following example, there is a chiral center, but it does not undergo inversion:

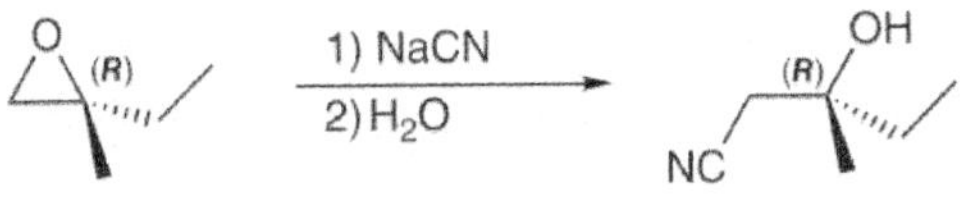

It is true that the reaction proceeds via an S$_N$2 process, but the nucleophile is not attacking the chiral center in this case:

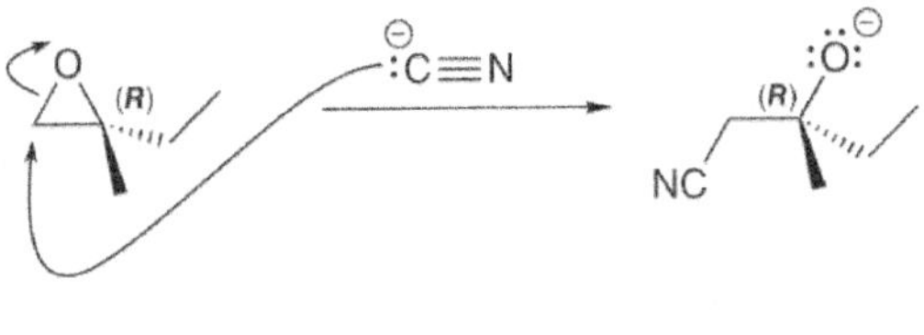

EXERCISE 14.26   Predict the expected product for each of the following reactions:

(a)

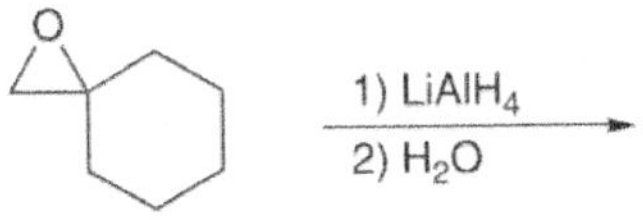

(b)

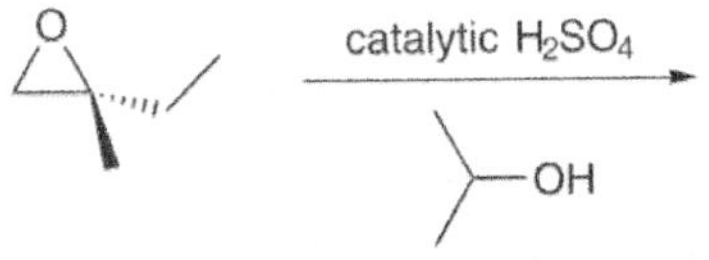

Answer
(a) The starting epoxide is unsymmetrical, so we must consider the regiochemical outcome. That is, we must decide whether the nucleophile attacks the more-substituted or the less-substituted side of the epoxide. The reagents do not indicate the presence of an acid (and we know that there cannot be any acid present because LiAlH$_4$ is not compatible with acidic conditions), so we expect the nucleophile to attack the less-substituted position:

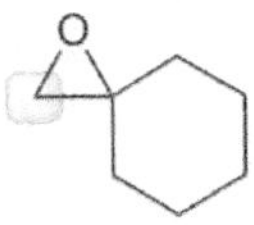

The nucleophile (H$^-$) attacks this position, opening the epoxide:

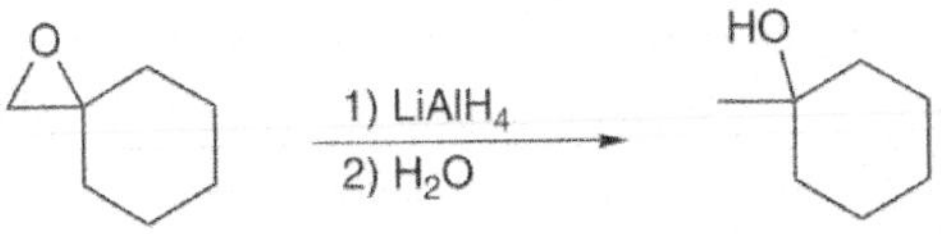

Neither the starting epoxide nor the product has a chiral center, so we did not have to consider the stereochemical outcome in this case.

(b) The starting epoxide is unsymmetrical, so we must consider the regiochemical outcome. The reagents indicate the presence of an acid, so we expect the nucleophile to attack the more-substituted position:

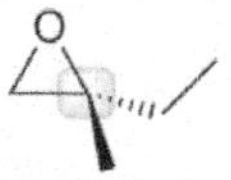
O

The nucleophile (isopropanol) attacks this position, opening the epoxide and installing an isopropoxy group. The starting epoxide has a chiral center, and the S_N2 process does indeed occur at that location, so we expect inversion of configuration:

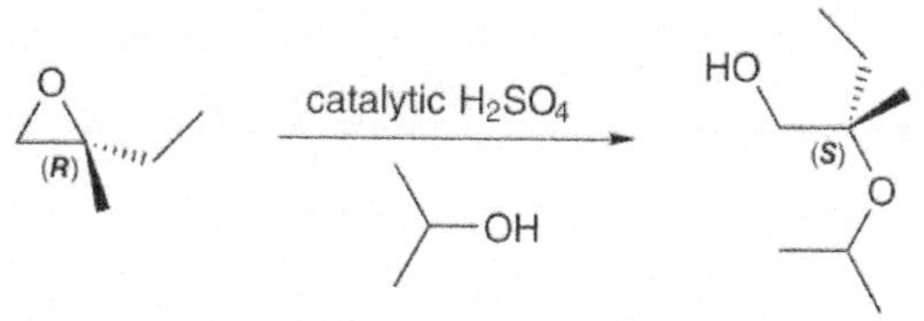
O
(R)
catalytic H_2SO_4
OH
HO
(S)
O

PROBLEMS Predict the expected product for each of the following reactions:

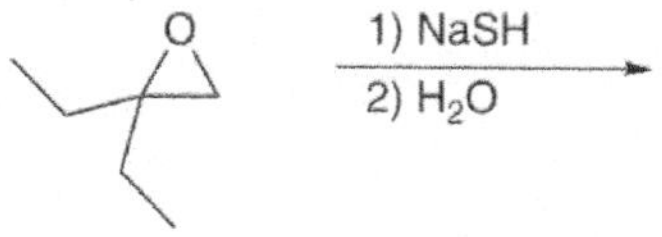
O
1) NaSH
2) H_2O

14.27

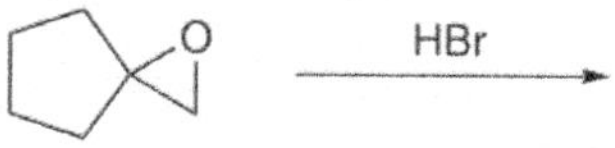
O
HBr

14.28

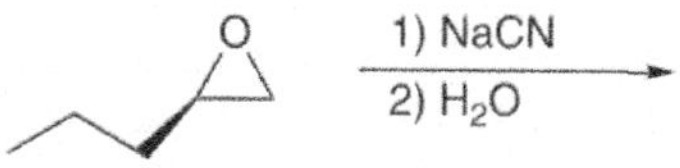
O
1) NaCN
2) H_2O

14.29

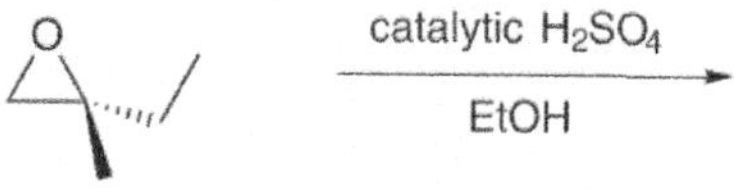
O
catalytic H_2SO_4
EtOH

14.30

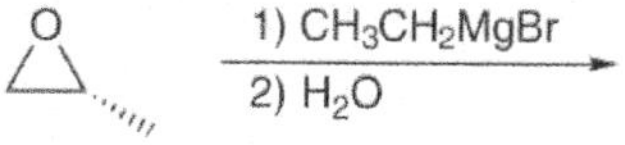
O
1) CH_3CH_2MgBr
2) H_2O

14.31

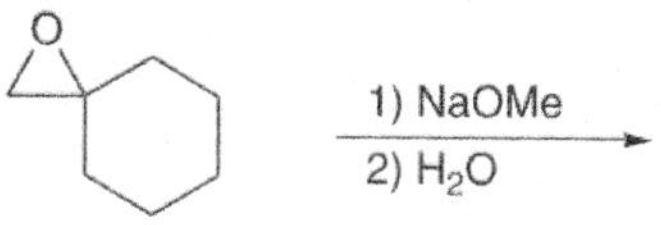
O
1) NaOMe
2) H_2O

14.32

SYNTHESIS

Synthesis is really just the flipside of predicting products. In any reaction, there are three groups of chemicals involved: the starting material, the reagents, and the products:

$$\text{Starting material} \xrightarrow{\text{Reagents}} \text{Products}$$

When the products are not shown, then you have a "predict the product" problem:

$$\text{Starting material} \xrightarrow{\text{Reagents}} \quad ?$$

When the reagents are not shown, then you have a synthesis problem:

$$\text{Starting material} \xrightarrow{?} \text{Products}$$

   Synthesis problems can be easy (if they are only one step) or they can be difficult (if they are more than one step). When you begin learning reactions in your course, you will start to encounter synthesis problems in your textbook. At first, you will get one-step problems, and as the course progresses, you will see multistep syntheses. In a multistep synthesis, you can often end up with a product that looks very different from the starting material. For example, look at the following series of reactions below. Don't concentrate on how the changes were made. For now, just focus on the fact that each reaction changes the compound only slightly, but in the end, we end up with a product completely different from the starting material:

Starting Material

Product

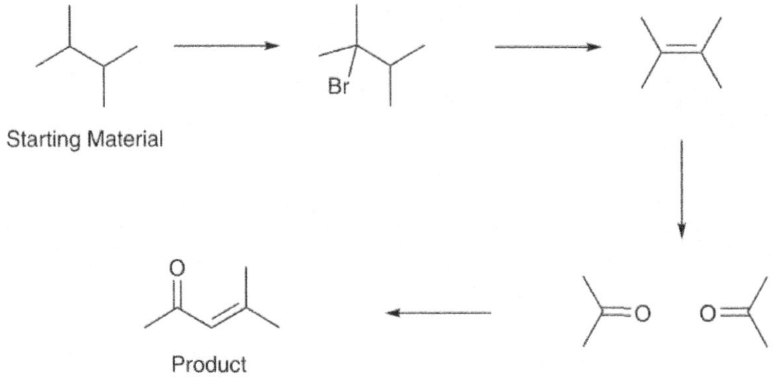

It can only take three or four steps before the problem can get quite difficult. If you convert the sequence above into a synthesis problem, it would look like this:

If you are having trouble with synthesis problems when you first encounter them, the worst thing you can do is to give up and say: "Oh, well, I'm not good at synthesis problems." As the course moves on, this attitude will slowly kill your grade in the course. To see why this is so, let's compare organic chemistry to a game of chess.

Imagine that you are learning how to play chess. You first learn about the pieces: how they are named, how to set up the board, and so on. Then you learn how each piece moves and how they capture each other. When you start playing your first game, you realize that there is quite a bit of strategy involved. Most strategies involve thinking more than just one move in advance. It is not good enough to know only how to move the pieces. You also need to think about how to plan out the next few moves so that you can coordinate an attack on your opponent's pieces. Imagine how silly it would be to take the time to learn how to move the pieces, but to then say to yourself that you are not good at strategy. Imagine thinking that you will keep playing chess, but you just won't be good at that one aspect of the game. That would be silly, because that one aspect of the game is the whole game itself. You either need to learn how to strategize, or just don't play chess. There is no in-between.

Organic chemistry is very much the same. Synthesis is all about strategizing. You need to think a few moves ahead, and you must learn how to do this. You cannot tell yourself that you are not good at synthesis problems, and therefore you will just focus on the other aspects of organic chemistry. Synthesis *is* organic chemistry. The second half of the course is all about learning reactions and applying them in syntheses. Everything that you have learned so far has prepared you for synthesis. The only way to become proficient at synthesis is to *practice*. Don't be lazy, and don't think that you can get through the course without learning how to propose syntheses. If you do, you will find that your performance in the course will spiral down to a point that will make you very unhappy.

There are a few techniques that will make you feel more comfortable with synthesis problems, and there are exercises that you can go through to increase your proficiency in approaching synthesis problems.

15.1 ONE-STEP SYNTHESES

As we mentioned earlier, one-step syntheses are the first synthesis problems you will encounter. These problems are not more difficult than predicting products. Before you can move on to multistep syntheses, you first need to feel comfortable with one-step syntheses.

To do this, we need to make a list of reactions, but we will leave out the reagents, so that we can repeatedly photocopy the list and get practice filling in the reagents.

As you learn more and more reactions, this list will grow. With every five new reactions, you should photocopy all of the reactions that you have recorded here. Then, start filling in the reagents on the photocopy. Repeat this procedure whenever you have entered five new reactions.

If you keep up with this exercise as the course progresses, you will be in very good shape for solving one-step synthesis problems. The hardest challenge that you will face is keeping up with the work and not waiting until the night before the exam. If you wait (as most students do), you will find it very difficult to spend the time that it takes to master this material. Don't make that mistake. The secret to success in this course is to do a little bit every night (rather than cramming on the night before the exam). Cramming might work well for other courses, but it doesn't work well in organic chemistry.

Begin your list on the next page.

For now, skip forward a few pages. We have some techniques to go over that will help you solve synthesis problems.

Remember not to fill in the reagents or the mechanisms. For each reaction, just draw the starting material in front of the arrow and the products after the arrow. Leave the space above the arrow empty. You will fill in the reagents when you photocopy these pages:

Now photocopy this page, and try to fill in the reagents on your photocopied page.

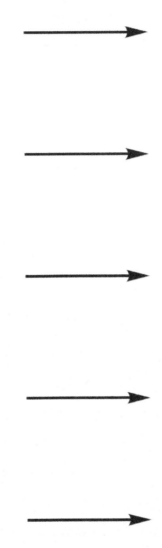

Now photocopy this page again, and fill in the reagents for every reaction on this page.

Now photocopy this page AND the previous pages, and fill in all of the reagents.

Now photocopy this page AND the previous pages, and fill in all of the reagents.

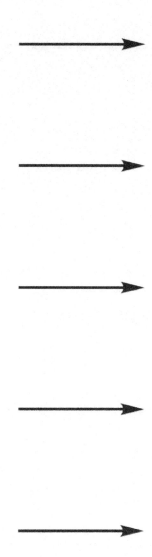

Now photocopy this page AND the previous pages, and fill in all of the reagents.

Now photocopy this page AND the previous pages, and fill in all of the reagents.

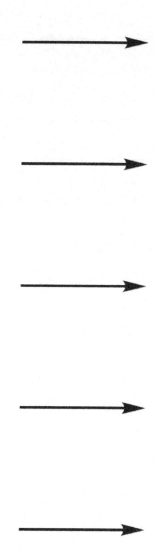

Now photocopy this page AND the previous pages, and fill in all of the reagents.

Now photocopy this page AND the previous pages, and fill in all of the reagents.

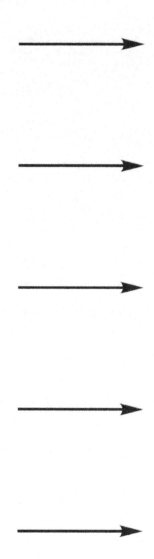

Now photocopy this page AND the previous pages, and fill in all of the reagents.

Now photocopy this page AND the previous pages, and fill in all of the reagents.

If you cover more than 30 reactions and need more space to continue, then you can just use a regular piece of paper to keep your list going.

15.2   MULTISTEP SYNTHESES

To prepare yourself for solving multistep syntheses, you need to learn how to think in more than one move. If you carefully review your list of reactions, you will find that the products of some reactions are the starting materials for other reactions. For example, you will find that some reactions are used to form double bonds, and other reactions add reagents across double bonds. So if you pair up all of the possibilities, you will create a list of many two-step syntheses. By studying these two-step possibilities, you will begin to get familiar with seeing syntheses that are more than one step.

Let's see an example of what we mean. Below is one reaction that forms a *cis*-alkene from an alkyne.

Now consider a reaction in which this alkene is the starting material, for example:

If we put these two reactions together into a two-step synthesis, we get the following:

You should now get some practice with this. You will probably learn around five methods for making alkenes and probably around 10 reactions that involve alkenes as starting materials. If you put together all of the possibilities, you will find that there are around 50 possibilities, depending on exactly how many reactions you learn. Clearly, you cannot keep a list like this as you go through the course. The list would be too long to study. And if you try to consider three-step syntheses, you will find that the number of permutations is too large to even compile such a list. It's just like our analogy to a game of chess.

In chess, you cannot possibly memorize every possible orientation of all of the pieces and then memorize the best move for each of those possibilities. There are too many permutations. Instead, you learn how to analyze each situation and as time goes on, you get better and better at it. By familiarizing yourself with certain permutations, you will get better at figuring things out as you go along. So let's start with the list that we talked about above—the approximately 50 possible two-step syntheses that involve forming an alkene and then doing something with that alkene.

Again, you should not try to make a list like this throughout your entire course. This task would be impractical. But if you make this first list of roughly 50 syntheses, you will learn how to start thinking in more than one step. It is important for you to get accustomed to thinking this way. Take a separate sheet of paper and try to create this list using the reactions in the beginning of your course. If you do not get a chance to write down all 50, that's OK. As long as you begin the process and draw at least 10 or 20 of them, then you will start to understand what it is like to think in more than one step.

After you have done this, we can start focusing on the main techniques for analyzing problems that display permutations that you have never seen. That is what the next section is all about.

15.3 RETROSYNTHETIC ANALYSIS

When you see a synthesis problem for the first time, you are not expected to immediately know the answer. I cannot stress this enough. It is so common for students to get overly anxious when they see synthesis problems that they cannot solve. Get used to it. This is the way it is supposed to be. Going back to our chess analogy, you don't need to make a move as soon as it is your turn. You are allowed to think about it first. In fact, you are supposed to think about it first. So, how do you begin thinking about a multistep synthesis problem where you do not immediately see the solution? The most powerful technique is called *retrosynthetic analysis*. This means that you analyze the problem backward. Let's see how this works with an example:

The synthesis problem above is a multistep synthesis problem, because we do not have a single reaction that allows us to do this transformation in just one step. So the best way to start is to first look at the product and work our way backward.

We see that the product is a dibromide. So we ask ourselves: Do we know any way of making a dibromide? You can see that to answer this question, you must have first mastered one-step syntheses. If you have not yet done this for all of the reactions that you have learned so far, you will need to go back to the beginning of this chapter and do that first (if you are a student in this situation, continue reading for now, so you can see where this is all going).

So we should be able to recognize that we know how to make dibromides from double bonds. We draw the alkene that would have been used to form the product:

Now we are one step closer to solving this problem. The next step is to ask if there is a way to turn the starting material into this double bond. And there is. We can perform an elimination reaction to give the alkene. So now we have solved our synthesis by working backward:

Notice that the stereochemistry and regiochemistry need to work out for every step. You cannot use a step that has the wrong stereochemistry or regiochemistry. I suppose you could have memorized all possible two-step syntheses from the reactions in your textbook, and then you would have gotten this problem right away (maybe ...), but that is not a practical approach. What will you do for three-step or four-step syntheses? You need to get accustomed to thinking backward. The more practice you can get, the better off you will be.

Here is where we run into a big problem. There is no way for me to give you problems that are appropriate. Every course goes at its own pace, in its own order, and with exams at different points in the course. I cannot give problems that will be perfectly appropriate for every student everywhere. So, how are we going to get practice? Very simply. You are going to make your own problems, as described in the next section.

15.4 CREATING YOUR OWN PROBLEMS

Creating your own problems is easier than it sounds. You just choose any reaction from the list that you prepared earlier in this chapter. Then look at the product of that reaction and choose another reaction that you have learned that will transform that compound into something else. We work backward to solve synthesis problems, but we work forward to create a synthesis problem. At each step, draw the product of that reaction and then move on to the next step. After you have gone two or three or four steps, erase everything in the middle. Just draw the very first compound and the final product. Draw an arrow between them, and you have a synthesis problem.

There is one catch. You will not find that problem to be very challenging, because you are the one who made it. So here is what you should do. Find a friend in the course, and each of you should make up 10 or 20 problems. Then you switch off with each other. You will find that this is a very effective method for studying. A study group of 3-4 people is even better, so don't be shy. You will need to work with a friend to get the practice that you need, not to mention the valuable peer support. If you are reading this book, then chances are that other students in your course have this book also. They will have the same need that you do. Team up with them.

Even if you cannot find a friend with whom you can swap problems, it will still be a useful exercise to create your own problems. The process of creating problems by itself is a worthwhile process. It will help you get accustomed to thinking in multiple steps for synthesis problems.

To summarize, these are the keys to becoming proficient at solving synthesis problems:

1. Master the one-step syntheses by constant review.

2. Train yourself to work backward when solving a problem.

3. And, finally, get lots of practice.

DETAILED SOLUTIONS

CHAPTER 1

1.2 Each corner represents a carbon atom, and each endpoint represents a carbon atom. So this compound has a total of six carbon atoms (and we will soon see that each carbon atom must be connected to enough hydrogen atoms to have exactly four bonds):

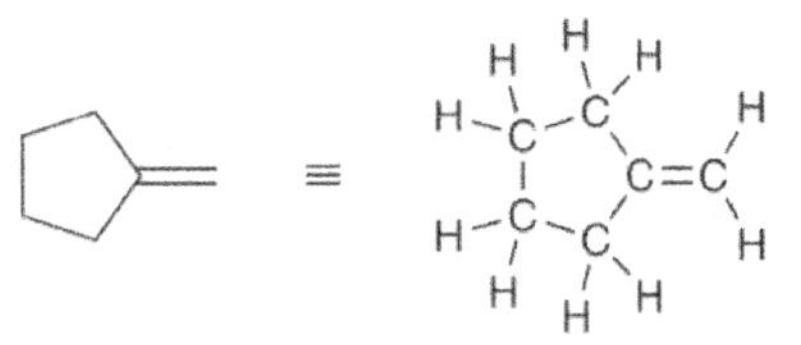

1.3 Each corner represents a carbon atom, and each endpoint represents a carbon atom. So this compound has a total of six carbon atoms (and we will soon see that each carbon atom must be connected to enough hydrogen atoms to have exactly four bonds):

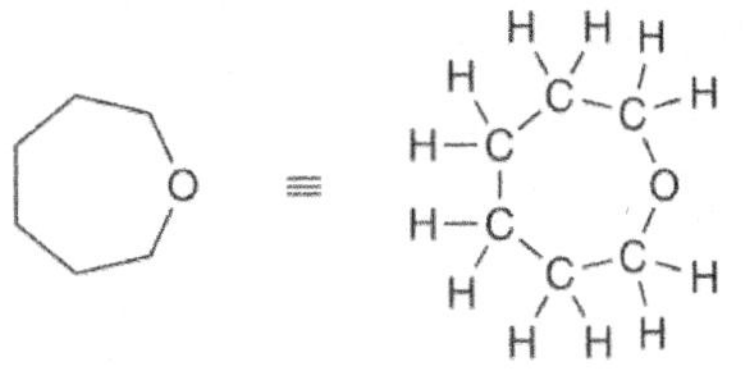

1.4 Each corner represents a carbon atom, and each endpoint represents a carbon atom. So this compound has a total of five carbon atoms (and we will soon see that each carbon atom must be connected to enough hydrogen atoms to have exactly four bonds):

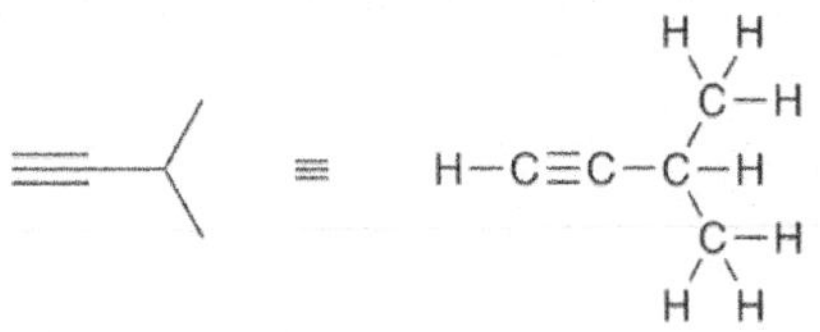

1.5 Each corner represents a carbon atom, and each endpoint represents a carbon atom. So this compound has a total of six carbon atoms (and we will soon see that each carbon atom must be connected to enough hydrogen atoms to have exactly four bonds):

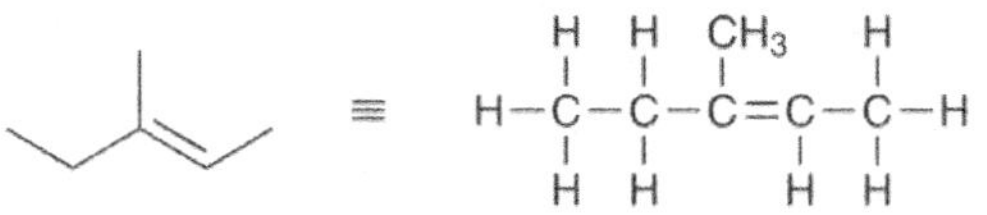

1.6 Each corner represents a carbon atom, and each endpoint represents a carbon atom. So this compound has a total of six carbon atoms (and we will soon see that each carbon atom must be connected to enough hydrogen atoms to have exactly four bonds):

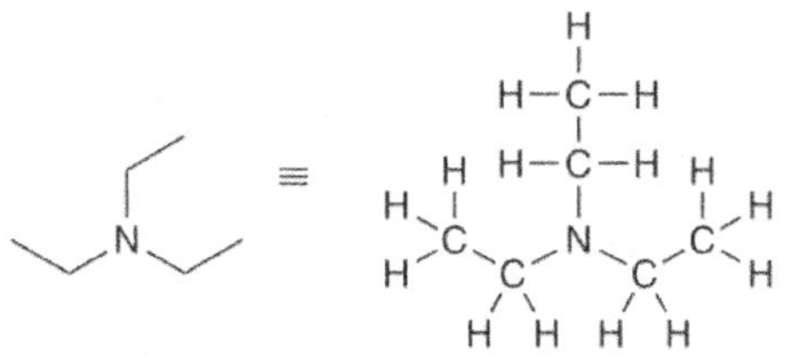

1.7 Each corner represents a carbon atom, and each endpoint represents a carbon atom. So this compound has a total of eight carbon atoms (and we will soon see that each carbon atom must be connected to enough hydrogen atoms to have exactly four bonds):

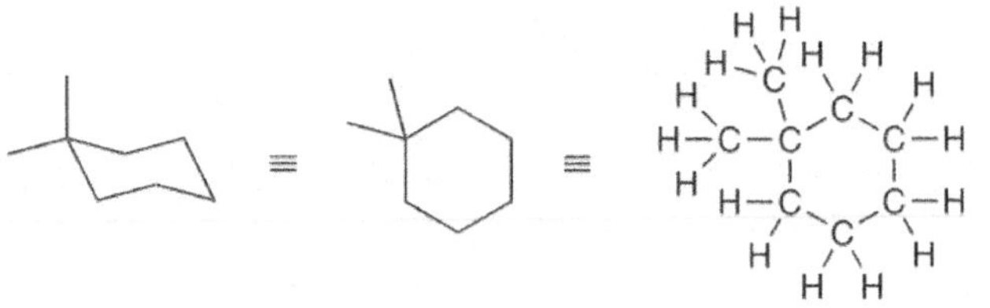

1.8 Each corner represents a carbon atom, and each endpoint represents a carbon atom. So this compound has a total of four carbon atoms (and we will soon see that each carbon atom must be connected to enough hydrogen atoms to have exactly four bonds):

1.9 Each corner represents a carbon atom, and each endpoint represents a carbon atom. So this compound has a total of nine carbon atoms (and we will soon see that each carbon atom must be connected to enough hydrogen atoms to have exactly four bonds):

1.10 Each corner represents a carbon atom, and each endpoint represents a carbon atom. So this compound has a total of nine carbon atoms (and we will soon see that each carbon atom must be connected to enough hydrogen atoms to have exactly four bonds):

1.11 Each corner represents a carbon atom, and each endpoint represents a carbon atom. So this compound has a total of ten carbon atoms (and we will soon see that each carbon atom must be connected to enough hydrogen atoms to have exactly four bonds):

1.13 Each carbon atom should have four bonds. We therefore draw enough hydrogen atoms in order to give each carbon atom a total of four bonds. Any carbon atoms that already have four bonds will not have any hydrogen atoms:

1.14 Each carbon atom should have four bonds. We therefore draw enough hydrogen atoms in order to give each carbon atom a total of four bonds. Any carbon atoms that already have four bonds will not have any hydrogen atoms:

1.15 Each carbon atom should have four bonds. We therefore draw enough hydrogen atoms in order to give each carbon atom a total of four bonds. Any carbon atoms that already have four bonds will not have any hydrogen atoms:

1.16 Each carbon atom should have four bonds. We therefore draw enough hydrogen atoms in order to give each carbon atom a total of four bonds. Any carbon atoms that already have four bonds will not have any hydrogen atoms:

1.17 Each carbon atom should have four bonds. We therefore draw enough hydrogen atoms in order to give each carbon atom a total of four bonds. Any carbon atoms that already have four bonds will not have any hydrogen atoms:

1.18 Each carbon atom should have four bonds. We therefore draw enough hydrogen atoms in order to give each carbon atom a total of four bonds. Any carbon atoms that already have four bonds will not have any hydrogen atoms:

1.19 Each carbon atom should have four bonds. We therefore draw enough hydrogen atoms in order to give each carbon atom a total of four bonds. Any carbon atoms that already have four bonds will not have any hydrogen atoms:

1.20 Each carbon atom should have four bonds. We therefore draw enough hydrogen atoms in order to give each carbon atom a total of four bonds. Any carbon atoms that already have four bonds will not have any hydrogen atoms:

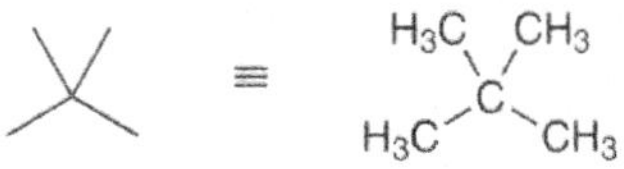

1.21 In a bond-line drawing, each corner represents a carbon atom, and each endpoint represents a carbon atom. This compound has five carbon atoms and can be drawn like this:

1.22 In a bond-line drawing, each corner represents a carbon atom, and each endpoint represents a carbon atom. This compound has seven carbon atoms and is drawn like this:

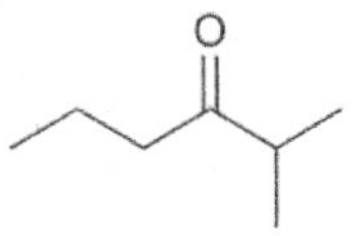

1.23 In a bond-line drawing, each corner represents a carbon atom, and each endpoint represents a carbon atom. This compound has six carbon atoms and is drawn like this:

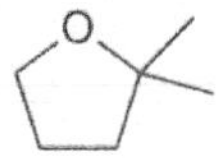

1.24 In a bond-line drawing, each corner represents a carbon atom, and each endpoint represents a carbon atom. This compound has three carbon atoms and can be drawn like this:

1.25 In the transformation shown below, Cl has been replaced by an OH group.

1.26 In the transformation shown below, two OH groups have been installed, and a double bond has been converted into a single bond.

1.27 In the transformation shown below, H and Cl (highlighted) have been removed, and a single bond has been converted into a double bond.

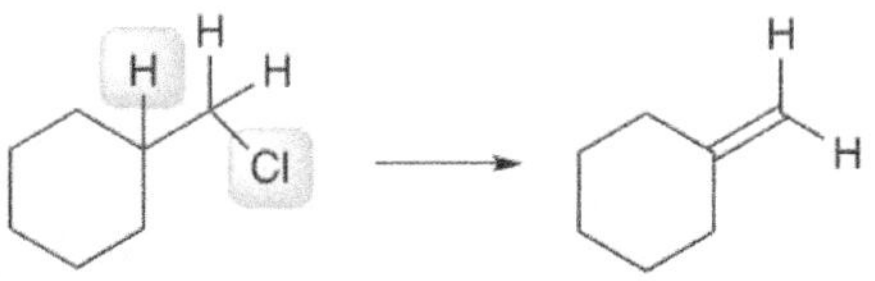

1.28 In the transformation shown below, two bromine atoms have been installed, and a double bond has been converted into a single bond.

1.29 In the transformation shown below, two hydrogen atoms (highlighted) have been removed, and a single bond has been converted into a double bond.

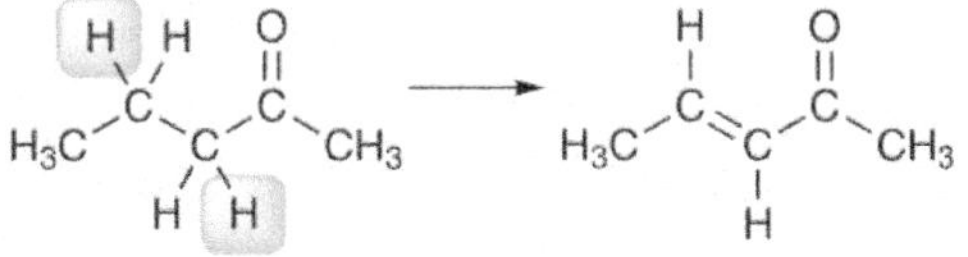

1.30 In the transformation shown below, iodine has been replaced by an SH group.

1.31 In the transformation shown, two hydrogen atoms (highlighted) have been removed, and a double bond has been converted into a triple bond.

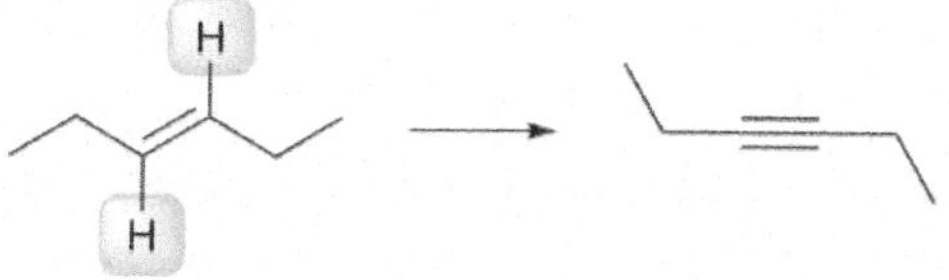

1.32 In the transformation shown below, two hydrogen atoms (highlighted) have been installed, and a triple bond has been converted into a double bond.

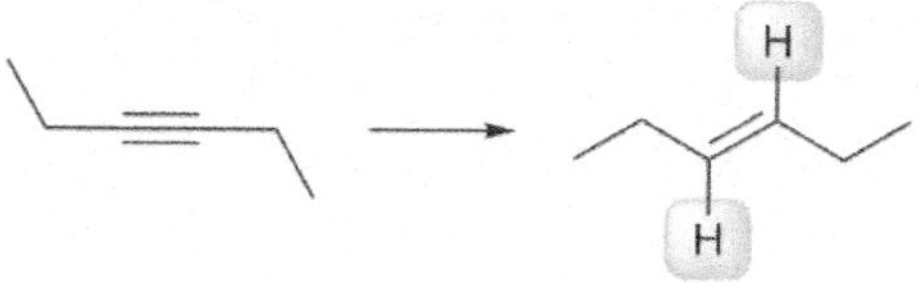

1.34 Oxygen is in Column 6A of the periodic table, so an oxygen atom is supposed to have six valence electrons. Now let's consider the oxygen atom in this case, and let's count how many valence electrons are *actually* around it.

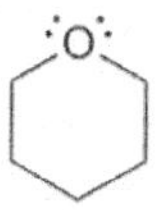

This oxygen atom has two bonds and two lone pairs. Each bond requires one valence electron from oxygen, and each lone pair requires two valence electrons from oxygen, for a total of six valence electrons. This is exactly the number of valence electrons that an oxygen atom is supposed to have, so this oxygen atom has no formal charge.

1.35 Oxygen is in Column 6A of the periodic table, so an oxygen atom is supposed to have six valence electrons. Now let's consider the oxygen atom in this case, and let's count how many valence electrons are *actually* around it.

This oxygen atom (shown below) has three bonds and one lone pair. Each bond requires one valence electron from oxygen, and the lone pair requires two valence electrons from oxygen, for a total of five valence electrons. But oxygen is supposed to have six valence electrons, so this oxygen atom is missing an electron. Therefore, this oxygen atom will have a positive charge.

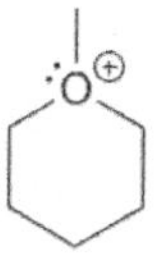

1.36 Nitrogen is in Column 5A of the periodic table, so a nitrogen atom is supposed to have five valence electrons. Now let's consider the nitrogen atom in this case, and let's count how many valence electrons are *actually* around it.

In this case, the nitrogen atom has two bonds and two lone pairs. Each bond requires one valence electron from nitrogen, and each

lone pair requires two valence electrons from nitrogen, for a total of six valence electrons. But nitrogen is supposed to have only five valence electrons, so this nitrogen atom has an extra electron. Therefore, this nitrogen atom will have a negative charge.

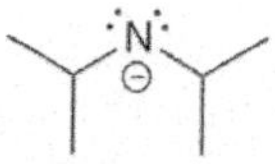

1.37 Nitrogen is in Column 5A of the periodic table, so a nitrogen atom is supposed to have five valence electrons. Now let's consider the nitrogen atom in this case, and let's count how many valence electrons are *actually* around it.

This nitrogen atom (shown below) has three bonds and one lone pair. Each bond requires one valence electron from nitrogen, and the lone pair requires two valence electrons from nitrogen, for a total of five valence electrons. This is exactly the number of valence electrons that a nitrogen atom is supposed to have, so this nitrogen atom has no formal charge.

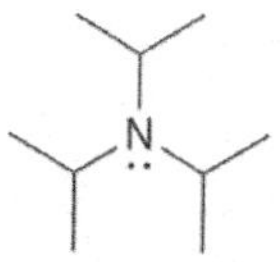

1.38 Nitrogen is in Column 5A of the periodic table, so a nitrogen atom is supposed to have five valence electrons. Now let's consider the nitrogen atom in this case, and let's count how many valence electrons are *actually* around it.

This nitrogen atom (shown below) has four bonds and no lone pairs. Each bond requires one valence electron from nitrogen, for a total of four valence electrons. But nitrogen is supposed to have five valence electrons, so this nitrogen is missing an electron. Therefore, this nitrogen atom will have a positive charge.

1.39 Oxygen is in Column 6A of the periodic table, so an oxygen atom is supposed to have six valence electrons. Now let's consider the oxygen atom in this case, and let's count how many valence electrons are *actually* around it.

This oxygen atom (shown below) has one bond and three lone pairs. The bond requires one valence electron from oxygen, and each lone pair requires two valence electrons from oxygen, for a total of seven valence electrons. But oxygen is supposed to have only six valence electrons, so this oxygen atom has an extra electron. Therefore, this oxygen atom will have a negative charge.

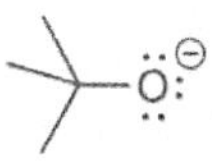

1.40 Oxygen is in Column 6A of the periodic table, so an oxygen atom is supposed to have six valence electrons. Now let's consider the oxygen atom in this case, and let's count how many valence electrons are *actually* around it.

This oxygen atom (shown below) has three bonds and one lone pair. Each bond requires one valence electron from oxygen, and the lone pair requires two valence electrons from oxygen, for a total of five valence electrons. But oxygen is supposed to have six valence electrons, so this oxygen atom is missing an electron. Therefore, this oxygen atom will have a positive charge.

1.41 Nitrogen is in Column 5A of the periodic table, so a nitrogen atom is supposed to have five valence electrons. Now let's consider the nitrogen atom in this case, and let's count how many valence electrons are *actually* around it.

This nitrogen atom (shown below) has two bonds and two lone pairs. Each bond requires one valence electron from nitrogen, and each lone pair requires two valence electrons from nitrogen, for a total of six valence electrons. But nitrogen is supposed to have only five valence electrons, so this nitrogen atom has an extra electron. Therefore, this nitrogen atom will have a negative charge.

1.42 Nitrogen is in Column 5A of the periodic table, so a nitrogen atom is supposed to have five valence electrons. Now let's consider the nitrogen atom in this case, and let's count how many valence electrons are *actually* around it.

This nitrogen atom (shown below) has three bonds and one lone pair. Each bond requires one valence electron from nitrogen, and the lone pair requires two valence electrons from nitrogen, for a total of five valence electrons. This is exactly the number of valence electrons that a nitrogen atom is supposed to have, so this nitrogen atom has no formal charge.

1.43 Nitrogen is in Column 5A of the periodic table, so a nitrogen atom is supposed to have five valence electrons. Now let's consider

the nitrogen atom in this case, and let's count how many valence electrons are *actually* around it.

This nitrogen atom (shown below) has four bonds and no lone pairs. Each bond requires one valence electron from nitrogen, for a total of four valence electrons. But nitrogen is supposed to have five valence electrons, so this nitrogen is missing an electron. Therefore, this nitrogen atom will have a positive charge.

1.44 Oxygen is in Column 6A of the periodic table, so an oxygen atom is supposed to have six valence electrons. Now let's consider the oxygen atom in this case, and let's count how many valence electrons are *actually* around it.

This oxygen atom (shown below) has two bonds and two lone pairs. Each bond requires one valence electron from oxygen, and each lone pair requires two valence electrons from oxygen, for a total of six valence electrons. This is exactly the number of valence electrons that an oxygen atom is supposed to have, so this oxygen atom has no formal charge.

1.45 Nitrogen is in Column 5A of the periodic table, so a nitrogen atom is supposed to have five valence electrons. Now let's consider the nitrogen atom in this case, and let's count how many valence electrons are *actually* around it.

This nitrogen atom (shown below) has three bonds and one lone pair. Each bond requires one valence electron from nitrogen, and the lone pair requires two valence electrons from nitrogen, for a total of five valence electrons. This is exactly the number of valence electrons that a nitrogen atom is supposed to have, so this nitrogen atom has no formal charge.

$$H_3C-C\equiv N:$$

1.47 The oxygen atom shown below has two bonds and no formal charge. Therefore, this oxygen atom has two lone pairs.

We will now count electrons to confirm this answer. Oxygen is in Column 6A of the periodic table, so it is supposed to have six valence electrons. Since this oxygen atom lacks a formal charge, this oxygen atom will actually be using six valence electrons. Each bond requires one valence electron from the oxygen atom. This leaves four valence electrons, corresponding to two lone pairs.

1.48 The oxygen atom shown below has two bonds and no formal charge. Therefore, this oxygen atom has two lone pairs.

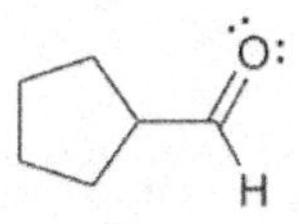

We will now count electrons to confirm this answer. Oxygen is in Column 6A of the periodic table, so it is supposed to have six valence electrons. Since this oxygen atom lacks a formal charge, this oxygen atom will actually be using six valence electrons. Each bond requires one valence electron from the oxygen atom. This leaves four valence electrons, corresponding to two lone pairs.

1.49 Each of the oxygen atoms shown below has two bonds and no formal charge. Therefore, each of these oxygen atoms has two lone pairs.

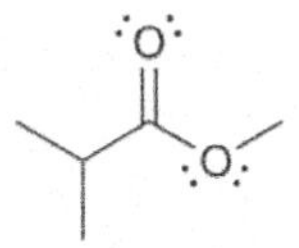

We will now count electrons to confirm this answer. Oxygen is in Column 6A of the periodic table, so each oxygen atom is supposed to have six valence electrons. Since each oxygen atom lacks a formal charge, each of these oxygen atoms must have exactly six valence electrons around it. Each bond requires one valence electron. So each oxygen atom has four remaining valence electrons that must comprise lone pairs. In other words, each oxygen atom has two lone pairs.

1.50 Each of the oxygen atoms shown below has two bonds and no formal charge. Therefore, each of these oxygen atoms has two lone pairs.

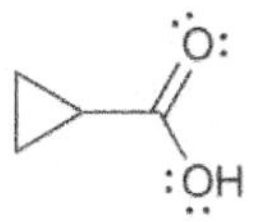

We will now count electrons to confirm this answer. Oxygen is in Column 6A of the periodic table, so each oxygen atom is supposed to have six valence electrons. Since each oxygen atom lacks a formal charge, each oxygen atom must have exactly six valence electrons around it. Each bond requires one valence electron. So each oxygen atom has four remaining valence electrons that must comprise lone pairs. In other words, each oxygen atom has two lone pairs.

1.51 The oxygen atom shown has three bonds and a positive formal charge. Therefore, this oxygen atom has one lone pair.

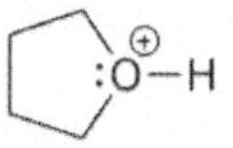

We will now count electrons to confirm this answer. Oxygen is in Column 6A of the periodic table, so it is supposed to have six valence electrons. Since this oxygen atom has a positive formal charge, it is missing a valence electron, which means that it has only five valence electrons around it. Each bond requires one valence electron from the oxygen atom. This leaves two valence electrons, corresponding to one lone pair.

1.52 The left-most oxygen atom shown below has two bonds and no formal charge, while the right-most oxygen atom shown below has one bond and a negative formal charge. Therefore, the left-most oxygen atom has two lone pairs, and the right-most oxygen atom has three lone pairs:

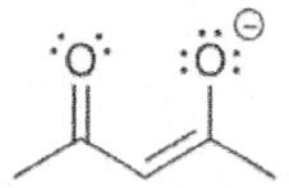

We will now count electrons to confirm this answer. Oxygen is in Column 6A of the periodic table, so it is supposed to have six valence electrons. The left-most oxygen atom has no formal charge, so it must have exactly six valence electrons around it. Each bond requires one valence electron from the oxygen atom. This leaves four valence electrons, corresponding to two lone pairs.

The right-most oxygen atom has a negative formal charge so it has one extra electron. Therefore, this oxygen atom must have exactly seven valence electrons around it. The one bond with this oxygen atom requires one valence electron from the oxygen atom. This leaves six valence electrons, corresponding to three lone pairs.

1.54 The nitrogen atom shown below has three bonds and no formal charge. Therefore, this nitrogen atom has one lone pair.

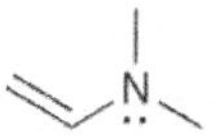

We will now count electrons to confirm this answer. Nitrogen is in Column 5A of the periodic table, so it is supposed to have five valence electrons. Since this nitrogen atom lacks a formal charge, this nitrogen atom must have five valence electrons around it. Each bond requires one valence electron from the nitrogen atom. This leaves two valence electrons, corresponding to one lone pair.

1.55 The nitrogen atom shown below has four bonds and a positive charge. Therefore, this nitrogen atom has no lone pairs.

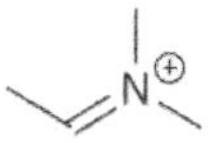

We will now count electrons to confirm this answer. Nitrogen is in Column 5A of the periodic table, so it is supposed to have five valence electrons. This nitrogen atom has a positive charge (it is missing a valence electron), so it must have exactly four valence electrons around it. Each bond requires one valence electron from the nitrogen atom. This leaves no valence electrons for lone pairs.

1.56 The nitrogen atom shown below has three bonds and no formal charge. Therefore, this nitrogen atom has one lone pair.

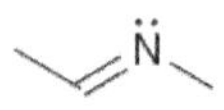

We will now count electrons to confirm this answer. Nitrogen is in Column 5A of the periodic table, so it is supposed to have five valence electrons. Since this nitrogen atom lacks a formal charge, this nitrogen atom must have exactly five valence electrons around it. Each bond requires one valence electron from the nitrogen atom. This leaves two valence electrons, corresponding to one lone pair.

1.57 The nitrogen atom shown below has two bonds and a negative charge. Therefore, this nitrogen atom has two lone pairs.

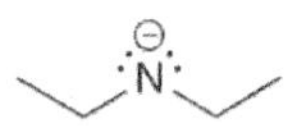

We will now count electrons to confirm this answer. Nitrogen is in Column 5A of the periodic table, so it is supposed to have five valence electrons. This nitrogen atom has a negative charge (it has an extra valence electron), so it must have exactly six valence electrons around it. Each bond requires one valence electron from the nitrogen atom. This leaves four valence electrons, corresponding to two lone pairs.

1.58 The nitrogen atom shown below has three bonds and no formal charge. Therefore, this nitrogen atom has one lone pair.

We will now count electrons to confirm this answer. Nitrogen is in Column 5A of the periodic table, so it is supposed to have five valence electrons. Since this nitrogen atom lacks a formal charge, this nitrogen atom must have exactly five valence electrons around it. Each bond requires one valence electron from the nitrogen atom. This leaves two valence electrons, corresponding to one lone pair.

1.59 The nitrogen atom shown below has four bonds and a positive charge. Therefore, this nitrogen atom has no lone pairs.

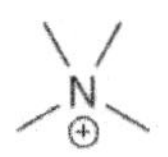

We will now count electrons to confirm this answer. Nitrogen is in Column 5A of the periodic table, so it is supposed to have five valence electrons. This nitrogen atom has a positive charge (it is missing a valence electron), so it must have only four valence electrons around it. Each bond requires one valence electron from the nitrogen atom. This leaves no valence electrons for lone pairs.

1.60 This structure has a carbon atom with a negative charge. Whenever a carbon atom has a negative charge, that carbon atom will have three bonds (not four) and one lone pair:

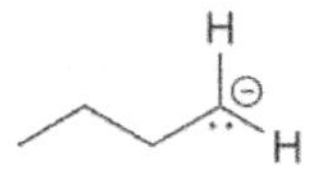

1.61 This structure has a carbon atom with a positive charge. Whenever a carbon atom has a positive charge, that carbon atom will have three bonds (not four) and no lone pairs. So this structure does not have any lone pairs:

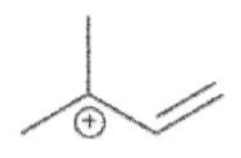

1.62 This structure has a carbon atom with a negative charge. Whenever a carbon atom has a negative charge, that carbon atom will have three bonds (not four) and one lone pair:

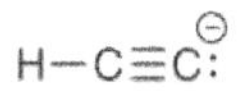

1.63 In the structure shown below, the nitrogen atom has three bonds and no formal charge, so this nitrogen atom has one lone pair. The right-most oxygen atom has one bond and a negative charge, so it must have three lone pairs. The other oxygen atom has two bonds and no formal charge, so it has two lone pairs.

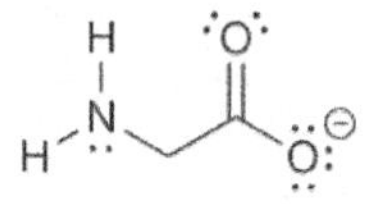

1.64 In the structure shown below, the oxygen atom has two bonds and no formal charge, so it has two lone pairs. Also, the right-most carbon atom bears a negative charge, and this carbon atom must have three bonds and one lone pair (rather than four bonds):

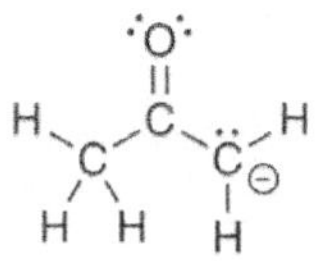

1.65 In the structure shown below, the nitrogen atom has three bonds and no formal charge, so this nitrogen atom has one lone pair. Each oxygen atom has two bonds and no charge, so each oxygen atom has two lone pairs.

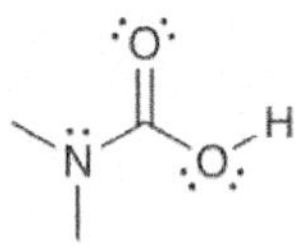

1.66 In the structure shown below, each nitrogen atom has three bonds and no formal charge, so each nitrogen atom has one lone pair. The oxygen atom has three bonds and a positive charge, so this oxygen atom has one lone pair.

1.67 Each nitrogen atom has three bonds and no formal charge, so each nitrogen atom has one lone pair.

1.68 In the structure given below, each nitrogen atom has three bonds and no formal charge, so each nitrogen atom has one lone pair.

CHAPTER 2

2.2 This curved arrow violates the octet rule (the second commandment) because nitrogen cannot have five bonds.

2.3 This curved arrow violates the octet rule (the second commandment) because nitrogen cannot have four bonds and one lone pair.

2.4 This curved arrow violates the octet rule (the second commandment) because oxygen cannot have three bonds and two lone pairs.

2.5 This curved arrow does not violate either of the two commandments.

2.6 This curved arrow violates the octet rule (the second commandment) because carbon cannot have five bonds.

2.7 This curved arrow violates the first commandment because the tail of the curved arrow has been placed on a single bond.

2.8 This curved arrow violates the octet rule (the second commandment) because carbon cannot have five bonds.

2.9 This curved arrow does not violate either of the two commandments.

2.10 This curved arrow violates the first commandment because the tail of the curved arrow has been placed on a single bond.

2.11 This curved arrow violates the octet rule (the second commandment) because carbon cannot have five bonds.

2.12 This curved arrow does not violate either of the two commandments.

2.14 The difference between the two structures is the positions of the double bonds. We will need one curved arrow to move each double bond, so we will need a total of three curved arrows.

These curved arrows can be drawn in a clockwise fashion, or they can be drawn in a counterclockwise fashion, as shown here:

2.15 We look for any double bonds or lone pairs that are appearing or disappearing. In this case, the C=N double bond becomes a single bond (so the tail of a curved arrow must be placed on this double bond), and the nitrogen atom receives a lone pair (so the head of the curved arrow must be placed on the nitrogen atom). Only one curved arrow is required here, pushing the electrons of the double bond up onto the nitrogen atom:

2.16 We look for any double bonds or lone pairs that are appearing or disappearing. In this case, the C=C double bond becomes a single bond (so the tail of a curved arrow must be placed on this double bond), and the C—C single bond becomes a double bond (so the head of the curved arrow must be placed on the C—C bond). Only one curved arrow is required here, pushing the electrons of the double bond over to the left:

2.17 We look for any double bonds or lone pairs that are appearing or disappearing. In this case, a C=C double bond becomes a single bond (so the tail of a curved arrow must be placed on this double bond), and a C—C single bond becomes a double bond (so the head of the curved arrow must be placed on this C—C single bond). Only one curved arrow is required here, pushing the electrons of the double bond over to the right:

2.18 We look for any double bonds or lone pairs that are appearing or disappearing. In this case, the oxygen atom at the top of the structure is losing a lone pair but gaining a double bond (requiring one curved arrow). The other oxygen atom is losing a double bond but gaining a lone pair (requiring another curved arrow). So, we will use two curved arrows in the case. The first curved arrow is drawn showing a lone pair (at the top of the structure) being pushed down to form a double bond, and the second curved arrow shows a double bond being pushed down to become a lone pair:

2.19 Recall that a negatively charged carbon atom will have a lone pair. In this case, there is a carbon atom with a negative charge, so that carbon atom must have a lone pair, even though it has not been drawn. That lone pair is being converted into a double bond (which requires a curved arrow). Also, the C≡N double bond is being pushed onto the nitrogen atom to form a lone pair, and this requires a second curved arrow. Both curved arrows are shown below:

2.21 In this case, the double bond is being pushed over. Notice the change in location of the positive formal charge.

Think of a C⁺ as a "hole," where electron density is missing. By moving the double bond, the hole is filled, but a new hole is created nearby.

2.22 In this case, a lone pair is being pushed to form a bond, and a double bond is pushed to form a lone pair.

Notice the change in location of the negative charge, which has been pushed as the result of two curved arrows.

2.23 In this case, there are three curved arrows, each of which is described here:

1. a lone pair on nitrogen is pushed to form a C≡N bond (giving a positive charge on nitrogen), and

2. a double bond is pushed to form a new double bond, and

3. a double bond is pushed to form a lone pair on a carbon atom (giving a negative charge on carbon).

Notice that the original structure has no net charge, so the resulting resonance structure must also have no net charge (it has one positive charge and one negative charge). Also notice the position of the positive and negative charges. Nitrogen originally had a lone pair of electrons, but these two electrons are being shared in the resonance structure. In going from the first resonance structure to the second resonance structure, nitrogen has lost an electron, giving nitrogen a positive formal charge. A negative charge results at the bottom of the structure (read the curved arrows to convince yourself that electron density has been pushed from the top of the structure to the bottom of the structure).

2.24 In this case, there are two curved arrows, each of which is described here:

1. a C≡C double bond is pushed to form a new C≡C double bond (giving a positive charge on the carbon atom that lost a bond) and

2. a C≡O double bond is pushed to form a lone pair on an oxygen atom (giving a negative charge on oxygen).

Notice that the original structure has no net charge, so the resulting resonance structure must also have no net charge (it has one positive charge and one negative charge). Also notice the position of the positive and negative charges. Read the curved arrows to convince yourself that electron density has been pushed from the bottom of the structure to the top of the structure.

2.25 In this case, there are two curved arrows, each of which is described here:

1. a lone pair on oxygen is pushed to form an N=O double bond and

2. an N=O double bond is pushed to form a lone pair on an oxygen atom.

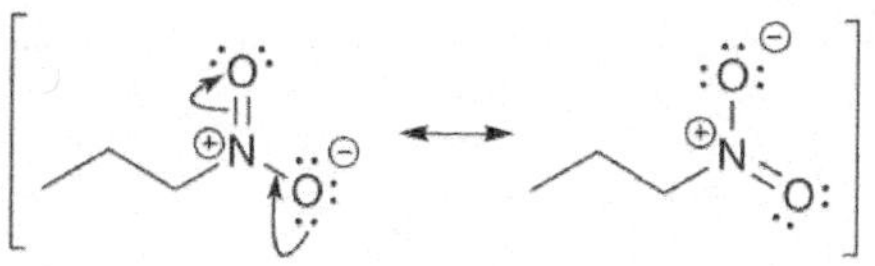

Notice that the original structure has no net charge, so the resulting resonance structure must also have no net charge. Also notice the position of the negative charge in both structures. Read the curved arrows to convince yourself that electron density has been pushed from one oxygen atom to another oxygen atom.

2.26 In this case, there are two curved arrows, each of which is described here:

1. a C=C double bond is pushed to form a C=N double bond and

2. an N=O double bond is pushed to form a lone pair on an oxygen atom.

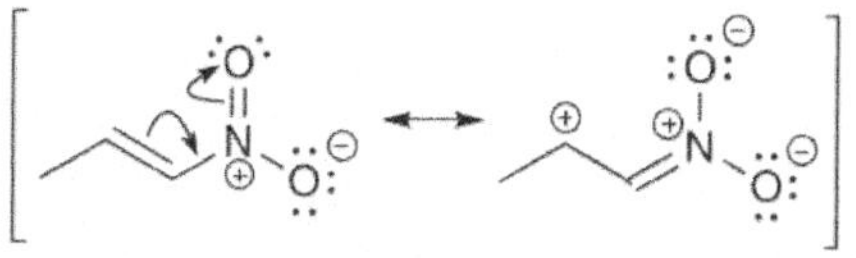

Notice that the original structure has no net charge (one positive charge on nitrogen and one negative charge on oxygen), so the resulting resonance structure must also have no net charge. Also notice the position of the positive and negative charges. Read the curved arrows to convince yourself that electron density has been pushed away from a carbon atom and onto an oxygen atom.

2.27 In this case, there are two curved arrows, each of which is described here:

1. a lone pair on nitrogen is pushed to form a C=N double bond and

2. a C=O double bond is pushed to form a lone pair on an oxygen atom.

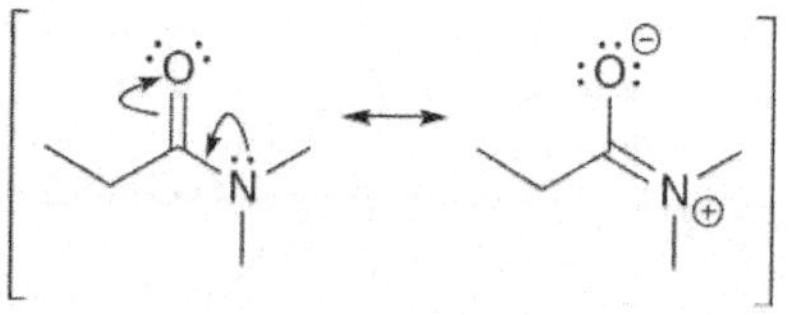

Notice that the original structure has no net charge, so the resulting resonance structure must also have no net charge (it has one positive charge and one negative charge). Also notice the position of the positive and negative charges. Read the curved arrows to convince yourself that electron density has been pushed from the nitrogen atom up onto the oxygen atom.

2.28 In this case, there are two curved arrows, each of which is described here:

1. a lone pair on carbon is pushed to form a C=C double bond and

2. a C=N double bond is pushed to form a lone pair on a nitrogen atom.

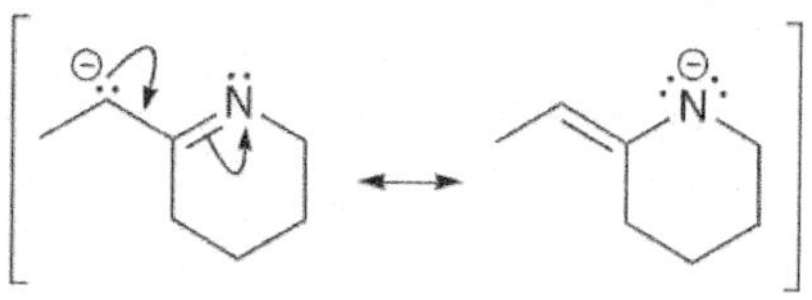

Notice the change in the location of the negative charge, which has been pushed from a carbon atom onto a nitrogen atom.

2.30 Let's begin by looking for electrons that can be pushed via resonance. We are specifically looking for pi electrons (i.e., double bonds) and lone pairs. The oxygen atom has lone pairs, as does the nitrogen atom, and there is also a C=O bond. If we try to push one of the lone pairs on oxygen down to form a triple bond between C and O, that would violate the octet rule,

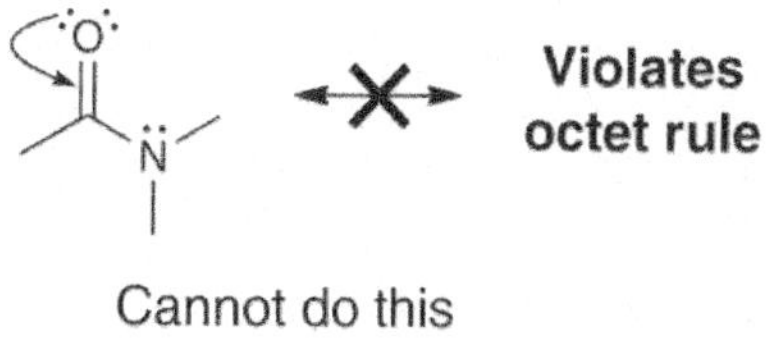

Cannot do this

so the lone pairs on the oxygen atom cannot be pushed.

Now let's consider the C=O bond. This bond can be moved to give a lone pair on O, generating C⁺ and O⁻, as shown here:

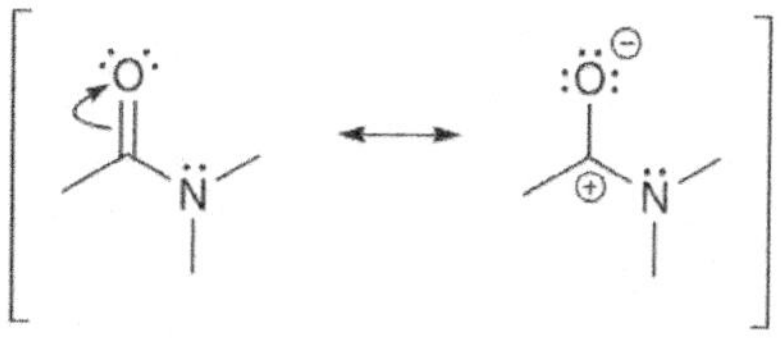

The resulting resonance structure has a positive charge on the carbon atom that is adjacent to nitrogen atom, so the lone pair on nitrogen can be pushed (without violating the octet rule), as shown below. This generates a third resonance structure in which a positive formal charge is on nitrogen, and a negative formal charge is on oxygen.

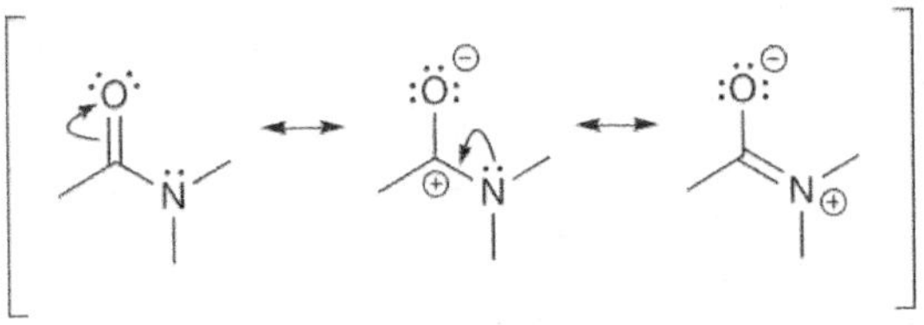

In the upcoming sections of this chapter, we will develop specific skills that can be used to draw the above three structures more methodically. Do not be discouraged if you struggled with this problem. Clarity is coming!

2.32 The structure shown has a lone pair next to a pi bond, so we draw two curved arrows: one from a lone pair on oxygen to form a new pi bond, and one from the existing pi bond to form a lone pair on oxygen:

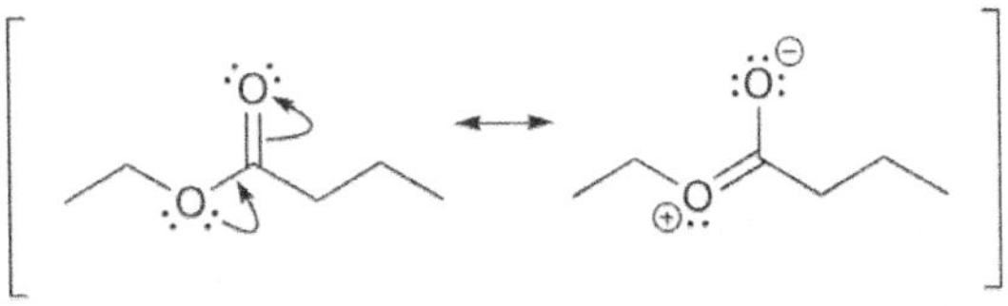

Look carefully at the charges. Both structures have no net charge, but the second resonance structure has charge separation, with a negative charge on one oxygen atom, and a positive charge on the other oxygen atom.

Note: There is a third resonance structure that has not been shown here. This problem is focused on helping you identify the pattern "lone pair next to a pi bond," but there is another pattern here, which we will soon learn. If you are coming back to review this problem (after finishing the chapter), please do not be alarmed that there is a third resonance structure that is not shown above.

2.33 The structure shown has a lone pair next to a pi bond, so we draw two curved arrows: one from the lone pair to form a pi bond, and one from the pi bond to form a lone pair:

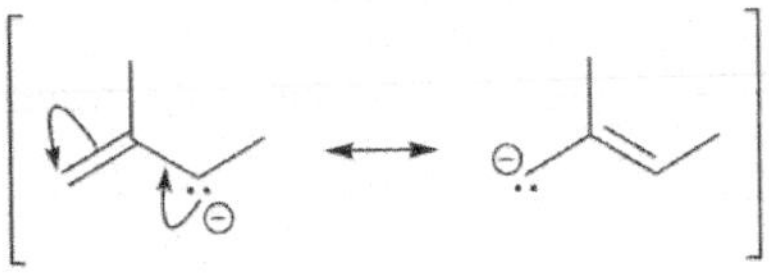

Notice that the location of the negative charge has moved.

2.34 The structure shown has a lone pair next to a pi bond, so we draw two curved arrows: one from a lone pair on oxygen to form a

pi bond, and one from the pi bond to form a lone pair on a carbon atom:

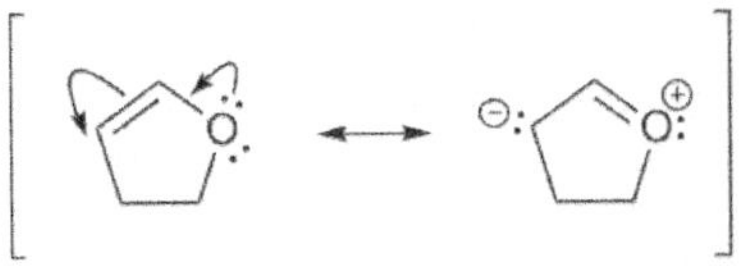

Look carefully at the charges. Both structures have no net charge, but the second resonance structure has charge separation, with a negative charge on a carbon atom, and a positive charge on the oxygen atom.

2.35 The structure shown has a lone pair next to a pi bond, so we draw two curved arrows: one from a lone pair on nitrogen to form a C=N pi bond, and one from the pi bond to form a lone pair on a carbon atom:

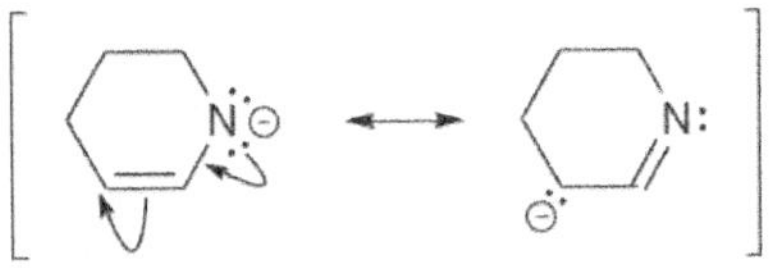

Notice that the negative charge has been pushed from nitrogen to carbon.

2.36 The structure shown has a lone pair next to a pi bond, so we draw two curved arrows: one from the lone pair on nitrogen to form a C=N pi bond, and one from the pi bond to form a lone pair on a carbon atom:

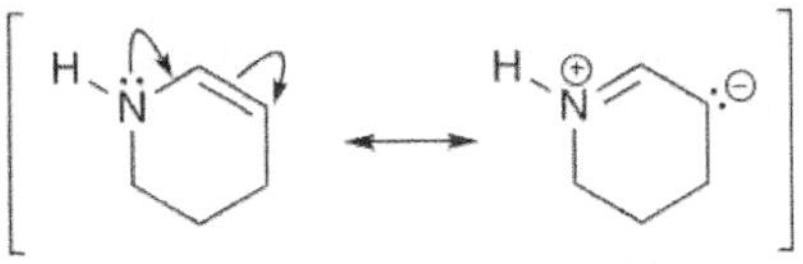

Look carefully at the charges. Both structures have no net charge, but the second resonance structure has charge separation, with a negative charge on a carbon atom, and a positive charge on the nitrogen atom.

2.37 The structure shown has a lone pair next to a pi bond, so we draw two curved arrows: one from a lone pair on oxygen to form a pi bond, and one from the pi bond to form a lone pair on an oxygen atom:

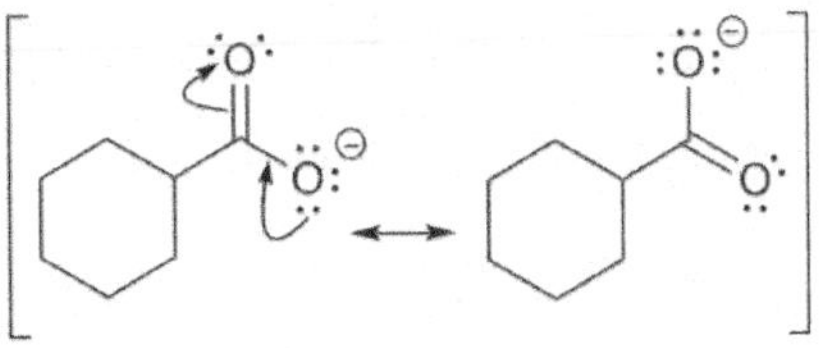

Notice that the negative charge has been pushed from one oxygen atom to the other.

2.38 The structure shown has a lone pair next to a pi bond, so we draw two curved arrows: one from a lone pair on oxygen to form a pi bond, and one from the pi bond to form a lone pair on a carbon atom:

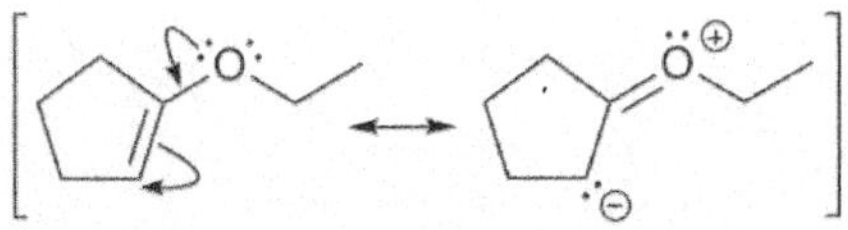

Look carefully at the charges. Both structures have no net charge, but the second resonance structure has charge separation, with a negative charge on a carbon atom, and a positive charge on the oxygen atom.

2.39 The structure shown has a lone pair next to a pi bond, so we draw two curved arrows: one from a lone pair on oxygen to form a pi bond, and one from the pi bond to form a lone pair on an oxygen atom:

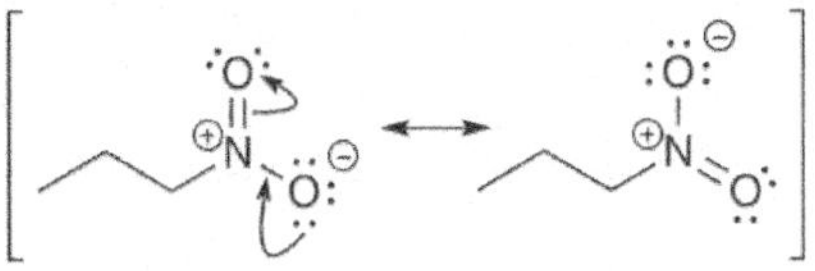

Look carefully at the charges. Both structures have charge separation but no net charge. These resonance structures differ in the location of the negative charge.

2.40 The structure shown has a lone pair next to a positive charge, so we draw one curved arrow that converts the lone pair into a pi bond:

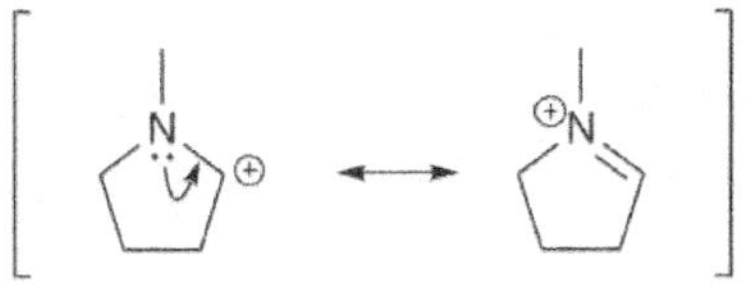

Notice that the position of the positive charge has moved from carbon to nitrogen.

2.41 The structure shown has a lone pair next to a positive charge, so we draw one curved arrow that converts the lone pair into a pi bond:

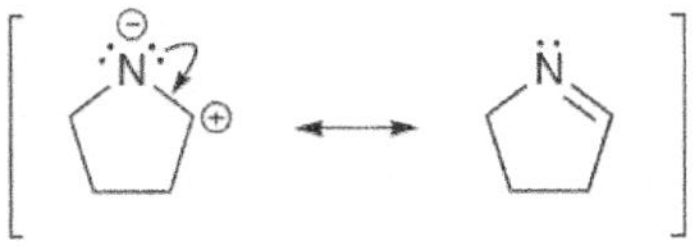

Look carefully at the charges. Both structures have no net charge, but the first resonance structure has charge separation, while the second resonance structure does not have charge separation.

2.42 The structure shown has a lone pair next to a positive charge, so we draw one curved arrow that converts a lone pair into a pi bond:

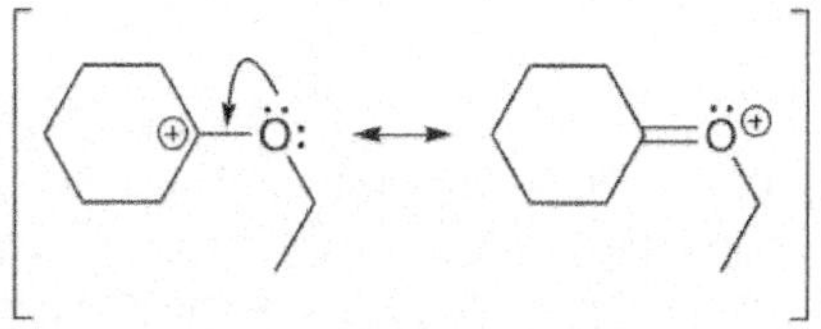

Notice that the position of the positive charge has moved from carbon to oxygen.

2.43 The structure shown has a lone pair next to a positive charge, so we draw one curved arrow that converts the lone pair into a pi bond:

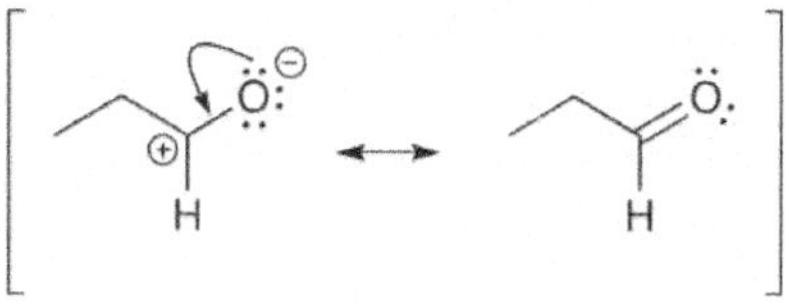

Look carefully at the charges. Both structures have no net charge, but the first resonance structure has charge separation, while the second resonance structure does not have charge separation.

2.44 The structure shown has a pi bond next to a positive charge, so we draw one curved arrow that pushes the pi bond over:

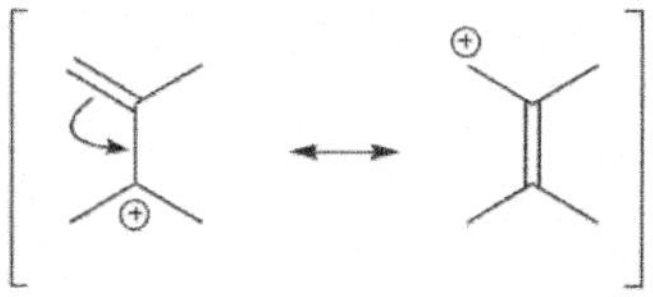

Notice that the position of the positive charge has moved from one carbon atom to another.

2.45 The structure shown has a pi bond next to a positive charge, so we draw one curved arrow that pushes the pi bond over:

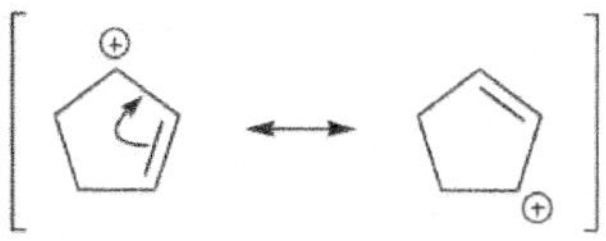

Notice that the position of the positive charge has moved from one carbon atom to another.

2.46 The structure shown has a pi bond next to a positive charge, so we draw one curved arrow that pushes the pi bond over:

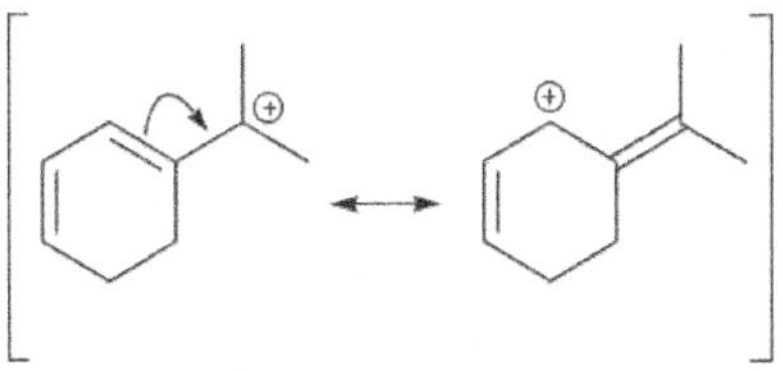

Notice that the position of the positive charge has moved from one carbon atom to another. Also notice that the new positive charge is once again next to a pi bond, so we can draw a third resonance structure, as shown below:

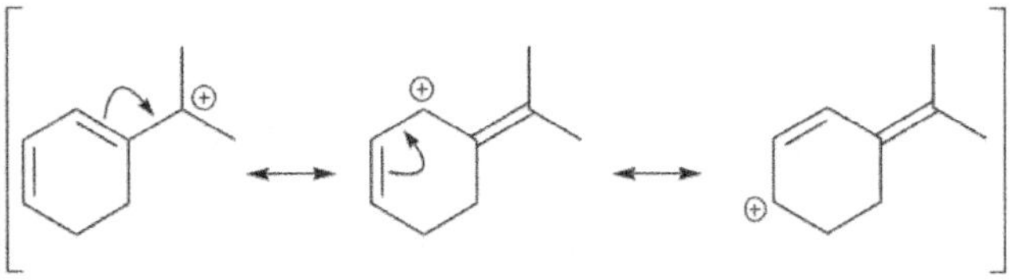

2.47 The structure shown has a C=N pi bond, so we draw one curved arrow that pushes the pi electrons onto the nitrogen atom, generating a positive charge on carbon and a negative charge on nitrogen:

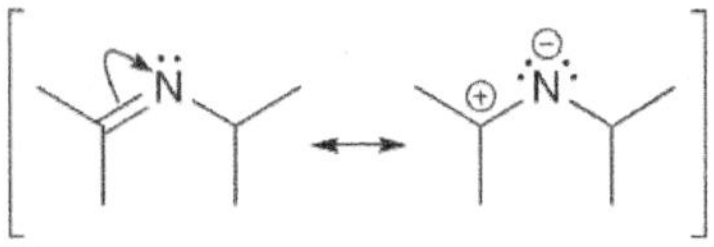

Look carefully at the charges. Both structures have no net charge, but the second resonance structure has charge separation, with the negative charge on the more electronegative atom.

2.48 The structure shown has a C=O pi bond, so we draw one curved arrow that pushes the pi electrons onto the oxygen atom, generating a positive charge on carbon and a negative charge on oxygen:

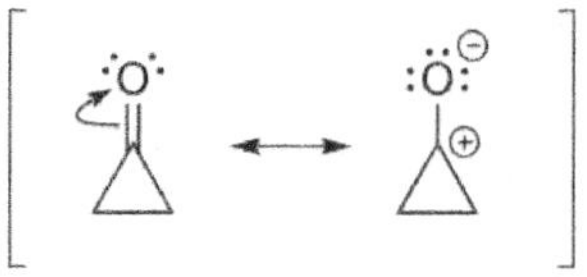

Look carefully at the charges. Both structures have no net charge, but the second resonance structure has charge separation, with the negative charge on the more electronegative atom.

2.49 The structure shown has a C=O pi bond, so we draw one curved arrow that pushes the pi electrons onto the oxygen atom, generating a positive charge on carbon and a negative charge on oxygen:

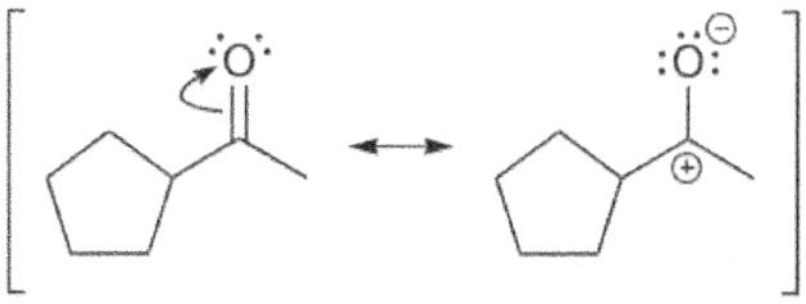

Look carefully at the charges. Both structures have no net charge, but the second resonance structure has charge separation, with the negative charge on the more electronegative atom.

2.50 This structure has a C=N bond, so we can draw a curved arrow that pushes the pi electrons up onto the nitrogen atom, generating a positive charge on a carbon atom and a negative charge on a nitrogen atom. The resulting resonance structure has a lone pair next to a positive charge, so we draw one curved arrow to generate the third resonance structure below:

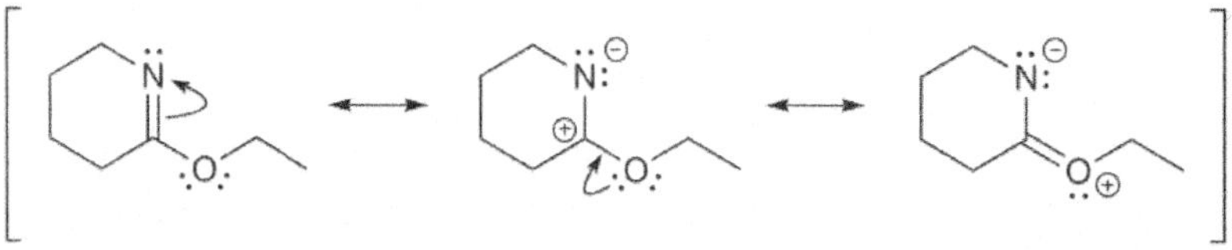

Alternatively, you might have noticed that the structure has a lone pair (on oxygen) next to a pi bond, so you could have drawn two curved arrows which would have converted the first resonance structure above directly into the third resonance structure above. If you had done this, you would still be able to generate the missing resonance structure, by recognizing that you now have a C=O bond, and the pi electrons can be pushed to generate the missing resonance structure:

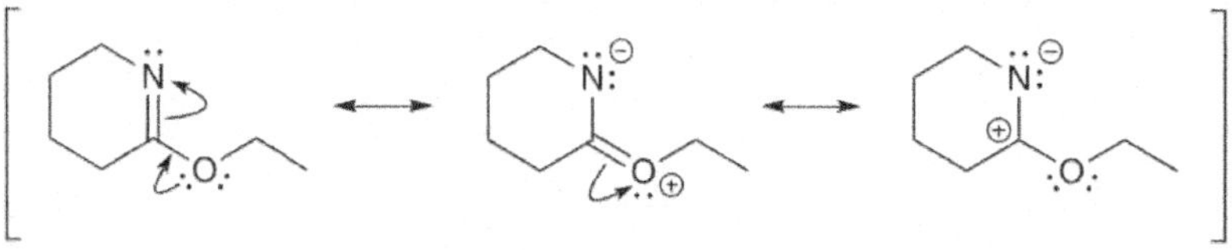

Either way, you can generate all three resonance structures by continuing to look for any arrow-pushing patterns until you cannot find any more patterns to explore.

2.51 The structure shown has a pi bond next to a positive charge, so we draw one curved arrow that pushes the pi bond over. The new positive charge is once again next to a pi bond, so we can draw a third resonance structure, as shown below:

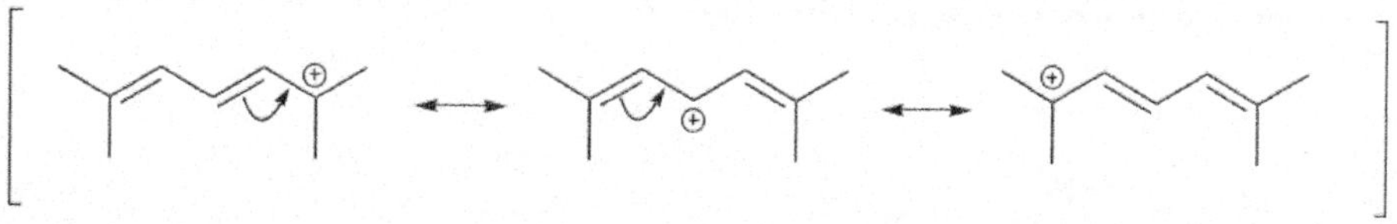

2.52 This structure has a positive charge next to two different atoms that each has a lone pair (nitrogen and oxygen). We can start by pushing a lone pair either from oxygen or from nitrogen. Either way, we can still generate two additional resonance structures. Below, we will start by pushing the lone pair on nitrogen, which generates a resonance structure with a positive charge on nitrogen. This resonance structure now has a lone pair (on oxygen) next to a pi bond, so we draw two curved arrows, generating a resonance structure in which the positive charge is on oxygen:

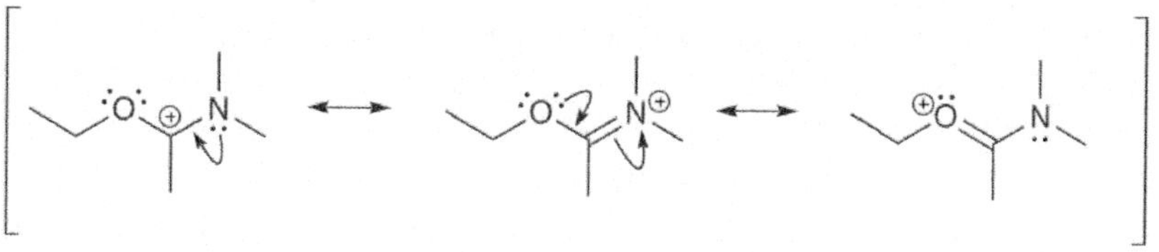

2.53 The structure shown has alternating single and double bonds in a ring, so we can push all of the pi bonds to generate a new resonance structure. This structure (like the previous structure) has a pi bond next to a positive charge, so we draw one curved arrow that pushes the pi bond over. The new positive charge is once again next to a pi bond, so we can draw another resonance structure. This process is repeated one more time to generate the fifth and final resonance structure:

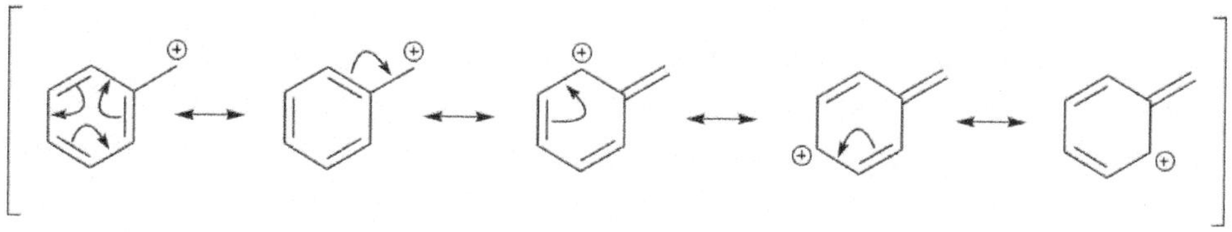

2.54 The structure shown has a lone pair next to a positive charge, so we draw one curved arrow that converts a lone pair into a pi bond. Notice that the position of the positive charge has moved from carbon to nitrogen.

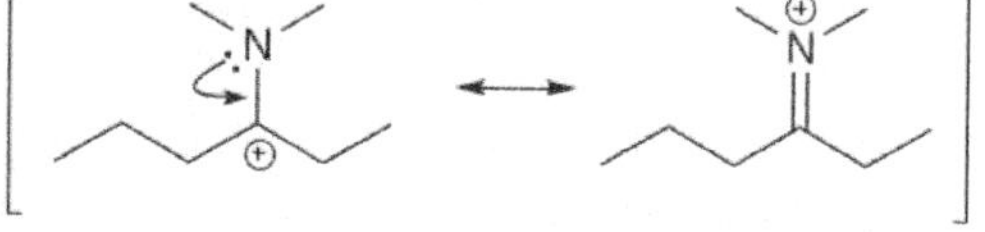

2.55 This structure has a C=O bond, so we can draw a curved arrow that pushes the pi electrons up onto the oxygen atom, generating a positive charge on a carbon atom and a negative charge on an oxygen atom. The resulting resonance structure has a lone pair next to a positive charge, so we draw one curved arrow to generate the third resonance structure below:

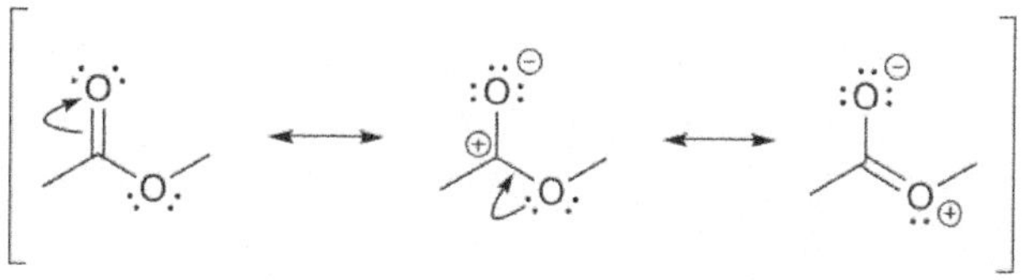

Alternatively, you might have noticed that the structure has a lone pair (on oxygen) next to a pi bond, so you could have drawn two curved arrows that would have converted the first resonance structure above directly into the third resonance structure above. If you had done this, you would still be able to generate the missing resonance structure, by recognizing that you now have a C=O bond, and the pi electrons can be pushed to generate the missing resonance structure:

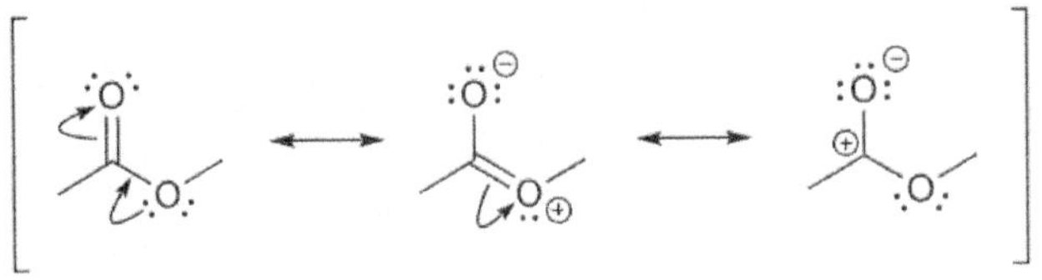

Either way, you can generate all three resonance structures by continuing to look for any arrow-pushing patterns until you cannot find any more patterns to explore.

2.56 This structure has a C=O bond, so we can draw a curved arrow that pushes the pi electrons up onto the oxygen atom, generating a positive charge on a carbon atom and a negative charge on an oxygen atom. The resulting resonance structure has a lone pair next to a positive charge, so we draw one curved arrow to generate the third resonance structure below:

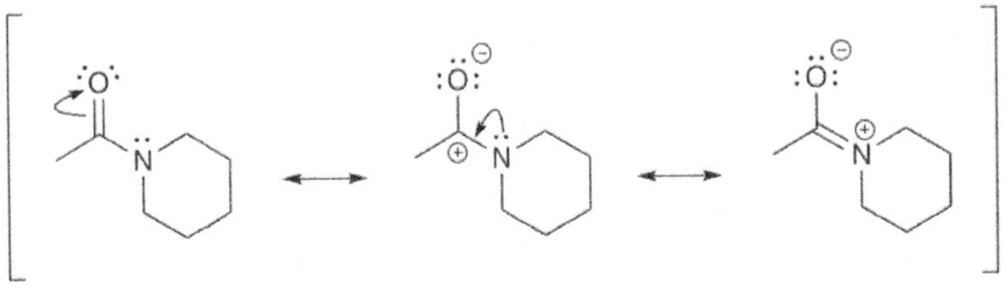

Alternatively, you might have noticed that the structure has a lone pair (on nitrogen) next to a pi bond, so you could have drawn two curved arrows that would have converted the first resonance structure above directly into the third resonance structure above. If you had done this, you would still be able to generate the missing resonance structure, by recognizing that you now have a C=N bond, and the pi electrons can be pushed to generate the missing resonance structure:

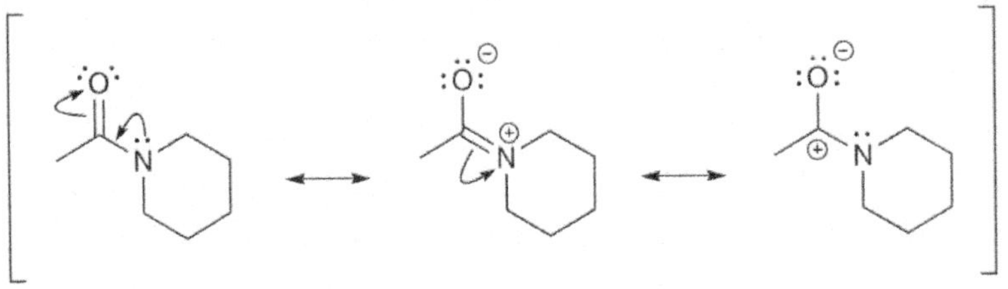

Either way, you can generate all three resonance structures by continuing to look for any arrow-pushing patterns until you cannot find any more patterns to explore.

2.57 The structure shown has a lone pair next to a positive charge, so we draw one curved arrow that converts a lone pair into a pi bond. Notice that the position of the positive charge has moved from carbon to oxygen.

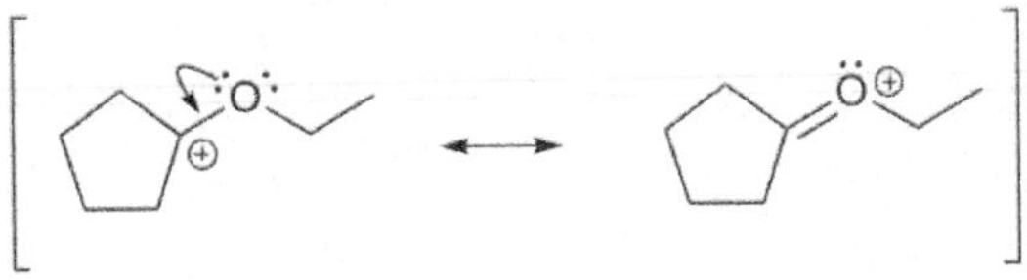

2.58 The structure shown has a lone pair next to a pi bond, so we draw two curved arrows, thereby moving the location of the negative charge. The resulting resonance structure also has a lone pair next to a pi bond, so we draw that arrow-pushing pattern again (two curved arrows) to give the third resonance structure shown below:

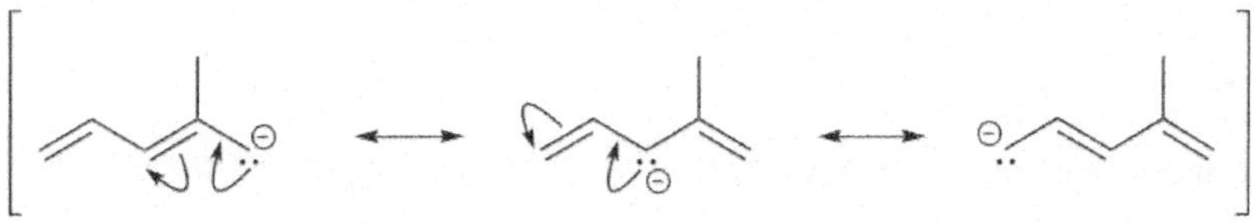

2.59 The structure shown has a lone pair next to a positive charge, so we draw one curved arrow that converts a lone pair into a pi bond. Notice that the position of the positive charge has moved from carbon to oxygen.

2.60 This structure has a C=O bond, so we can draw a curved arrow that pushes the pi electrons up onto the oxygen atom, generating a positive charge on a carbon atom and a negative charge on an oxygen atom. The resulting resonance structure has a lone pair (on Cl) next to a positive charge, so we draw one curved arrow to generate the third resonance structure below:

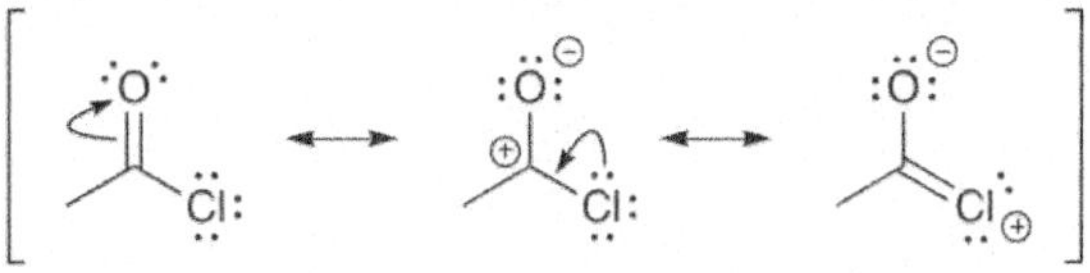

Alternatively, you might have noticed that the structure has a lone pair (on chlorine) next to a pi bond, so you could have drawn two curved arrows which would have converted the first resonance structure above directly into the third resonance structure above. If you had done this, you would still be able to generate the missing resonance structure, by recognizing that you now have a C=Cl bond, and the pi electrons can be pushed to generate the missing resonance structure, as shown here:

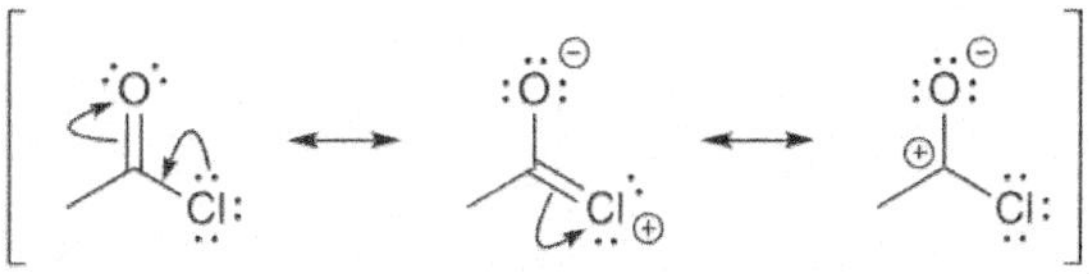

Either way, you can generate all three resonance structures by continuing to look for any arrow-pushing patterns until you cannot find any more patterns to explore.

2.62 This structure has a C=N pi bond, so we draw a curved arrow, giving a resonance structure with charge separation (C$^+$ and N$^-$).

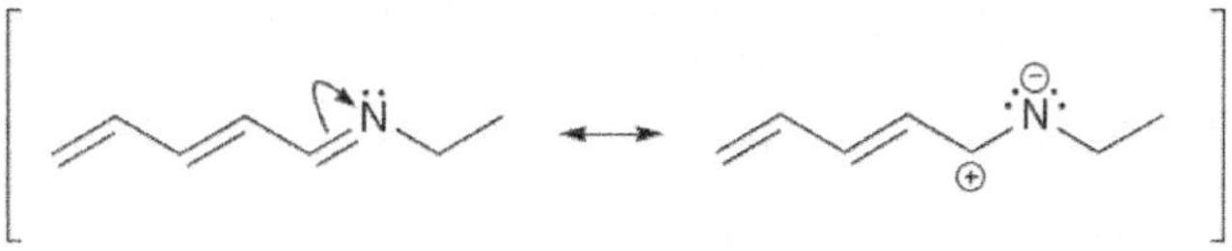

This resonance structure has a positive charge next to a pi bond, so we can push the pi bond over, thereby moving the location of the positive charge. The resulting resonance structure (as seen below) exhibits the same pattern again (positive charge next to pi bond), so we push a pi bond over one more time, giving the final resonance structure. In total, there are four significant resonance structures, all shown below. The first structure below is the major contributor to the resonance hybrid because all atoms in this structure have filled octets and there is no charge separation.

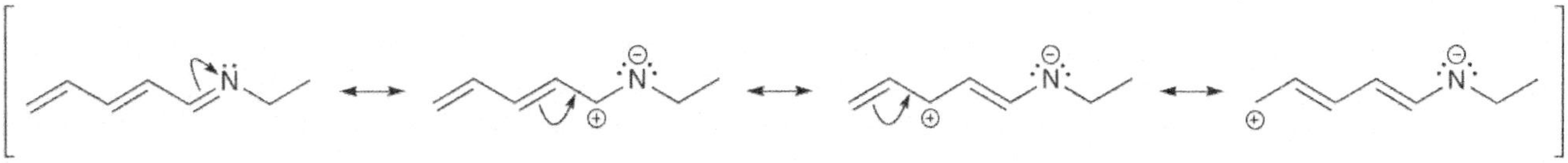

2.63 This structure has a lone pair (on nitrogen) next to a pi bond (C=C), so we draw two curved arrows, giving a resonance structure in which the negative charge is on a carbon atom. The resulting negatively charged carbon atom has a lone pair, so there is a lone pair next to a pi bond, and we can once again draw two curved arrows, giving a third resonance structure. All three resonance structures are significant. The first structure below is the major contributor to the resonance hybrid because the negative charge is on the more electronegative atom (nitrogen, rather than carbon):

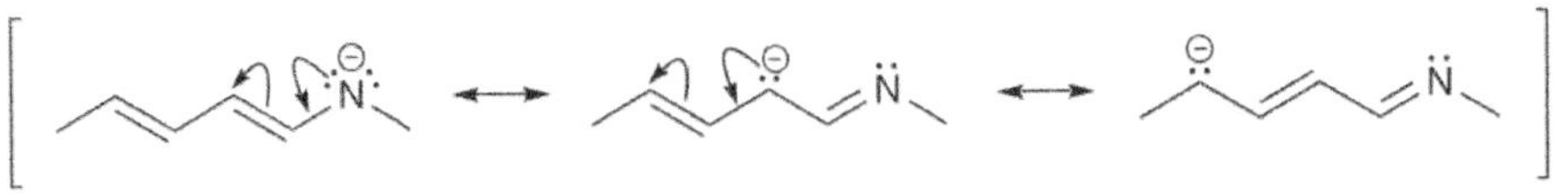

2.64 This structure has a lone pair (on oxygen) next to a pi bond (C=C), so we draw two curved arrows, giving a resonance structure with charge separation (O⁺ and C⁻). The negatively charged carbon atom has a lone pair, so there is a lone pair next to a pi bond, and we can once again draw two curved arrows, giving a third resonance structure. All three resonance structures (shown below) are significant. The first resonance structure is the major contributor to the resonance hybrid because it has no charge separation:

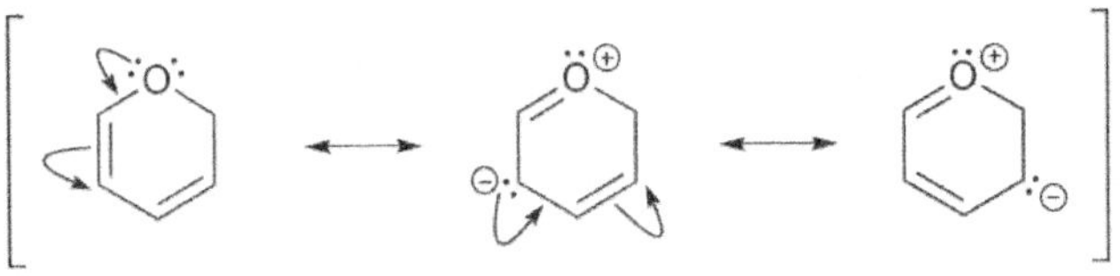

2.65 This structure has a positive charge next to a pi bond, so we draw one curved arrow, thereby generating a resonance structure in which the location of the positive charge has moved. This resonance structure exhibits the same pattern (positive charge next to a pi bond), so we draw another curved arrow to generate another resonance structure. We repeat the process until we have generated all seven resonance structures, as shown below. All seven structures are significant:

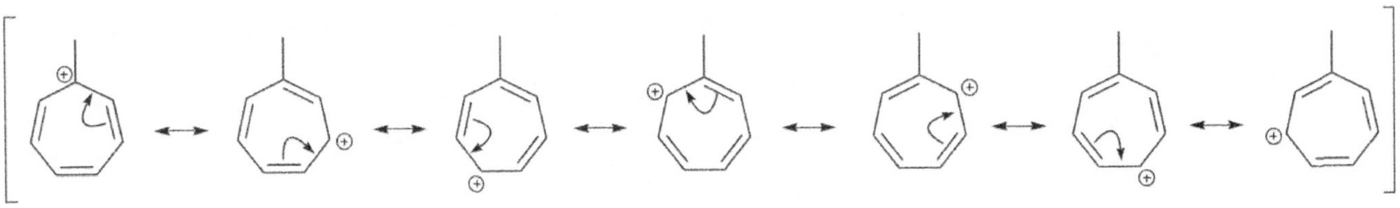

2.66 This structure has a C=O bond, so we draw one curved arrow, generating charge separation (C⁺ and O⁻). This resonance structure has a positive charge next to a pi bond, so we can push the pi bond over, thereby moving the location of the positive charge. The resulting resonance structure (as seen below) exhibits the same pattern again (positive charge next to pi bond), so we push a pi bond over one more time, giving the final resonance structure. In total, there are four significant resonance structures, all shown below. The first structure below is the major contributor to the resonance hybrid because all atoms in this structure have filled octets and there is no charge separation.

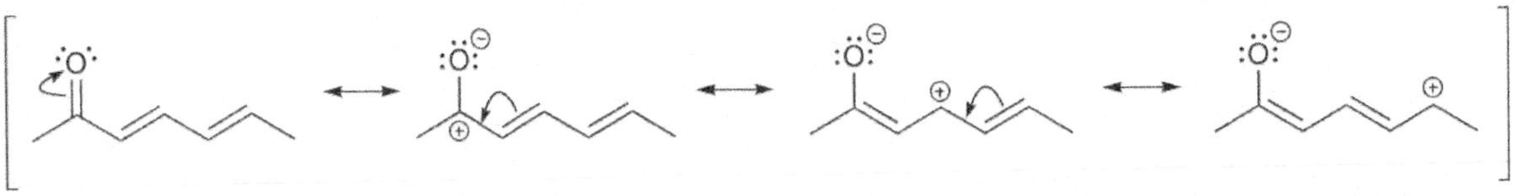

2.67 Avoid drawing resonance structures with too many charges. This structure already has a charge, so we don't want to create more charges; our goal is to push the charge.

This structure has a lone pair (on carbon) next to a pi bond (actually, there are two pi bonds in a C≡N group), so we draw two curved arrows, generating a resonance structure in which the negative charge is on nitrogen, as shown below. The second resonance structure is the major contributor to the resonance hybrid because the negative charge is on the more electronegative atom (nitrogen, rather than carbon).

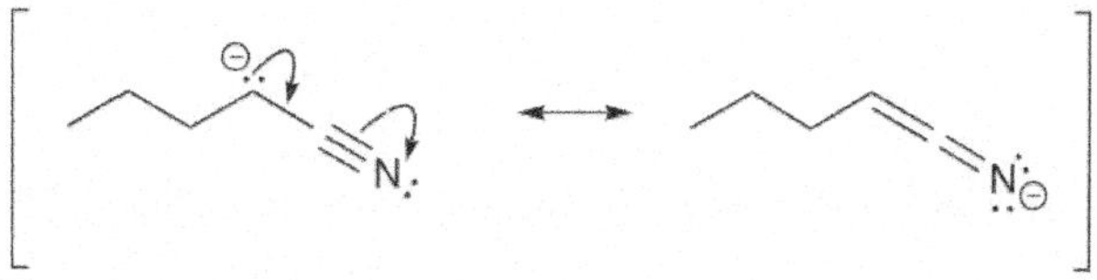

2.68 This structure has a C=O bond, so we can draw a curved arrow that pushes the pi electrons up onto the oxygen atom, generating a positive charge on a carbon atom and a negative charge on an oxygen atom. The resulting resonance structure has a lone pair (on oxygen) next to a positive charge, so we draw one curved arrow to generate the third resonance structure below:

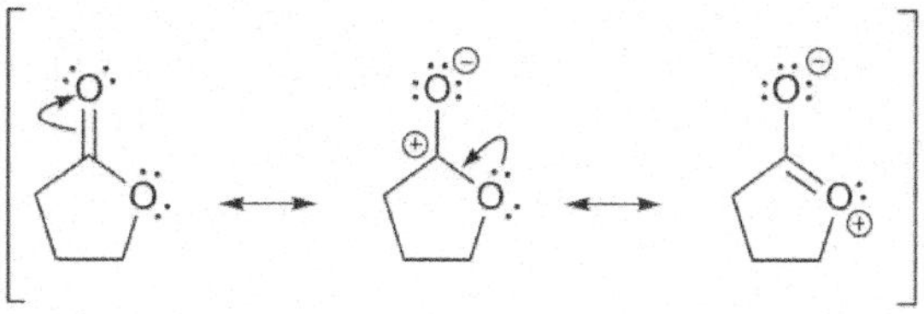

Alternatively, you might have noticed that the structure has a lone pair (on oxygen) next to a pi bond, so you could have drawn two curved arrows which would have converted the first resonance structure above directly into the third resonance structure above. If you had done this, you would still be able to generate the missing resonance structure, by recognizing that you now have a C=O bond, and the pi electrons can be pushed to generate the missing resonance structure, as shown here:

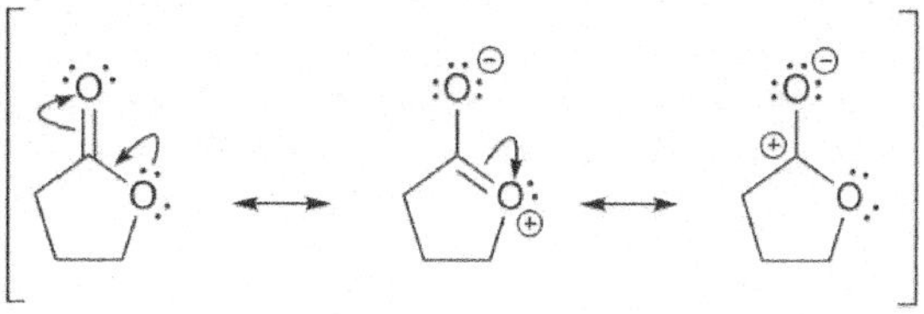

All three resonance structures are significant, although the first is the major contributor to the resonance hybrid because it has filled octets and no separation of charge. The resonance structure with C+ is the least contributing because it is missing an octet.

2.69 Avoid drawing resonance structures with too many charges. This structure already has a charge, so we don't want to create more charges; our goal is to push the charge.

This structure has a lone pair (on carbon) next to a pi bond (C=C), so we draw two curved arrows, giving a resonance structure in which the location of the negative charge has moved. The resulting negatively charged carbon atom has a lone pair, so there is a lone pair next to a pi bond, and we can once again draw two curved arrows, giving a third resonance structure in which the negative charge is on oxygen. All three resonance structures are shown below:

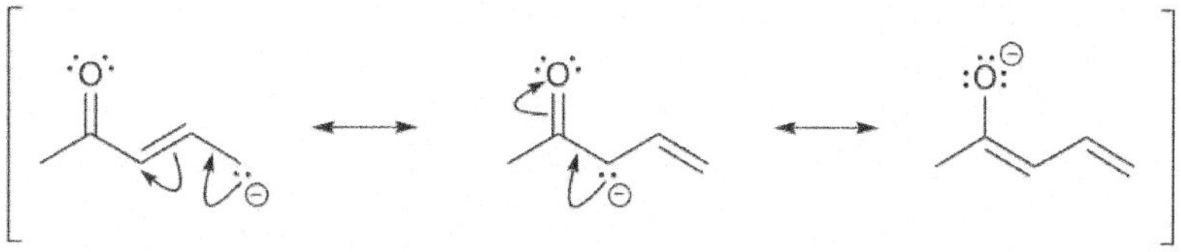

All three resonance structures are significant, although the third structure is the major contributor to the overall resonance hybrid because the negative charge is on oxygen, which is more electronegative than carbon.

2.70 This structure has a lone pair (on oxygen) next to a pi bond (N=O) so we draw two curved arrows. The resulting resonance structure exhibits the same pattern (a lone pair next to a pi bond), so we draw two curved arrows again, generating the third resonance structure shown below. All three resonance structures are significant, although the third structure is the major contributor to the overall resonance hybrid because both negative charges are on oxygen atoms (and oxygen is more electronegative than carbon).

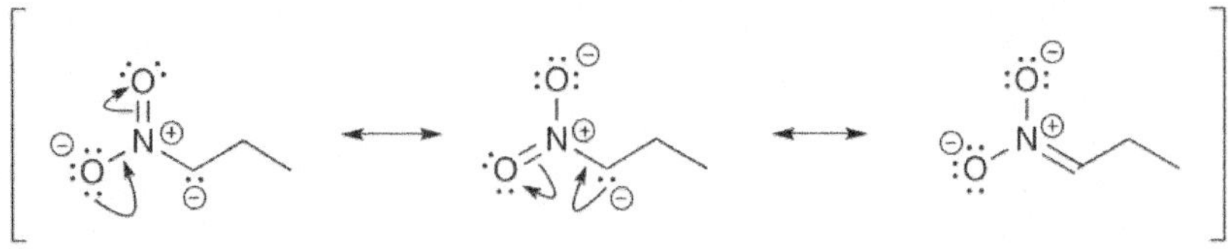

2.71 Avoid drawing resonance structures with too many charges. This structure already has a charge, so we don't want to create more charges; our goal is to push the charge.

This structure has a lone pair (on oxygen) next to a pi bond (C=C), so we draw two curved arrows, giving a resonance structure in which the location of the negative charge has moved from oxygen to carbon. The resulting negatively charged carbon atom has a lone pair, so there is a lone pair next to a pi bond, and we can once again draw two curved arrows, giving a third resonance structure in which the negative charge is on an oxygen atom. All three resonance structures are significant. The first and third structures are equivalent and represent the major contributors to the resonance hybrid because the negative charge is on the more electronegative atom (oxygen, rather than carbon):

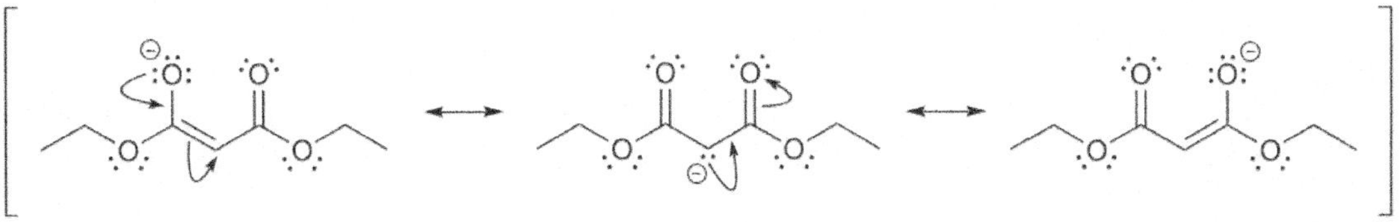

2.72 This compound has a C=O bond, so we can draw one curved arrow, giving a resonance structure with charge separation. This resonance structure exhibits a positive charge next to a lone pair (on either side), so we draw one curved arrow and the resulting resonance structure. This third resonance structure has a lone pair next to a pi bond, so we draw two curved arrows to give the final resonance structure. In total, there are four significant resonance structures, as shown below. The first structure is the major contributor because it has no charge separation, and the second resonance structure is the least contributing because it has an atom that lacks an octet (C^+).

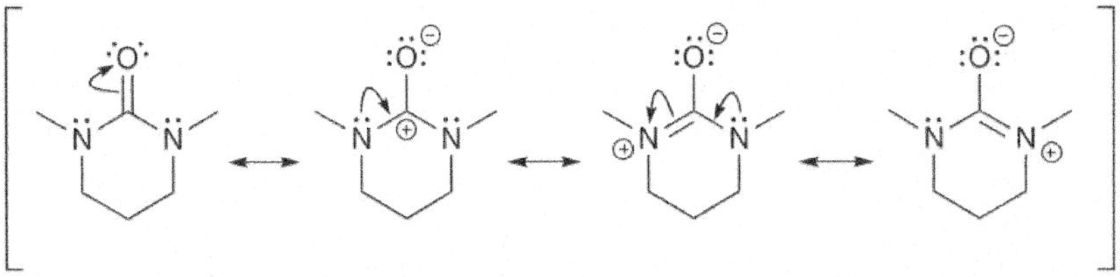

Note: You may have started by noticing that the compound has a lone pair (on nitrogen) next to a pi bond, and you might have drawn two curved arrows, going directly to either the third or fourth resonance structure above. If you did that, it is ok, just keep looking for more patterns. You should still be able to draw all four resonance structures, albeit in a different order. There is nothing magical about the order above. As long as you draw all four significant resonance structures, you can draw them in any order.

2.73 This structure has a C=O bond, so we can draw one curved arrow, generating a resonance structure with charge separation (C^+ and O^-). This resonance structure has a positive charge next to a pi bond, so we can draw one curved arrow, pushing the pi bond over. The resulting resonance structure has a positive charge next to a lone pair (on oxygen), so we draw one curved arrow and the resulting resonance structure (in which the positive charge is now on oxygen). Finally, there is a lone pair (on oxygen) next to a pi bond (C=C), so we can draw two curved arrows and the resulting resonance structure. In total, there are five resonance structures, as shown below.

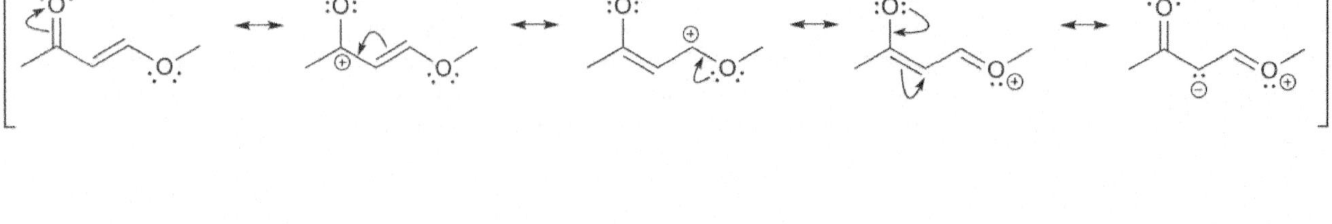

The first resonance structure is the major contributor to the resonance hybrid because this resonance structure has full octets and does not have charge separation. Among the remaining four resonance structures, the second-to-last structure contributes more character than the others because it has full octets, and because the negative charge is on an oxygen atom rather than a carbon atom. The second and third resonance structures contribute the least character to the resonance hybrid because those structures do not have full octets (they have C⁺).

Note: You may have started by noticing that the compound has a lone pair (on oxygen) next to a pi bond (C=C), and you might have drawn two curved arrows, going directly to the final resonance structure above. If you did that, it is ok, just keep looking for more patterns. You should still be able to draw all five resonance structures, albeit in a different order. There is nothing magical about the order above. As long as you draw all five resonance structures, you can draw them in any order.

After drawing the final resonance structure above, make sure to avoid drawing the following insignificant resonance structure:

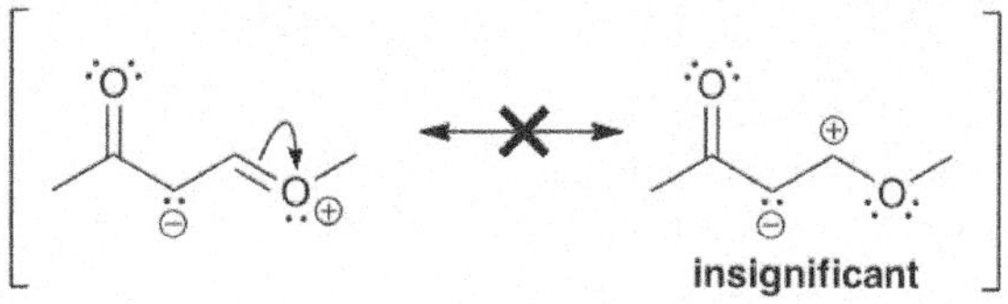

This resonance structure is insignificant because it suffers from two major deficiencies: 1) it does not have filled octets and 2) it has a negative charge on a carbon atom (which is not an electronegative atom). Either of these deficiencies alone would render the resonance structure a minor contributor. But with both deficiencies together (C⁺ and C⁻), this resonance structure is insignificant. The same is generally true for any resonance structure that has both C⁺ and C⁻.

2.74 This structure has a lone pair (on nitrogen) next to a pi bond (C=C), so we draw two curved arrows, giving a resonance structure that has a separation of charge (N⁺ and C⁻). The resulting negatively charged carbon atom has a lone pair, so there is a lone pair next to a pi bond, and we can once again draw two curved arrows, giving a third resonance structure. Finally, we draw the same pattern one last time to generate a fourth resonance structure, as shown below. All four resonance structures are significant, although the first is the major contributor because it lacks charge separation.

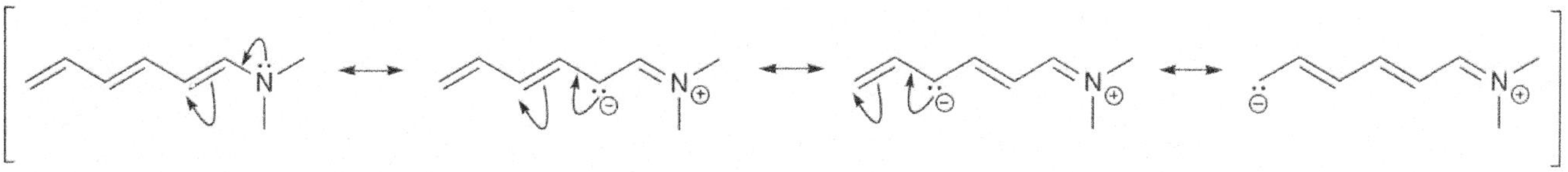

CHAPTER 3

3.2 In order to compare the acidity of the two highlighted protons, we remove each of them (one at a time) and draw the resulting conjugate base in each case. This gives two possible conjugate bases:

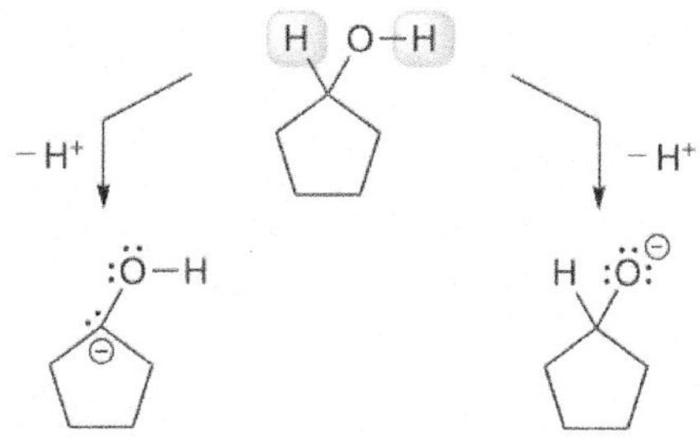

Conjugate base #1   Conjugate base #2

Next, we compare these conjugate bases and consider which one is more stable. In this case, conjugate base #2 is more stable than

conjugate base #1 because a negative charge on an oxygen atom is more stable than a negative charge on a carbon atom (oxygen is more electronegative than carbon). Since conjugate base #2 is more stable, we conclude that the highlighted proton shown below is the more acidic proton because removing this proton generates the more stable conjugate base:

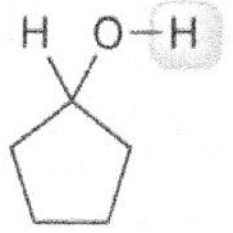

3.3 In order to compare the acidity of the two highlighted protons, we remove each of them (one at a time) and draw the resulting conjugate base in each case. This gives two possible conjugate bases:

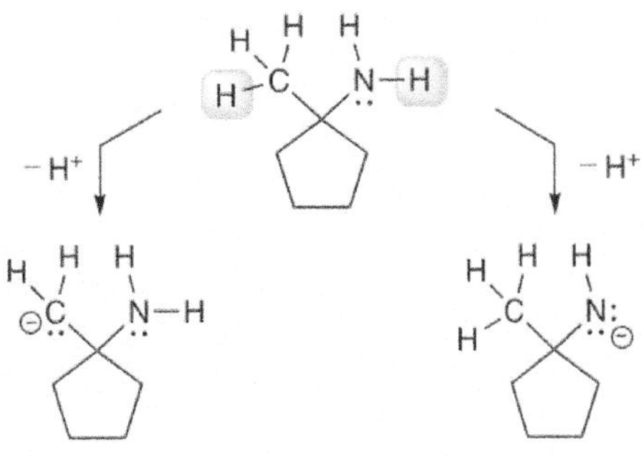

Conjugate base #1   Conjugate base #2

Next, we compare these conjugate bases and consider which one is more stable. In this case, conjugate base #2 is more stable than conjugate base #1 because a negative charge on a nitrogen atom is more stable than a negative charge on a carbon atom (nitrogen is more electronegative than carbon). Since conjugate base #2 is more stable, we conclude that the highlighted proton shown below is the more acidic proton because removing this proton generates the more stable conjugate base:

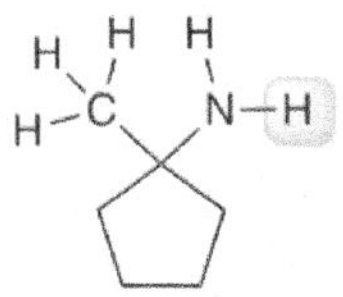

3.4 In order to compare the acidity of the two highlighted protons, we remove each of them (one at a time) and draw the resulting conjugate base in each case. This gives two possible conjugate bases:

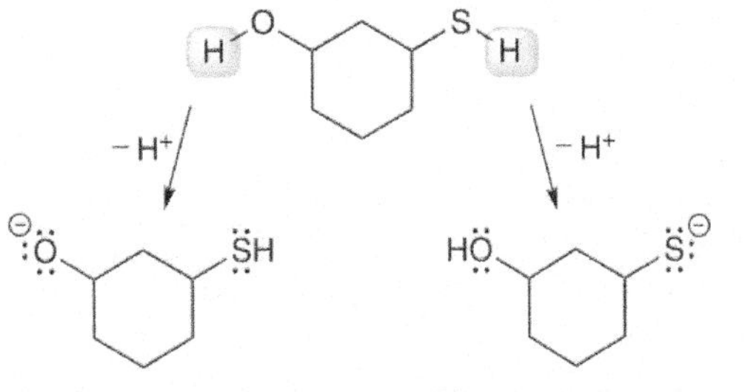

Conjugate base #1   Conjugate base #2

Next, we compare these conjugate bases and consider which one is more stable. In this case, conjugate base #2 is more stable than conjugate base #1 because a negative charge on a sulfur atom is more stable than a negative charge on an oxygen atom (sulfur is larger in size than oxygen). Since conjugate base #2 is more stable, we conclude that the highlighted proton shown below is the more acidic proton because removing this proton generates the more stable conjugate base:

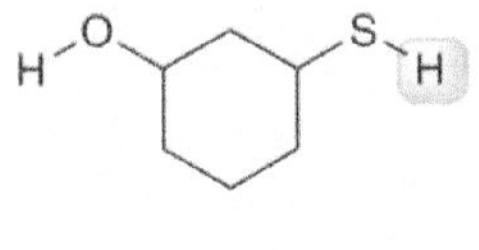

3.5 In order to compare the acidity of the two highlighted protons, we remove each of them (one at a time) and draw the resulting conjugate base in each case. This gives two possible conjugate bases:

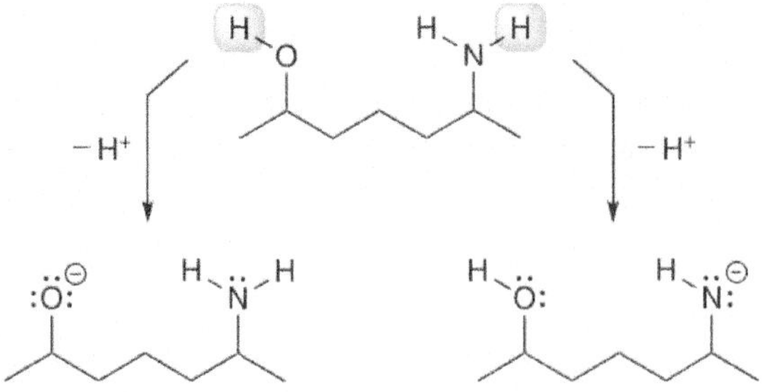

Conjugate base #1   Conjugate base #2

Next, we compare these conjugate bases and consider which one is more stable. In this case, conjugate base #1 is more stable than conjugate base #2 because a negative charge on an oxygen atom is more stable than a negative charge on a nitrogen atom (oxygen is more electronegative than nitrogen). Since conjugate base #1 is more stable, we conclude that the highlighted proton shown below is the more acidic proton because removing this proton generates the more stable conjugate base:

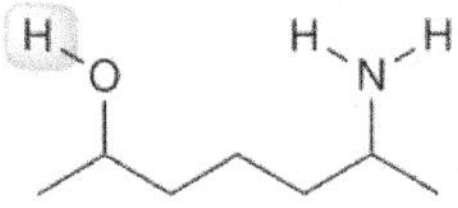

3.7 In order to compare the acidity of the two highlighted protons, we remove each of them (one at a time) and draw the resulting conjugate base in each case. This gives two possible conjugate bases:

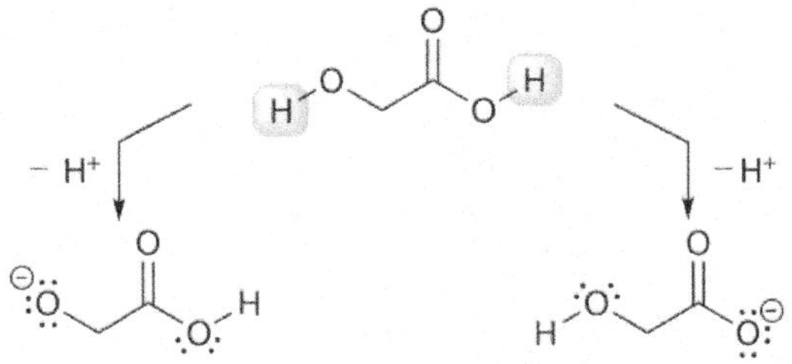

Conjugate base #1   Conjugate base #2

Next, we compare these conjugate bases and consider which one is more stable. In both structures, the negative charge is on an oxygen atom. But the second structure is more stable because its negative charge is delocalized by resonance:

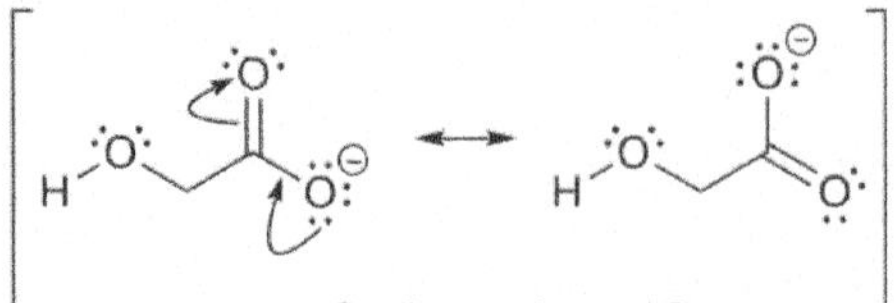

Conjugate base #2

Since conjugate base #2 is more stable, we conclude that the highlighted proton shown below is the more acidic proton because removing this proton generates the more stable conjugate base:

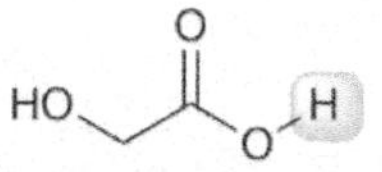

3.8 In order to compare the acidity of the two highlighted protons, we remove each of them (one at a time) and draw the resulting conjugate base in each case. This gives two possible conjugate bases:

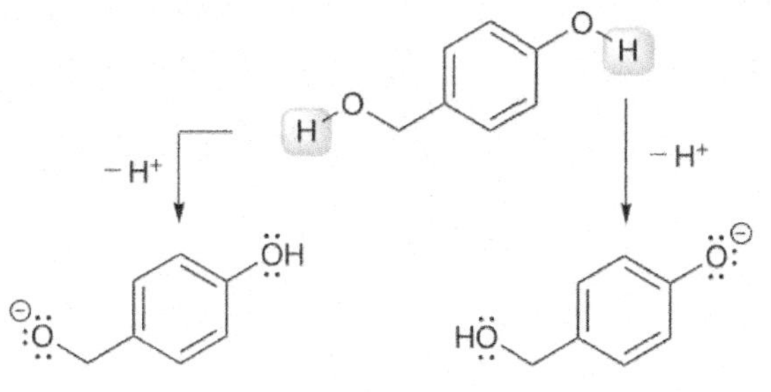

Conjugate base #1   Conjugate base #2

Next, we compare these conjugate bases and consider which one is more stable. In both structures, the negative charge is on an oxygen atom. But the second structure is more stable because its negative charge is delocalized by resonance:

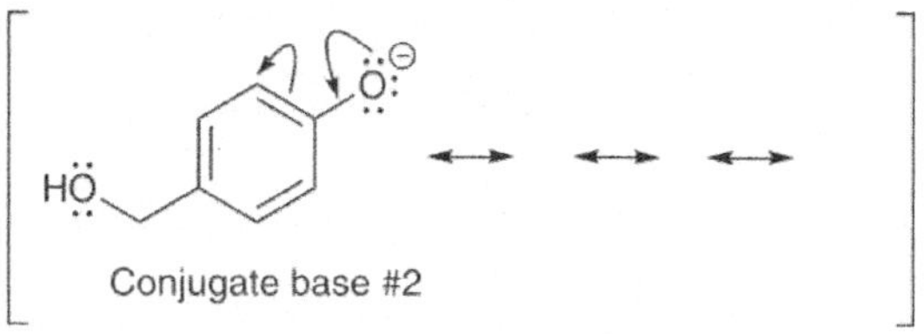

Conjugate base #2

Since conjugate base #2 is more stable, we conclude that the highlighted proton shown below is the more acidic proton because removing this proton generates the more stable conjugate base:

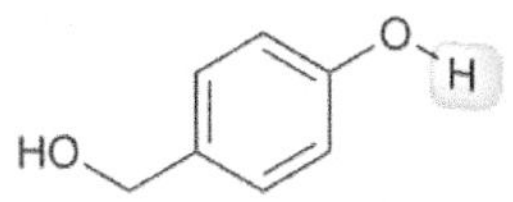

3.9 In order to compare the acidity of the two highlighted protons, we remove each of them (one at a time) and draw the resulting conjugate base in each case. This gives two possible conjugate bases:

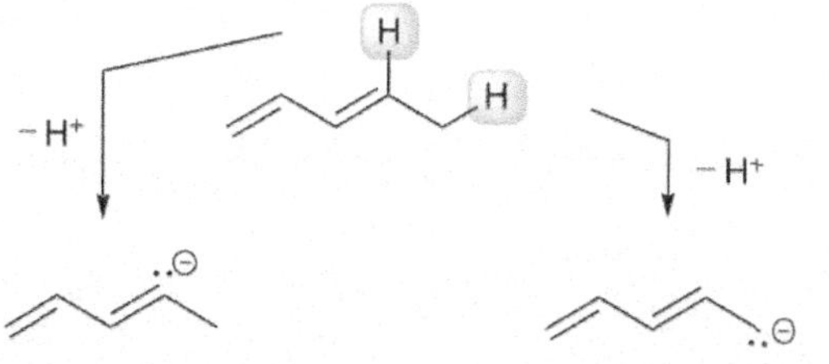

Conjugate base #1   Conjugate base #2

Next, we compare these conjugate bases and consider which one is more stable. In both structures, the negative charge is on a carbon atom. Conjugate base #1 has a localized charge (not stabilized by resonance), while conjugate base #2 has a negative charge that is delocalized (stabilized by resonance):

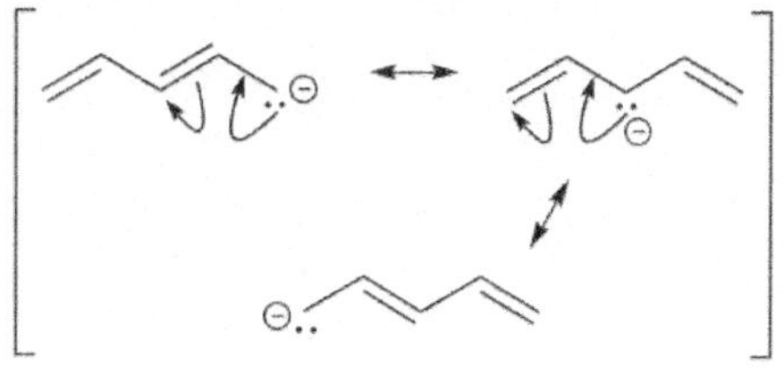

Delocalization stabilizes a negative charge, so conjugate base #2 is more stable than conjugate base #1. As a result, we conclude that the highlighted proton shown below is the more acidic proton because removing this proton generates the more stable conjugate base:

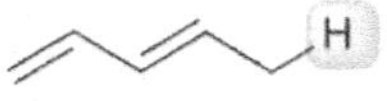

3.10 In order to compare the acidity of the two highlighted protons, we remove each of them (one at a time) and draw the resulting conjugate base in each case. This gives two possible conjugate bases:

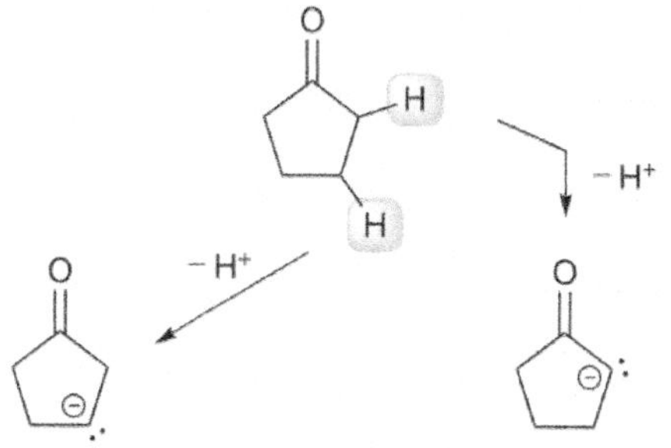

Conjugate base #1   Conjugate base #2

Next, we compare these conjugate bases and consider which one is more stable. Conjugate base #1 has a negative charge that is localized (not stabilized by resonance), while conjugate base #2 has a negative charge that is delocalized (stabilized by resonance):

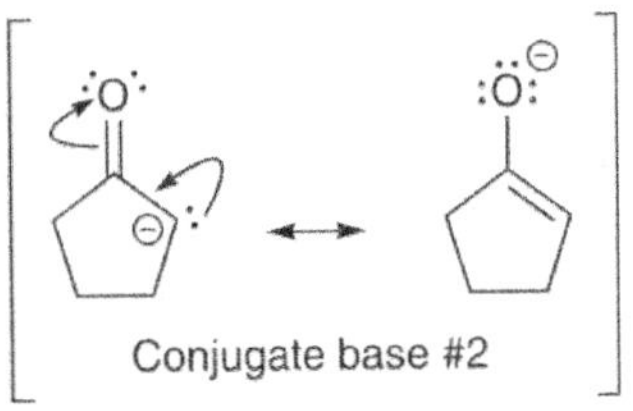

Conjugate base #2

Conjugate base #2 is more stable. As a result, we conclude that the highlighted proton shown below is the more acidic proton because removing this proton generates the more stable conjugate base:

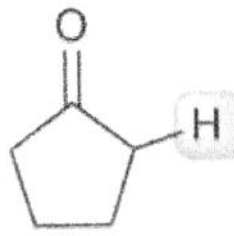

3.11 In order to compare the acidity of the two highlighted protons, we remove each of them (one at a time) and draw the resulting conjugate base in each case. This gives two possible conjugate bases:

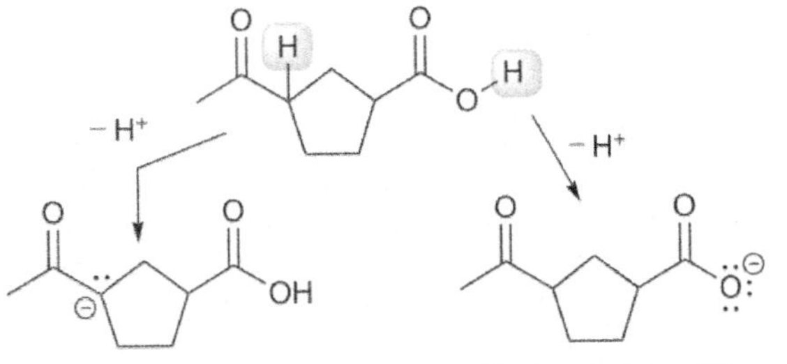

Conjugate base #1   Conjugate base #2

Next, we compare these conjugate bases and consider which one is more stable. Both conjugate bases are stabilized by resonance. Here are the resonance structures for the first conjugate base,

Conjugate base #1

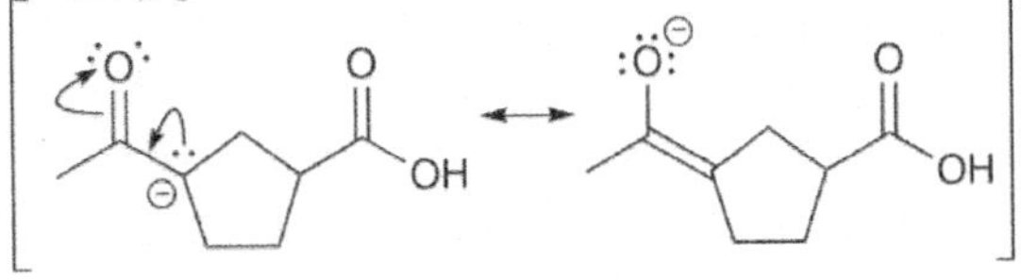

and here are the resonance structures for the second conjugate base:

Conjugate base #2

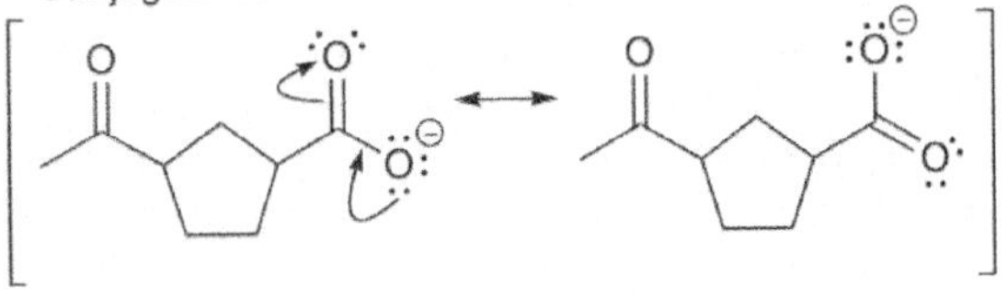

Now compare the location of the negative charge in each case. In the first conjugate base, the negative charge is spread over a carbon atom and an oxygen atom. In contrast, the second conjugate base has the negative charge spread over two oxygen atoms. Oxygen is more electronegative than carbon, so conjugate base #2 is more stable than conjugate base #1. Therefore, we conclude that the highlighted proton shown below is the more acidic proton because removing this proton generates the more stable conjugate base:

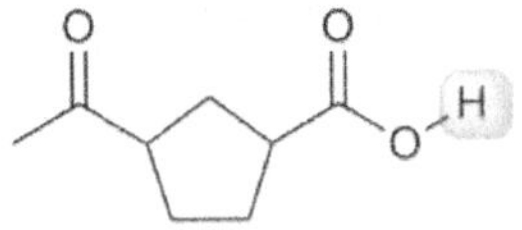

3.12 In order to compare the acidity of the two highlighted protons, we remove each of them (one at a time) and draw the resulting conjugate base in each case. This gives two possible conjugate bases:

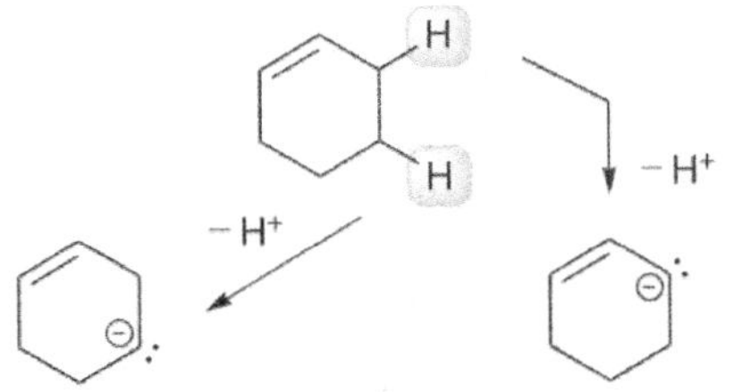

Conjugate base #1   Conjugate base #2

Next, we compare these conjugate bases and consider which one is more stable. In both structures, the negative charge is on a carbon atom. Conjugate base #1 has a localized charge (not stabilized by resonance), while conjugate base #2 has a negative charge that is delocalized (stabilized by resonance):

Conjugate base #2

Delocalization stabilizes a negative charge, so conjugate base #2 is more stable than conjugate base #1. As a result, we conclude that the highlighted proton below is the more acidic proton because removing this proton generates the more stable conjugate base:

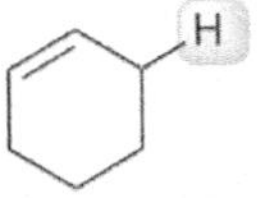

3.14 In order to compare the acidity of the two highlighted protons, we remove each of them (one at a time) and draw the resulting conjugate base in each case. This gives two possible conjugate bases:

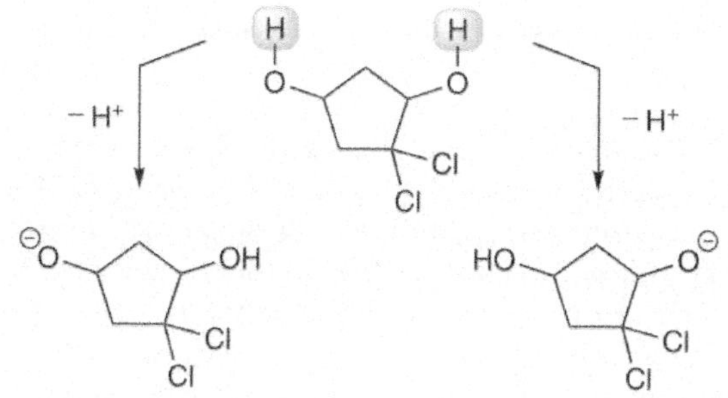

Conjugate base #1   Conjugate base #2

Next, we compare these conjugate bases and consider which one is more stable. In both structures, the negative charge is on an oxygen atom; and in both structures, the charge is localized. The difference between the two structures is the proximity of the charge to the electron-withdrawing chlorine atoms. In conjugate base #2, the charge is closer to the chlorine atoms (as compared with conjugate base #1). The inductive effects of the chlorine atoms serve to stabilize the negative charge by withdrawing some electron density away from the negative charge. These stabilizing inductive effects are stronger with closer proximity, so conjugate base #2 is more stable than conjugate base #1. Therefore, we conclude that the highlighted proton shown below is the more acidic proton because removing this proton generates the more stable conjugate base:

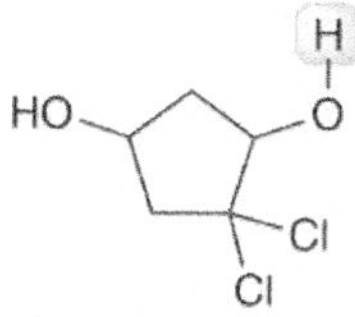

3.15 In order to compare the acidity of the two highlighted protons, we remove each of them (one at a time) and draw the resulting conjugate base in each case. This gives two possible conjugate bases:

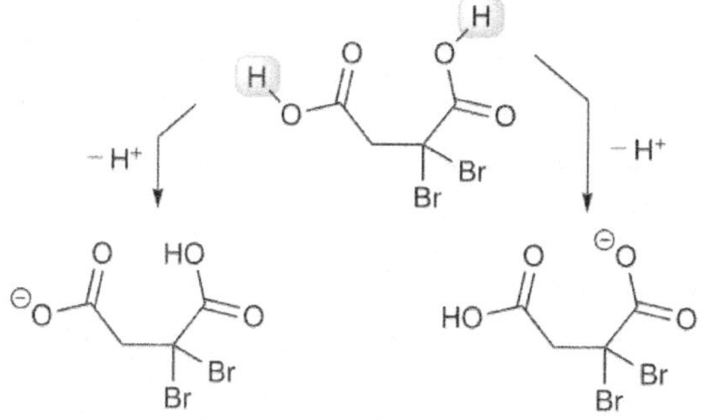

Conjugate base #1   Conjugate base #2

Next, we compare these conjugate bases and consider which one is more stable. In both structures, the negative charge is on an

oxygen atom; and in both structures, the charge is stabilized by resonance. The difference between the two structures is the proximity of the charge to the electron-withdrawing bromine atoms. In conjugate base #2, the charge is closer to the bromine atoms (as compared with conjugate base #1). The inductive effects of the bromine atoms serve to stabilize the negative charge by withdrawing some electron density away from the negative charge. These stabilizing inductive effects are stronger with closer proximity, so conjugate base #2 is more stable than conjugate base #1. Therefore, we conclude that the highlighted proton shown below is the more acidic proton because removing this proton generates the more stable conjugate base:

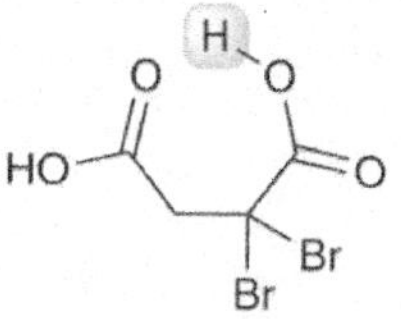

3.16 In order to compare the acidity of the two highlighted protons, we remove each of them (one at a time) and draw the resulting conjugate base in each case. This gives two possible conjugate bases:

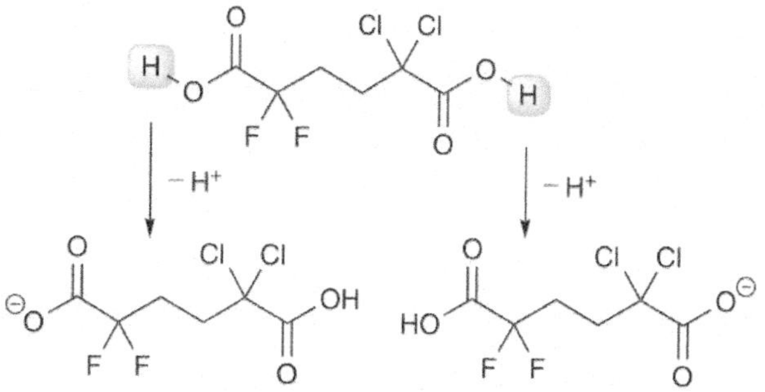

Conjugate base #1   Conjugate base #2

Next, we compare these conjugate bases and consider which one is more stable. In both structures, the negative charge is on an oxygen atom; and in both structures, the charge is stabilized by resonance. Furthermore, each structure has electron-withdrawing groups (Cl or F), which stabilize the charge via induction. The difference between these structures is the identity of the electron-withdrawing groups. The inductive effects of fluorine atoms are stronger than the inductive effects of chlorine atoms (because fluorine is the most electronegative element). So conjugate base #1 is more stable than conjugate base #2. Therefore, we conclude that the highlighted proton shown below is the more acidic proton because removing this proton generates the more stable conjugate base:

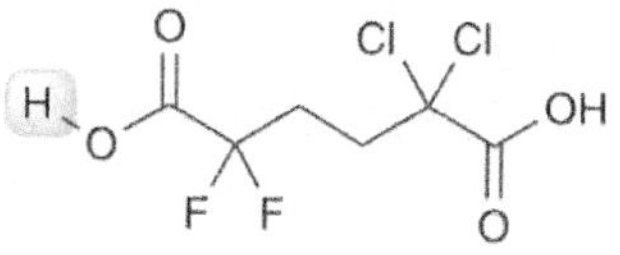

3.19 In order to compare the acidity of the two highlighted protons, we remove each of them (one at a time) and draw the resulting conjugate base in each case. This gives two possible conjugate bases:

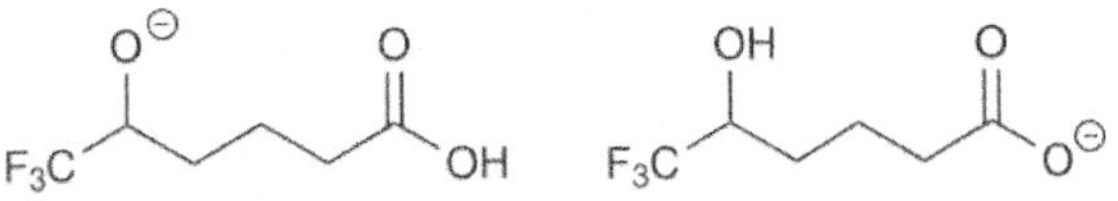

Conjugate base #1   Conjugate base #2

Next, we compare these conjugate bases and consider which one is more stable. Conjugate base #1 is stabilized by the inductive effects of the three fluorine atoms, while conjugate base #2 is stabilized by resonance. In general, resonance is a stronger effect than induction (R has priority over I in ARIO), so we conclude that conjugate base #2 is more stable. Therefore, the highlighted proton shown below is the more acidic proton because removing this proton generates the more stable conjugate base:

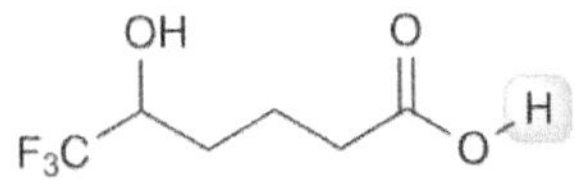

3.20 In order to compare the acidity of the two highlighted protons, we remove each of them (one at a time) and draw the resulting conjugate base in each case. This gives two possible conjugate bases:

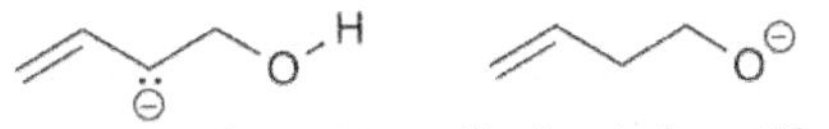

Conjugate base #1   Conjugate base #2

Next, we compare these conjugate bases and consider which one is more stable. Conjugate base #1 is stabilized by resonance (the charge is spread over two carbon atoms), while conjugate base #2 is stabilized by the charge being on an electronegative atom (O). In general, the identity of the atom (bearing the charge) is more important than the presence of resonance (A has priority over R in ARIO), so we conclude that conjugate base #2 is more stable. Therefore, the highlighted proton shown below is the more acidic proton because removing this proton generates the more stable conjugate base:

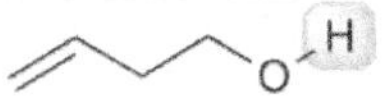

3.21 In order to compare the acidity of the two highlighted protons, we remove each of them (one at a time) and draw the resulting conjugate base in each case. This gives two possible conjugate bases:

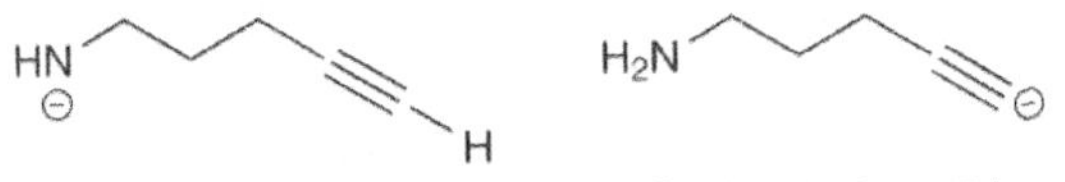

Conjugate base #1   Conjugate base #2

Next, we compare these conjugate bases and consider which one is more stable. Conjugate base #1 is stabilized by the charge being on an electronegative atom (N), while conjugate base #2 is stabilized by the charge being associated with an sp hybridized orbital. This represents an exception to the **ARIO** prioritization (we might expect Atom to beat Orbital, but this is the exception). It is actually more of a stabilizing effect for the charge to be associated with an sp hybridized orbital. Note: This exception only holds true when comparing a negative charge on an sp hybridized C with a negative charge on an sp^3 hybridized N.

We conclude that conjugate base #2 is more stable. Therefore, the highlighted proton shown below is the more acidic proton because removing this proton generates the more stable conjugate base:

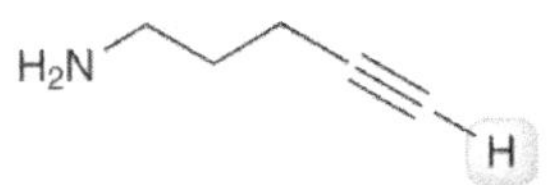

3.22 In order to compare the acidity of the two highlighted protons, we remove each of them (one at a time) and draw the resulting conjugate base in each case. This gives two possible conjugate bases:

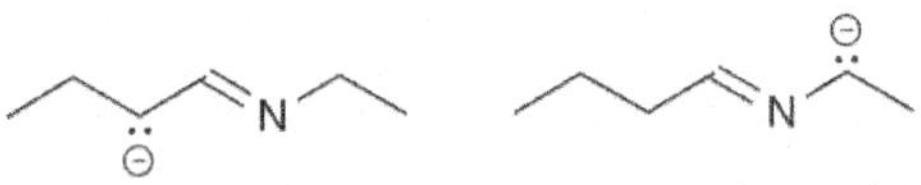

Conjugate base #1   Conjugate base #2

Next, we compare these conjugate bases and consider which one is more stable. Both conjugate bases are stabilized by resonance. Here are the resonance structures for the first conjugate base,

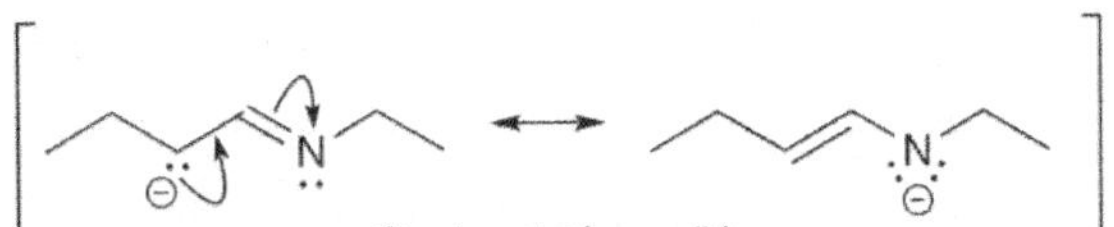

Conjugate base #1

and here are the resonance structures for the second conjugate base:

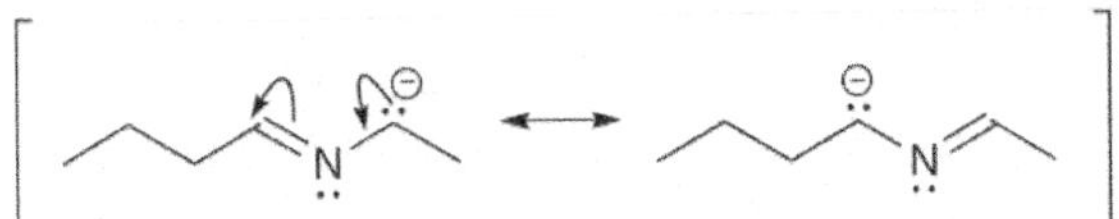

Conjugate base #2

Now compare the location of the negative charge in each case. In the first conjugate base, the negative charge is spread over a carbon atom and a nitrogen atom. In contrast, the second conjugate base has the negative charge spread over two carbon atoms. Nitrogen is more electronegative than carbon, so conjugate base #1 is more stable than conjugate base #2. Therefore, we conclude that the highlighted proton shown below is the more acidic proton because removing this proton generates the more stable conjugate base:

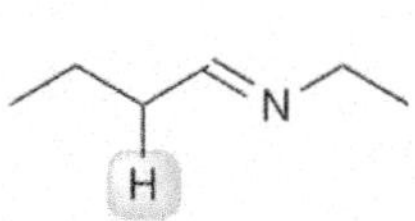

3.23 In order to compare the acidity of the two highlighted protons, we remove each of them (one at a time) and draw the resulting conjugate base in each case. This gives two possible conjugate bases:

Conjugate base #1   Conjugate base #2

Next, we compare these conjugate bases and consider which one is more stable. Both conjugate bases are stabilized by resonance. Here are the resonance structures for the first conjugate base,

Conjugate base #1

and here are the resonance structures for the second conjugate base:

Conjugate base #2

Now compare the location of the negative charge in each case. In the first conjugate base, the negative charge is spread over two nitrogen atoms. In contrast, the second conjugate base has the negative charge spread over two carbon atoms. Nitrogen is more electronegative than carbon, so conjugate base #1 is more stable than conjugate base #2. Therefore, we conclude that the highlighted proton below is the more acidic proton because removing this proton generates the more stable conjugate base:

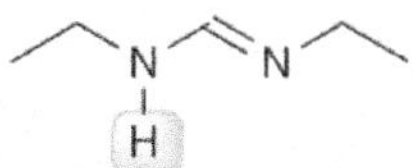

3.24 In order to compare the acidity of the two highlighted protons, we remove each of them (one at a time) and draw the resulting conjugate base in each case. This gives two possible conjugate bases:

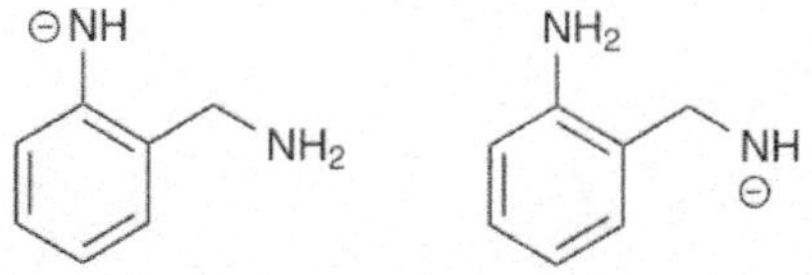

Conjugate base #1   Conjugate base #2

Next, we compare these conjugate bases and consider which one is more stable. In both structures, the negative charge is on a nitrogen atom. Conjugate base #1 has a delocalized negative charge (stabilized by resonance), while conjugate base #2 has a negative charge that is localized (not stabilized by resonance):

Conjugate base #1

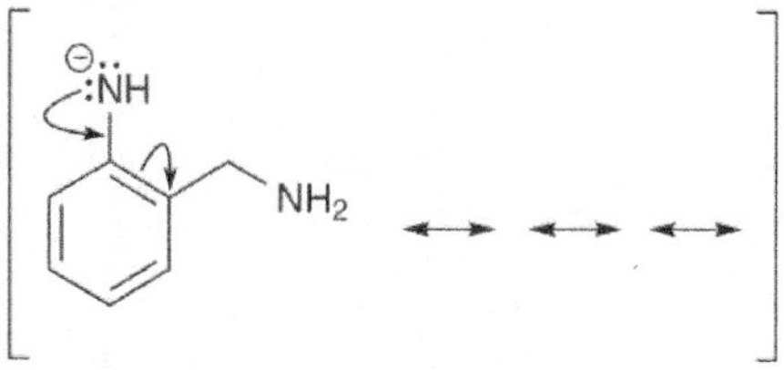

Delocalization is a stabilizing effect, so conjugate base #1 is more stable than conjugate base #2. As a result, we conclude that the highlighted proton shown below is the more acidic proton because removing this proton generates the more stable conjugate base:

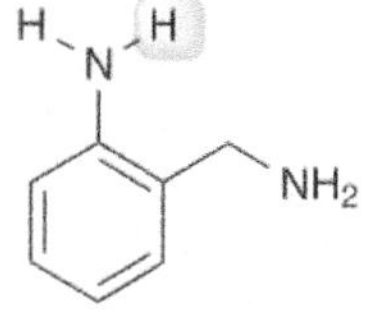

3.25 In order to compare the acidity of the two highlighted protons, we remove each of them (one at a time) and draw the resulting conjugate base in each case. This gives two possible conjugate bases:

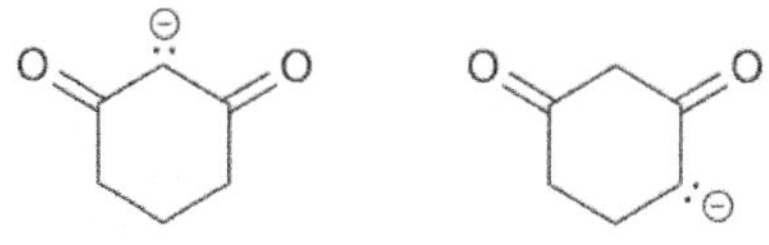

Conjugate base #1   Conjugate base #2

Next, we compare these conjugate bases and consider which one is more stable. Each of these conjugate bases is stabilized by resonance, so let's draw the resonance structures. Here are the resonance structures of conjugate base #1,

Conjugate base #1

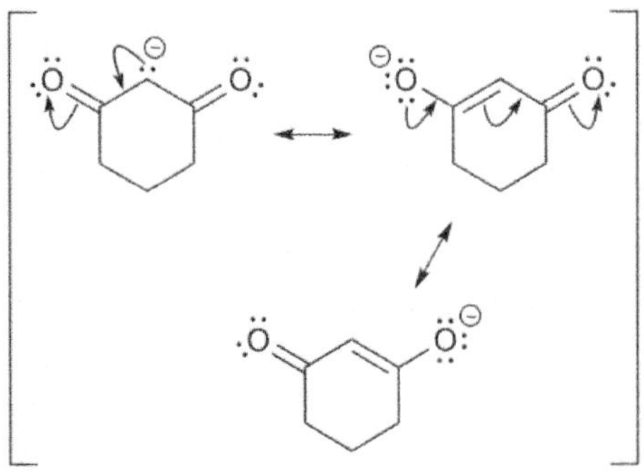

and here are the resonance structures of conjugate base #2:

Conjugate base #2

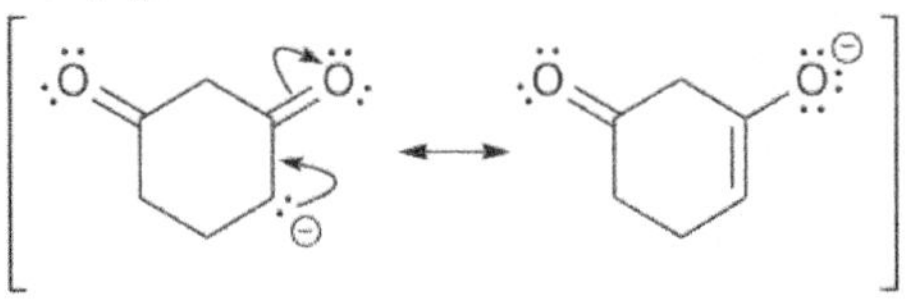

Notice that conjugate base #1 has three resonance structures (with the negative charge spread over two oxygen atoms and one carbon atom), while conjugate base #2 has only two resonance structures (with the negative charge spread over one oxygen atom and one carbon atom). Because of its additional resonance structure, conjugate base #1 is more stable than conjugate base #2.

Therefore, the highlighted proton shown below is the more acidic proton because removing this proton generates the more stable conjugate base:

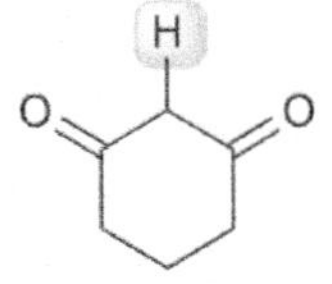

3.26 In order to compare the acidity of the two highlighted protons, we remove each of them (one at a time) and draw the resulting conjugate base in each case. This gives two possible conjugate bases:

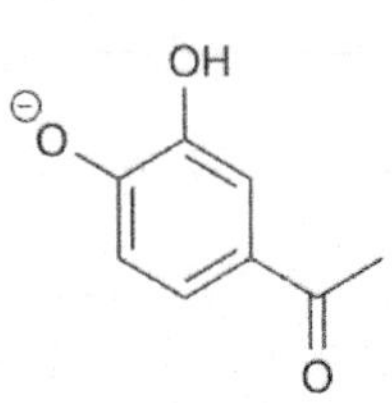

Conjugate base #1 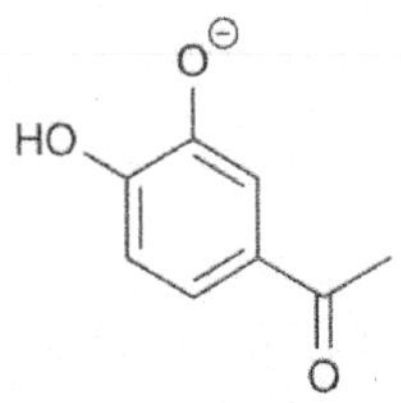   Conjugate base #2

Next, we compare these conjugate bases and consider which one is more stable. Each of these conjugate bases is stabilized by resonance, so let's draw the resonance structures. Here are the resonance structures of conjugate base #1.

Conjugate base #1

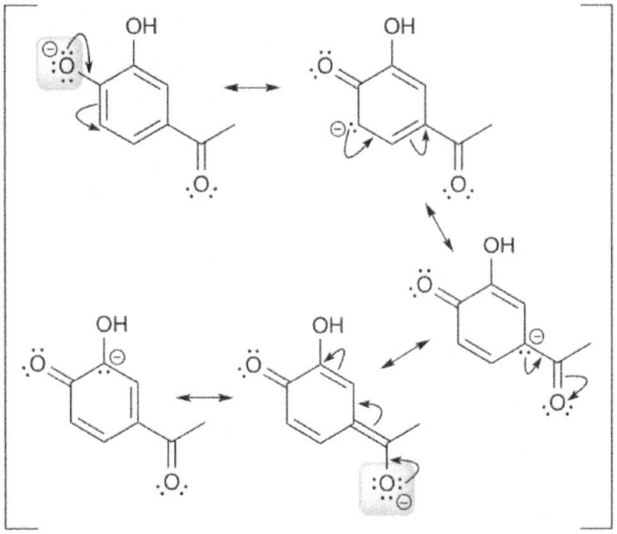

and here are the resonance structures of conjugate base #2:

Conjugate base #2

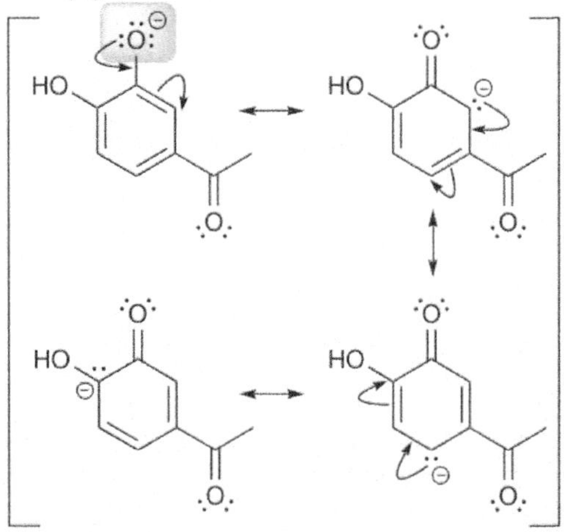

In each case, there are many resonance structures, but conjugate base #1 has one extra resonance structure that conjugate base #2 lacks. Notice the highlighted regions, which indicate that the negative charge is spread over two different oxygen atoms in conjugate base #1, but the negative charge can only be pushed onto one oxygen atom in conjugate base #2. Because of its additional resonance structure, conjugate base #1 is more stable than conjugate base #2.

Therefore, the highlighted proton shown below is the more acidic proton because removing this proton generates the more stable conjugate base:

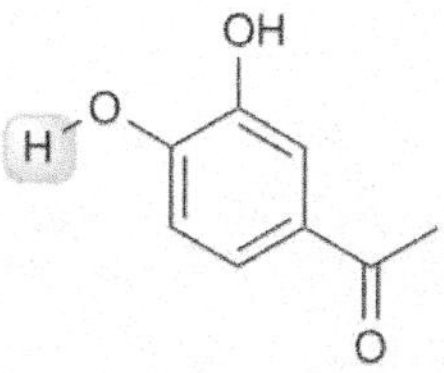

3.27 In order to compare the acidity of HI and HBr, we deprotonate each of them and compare the resulting conjugate bases. Iodide (I⁻) is more stable (because it is larger) than bromide (Br⁻), and therefore, HI is a stronger acid than HBr.

3.28 In order to compare the acidity of CH_3SH and CH_3OH, we remove the most acidic proton from each structure and compare the resulting conjugate bases. CH_3S^- is more stable than CH_3O^- because S is larger than O and can better stabilize a negative charge. Therefore, CH_3SH is a stronger acid than CH_3OH.

3.29 In order to compare the acidity of NH_3 and H_2O, we deprotonate each of them and compare the resulting conjugate bases. HO^- is more stable than H_2N^- because O is more electronegative than N and can better stabilize a negative charge. Therefore, H_2O is a stronger acid than NH_3.

3.30 In order to compare the acidity of the two compounds, we remove the most acidic proton from each structure and compare the resulting conjugate bases.

Conjugate base #1   Conjugate base #2

In both conjugate bases, the negative charge is on a carbon atom, and neither conjugate base is stabilized by resonance. Inductive effects are not relevant here. The difference between these two conjugate bases is the type of orbital that the negative charge occupies. In conjugate base #1, the negative charge occupies an *sp* hybridized orbital, and in conjugate base #2, the negative charge occupies an sp^2 hybridized orbital. Conjugate base #1 is more stable because the negative charge is held more closely to the positively charged nucleus. Therefore, the compound shown below is the more acidic compound (deprotonation of this compound produces the more stable conjugate base).

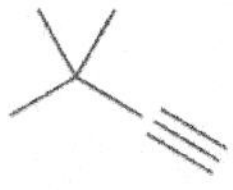

3.31 In order to compare the acidity of the two compounds, we remove the most acidic proton from each structure and compare the resulting conjugate bases.

Conjugate base #1   Conjugate base #2

In conjugate base #1, the charge is on a nitrogen atom. In conjugate base #2, the charge is on a sulfur atom. Nitrogen and sulfur are neither in the same row nor in the same column of the periodic table. Nevertheless, these two atoms can still be compared by considering a similar structure with a negative charge on an oxygen atom (see the middle structure below):

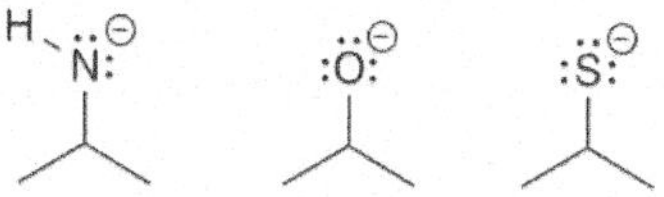

We can compare a negative charge on N with a negative charge on O because N and O are in the same row of the periodic table. When we compare them, we find that a negative charge is more stable on an oxygen atom than on a nitrogen atom (because of electronegativity).

We can also compare a negative charge on S with a negative charge on O because S and O are in the same column of the periodic table. When we compare them, we find that a negative charge is more stable on a sulfur atom than on an oxygen atom (because of size).

In summary, the three anions above (with negative charges on N, O, and S) are arranged in order of increasing stability, with the most stable anion on the right. Since that conjugate base is the more stable conjugate base, we therefore conclude that the following compound is the stronger acid.

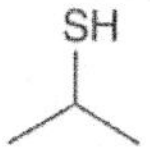

3.32 In order to compare the acidity of the two compounds, we deprotonate each of them and compare the resulting conjugate bases.

Conjugate base #1   Conjugate base #2

Next, we compare these conjugate bases and consider which one is more stable. In both structures, the negative charge is on an oxygen atom; and neither structure is stabilized by resonance. Furthermore, each structure has electronegative atoms (Cl or F), which stabilize the charge via induction. The inductive effects of fluorine atoms are stronger than the inductive effects of chlorine atoms (because fluorine is the most electronegative element). So conjugate base #2 is more stable than conjugate base #1. Therefore, we conclude that the compound shown below is the more acidic compound because deprotonating this compound generates the more stable conjugate base:

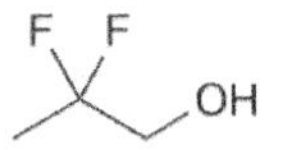

3.34 We compare the base on either side of the equilibrium. On the left side of the equilibrium, there is a negative charge on a carbon atom. On the right side of the equilibrium, there is a negative charge on an oxygen atom. A negative charge is more stable on an oxygen atom than on a carbon atom (because oxygen is more electronegative than carbon), so the equilibrium favors the right (the products).

3.35 We compare the base on either side of the equilibrium. In each case, the negative charge is on an oxygen atom. But the base on the left side of the equilibrium is stabilized by resonance:

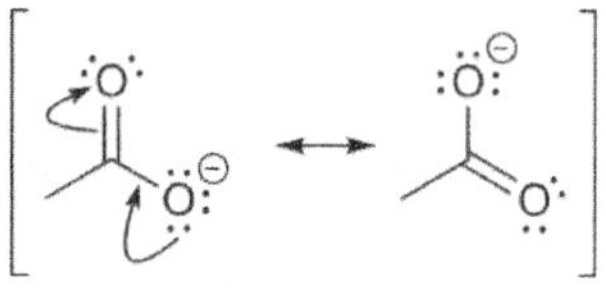

In contrast, the base on the right side of the equilibrium (HO⁻) is not stabilized by resonance. The equilibrium will favor the more stable base, so the equilibrium will favor the left (the reactants).

3.36 We compare the base on either side of the equilibrium. In each case, the negative charge is on an oxygen atom. But the base on the right side of the equilibrium is stabilized by resonance:

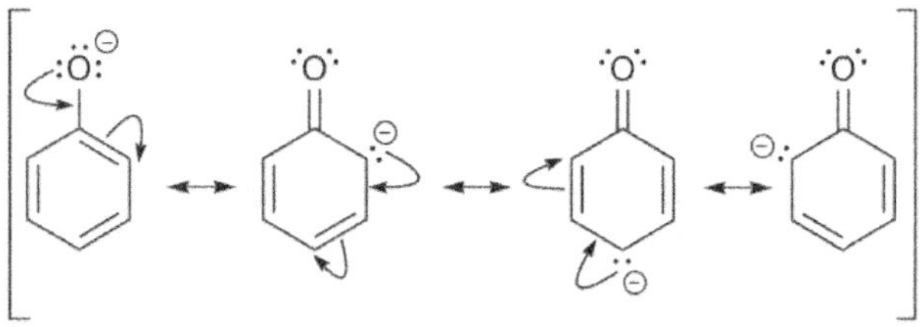

In contrast, the base on the left side of the equilibrium is not stabilized by resonance. The equilibrium will favor the more stable base, so the equilibrium will favor the right (the products).

3.38 We must draw two curved arrows to show the transfer of a proton. The first curved arrow is drawn from the base to the acidic proton, and the second curved arrow is drawn showing the breaking of the O—H bond to give a negative charge on an oxygen atom:

3.39 We must draw two curved arrows to show the transfer of a proton. The first curved arrow is drawn from the base to the acidic proton, and the second curved arrow is drawn showing the breaking of the N—H bond to give a negative charge on a nitrogen atom:

3.40 Hydroxide is a strong base, and it will remove the most acidic proton in CH_3SH (the proton connected to the sulfur atom). We must draw two curved arrows to show the transfer of this proton. The first curved arrow is drawn from the base (hydroxide) to the acidic proton, and the second curved arrow is drawn showing the breaking of the S—H bond to give a negative charge on sulfur:

3.41 Hydroxide is a strong base, and it will remove the most acidic proton in the compound. The most acidic proton is connected to the central carbon atom because removal of this proton generates a conjugate base with three resonance structures (two of which have the negative charge on an oxygen atom).

To draw a mechanism for this reaction, we must draw two curved arrows to show the transfer of the acidic proton. The first curved arrow is drawn from the base (hydroxide) to the acidic proton, and the second curved arrow is drawn showing the breaking of the C—H bond to give a resonance-stabilized conjugate base:

3.42 Hydroxide is a strong base, and it will remove the most acidic proton in the compound. The most acidic proton is the one that can be removed to generate the most stable conjugate base. In this case, the proton shown below is the most acidic because removal of this proton generates a conjugate base with three resonance structures (one of which has the negative charge on an oxygen atom).

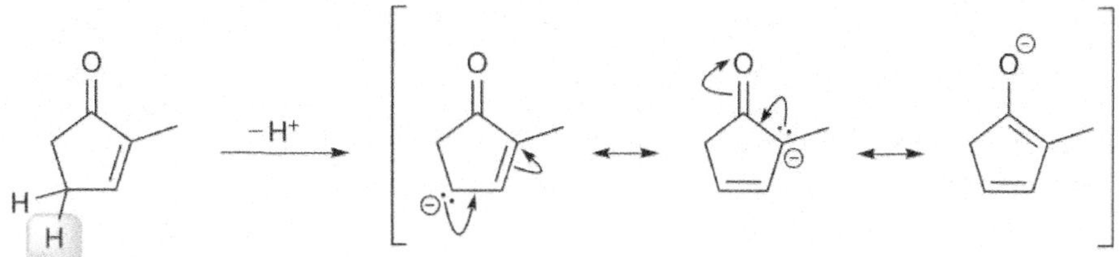

To draw a mechanism for this acid–base reaction, we must draw two curved arrows to show the transfer of the acidic proton. The first curved arrow is drawn from the base (hydroxide) to the acidic proton, and the second curved arrow is drawn showing the breaking of the C—H bond to give a resonance-stabilized conjugate base:

3.43 H$_2$N⁻ is a very strong base, and it will remove the most acidic proton in the compound. The most acidic proton is the one that can be removed to generate the most stable conjugate base. In this case, the proton shown below is the most acidic because removal of this proton generates a conjugate base in which the negative charge occupies an *sp* hybridized orbital.

To draw a mechanism for this reaction, we must draw two curved arrows to show the transfer of the acidic proton. The first curved arrow is drawn from the base (H$_2$N⁻) to the acidic proton, and the second curved arrow is drawn showing the breaking of the C—H bond to give the corresponding conjugate base:

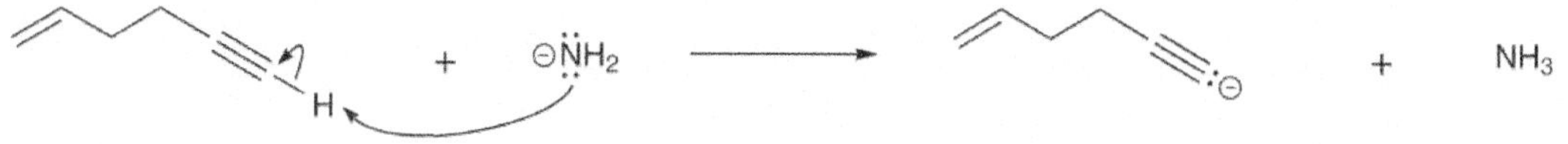

3.44 H$_2$N⁻ is a very strong base, and it will remove the most acidic proton in the compound. The most acidic proton is the one that can be removed to generate the most stable conjugate base. In this case, the proton shown below is the most acidic because removal of this proton generates a conjugate base in which the negative charge is on an oxygen atom.

To draw a mechanism for this reaction, we must draw two curved arrows to show the transfer of the acidic proton. The first curved arrow is drawn from the base (H$_2$N⁻) to the acidic proton, and the second curved arrow is drawn showing the breaking of the O—H bond to give the corresponding conjugate base:

CHAPTER 4

4.2 This carbon atom is connected to three other atoms and has no lone pairs. The sum is $3 + 0 = 3$, which means that this carbon atom is sp^2 hybridized.

4.3 This carbon atom is connected to two other atoms and has no lone pairs. The sum is $2 + 0 = 2$, which means that this carbon atom is sp hybridized.

4.4 This carbon atom is connected to four other atoms and has no lone pairs. The sum is $4 + 0 = 4$, which means that this carbon atom is sp^3 hybridized.

4.5 The central carbon atom is connected to three other atoms and has no lone pairs. The sum is $3 + 0 = 3$, which means that this carbon atom is sp^2 hybridized.

4.6 The central carbon atom is connected to three other atoms and has one lone pair (recall that a carbon atom with a negative charge will have a lone pair and three bonds). The sum is $3 + 1 = 4$, which means that this carbon atom is sp^3 hybridized.

4.7 The central carbon atom is connected to two other atoms and has no lone pairs. The sum is $2 + 0 = 2$, which means that this carbon atom is sp hybridized.

4.8 The carbon atoms with double bonds are sp^2 hybridized, and the carbon atoms with triple bonds are sp hybridized. The remaining carbon atoms have only single bonds (no double bonds or triple bonds), so they are sp^3 hybridized:

a = sp^3
b = sp^2
c = sp

4.10 Each of the six carbon atoms bears a double bond so each carbon atom in this compound is sp^2 hybridized. Carbon atoms that are sp^2 hybridized (with three bonds and no lone pairs) have trigonal planar geometry.

4.11 The two carbon atoms bearing the double bond are sp^2 hybridized. Carbon atoms that are sp^2 hybridized (with three bonds and no lone pairs) have trigonal planar geometry.

The remaining carbon atoms have only single bonds (no double bonds or triple bonds), so they are sp^3 hybridized, and have tetrahedral geometry:

a = sp^2, trigonal planar
b = sp^3, tetrahedral

4.12 The carbon atoms with double bonds are sp^2 hybridized and have trigonal planar geometry. The carbon atoms with triple bonds are sp hybridized and are linear. The remaining carbon atoms have only single bonds (no double bonds or triple bonds), so they are sp^3 hybridized, with tetrahedral geometry:

a = sp^3, tetrahedral
b = sp^2, trigonal planar
c = sp, linear

4.13 Each of the carbon atoms bearing a double bond is sp^2 hybridized and has trigonal planar geometry. All of the other carbon atoms have only single bonds (no double bonds or triple bonds), so they are sp^3 hybridized, with tetrahedral geometry. The nitrogen atom has three bonds and one lone pair, so it is sp^3 hybridized, with trigonal pyramidal geometry:

a = sp^3, tetrahedral
b = sp^2, trigonal planar
c = sp^3, trigonal pyramidal

4.14 In this compound, all of the carbon atoms have only single bonds (no double bonds or triple bonds), so they are sp^3 hybridized, with tetrahedral geometry. The nitrogen atom has three bonds and one lone pair, so it is sp^3 hybridized, with trigonal pyramidal geometry. The oxygen atom has two bonds and two lone pairs, so it has bent geometry.

a = tetrahedral
b = trigonal pyramidal
c = bent

4.15 The carbon atoms with double bonds are sp^2 hybridized and have trigonal planar geometry. The carbon atoms with triple bonds are sp hybridized and are linear. The remaining carbon atoms have only single bonds (no double bonds or triple bonds), so they are sp^3 hybridized, with tetrahedral geometry:

a = tetrahedral
b = trigonal planar
c = linear

4.16 The carbon atoms with double bonds are sp^2 hybridized and have trigonal planar geometry. The remaining carbon atoms have only single bonds (no double bonds or triple bonds), so they are sp^3 hybridized, with tetrahedral geometry:

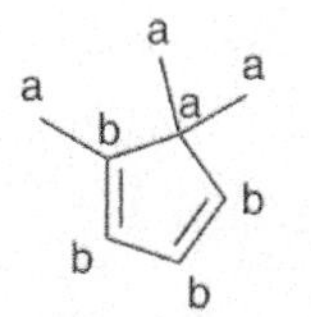

a = tetrahedral
b = trigonal planar

4.17 The carbon atoms with double bonds are sp^2 hybridized and have trigonal planar geometry. The remaining carbon atoms have only single bonds (no double bonds or triple bonds), so they are sp^3 hybridized, with tetrahedral geometry:

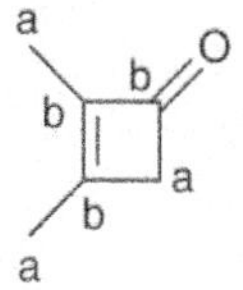

a = tetrahedral
b = trigonal planar

4.18 This nitrogen atom has a lone pair that is participating in resonance:

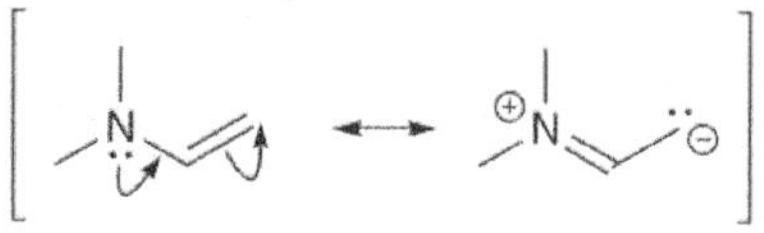

Therefore, this nitrogen atom is sp^2 hybridized (rather than sp^3 hybridized) and has trigonal planar geometry (rather than trigonal pyramidal geometry).

4.19 This nitrogen atom has a lone pair that is localized (not participating in resonance) because it is not close enough to the double bond (there is an sp^3 hybridized carbon atom that is in between the nitrogen atom and the double bond).

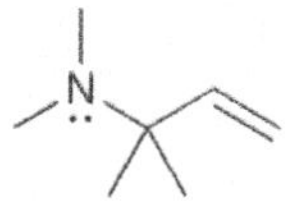

So this nitrogen atom is sp^3 hybridized and has trigonal pyramidal geometry.

4.20 This nitrogen atom has a lone pair that is participating in resonance:

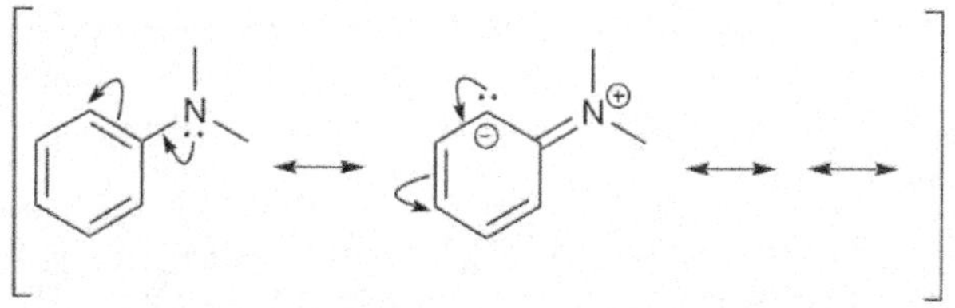

Therefore, this nitrogen atom is sp^2 hybridized (rather than sp^3 hybridized) and has trigonal planar geometry (rather than trigonal pyramidal geometry).

4.21 This nitrogen atom has a lone pair that is localized (not participating in resonance),

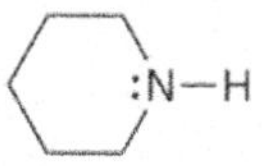

so this nitrogen atom is sp^3 hybridized and has trigonal pyramidal geometry.

4.22 This nitrogen atom has a lone pair that is participating in resonance:

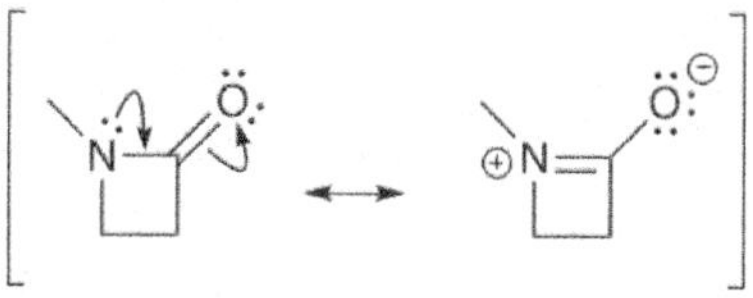

Therefore, this nitrogen atom is sp^2 hybridized (rather than sp^3 hybridized) and has trigonal planar geometry (rather than trigonal pyramidal geometry).

4.23 This nitrogen atom has a lone pair that is localized (not participating in resonance),

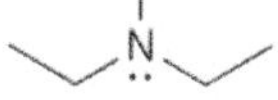

so this nitrogen atom is sp^3 hybridized and has trigonal pyramidal geometry.

CHAPTER 5

5.2 This compound has a $C=O$ bond that is flanked by two alkyl (R) groups, so this compound is a ketone. When naming this compound, the correct suffix is "−one."

5.3 This compound has a $C=O$ bond that is flanked by an alkyl (R) group and an alkoxy (OR) group, so this compound is an ester. When naming this compound, the correct suffix is "−oate."

5.4 This compound has a C=O bond that is flanked by a hydrogen atom and an alkyl (R) group, so this compound is an aldehyde. When naming this compound, the correct suffix is "–al."

5.5 This compound has an NH₂ group, so this compound is an amine. When naming this compound, the correct suffix is "–amine." The chlorine atom will be named as a substituent in the name of the compound.

5.6 This compound has an OH group, so this compound is an alcohol. When naming this compound, the correct suffix is "–ol." The bromine atom will be named as a substituent in the name of the compound.

5.7 This compound has an OH group, so this compound is an alcohol. When naming this compound, the correct suffix is "–ol." The fluorine atoms will be named as substituents in the name of the compound.

5.8 This compound has a C=O bond that is flanked by a hydrogen atom and an alkyl (R) group, so this compound is an aldehyde. There is also an OH group, so the compound is also an alcohol, but the aldehyde group takes priority, and the compound is named as an aldehyde (the OH group will be listed as a substituent). Therefore, when naming this compound, the correct suffix is "–al."

5.9 This compound has a C=O bond that is flanked by two alkyl (R) groups, so this compound is a ketone. There is also an OH group, so the compound is also an alcohol, but the ketone group takes priority, and the compound is named as a ketone (the OH group will be listed as a substituent). Therefore, when naming this compound, the correct suffix is "–one."

5.10 This compound has a C=O bond that is flanked by an alkyl (R) group and an OH group, so this compound is a carboxylic acid. There is also an NH₂ group, so the compound is also an amine, but the carboxylic acid group takes priority, and the compound is named as a carboxylic acid (the NH₂ group will be listed as a substituent). Therefore, when naming this compound, the correct suffix is "–oic acid."

5.12 This compound has one double bond, so the unsaturation for this compound is –en–.

5.13 This compound has one triple bond, so the unsaturation for this compound is –yn–.

5.14 This compound has two double bonds, so the unsaturation for this compound is –dien–.

5.15 This compound has three double bonds, so the unsaturation for this compound is –trien–.

5.16 This compound has three double bonds, so the unsaturation for this compound is –trien–.

5.17 This compound has one double bond and two triple bonds, so the unsaturation for this compound is –endiyn–.

5.19 This compound has no functional groups and no double bonds or triple bonds. So, we simply choose the longest chain, which has six carbon atoms. Therefore, the parent is "hex–."

5.20 This compound has a double bond, so we choose the longest chain that includes the double bond. In this case, the longest chain (including the double bond) has seven carbon atoms. Therefore, the parent is "hept–."

5.21 This compound has a double bond, so we choose the longest chain that includes the double bond. In this case, the longest chain (including both carbon atoms of the double bond) has six carbon atoms. Therefore, the parent is "hex–."

5.22 This compound has no functional groups and no double bonds or triple bonds. So, we simply choose the longest chain. In this case, the longest chain has nine carbon atoms. Therefore, the parent is "non–."

5.23 This compound has no functional groups and no double bonds or triple bonds. So, we simply choose the longest chain. In this case, the longest chain has eight carbon atoms. Therefore, the parent is "oct–."

5.24 The bromine atom will be listed as a substituent, and the suffix of the compound will be "–e." This compound has no double bonds or triple bonds, so we simply choose the longest chain. In this case, the longest chain has six carbon atoms. Therefore, the parent is "hex–."

5.25 This compound has a functional group (OH), a double bond, and a triple bond. The first two of these features (functional group and double bond) are the most important when determining the parent chain. We must choose the longest chain that includes the two carbon atoms of the double bond as well as the carbon atom connected to the OH group. This chain has six carbon atoms. Therefore, the parent is "hex–."

5.26 This compound has a functional group (OH) but no double bonds or triple bonds. So we choose the longest chain that includes the carbon atom connected to the OH group. This chain has six carbon atoms. Therefore, the parent is "hex–."

5.27 This compound has a functional group (a ketone) and a double bond. We must choose the longest chain that includes the two carbon atoms of the double bond as well as the carbon atom of the C=O group. This chain has five carbon atoms (it includes the two carbon atoms to the right of the C=O group). Therefore, the parent is "pent–."

5.29 The parent in this case is propane, and there are two chlorine atoms connected to the parent. Each of these chlorine atoms will be named as a "chloro" substituent. With two of them, the term "dichloro" will appear in the name of the compound.

5.30 The parent in this case is propane, and there is a bromine atom and an iodine atom connected to the parent. Each of these is considered to be a substituent. The bromine atom will appear as "bromo" in the name of the compound, while the iodine atom will appear as "iodo" in the name of the compound.

5.31 The parent in this case is octane, and there are five methyl groups connected to the parent. With five of them, the term "pentamethyl" will appear in the name of the compound.

5.32 The parent in this case is ethane, and there are six fluorine atoms connected to the parent. Each of these fluorine atoms will be named as a "fluoro" substituent. With six of them, the term "hexafluoro" will appear in the name of the compound.

5.33 The parent in this case has five carbon atoms. The OH group is the functional group, and the compound will be named as an alcohol (with the suffix "ol"). So, there is only one substituent, called a methyl group.

5.34 The longest chain has seven carbon atoms, so the parent in this case is heptane. There are two substituents connected to this parent. These substituents are called "chloro" and "*tert*-butyl."

5.35 The parent has four carbon atoms, and the carboxylic acid group is the functional group. All of the other groups will be listed as substituents, including NH_2 (amino), Br (bromo), Cl (chloro), and F (fluoro).

5.36 The parent has six carbon atoms, and there are three substituents attached to this parent. These three substituents are I (iodo), F (fluoro), and Br (bromo).

5.37 The parent has eight carbon atoms, and there is only one substituent connected to this parent. This substituent is called an isopropyl group.

5.38 The parent has six carbon atoms, and the carboxylic acid group takes priority as the functional group. So the OH group must be listed as a substituent, called a hydroxy group. There is also an ethyl group connected to the parent.

5.39 The methyl groups are on opposite sides of the double bond, so this double bond has a *trans* configuration.

5.40 The two fluoro substituents are on the same side of the double bond, so this double bond has a *cis* configuration.

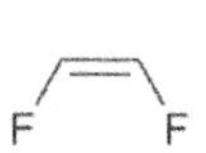

5.41 This double bond has two hydrogen atoms that are on opposite sides of the double bond, so this double bond has a *trans* configuration.

5.42 This double bond has two ethyl groups that are on the same side of the double bond, so this double bond has a *cis* configuration.

5.43 This double bond has two hydrogen atoms that are on opposite sides of the double bond, so this double bond has a *trans* configuration.

5.45 This compound has a triple bond, and the parent is the longest chain that contains the two carbon atoms of the triple bond. The parent is numbered in the direction that gives the triple bond the lower possible number (the triple bond is at position 2, rather than position 6), as shown:

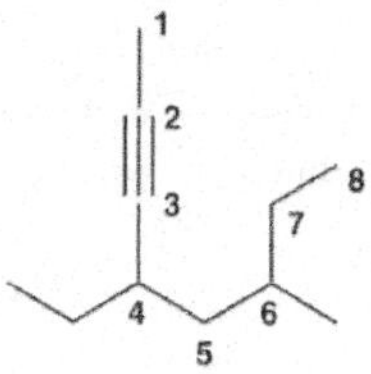

5.46 This compound has a double bond, and the parent is the longest chain that contains the two carbon atoms of the double bond. The parent is numbered in the direction that gives the double bond the lower possible number (the double bond is at position 2, rather than position 4), as shown:

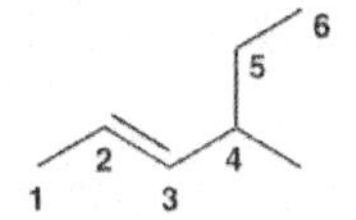

5.47 This compound is a ketone, and the parent is the longest chain that contains the carbon atom of the C=O bond. The parent is numbered in the direction that gives the functional group the lower possible number (at position 4, rather than position 6), as shown:

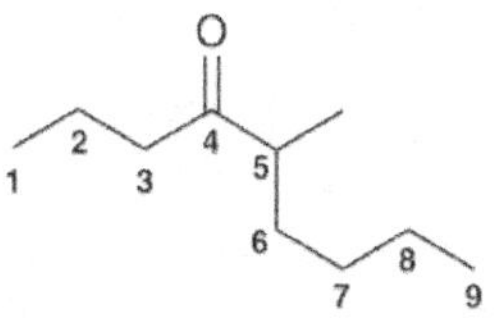

5.48 This compound contains an OH group and a C=C double bond. The parent is the longest chain that contains the carbon atom connected to the functional group (OH) and the two carbon atoms of the double bond. This parent has five carbon atoms (not seven). The parent is numbered in the direction that gives the functional group the lower possible number (at position 1, rather than position 5), as shown:

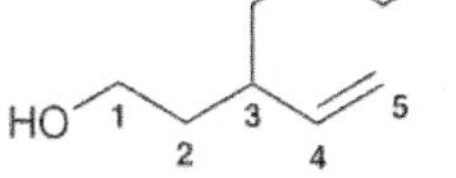

5.49 This compound has no functional group, no double bond, and no triple bond. So the parent is simply the longest chain. This chain is numbered in the direction that gives the lowest possible number to the first substituent.

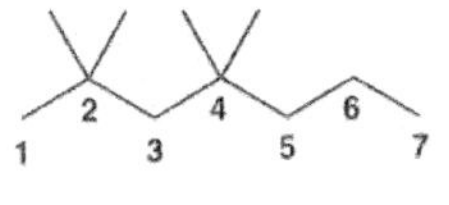

5.50 This compound has a triple bond, and the parent is the longest chain that contains the two carbon atoms of the triple bond. The parent is numbered in the direction that gives the triple bond the lower possible number (the double bond is at position 3, rather than position 4), as shown here:

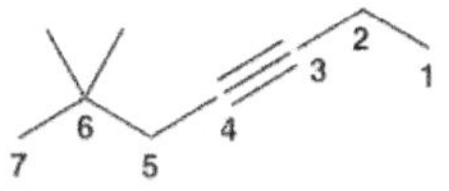

5.51 This compound has no functional group, no double bond, and no triple bond. So the parent is simply the longest chain. Regardless of the direction in which we number the parent, the first substituent will be at position 2. So we consider the number of the second substituent. In this case, we number the parent in the direction that gives the lower possible number to the second substituent (at position 2, rather than position 4).

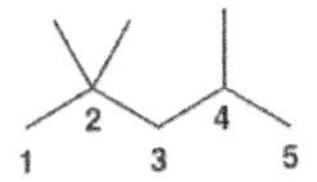

5.52 This compound contains an OH group, and the parent is the longest chain that contains the carbon atom connected to this functional group. The parent is numbered in the direction that gives the functional group the lower possible number (at position 2, rather than position 6), even though numbering the other way would give two substituents at position 2. When numbering the parent, the functional group takes priority over the substituents.

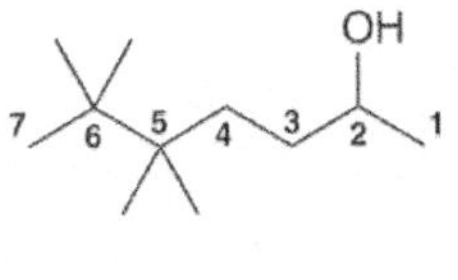

5.53 This compound has a double bond, and the parent is the longest chain that contains the two carbon atoms of the double bond. The parent is numbered in the direction that gives the double bond the lower possible number (the double bond is at position 1, rather than position 6), as shown:

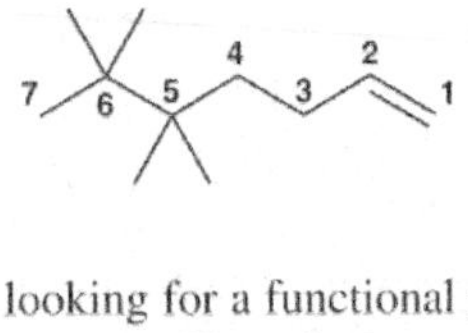

5.55 We begin by looking for a functional group that will determine the suffix of the name. There isn't one, so the suffix of the name will just be the letter "e." There is a double bond, which will

be named as the unsaturation (–en–). The parent has eight carbon atoms (–oct–), and there are two substituents (a methyl group and an ethyl group). The parent is numbered from left to right to give the lowest possible number to the double bond:

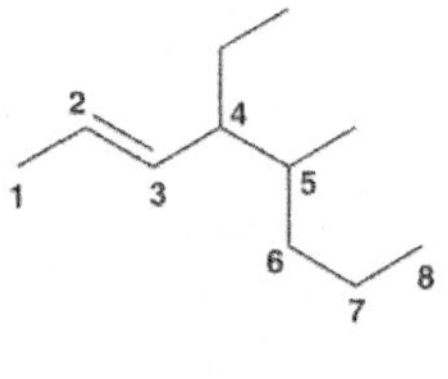

With this numbering system, the double bond is at position 2, and this must be indicated in the name (there are two different options for how this can be indicated, and both are shown below). According to this numbering system, the ethyl group is at position 4, and the methyl group is at position 5. Also, the double bond has the *trans* configuration, and this must be indicated at the beginning of the name, parenthetically. Putting it all together, we have the following name, in which the substituents are listed alphabetically:

(*trans*)-4-Ethyl-5-methyloct-2-ene
or
(*trans*)-4-Ethyl-5-methyl-2-octene

5.56 We begin by looking for a functional group that will determine the suffix of the name. In this case, there is an OH group, so this compound is an alcohol, and the suffix will be "–ol." There is no double bond or triple bond, so the letters "an" will appear just before the suffix to indicate that the compound lacks unsaturation. The parent is the longest chain that contains the carbon atom connected to the OH group. This parent has nine carbon atoms (–non–) and is numbered to give the OH group the lowest number (at position 3, rather than position 7):

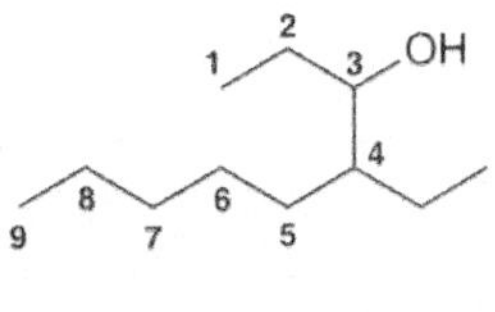

There is one substituent (an ethyl group) located at position 4. Putting it all together, we have the following name:

4-Ethylnonan-3-ol
or
4-Ethyl-3-nonanol

5.57 We begin by looking for a functional group that will determine the suffix of the name. There isn't one, so the suffix of the name will just be the letter "e." There is a triple bond, which will be named as the unsaturation (–yn–). The parent has six carbon atoms (–hex–), and there are two substituents (both methyl groups). The parent is numbered from left to right to give the lowest possible number to the triple bond.

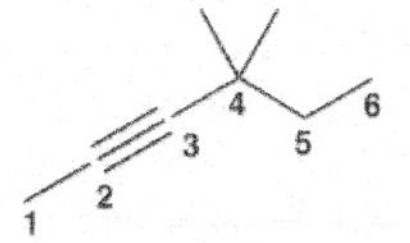

With this numbering system, the triple bond is at position 2, and this must be indicated in the name (there are two different options for how this can be indicated, and both are shown below). According to this numbering system, both methyl groups are at position 4 (notice that the location of each methyl group must be indicated separately, so it is 4,4-dimethyl, rather than just 4-dimethyl). Putting it all together, we have the following name:

4,4-Dimethylhex-2-yne

or

4,4-Dimethyl-2-hexyne

5.58 We begin by looking for a functional group that will determine the suffix of the name. This compound is a ketone, so the suffix will be "–one." There is no C═C bond or C≡C bond, so the letters "an" will appear just before the suffix to indicate that the compound lacks C═C and C≡C bonds. The parent is a six-membered ring, so it will be "cyclohex." There are two methyl groups, both located at the same position. The parent is numbered starting with the functional group.

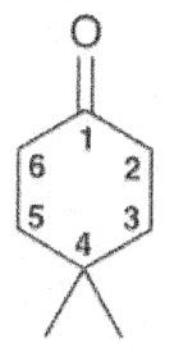

With this numbering system, both methyl groups are at position 4 (notice that the location of each methyl group must be indicated separately, so it is 4,4-dimethyl, rather than just 4-dimethyl). Putting it all together, we have the following name:

4,4-Dimethylcyclohexanone

Note that we don't indicate the location of the functional group (at position 1) because the functional group is understood to be at position 1 in a cyclic structure (the numbering always starts with the functional group, so by definition, the functional group is always at position 1).

5.59 We begin by looking for a functional group that will determine the suffix of the name. There isn't one, so the suffix of the name will just be the letter "e." There is no double bond or triple bond, so the letters "an" will appear just before the suffix ("–ane") to indicate that the compound lacks unsaturation. The parent is the longest chain, which has six carbon atoms (and therefore called –hex–) and is numbered from right to left to give the lowest number to the first substituent (at position 2, rather than the position 3):

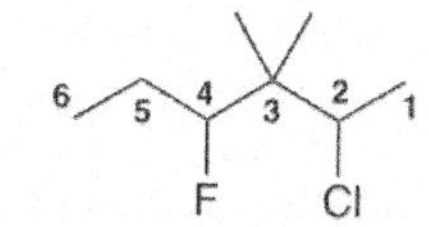

There are four substituents connected to this parent: a chloro substituent at position 2, two methyl substituents at position 3, and a fluoro substituent at position 4. These substituents appear in alphabetical order in the following name:

2-Chloro-4-fluoro-3,3-dimethylhexane

Notice that the location of each methyl group must be indicated separately, so it is 3,3-dimethyl, rather than just 3-dimethyl.

5.60 We begin by looking for a functional group that will determine the suffix of the name. In this case, there is an NH_2 group, so this compound is an amine, and the suffix will be "–amine." There is no double bond or triple bond, so the letters "an" will appear just before the suffix to indicate that the compound lacks unsaturation. The parent is the longest chain that contains the carbon atom connected to the NH_2 group. This parent has five carbon atoms (–pent–) and is numbered to give the NH_2 group the lowest number (at position 1, rather than position 5):

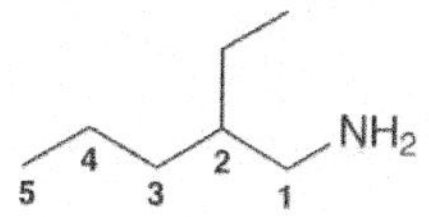

There is one substituent—an ethyl group located at position 2. Putting it all together, we have the following name:

2-Ethylpentan-1-amine

or

2-Ethyl-1-pentanamine

5.61 We begin by looking for a functional group that will determine the suffix of the name. In this case, there is a carboxylic acid group, so the suffix will be "–oic acid." There is no C═C bond or C≡C bond, so the letters "an" will appear just before the suffix to indicate that the compound lacks C═C and C≡C bonds. The parent is the longest chain that contains the carbon atom of the carboxylic acid group. This parent has only five carbon atoms (–pent–) and is numbered to give the lowest possible number to the carbon atom of the carboxylic acid group (at position 1, rather than position 5):

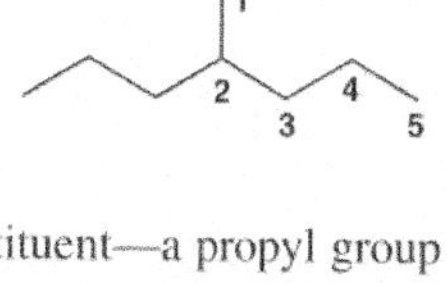

There is one substituent—a propyl group located at position 2. Putting it all together, we have the following name:

2-Propylpentanoic acid

Note that this name does not indicate the location of the functional group (at position 1). For a carboxylic acid, the functional group is understood to be at position 1 (the numbering always starts with the carbon atom of this functional group, so by definition, the carboxylic acid group is always at position 1).

5.62 We begin by looking for a functional group that will determine the suffix of the name. In this case, there is an OH group, so this compound is an alcohol, and the suffix will be "–ol." There is also a double bond, so the letters "en" will appear just before the suffix to indicate the presence of a double bond. The parent is the longest chain that contains the carbon atom connected to the OH group and the two carbon atoms of the double bond. This parent has eight carbon atoms (–oct–) and is numbered right to left to give the OH group the lowest number (at position 4, rather than position 5):

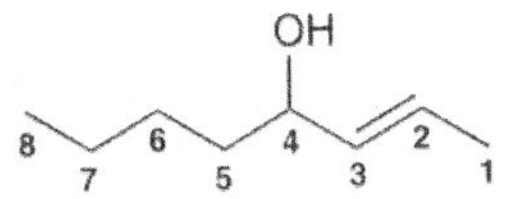

The double bond has the *trans* configuration, and this must be indicated at the beginning of the name, parenthetically. Putting it all together, we have the following name:

(*trans*)-Oct-2-en-4-ol
or
(*trans*)-2-Octen-4-ol

5.63 We begin by looking for a functional group that will determine the suffix of the name. There isn't one, so the suffix of the name will just be the letter "e." There is a double bond, which will be named as the unsaturation (–en–). The parent is the longest chain that contains the two carbon atoms of the double bond. This parent has eight carbon atoms (–oct–) and is numbered left to right to give the lowest number to the double bond (at position 2, rather than position 6):

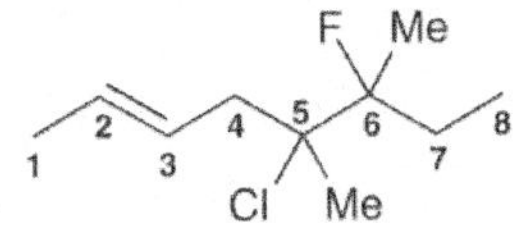

There are four substituents: a chloro substituent at position 5, a methyl group at position 5, a fluoro substituent at position 6, and a methyl group at position 6. With two methyl groups, the term "dimethyl" will appear in the name, although this term will be alphabetized as "m" rather than "d" when the substituents are listed in alphabetical order. Also, the double bond has the *trans* configuration, and this must be indicated at the beginning of the name, parenthetically. Putting it all together, we have the following name:

(*trans*)-5-Chloro-6-fluoro-5,6-dimethyloct-2-ene
or
(*trans*)-5-Chloro-6-fluoro-5,6-dimethyl-2-octene

CHAPTER 6

6.2 When viewed along the direction of the arrow, the front carbon atom is connected to three methyl groups, and the back carbon atom is connected to three methyl groups.

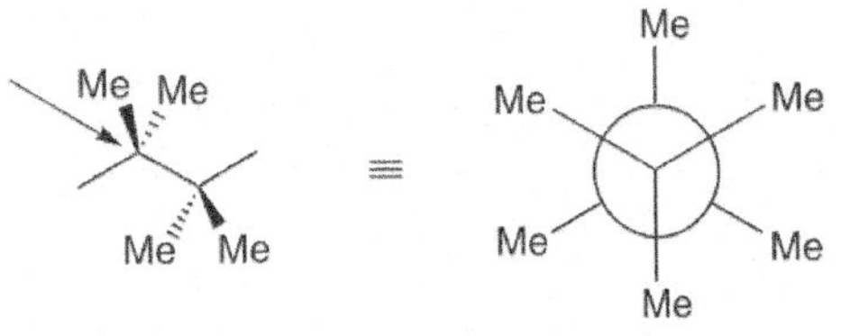

6.3 When viewed along the direction of the arrow, the front carbon atom is connected to a hydrogen atom and two methyl groups (one methyl group is straight down, and the other methyl group, on a dash, is up and to the left). The back carbon atom is also connected to a hydrogen atom and two methyl groups (one methyl group is straight up, and the methyl group on a wedge is down and to the right, when viewed from the direction indicated by the arrow).

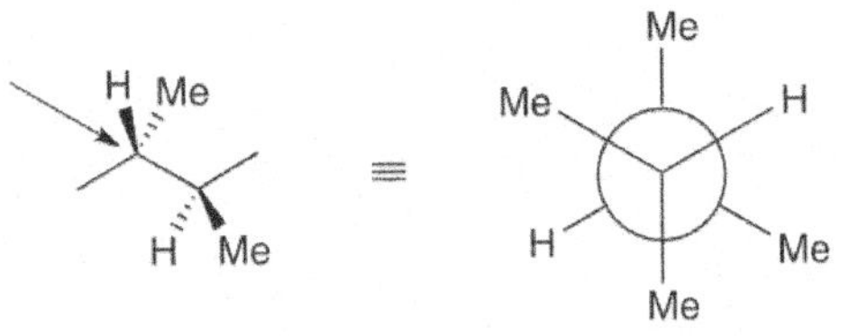

6.4 When viewed along the direction of the arrow, the front carbon atom is connected to two hydrogen atoms and one ethyl group (the ethyl group is pointing straight down). The back carbon atom is connected to two hydrogen atoms and a methyl group (straight up, when viewed from the direction indicated by the arrow).

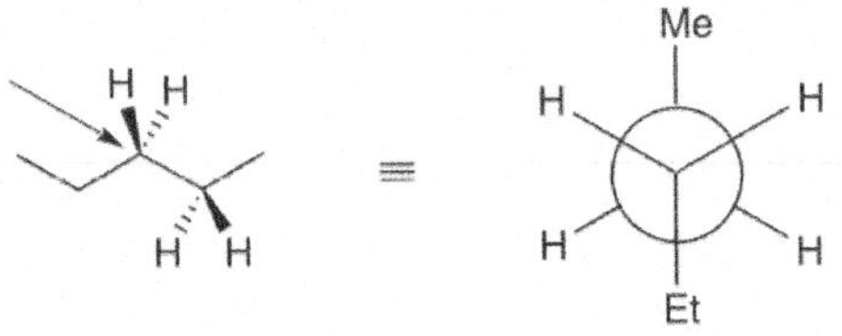

6.5   When viewed along the direction of the arrow, the front carbon atom is connected to two hydrogen atoms and one ethyl group (the ethyl group is pointing straight down). The back carbon atom is connected to a hydrogen atom, a methyl group (down and to the left), and an ethyl group (pointing up, when viewed from the direction indicated by the arrow).

6.6   When viewed along the direction of the arrow, the front carbon atom is connected to two hydrogen atoms and one isopropyl group (the isopropyl group is pointing straight down). The back carbon atom is also connected to two hydrogen atoms and an isopropyl group (straight up, when viewed from the direction indicated by the arrow).

6.7   When viewed along the direction of the arrow, the front carbon atom is connected to two methyl groups and one hydrogen atom (the hydrogen atom is pointing up and to the left). The back carbon atom is connected to two hydrogen atoms and a methyl group (this methyl group is pointing straight down, when viewed from the direction indicated by the arrow).

6.9   A Newman projection for this compound was first drawn in problem 6.2. The most stable conformation for this compound is a staggered conformation, and the least stable conformation is an eclipsed conformation.

Most stable
(staggered)

Least stable
(eclipsed)

6.10   A Newman projection for this compound was first drawn in problem 6.3. The most stable conformation for this compound is the staggered conformation that has only two Gauche interactions (the other staggered conformations have three Gauche interactions). The least stable conformation is the eclipsed conformation with two separate Me–Me eclipsing interactions.

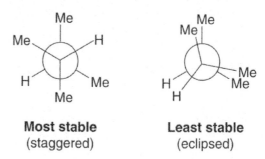

Most stable
(staggered)

Least stable
(eclipsed)

6.11   A Newman projection for this compound was first drawn in problem 6.4. The most stable conformation for this compound is the staggered conformation in which the methyl group and the ethyl group are anti to each other. The least stable conformation is the eclipsed conformation in which the methyl group and the ethyl group are eclipsing each other.

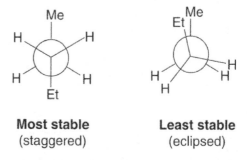

Most stable
(staggered)

Least stable
(eclipsed)

6.12   A Newman projection for this compound was first drawn in problem 6.5. The most stable conformation for this compound is the staggered conformation in which the two largest groups (the two ethyl groups) are anti to each other. The least stable conformation is the eclipsed conformation in which the two largest groups (the ethyl groups) are eclipsing each other.

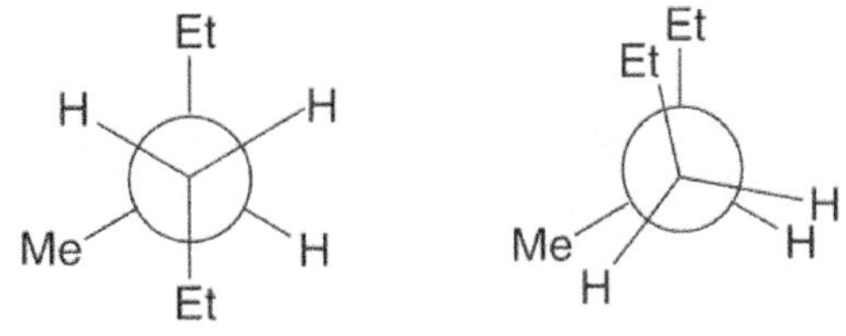

Most stable
(staggered)

Least stable
(eclipsed)

6.13 A Newman projection for this compound was first drawn in problem 6.6. The most stable conformation for this compound is the staggered conformation in which the two isopropyl groups are anti to each other. The least stable conformation is the eclipsed conformation in which the two isopropyl groups are eclipsing each other.

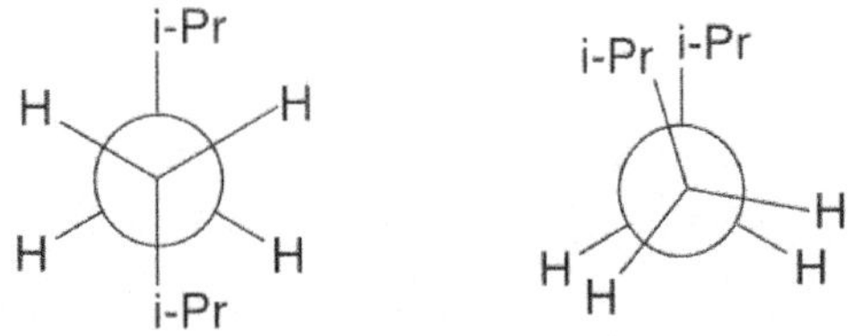

Most stable
(staggered)

Least stable
(eclipsed)

6.14 A Newman projection for this compound was first drawn in problem 6.7. The most stable conformation for this compound is a staggered conformation in which there is only one Gauche interaction (rather than a staggered conformation with two Gauche interactions). The least stable conformation is an eclipsed conformation in which two methyl groups are eclipsing each other.

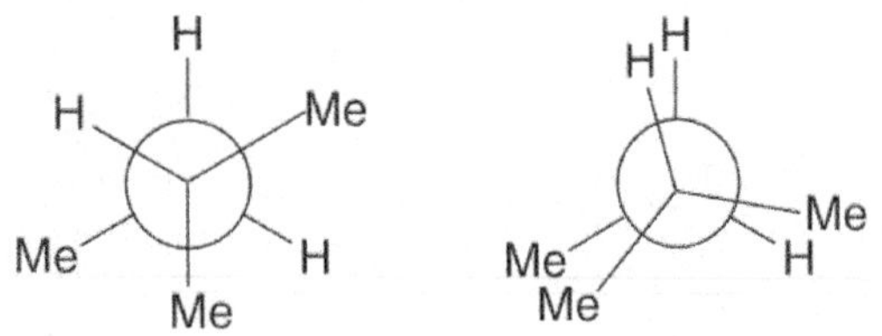

Most stable
(staggered)

Least stable
(eclipsed)

6.16 Begin by numbering the hexagon, starting from the first group (at the top) and then going clockwise. This puts the chloro

substituent at position 1 and the OH group at position 3. Then, we draw a chair, numbering it clockwise, and then place the substituents in the correct positions. The Cl is on a wedge, so it is up, and the OH is on a wedge, so it is up:

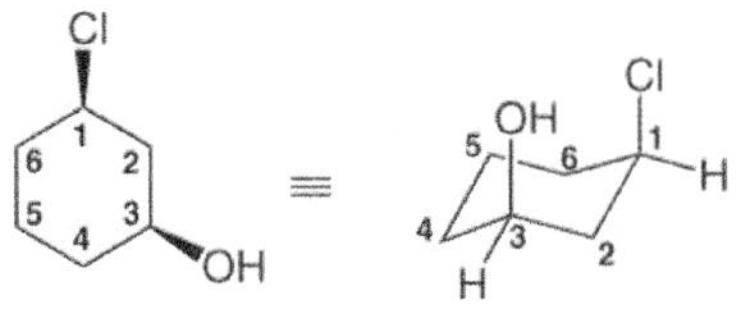

Note: These numbers are not in accordance with IUPAC rules, which would dictate that the numbering start at the carbon atom connected to the OH group (because the OH group has priority over Cl). However, we are not naming this compound right now. We are merely drawing a chair conformation, and we are using this numbering system as a tool to help us draw the chair conformation correctly. Once we are finished, we can delete the numbers, and draw only the chair, like this:

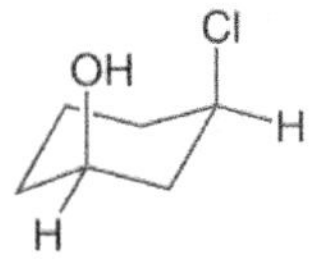

6.17 Begin by numbering the hexagon, starting from the first group (at the top) and then going clockwise. This puts the methyl group at position 1 and the ethyl group at position 4. Then, we draw a chair, numbering it clockwise, and then place the substituents in the correct positions. The methyl group is on a wedge, so it is up, and the ethyl group is on a dash, so it is down:

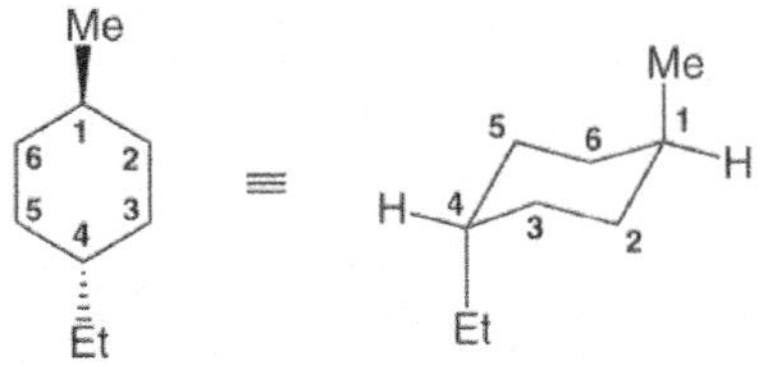

Note: These numbers are not in accordance with IUPAC rules, which would dictate that the numbering start at the carbon atom connected to the ethyl group. However, we are not naming this compound right now. We are merely drawing a chair conformation, and we are using this numbering system as a tool to help us draw the chair conformation correctly. Once we are finished, we can delete the numbers, and draw only the chair, just as we did in the solution to the previous problem.

6.18 Begin by numbering the hexagon, starting from the first group (at the top) and then going clockwise. This puts the Br at position 1 and the methyl group at position 4. Then, we draw a chair, numbering it clockwise, and then place the substituents in the correct positions. The Br is on a wedge, so it is up, and the methyl group is on a wedge, so it is up:

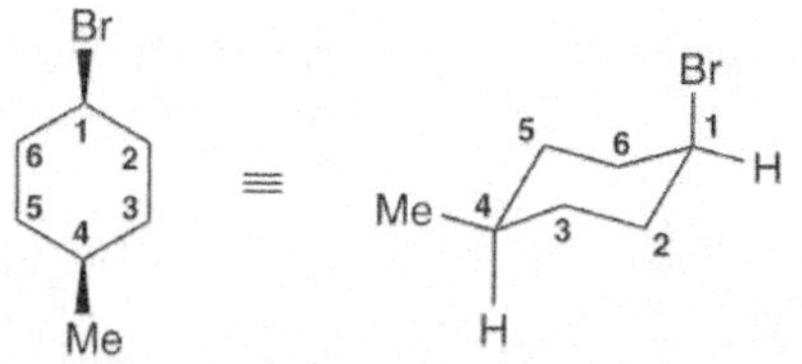

6.19 Begin by numbering the hexagon, starting from the first group (at the top) and then going clockwise. This puts the COOH group at position 1 and the OH group at position 3. Then, we draw a chair, numbering it clockwise, and then place the substituents in the correct positions. The COOH is on a dash, so it is down, and the OH group is on a wedge, so it is up:

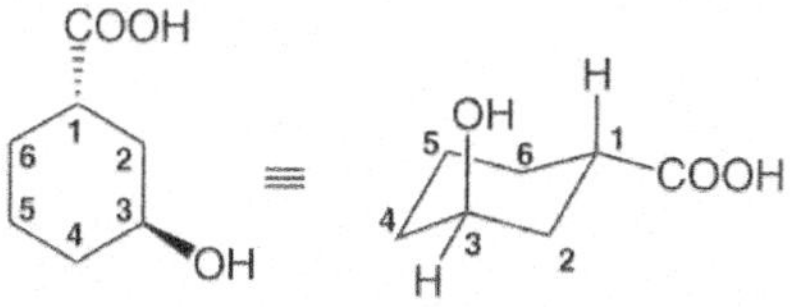

6.20 Begin by numbering the hexagon, starting from the first group (at the top) and then going clockwise. This puts the OH group at position 1 and the ethyl group at position 2. Then, we draw a chair, numbering it clockwise, and then place the substituents in the correct positions. The OH group is on a wedge, so it is up, and the ethyl group is on a dash, so it is down:

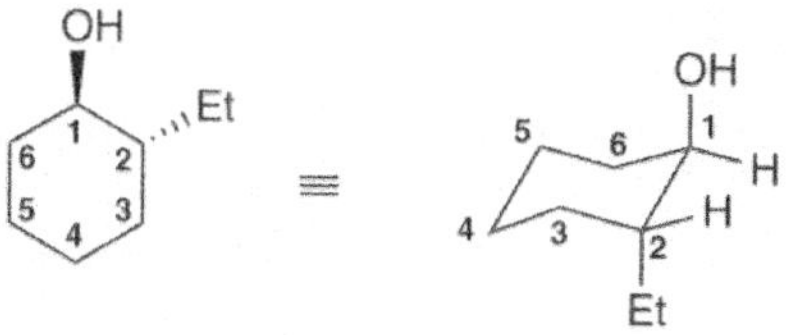

6.21 Begin by numbering the hexagon, starting from the first group (at the top) and then going clockwise. This puts the methyl groups at positions 1 and 2. Then, we draw a chair, numbering it clockwise, and then place the substituents in the correct positions. Both methyl groups are on wedges, so both are up:

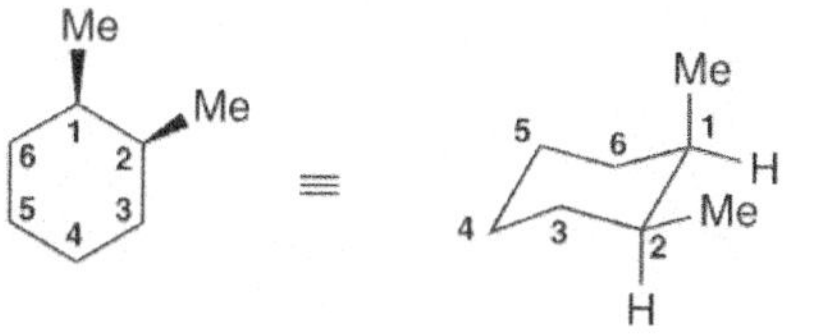

6.24 In the solution to problem 6.16, we already drew one of the two chair conformations for this compound, as shown again here.

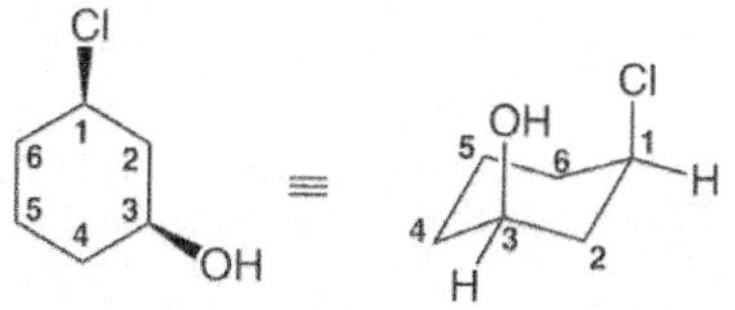

Note: These numbers are not in accordance with IUPAC rules, but that's okay, as explained in the solution to problem 6.16.

To draw the other chair conformation for this compound, we use a similar approach, beginning with a different chair skeleton. Both substituents are on wedges, so they are both up:

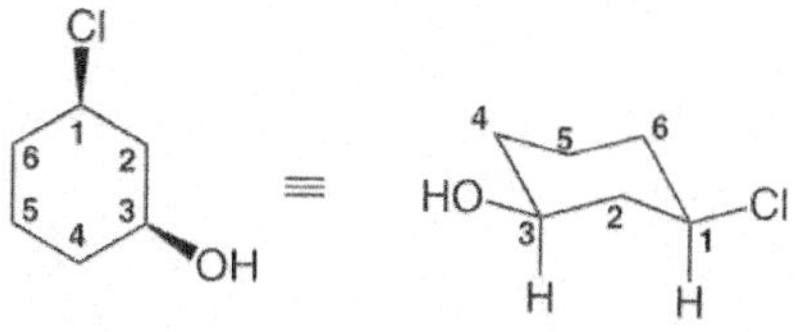

In summary, this compound has two chair conformations, as shown here:

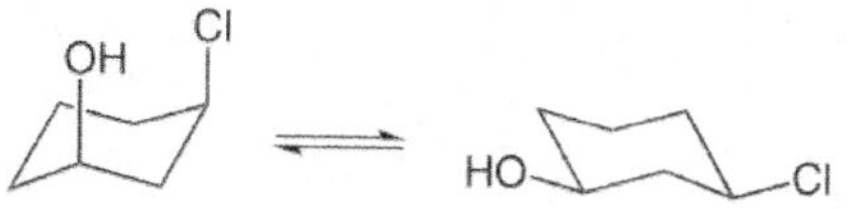

6.25 In the solution to problem 6.17, we already drew one of the two chair conformations for this compound, as shown again here.

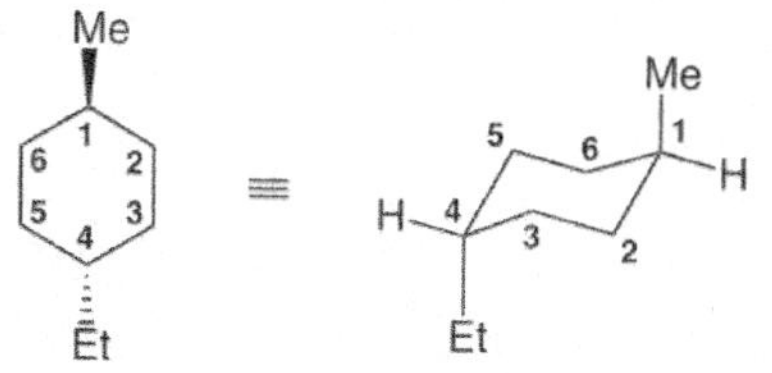

Note: These numbers are not in accordance with IUPAC rules, but that's okay, as explained in the solution to problem 6.17.

To draw the other chair conformation for this compound, we use a similar approach, beginning with a different chair skeleton. The methyl group is on a wedge, so it is up; while the ethyl group is on a dash, so it is down:

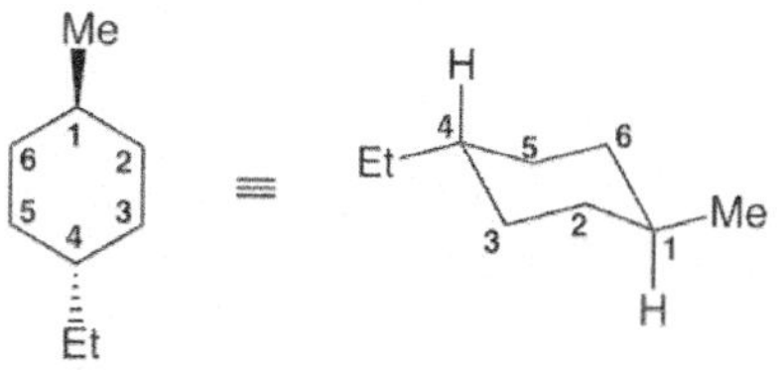

In summary, this compound has two chair conformations, as shown here:

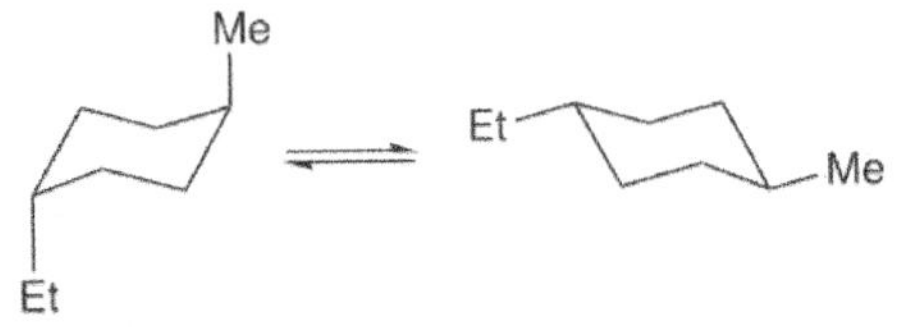

6.26 In the solution to problem 6.18, we already drew one of the two chair conformations for this compound, as shown again here:

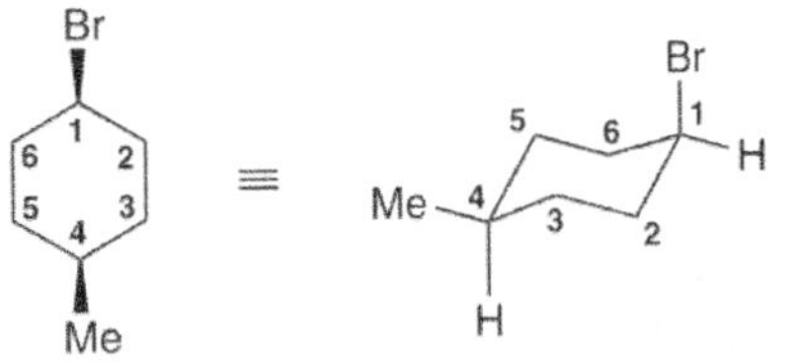

To draw the other chair conformation for this compound, we use a similar approach, beginning with a different chair skeleton. Both substituents are on wedges, so they are both up:

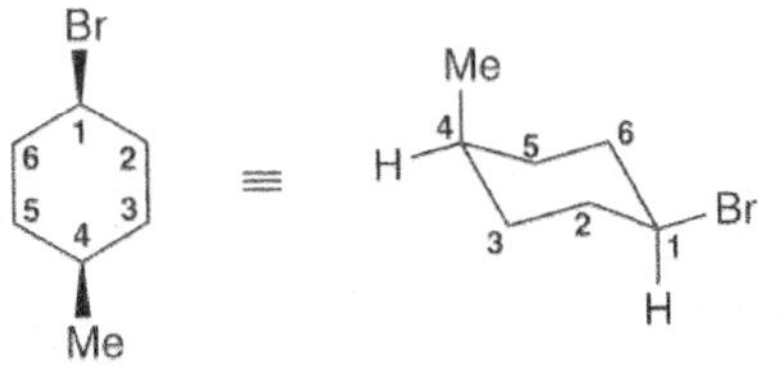

In summary, this compound has two chair conformations, as shown here:

6.27 In the solution to problem 6.19, we already drew one of the two chair conformations for this compound, as shown again here:

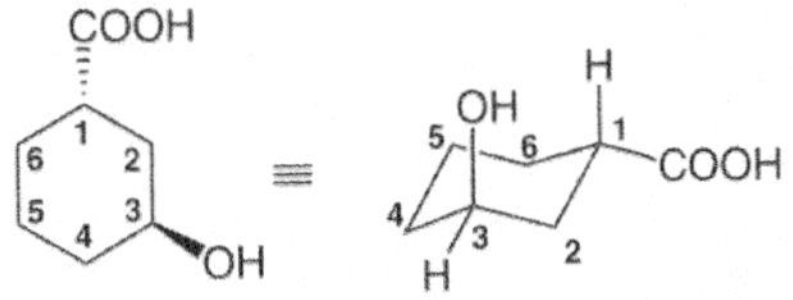

To draw the other chair conformation for this compound, we use a similar approach, beginning with a different chair skeleton. The COOH group is on a dash, so it is down; while the OH group is on a wedge, so it is up:

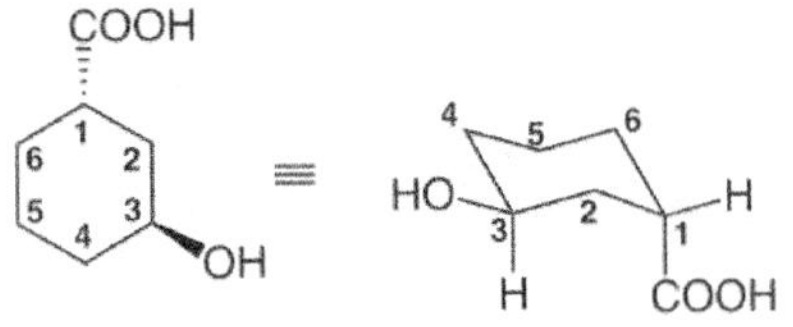

In summary, this compound has two chair conformations, as shown here:

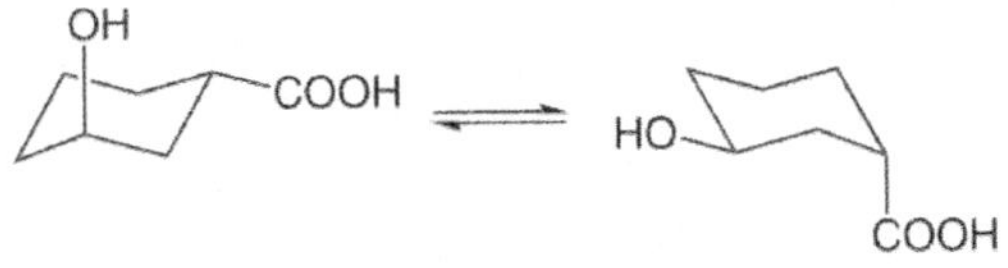

6.28 In the solution to problem 6.20, we already drew one of the two chair conformations for this compound, shown again here:

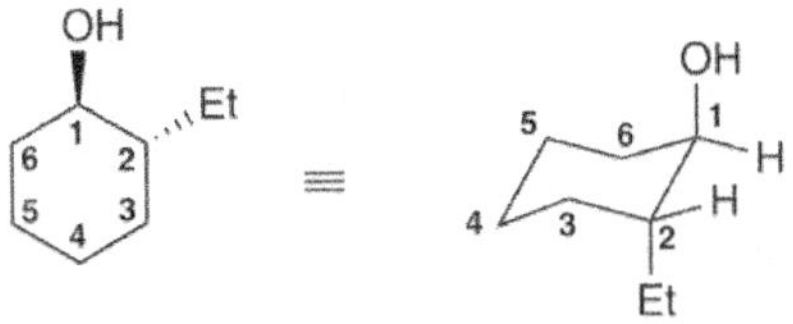

To draw the other chair conformation for this compound, we use a similar approach, beginning with a different chair skeleton. The OH group is on a wedge, so it is up; while the ethyl group is on a dash, so it is down:

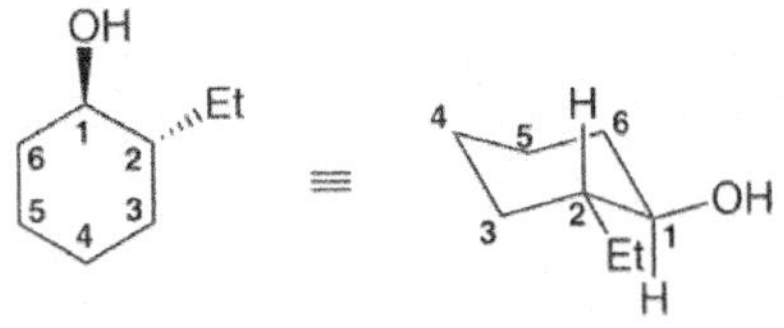

In summary, this compound has two chair conformations, as shown here:

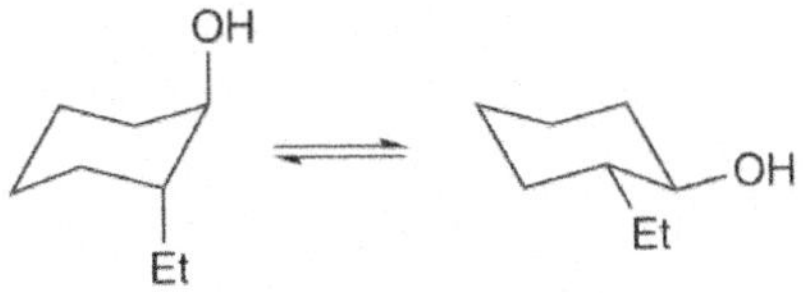

6.29 In the solution to problem 6.21, we already drew one of the two chair conformations for this compound, as shown again here.

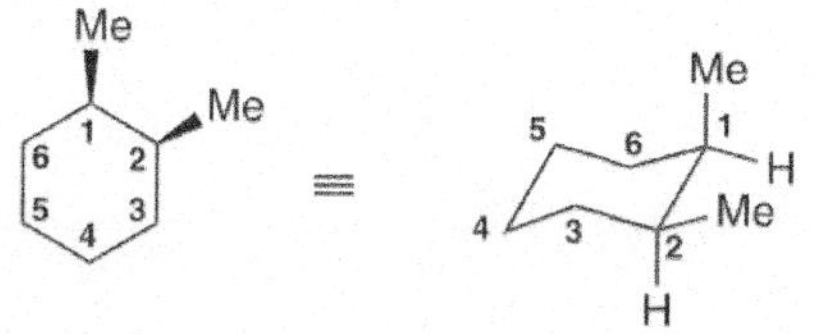

To draw the other chair conformation for this compound, we use a similar approach, beginning with a different chair skeleton. Both substituents are on wedges, so they are both up:

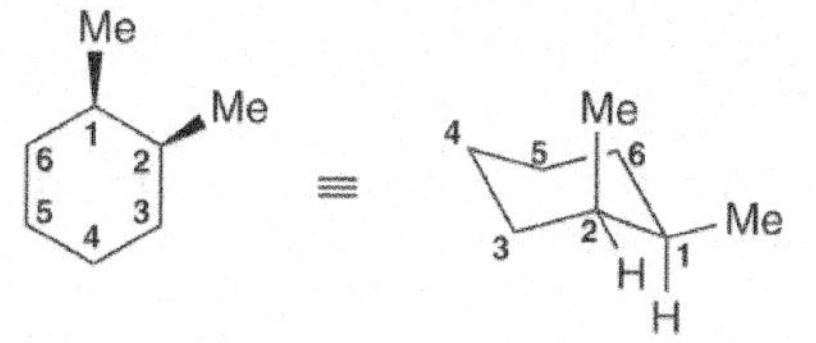

In summary, this compound has two chair conformations, as shown here:

6.31 We begin by placing numbers on the chair conformation that is shown in the problem statement. Start at the right side of the chair and continue numbering clockwise (shown below). With this numbering scheme, there is a methyl group pointing up at position 1, and there is a methyl group pointing up at position 2.

Next, draw the other chair skeleton, number it clockwise, starting on the right side, and then draw the two substituents—there is a methyl group pointing up at position 1, and there is a methyl group pointing up at position 2:

6.32 We begin by placing numbers on the chair conformation that is shown in the problem statement. Start at the right side of the chair and continue numbering clockwise (shown below). With this numbering scheme, there is a methyl group pointing down at position 1, and there is a methyl group pointing down at position 3.

Next, draw the other chair skeleton, number it clockwise, starting on the right side, and then draw the two substituents—there is a methyl group pointing down at position 1, and there is a methyl group pointing down at position 3:

6.33 We begin by placing numbers on the chair conformation that is shown in the problem statement. Start at the right side of the chair and continue numbering clockwise (shown below). With this numbering scheme, there is a methyl group pointing up at position 1, and there is a methyl group pointing down at position 2.

Next, draw the other chair skeleton, number it clockwise, starting on the right side, and then draw the two substituents—there is a methyl group pointing up at position 1, and there is a methyl group pointing down at position 2:

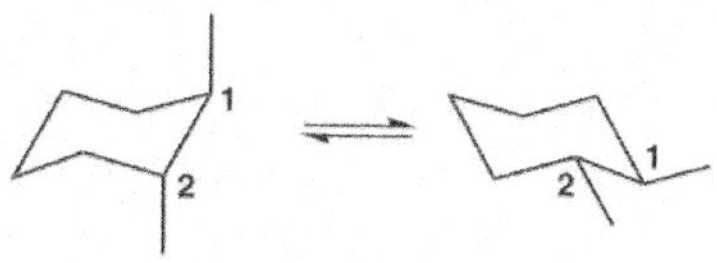

6.34 We begin by placing numbers on the chair conformation that is shown in the problem statement, starting with the carbon on the right-most side of the structure and continuing clockwise. With this numbering scheme, there is a methyl group pointing down at position 1, and there is a methyl group pointing up at position 6.

Next, draw the other chair skeleton, number it clockwise, and then draw the two substituents—there is a methyl group pointing down at position 1, and there is a methyl group pointing up at position 6:

Note: These numbers are not in accordance with IUPAC rules, which would dictate that this compound be called 1,2-dimethylcyclohexane, not 1,6-dimethylcyclohexane. Nevertheless, we are not naming this compound right now. We are merely drawing a chair conformation, and we are using this numbering system as a tool to help us draw the chair conformation correctly. Once we are finished, we can delete the numbers and draw only the chairs:

6.35 We begin by placing numbers on the chair conformation that is shown in the problem statement. Start at the right side of the chair and continue numbering clockwise (shown below). With this numbering scheme, there is a methyl group pointing up at position 1, and there is a methyl group pointing up at position 3.

Next, draw the other chair skeleton, number it clockwise, starting on the right side, and then draw the two substituents—there is a methyl group pointing up at position 1, and there is a methyl group pointing up at position 3:

6.36 We begin by placing numbers on the chair conformation that is shown in the problem statement. Start at the right side of the chair and continue numbering clockwise (shown below). With this numbering scheme, there is a methyl group pointing up at position 1, and there is a methyl group pointing down at position 3.

Next, draw the other chair skeleton, number it clockwise, starting on the right side, and then draw the two substituents—there is a methyl group pointing up at position 1, and there is a methyl group pointing down at position 3:

6.38 We begin by drawing both chair conformations:

The more stable chair conformation is the one in which both substituents occupy equatorial positions:

6.39 We begin by drawing both chair conformations:

The more stable chair conformation is the one in which the larger substituent (the isopropyl group) occupies an equatorial position:

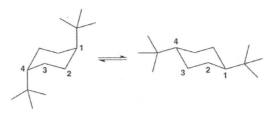

6.40 We begin by drawing both chair conformations:

The more stable chair conformation is the one in which both substituents occupy equatorial positions:

6.41 We begin by drawing both chair conformations:

The more stable chair conformation is the one in which both substituents occupy equatorial positions:

6.42 We begin by drawing both chair conformations:

The more stable chair conformation is the one in which both substituents occupy equatorial positions:

6.43 We begin by drawing both chair conformations:

The more stable chair conformation is the one in which the larger substituent (the isopropyl group) occupies an equatorial position:

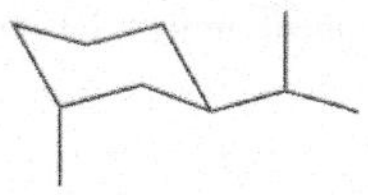

6.44 We begin by drawing both chair conformations:

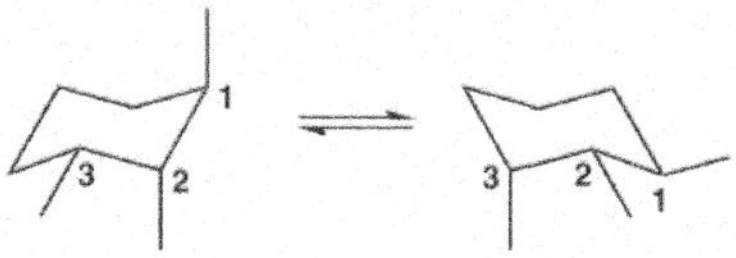

The more stable chair conformation is the one in which two of the three methyl groups occupy equatorial positions:

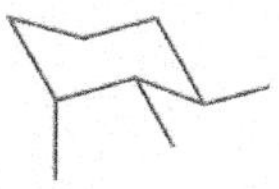

6.45 We begin by drawing both chair conformations:

The more stable chair conformation is the one in which two of the three methyl groups occupy equatorial positions.

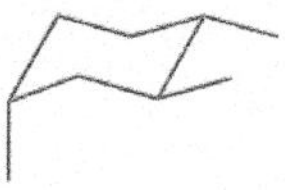

CHAPTER 7

7.2 The position highlighted below is a chiral center because it is connected to four different groups (a propyl group, an ethyl group, a methyl group, and H):

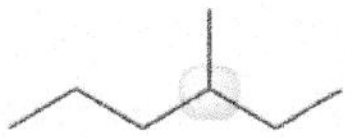

7.3 The position highlighted below is a chiral center because it is connected to four different groups (an isopropyl group, an ethyl group, a methyl group, and H):

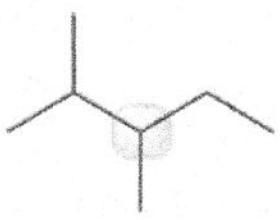

7.4 The position highlighted below is a chiral center because it is connected to four different groups (a cyclohexyl group, a methyl group, an OH group, and H):

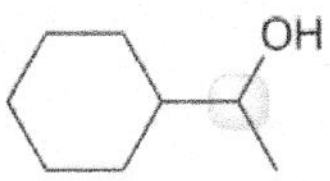

7.5 The position highlighted below is a chiral center because it is connected to four different groups (a phenyl group, an ethyl group, a methyl group, and H):

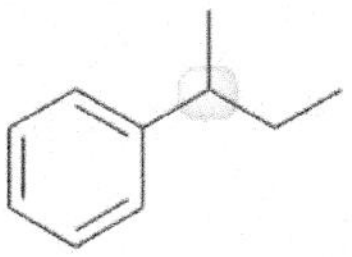

7.6 The position highlighted below is a chiral center because it is connected to four different groups:

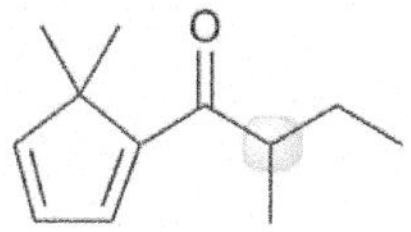

7.7 The position highlighted below is a chiral center because it is connected to four different groups:

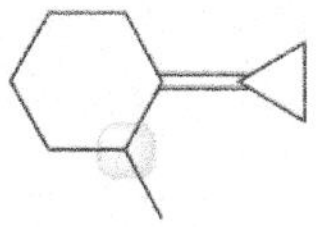

7.9 The positions highlighted below are chiral centers because each of them is connected to four different groups:

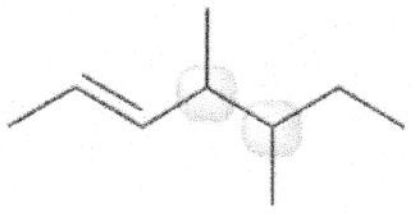

7.10 The positions highlighted below are chiral centers because each of them is connected to four different groups:

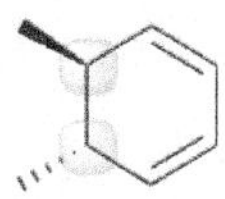

7.11 The positions highlighted below are chiral centers because each of them is connected to four different groups:

7.12 The position highlighted below is a chiral center because it is connected to four different groups:

7.13 The positions highlighted below are chiral centers because each of them is connected to four different groups:

7.14 The positions highlighted below are chiral centers because each of them is connected to four different groups:

7.15 The positions highlighted below are chiral centers because each of them is connected to four different groups:

7.17 The four groups are ranked according to atomic number. H has the lowest atomic number and is assigned the lowest priority (#4), while Cl has the highest atomic number and is assigned the highest priority (#1). The remaining two positions are both carbon atoms, so we make a list of the groups attached to each of these positions. The position on the right has the higher priority (#2) because (Cl, H, H) beats (C, H, H):

7.18 The four groups are ranked according to atomic number. H has the lowest atomic number and is assigned the lowest

priority (#4), while O has the highest atomic number and is assigned the highest priority (#1). The remaining two positions are both carbon atoms, so we make a list of the groups attached to each of these positions. The position on the left has the higher priority (#2) because (O, C, C) beats (C, C, C):

7.19 The four groups are ranked according to atomic number. H has the lowest atomic number and is assigned the lowest priority (#4), while S has the highest atomic number and is assigned the highest priority (#1). The remaining two positions are both carbon atoms, so we make a list of the groups attached to each of these positions. The position on the right has the higher priority (#2) because (C, C, C) beats (C, C, H):

7.21 The four groups are ranked according to atomic number. Br has the highest atomic number and is assigned the highest priority (#1), while Cl has the second-highest atomic number and is assigned the second-highest priority (#2). The remaining two positions are both carbon atoms, so we make a list of the groups attached to each of these positions. In this case, these lists are the same (C, H, H), so we continue to move farther away from the chiral center and repeat the process (making lists and comparing them). This analysis reveals that the position on the right has the higher priority (#3) because of the presence of the double bond: (C, C, H) beats (C, H, H):

7.22 The four groups are ranked according to atomic number. Cl has the highest atomic number and is assigned the highest priority (#1). The remaining three positions are carbon atoms, so we make a list of the groups attached to each of these positions. In this case, these lists are all the same (C, H, H), so we continue to move farther away from the chiral center and repeat the process (making lists and comparing them). This analysis reveals that the group on the dash has the higher priority (#2), while the group with the double bond has the next higher priority (#3) because the tie-breaking lists are as follows: (O, H, H) beats (C, C, H) which beats (C, H, H):

7.23 The four groups are ranked according to atomic number. H has the lowest atomic number and is assigned the lowest priority (#4), while O has the highest atomic number and is assigned the highest priority (#1). The remaining two positions are both carbon atoms, so we make a list of the groups attached to each of these positions. The position on the right has the higher priority (#2) because (C, C, H) beats (C, H, H):

7.24 The four groups are ranked according to atomic number. Br has the highest atomic number and is assigned the highest priority (#1). The remaining three positions are carbon atoms, so we make a list of the groups attached to each of these positions. The group on the wedge is assigned priority #2 because (C, C, H) beats (C, H, H) and (C, H, H). To compare the remaining two groups, we continue to move farther away from the chiral center and repeat the process (making lists and comparing them) until there is a difference. This analysis reveals that the group on the right has the higher priority (#3):

7.25 The four groups are ranked according to atomic number. H has the lowest atomic number and is assigned the lowest priority (#4). The remaining three positions are carbon atoms, so we make a list of the groups attached to each of these positions. When we compare these lists, we find that (C, C, H) beats (C, H, H), which beats (H, H, H). So the priorities are as follows:

7.26 The four groups are ranked according to atomic number. O has the highest atomic number and is assigned the highest

priority (#1). The remaining three positions are carbon atoms, so we make a list of the groups attached to each of these positions. The list for the methyl group is (H, H, H) so this group is assigned the lowest priority (#4). To compare the remaining two groups, we continue to move farther away from the chiral center and repeat the process (making lists and comparing them) until there is a difference. This analysis reveals that the group on the left has the higher priority (#2) because when we arrive at the double bond, (C, C, C) beats (C, C, H):

7.27 Group #4 is not on a dash, so we switch group #4 with group #2, which now places group #4 on the dash. Now we are ready to consider the sequence 1-2-3, which is clockwise here:

A clockwise sequence of 1-2-3 would be the *R* configuration, except that we switched two groups (which changed the configuration), and we must take this into account. So the answer must be *S*.

7.28 Group #4 is not on a dash, so we switch group #4 with group #3, which now places group #4 on the dash. Now we are ready to consider the sequence 1-2-3, which is counterclockwise here:

A counterclockwise sequence of 1-2-3 would be the *S* configuration, except that we switched two groups (which changed the configuration), and we must take this into account. So the answer must be *R*.

7.29 Group #4 is not on a dash, so we switch group #4 with group #1, which now places group #4 on the dash. Now we are ready to consider the sequence 1-2-3, which is clockwise here:

A clockwise sequence of 1-2-3 would be the **R** configuration, except that we switched two groups (which changed the configuration), and we must take this into account. So the answer must be **S**.

7.30 Group #4 is already on a dash, so we are ready to consider the sequence 1-2-3, which is counterclockwise here. A counterclockwise sequence of 1-2-3 indicates the **S** configuration.

7.31 Group #4 is not on a dash, so we switch group #4 with group #3, which now places group #4 on the dash. Now we are ready to consider the sequence 1-2-3, which is counterclockwise here.

A counterclockwise sequence of 1-2-3 would be the **S** configuration, except that we switched two groups (which changed the configuration), and we must take this into account. So the answer must be **R**.

7.32 Group #4 is not on a dash, so we switch group #4 with group #1, which now places group #4 on the dash. Now we are ready to consider the sequence 1-2-3, which is counterclockwise here.

A counterclockwise sequence of 1-2-3 would be the **S** configuration, Except that we switched two groups (which changed the configuration), and we must take this into account. So the answer must be **R**.

7.33 Group #4 is not on a dash, so we switch group #4 with group #1, which now places group #4 on the dash. Now we are ready to consider the sequence 1-2-3, which is counterclockwise here.

A counterclockwise sequence of 1-2-3 would be the **S** configuration, except that we switched two groups (which changed the configuration), and we must take this into account. So the answer must be **R**.

7.34 Group #4 is not on a dash, so we switch group #4 with group #3, which now places group #4 on the dash. Now we are ready to consider the sequence 1-2-3, which is counterclockwise here.

A counterclockwise sequence of 1-2-3 would be the **S** configuration, except that we switched two groups (which changed the configuration), and we must take this into account. So the answer must be **R**.

7.35 Group #4 is not on a dash, so we switch group #4 with group #3, which now places group #4 on the dash. Now we are ready to consider the sequence 1-2-3, which is clockwise here.

A clockwise sequence of 1-2-3 would be the **R** configuration, except that we switched two groups (which changed the configuration), and we must take this into account. So the answer must be **S**.

7.37 This compound has one chiral center. To assign the configuration, we rank the four groups (attached to the chiral center) according to atomic number. These priorities are shown below. Notice that priority #4 is already on a dash, so we simply consider the sequence 1-2-3, which is counterclockwise. A counterclockwise sequence of 1-2-3 indicates the **S** configuration.

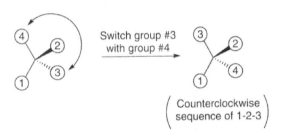

7.38 This compound has one chiral center. To assign the configuration, we rank the four groups (attached to the chiral center) according to atomic number. These priorities are shown below. Notice that priority #4 is already on a dash, so we simply consider the sequence 1-2-3, which is counterclockwise. A counterclockwise sequence of 1-2-3 indicates the *S* configuration.

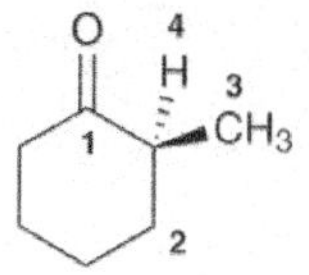

7.39 This compound has one chiral center. To assign the configuration, we rank the four groups (attached to the chiral center) according to atomic number. These priorities are shown below. Notice that priority #4 is already on a dash, so we simply consider the sequence 1-2-3, which is clockwise. A clockwise sequence of 1-2-3 indicates the *R* configuration.

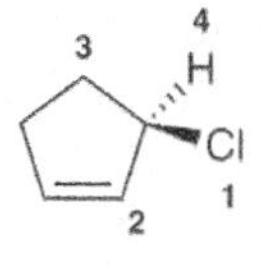

7.40 This compound has one chiral center. To assign the configuration, we rank the four groups (attached to the chiral center) according to atomic number. These priorities are shown below. Notice that priority #4 is already on a dash, so we simply consider the sequence 1-2-3, which is clockwise. A clockwise sequence of 1-2-3 indicates the *R* configuration.

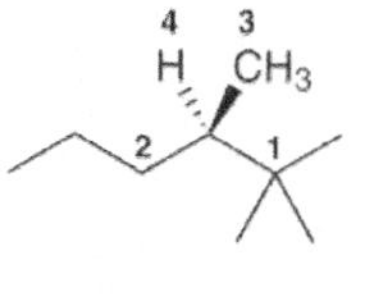

7.41 This compound has two chiral centers. To assign the configuration for each chiral center, we rank the four groups (attached to each chiral center) according to atomic number. The priorities for each chiral center are shown below. For the first chiral center, notice that priority #4 is already on a dash, so we simply consider the sequence 1-2-3, which is clockwise, indicating the *R* configuration. For the second chiral center, notice that priority #4 is not on a dash. So, we must switch it with the substituent that is on the dash (Br), giving a counterclockwise sequence of 1-2-3. A counterclockwise sequence of 1-2-3 would be the *S* configuration, except that we switched two groups (which changed the configuration), and we must take this into account. So the second chiral center has the *R* configuration.

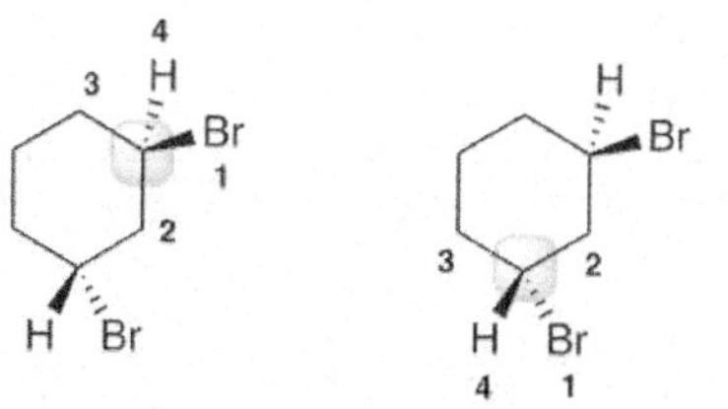

7.42 This compound has two chiral centers. To assign the configuration for each chiral center, we rank the four groups (attached to each chiral center) according to atomic number. The priorities for each chiral center are shown below. For each chiral center, notice that priority #4 is already on a dash, so we simply consider the sequence 1-2-3, which is clockwise in each case. This indicates that both chiral centers have the *R* configuration.

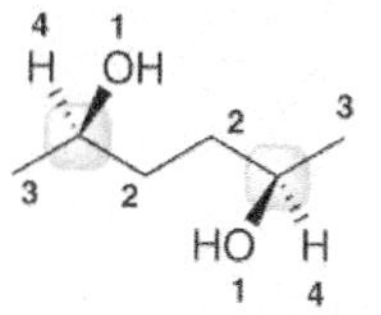

7.44 This compound has a stereoisomeric C=C double bond with the Z configuration. This configuration must be indicated at the beginning of the name of the compound, like this:

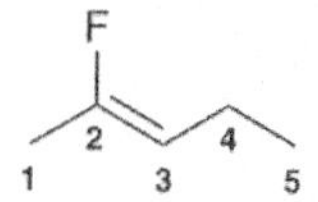

(*Z*)-2-fluoropent-2-ene

7.45 This compound has two chiral centers, and the configuration of each must be indicated at the beginning of the name of the compound, like this:

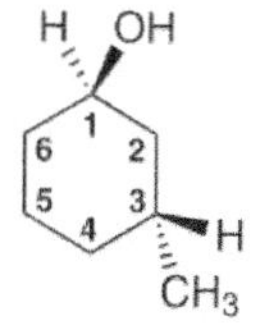

(1*R*,3*R*)-3-methylcyclohexanol

7.46 This compound has one chiral center, and the configuration must be indicated at the beginning of the name of the compound, like this:

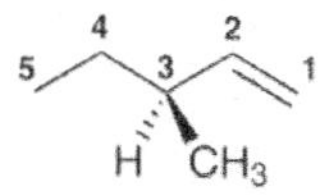

(*S*)-3-methylpent-1-ene

7.47 This compound has a stereoisomeric C=C double bond with the *E* configuration. This configuration must be indicated at the beginning of the name of the compound, like this:

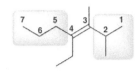

(*E*)-4-ethyl-2,3-dimethylhept-3-ene

7.48 This compound has two stereoisomeric C=C double bonds, and the configuration of each must be indicated at the beginning of the name of the compound, like this:

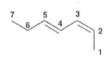

(2*Z*,4*E*)-hepta-2,4-diene

7.49 This compound has four stereoisomeric C=C double bonds and the configuration of each must be indicated at the beginning of the name of the compound, like this:

(2*E*,4*Z*,6*Z*,8*E*)-deca-2,4,6,8-tetraene

7.51 Redraw the molecule, but convert all wedges into dashes:

7.52 Redraw the molecule, but convert all wedges into dashes, and convert all dashes into wedges:

7.53 Redraw the molecule, but convert all wedges into dashes, and convert all dashes into wedges:

7.54 Redraw the molecule, but convert all wedges into dashes, and convert all dashes into wedges:

7.55 Redraw the molecule, but convert the wedge into a dash:

7.56 Redraw the molecule, but convert the wedge into a dash:

7.58 This type of drawing (a chair conformation) does not show any wedges or dashes. In such a case, the easiest way to draw the enantiomer is to draw a horizontal reflection of the original drawing, like this:

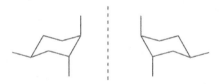

7.59 This type of drawing (of a bicyclic compound) does not show any wedges or dashes. In such a case, the easiest way to draw the enantiomer is to draw a horizontal reflection of the original drawing, like this:

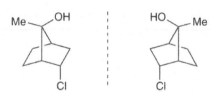

7.60 This type of drawing (of a bicyclic compound) does not show any wedges or dashes. In such a case, the easiest way to draw the

enantiomer is to draw a horizontal reflection of the original drawing, like this:

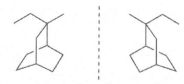

7.61 This type of drawing (a chair conformation) does not show any wedges or dashes. In such a case, the easiest way to draw the enantiomer is to draw a horizontal reflection of the original drawing, like this:

7.62 This type of drawing (of a bicyclic compound) does not show any wedges or dashes. In such a case, the easiest way to draw the enantiomer is to draw a horizontal reflection of the original drawing, like this:

7.63 This type of drawing (a chair conformation) does not show any wedges or dashes. In such a case, the easiest way to draw the enantiomer is to draw a horizontal reflection of the original drawing, like this:

7.65 These compounds are nonsuperimposable stereoisomers, and they are mirror images of each other. Therefore, they are enantiomers.

7.66 These compounds are nonsuperimposable stereoisomers, but they are not mirror images of each other. Therefore, they are diastereomers.

7.67 These compounds are nonsuperimposable stereoisomers, and they are mirror images of each other. Therefore, they are enantiomers.

7.68 These compounds are nonsuperimposable stereoisomers, and they are mirror images of each other. Therefore, they are enantiomers.

7.69 These compounds are nonsuperimposable stereoisomers, but they are not mirror images of each other. Therefore, they are diastereomers.

7.70 These compounds are nonsuperimposable stereoisomers, but they are not mirror images of each other. Therefore, they are diastereomers.

7.72 This compound has a plane of symmetry, chopping the molecule in half (vertically). The methyl group (at the top of the molecule) is chopped in half by the plane, and the methyl group at the bottom of the molecule is also chopped in half. The methyl groups on the sides reflect each other. Since this compound has chiral centers and also has a plane of symmetry, this compound is classified as a *meso* compound.

7.73 This compound lacks a plane of symmetry. If both Br substituents were on the same side of the ring (both on wedges, or both on dashes), then the compound would have a plane of symmetry. But this compound does not have such a plane. This compound is not a *meso* compound.

If you try to draw a plane of symmetry, chopping the five membered-ring in half (perpendicular to the plane of the ring), you should hopefully see that the bromine atoms do not reflect each other about that plane. If you are having trouble seeing this, I recommend building a molecular model of this compound (using any one of the many commercially available molecular model sets).

7.74 This compound has a plane of symmetry, chopping the molecule in half (vertically). The OH groups reflect each other, as do the methyl groups. Since this compound has chiral centers and also has a plane of symmetry, this compound is classified as a *meso* compound.

7.76 We redraw one of the horizontal lines as a wedge, and we redraw one of the vertical lines as a dash, and then we assign priorities (1–4):

Using these priorities, we determine that the chiral center has the *R* configuration.

The enantiomer is the mirror image and can be drawn by switching the location of the horizontal groups (H and Br):

$$
\begin{array}{cc}
\text{Et} & \text{Et} \\
\text{H}-\!\!\!\!-\text{Br} & \text{Br}-\!\!\!\!-\text{H} \\
\text{Me} & \text{Me} \\
R & S
\end{array}
$$

7.77 We redraw one of the horizontal lines as a wedge, and we redraw one of the vertical lines as a dash, and then we assign priorities (1–4):

$$
\begin{array}{ccc}
\text{CH}_2\text{OH} & & \overset{2}{\text{CH}_2\text{OH}} \\
\text{Me}-\!\!\!\!-\text{Br} & \equiv & \overset{4}{\text{Me}}\!\!\blacktriangleright\!\!-\!\!\!\!-\!\!\text{Br}\,{}^{1} \\
\text{Et} & & \underset{3}{\text{Et}}
\end{array}
$$

Using these priorities, we determine that the chiral center has the R configuration.

The enantiomer is the mirror image and can be drawn by switching the location of the horizontal groups (Me and Br):

$$
\begin{array}{cc}
\text{CH}_2\text{OH} & \text{CH}_2\text{OH} \\
\text{Me}-\!\!\!\!-\text{Br} & \text{Br}-\!\!\!\!-\text{Me} \\
\text{Et} & \text{Et} \\
R & S
\end{array}
$$

7.78 We redraw one of the horizontal lines as a wedge, and we redraw one of the vertical lines as a dash, and then we assign priorities (1–4):

$$
\begin{array}{ccc}
\text{O}\!\!=\!\!\text{CH} & & \text{O}\overset{2}{=}\!\!\text{CH}\,{}^{1} \\
\text{H}-\!\!\!\!-\text{OH} & \equiv & \overset{4}{\text{H}}\!\!\blacktriangleright\!\!-\!\!\!\!-\!\!\text{OH} \\
\text{CH}_2\text{OH} & & \underset{3}{\text{CH}_2\text{OH}}
\end{array}
$$

Using these priorities, we determine that the chiral center has the R configuration.

The enantiomer is the mirror image and can be drawn by switching the location of the horizontal groups (H and OH):

$$
\begin{array}{cc}
\text{O}\!\!=\!\!\text{CH} & \text{O}\!\!=\!\!\text{CH} \\
\text{H}-\!\!\!\!-\text{OH} & \text{HO}-\!\!\!\!-\text{H} \\
\text{CH}_2\text{OH} & \text{CH}_2\text{OH} \\
R & S
\end{array}
$$

7.79 We treat each chiral center separately. For each chiral center, we redraw one of the horizontal lines as a wedge, and we redraw one of the vertical lines as a dash, and then we assign priorities (1–4), and finally assign a configuration. This process is repeated for each chiral center, giving the following configurations:

$$
\begin{array}{c}
\text{COOH} \\
\text{H}-\!\!\overset{R}{-}\!\!-\text{OH} \\
\text{HO}-\!\!\overset{S}{-}\!\!-\text{H} \\
\text{CH}_2\text{OH}
\end{array}
$$

The enantiomer is the mirror image and can be drawn by switching the location of all horizontal groups, like this:

$$
\begin{array}{c}
\text{COOH} \\
\text{HO}-\!\!\!\!-\text{H} \\
\text{H}-\!\!\!\!-\text{OH} \\
\text{CH}_2\text{OH}
\end{array}
$$

7.80 We treat each chiral center separately. For each chiral center, we redraw one of the horizontal lines as a wedge, and we redraw one of the vertical lines as a dash, and then we assign priorities (1–4), and finally assign a configuration. This process is repeated for each chiral center, giving the following configurations:

$$
\begin{array}{c}
\text{COOH} \\
\text{H}-\!\!\overset{R}{-}\!\!-\text{OH} \\
\text{HO}-\!\!\overset{S}{-}\!\!-\text{H} \\
\text{H}-\!\!\overset{R}{-}\!\!-\text{OH} \\
\text{CH}_2\text{OH}
\end{array}
$$

The enantiomer is the mirror image and can be drawn by switching the location of all horizontal groups, like this:

$$
\begin{array}{c}
\text{COOH} \\
\text{HO}-\!\!\!\!-\text{H} \\
\text{H}-\!\!\!\!-\text{OH} \\
\text{HO}-\!\!\!\!-\text{H} \\
\text{CH}_2\text{OH}
\end{array}
$$

7.81 We treat each chiral center separately. For each chiral center, we redraw one of the horizontal lines as a wedge, and we redraw one of the vertical lines as a dash, and then we assign priorities (1–4), and finally assign a configuration. This process is repeated for each chiral center, giving the following configurations:

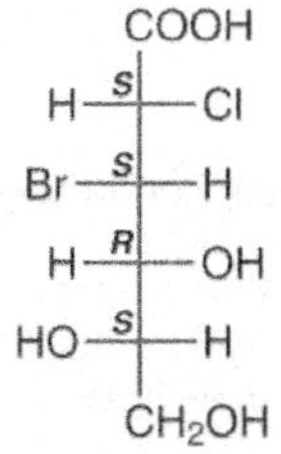

COOH

H ---^{S}--- Cl

Br ---^{S}--- H

H ---^{R}--- OH

HO ---^{S}--- H

CH$_2$OH

COOH

Cl --- H

H --- Br

HO --- H

H --- OH

CH$_2$OH

The enantiomer is the mirror image and can be drawn by switching the location of all horizontal groups, like this:

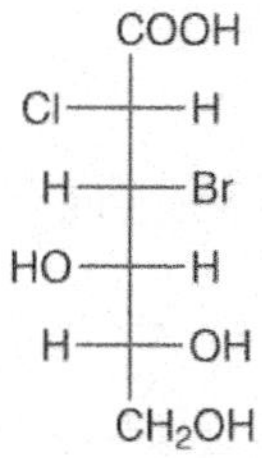

CHAPTER 8

8.2 To identify nucleophilic centers, we look for atoms containing either lone pairs or pi bonds. In this structure, the oxygen atom and the sulfur atom each has lone pairs and are therefore nucleophilic centers. Furthermore, the triple bond is a nucleophilic center because of its pi bonds. In summary, this compound has three nucleophilic centers, highlighted below.

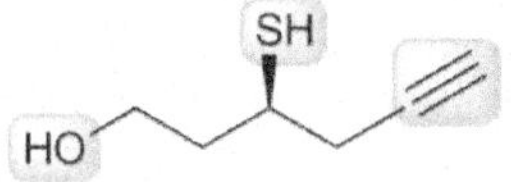

8.3 To identify nucleophilic centers, we look for atoms containing either lone pairs or pi bonds. This structure has one nucleophilic center—a carbon atom with a lone pair, highlighted below:

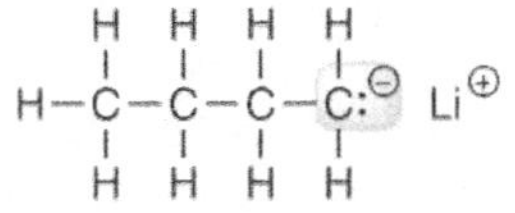

8.4 To identify nucleophilic centers, we look for atoms containing either lone pairs or pi bonds. Each of the oxygen atoms has lone pairs, so each oxygen atom is a nucleophilic center. Furthermore, the double bond is a nucleophilic center because of its pi bond. In summary, this compound has three nucleophilic centers, highlighted below:

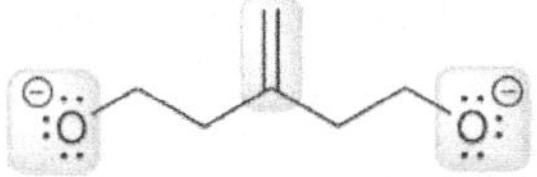

8.5 To identify nucleophilic centers, we look for atoms containing either lone pairs or pi bonds. Nitrogen has a lone pair, so it is a nucleophilic center. Furthermore, the double bond is a nucleophilic center because of its pi bond. In summary, this compound has two nucleophilic centers, highlighted below:

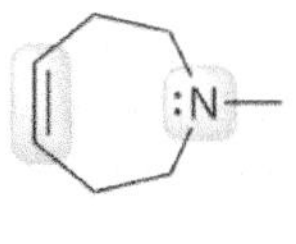

8.6 To identify nucleophilic centers, we look for atoms containing either lone pairs or pi bonds. The double bond is a nucleophilic center because of its pi bond, and the triple bond is also a nucleophilic center because of its pi bonds. In summary, this compound has two nucleophilic centers, highlighted below:

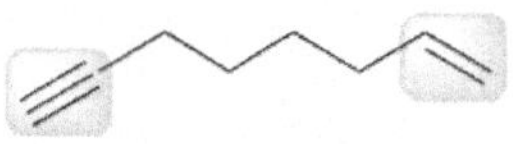

8.7 To identify nucleophilic centers, we look for atoms containing either lone pairs or pi bonds. In this structure, the sulfur atom has lone pairs. Furthermore, the double bond is a nucleophilic center because of its pi bond. In summary, this compound has two nucleophilic centers, highlighted below:

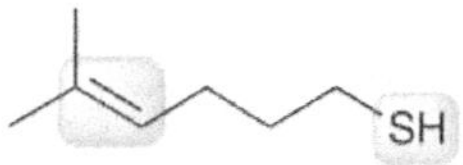

8.8 The positively charged carbon atom, highlighted, is an electrophilic center.

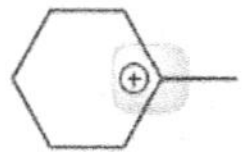

8.9 This structure has two electrophilic centers, both highlighted below. The carbon atom next to bromine is an electrophilic center because of the inductive effect of the neighboring electronegative bromine atom. And the carbon atom of the C=O group is also an electrophilic center because of induction and resonance.

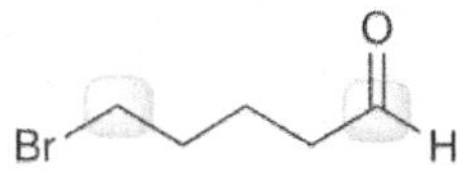

8.10 This structure has three electrophilic centers, highlighted below. Each of the carbon atoms connected to an oxygen atom is an electrophilic center because of the inductive effect of the oxygen atom. The central carbon atom (of the C=O bond) is also electrophilic because of resonance (in addition to the inductive effects of the neighboring oxygen atoms). This central carbon atom is the most electrophilic position because of resonance.

8.11 This structure has one electrophilic center, highlighted below. This position is electrophilic because of both induction and resonance.

8.12 This structure has two electrophilic centers, both highlighted below. The carbon atom from each C=O group is an electrophilic center because of both induction and resonance.

8.13 This structure has two electrophilic centers, both highlighted below. Each of the carbon atoms next to chlorine is an electrophilic center because of the inductive effect of the neighboring electronegative chlorine atom.

8.14 We begin by drawing the resonance structures for this compound:

Notice the highlighted regions above. These resonance structures reveal that this compound has two positions that are poor in electron density ($\delta+$). These two positions, highlighted below, are electrophilic centers:

8.15 In this case, hydroxide is attacking an electrophile, so hydroxide is functioning as a nucleophile.

8.16 In this case, hydroxide is removing a proton, so hydroxide is functioning as a base.

8.17 In this case, hydroxide is removing a proton, so hydroxide is functioning as a base.

8.18 In this case, hydroxide is attacking an electrophile, so hydroxide is functioning as a nucleophile.

8.19 In this case, hydroxide is removing a proton, so hydroxide is functioning as a base.

8.20 In this case, hydroxide is attacking an electrophile, so hydroxide is functioning as a nucleophile.

8.22 In the first step of this mechanism, hydroxide functions as a base and removes a proton, to generate a resonance-stabilized anion (proton transfer). Then, this anion functions as a base and removes a proton from water (proton transfer).

In summary, the sequence of arrow-pushing patterns is: 1) proton transfer, followed by 2) proton transfer.

8.23 In the first step of this mechanism, hydroxide functions as a nucleophile and attacks an electrophile (nucleophilic attack), with the simultaneous loss of a leaving group (the C—O bond is broken). So the first step of the mechanism is a concerted process. Then, in the second step of the mechanism, an alkoxide ion (RO^-) functions as a base and removes a proton from water (proton transfer).

In summary, the sequence of arrow-pushing patterns is: 1) nucleophilic attack and loss of a leaving group in a concerted fashion, followed by 2) proton transfer.

8.24 In the first step of this mechanism, the oxygen atom functions as a base and removes a proton from HCl (proton transfer). The resulting cation is then attacked by a chloride ion (nucleophilic attack), with the simultaneous loss of a leaving group (the C—O bond is broken), in a concerted fashion.

In summary, the sequence of arrow-pushing patterns is: 1) proton transfer, followed by 2) nucleophilic attack and loss of a leaving group in a concerted fashion.

8.25 In the first step of this mechanism, a cyanide ion ($N\equiv C^-$) functions as a nucleophile and attacks an electrophile (nucleophilic attack). Then, in the second step of the mechanism, an alkoxide ion (RO^-) functions as a base and removes a proton from HCN (proton transfer).

In summary, the sequence of arrow-pushing patterns is: 1) nucleophilic attack, followed by 2) proton transfer.

8.26 In the first step of this mechanism, the ester functions as a base and removes a proton (proton transfer). Then, in the second step, ethanol functions as a nucleophile and attacks an electrophile (nucleophilic attack). Then, in the third step, ethanol functions as a base and removes a proton (proton transfer). In the fourth step, an oxygen atom functions as a base and removes a proton (proton transfer). In the fifth step, methanol is expelled as a leaving group (loss of a leaving group). And finally, in the last step, ethanol functions as a base and removes a proton (proton transfer).

In summary, the sequence of arrow-pushing patterns is: 1) proton transfer, 2) nucleophilic attack, 3) proton transfer, 4) proton transfer, 5) loss of a leaving group, and 6) proton transfer.

8.27 In the first step of this mechanism, the alcohol functions as a base and removes a proton from HCl (proton transfer). Then, in the second step, water is expelled as a leaving group (loss of a leaving group). The third step is a carbocation rearrangement to give a resonance-stabilized intermediate. Finally, in the last step, the resonance-stabilized intermediate is attacked by a chloride ion (nucleophilic attack).

In summary, the sequence of arrow-pushing patterns is: 1) proton transfer, 2) loss of a leaving group, 3) carbocation rearrangement, and 4) nucleophilic attack.

8.29 The carbocation is secondary, and it can undergo a hydride shift, as shown below, to give a more stable, tertiary carbocation:

Hydride shift

Secondary → Tertiary

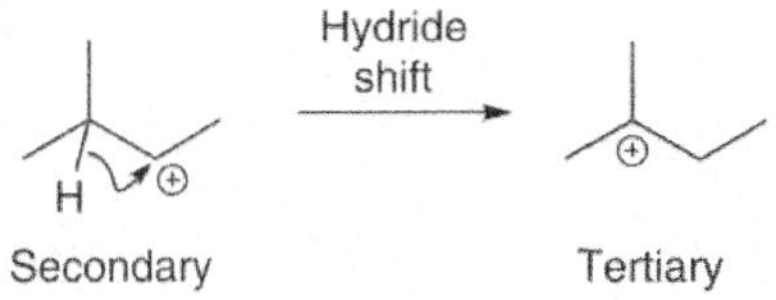

8.30 This carbocation is secondary, but it cannot undergo a carbocation rearrangement to generate a more stable carbocation. Indeed, a hydride shift would generate a less stable (primary) carbocation, which does not occur.

8.31 The carbocation is secondary, and it can undergo a methyl shift, as shown here, to give a more stable, tertiary carbocation:

Methyl shift

Secondary → Tertiary

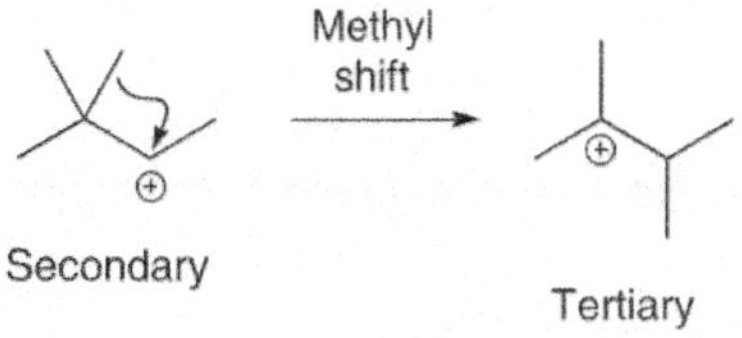

8.32 The carbocation is secondary, and it can undergo a hydride shift, as shown below, to give a more stable, secondary allylic carbocation, which is resonance-stabilized:

Hydride shift

Secondary → Secondary allylic

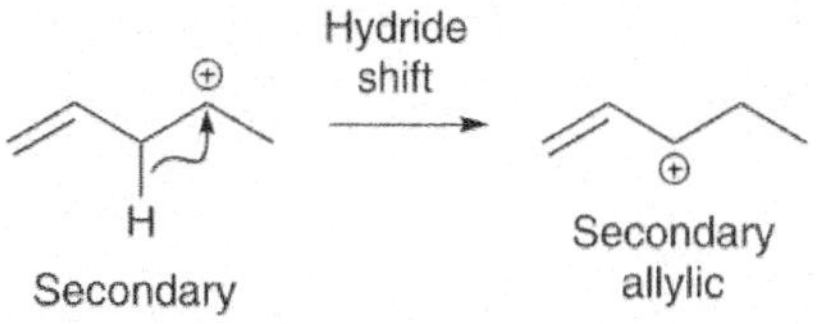

8.33 This carbocation is tertiary, and it cannot undergo a carbocation rearrangement to generate a more stable carbocation. Indeed, a hydride shift would generate a less stable (primary) carbocation, which does not occur.

8.34 The carbocation is secondary, and it can undergo a hydride shift, as shown below, to give a more stable, secondary benzylic carbocation, which is resonance-stabilized:

Hydride shift

Secondary → Secondary benzylic

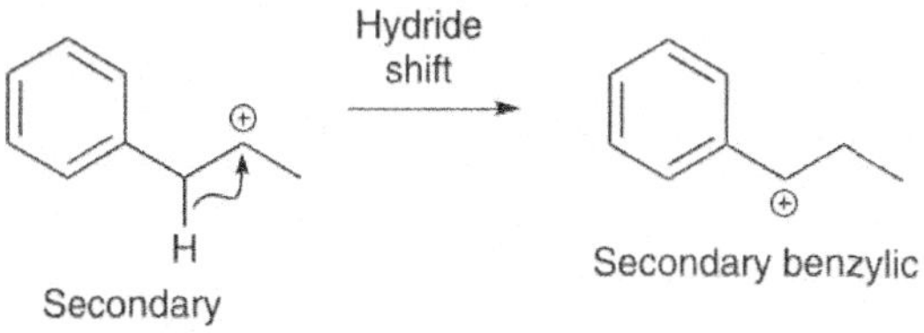

CHAPTER 9

9.2 The substrate is secondary, so we predict an S_N2 reaction.

9.3 The substrate is primary, so we predict an S_N2 reaction.

9.4 The substrate is secondary, so we predict an S_N2 reaction.

9.5 The substrate is tertiary, so we predict an S_N1 reaction.

9.7 The leaving group (Cl^-) is connected directly to the benzene ring, so the leaving group does not occupy a benzylic position. In order to be benzylic, there must be an sp^3 hybridized carbon between the benzene ring and the leaving group. Loss of the leaving group, in this case, will not generate a resonance-stabilized carbocation.

Not stabilized by resonance

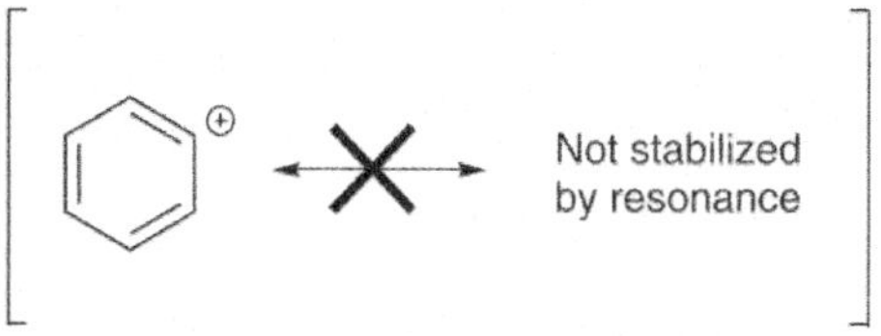

9.8 Loss of the leaving group (Cl^-) will generate a resonance-stabilized carbocation (a benzylic carbocation).

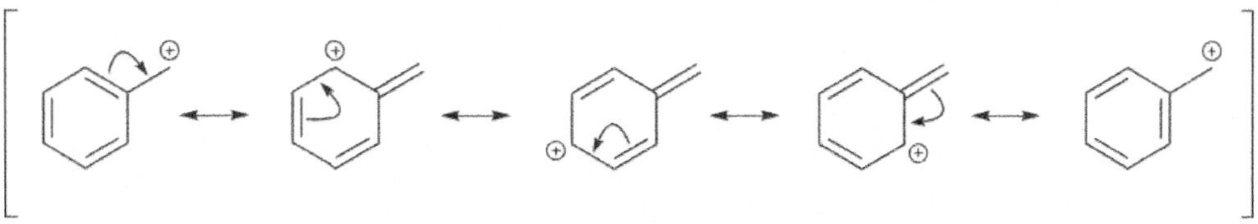

9.9 The leaving group (Br⁻) is connected directly to the double bond, so the leaving group does not occupy an allylic position. In order to be allylic, there must be an *sp³* hybridized carbon between the double bond and the leaving group. Loss of the leaving group, in this case, will not generate a resonance-stabilized carbocation.

Not stabilized by resonance

9.10 Loss of the leaving group (Br⁻) will generate a resonance-stabilized carbocation (an allylic carbocation):

9.12 This reagent is an alkoxide ion (RO⁻), which is a strong nucleophile, so we expect an S_N2 reaction.

9.13 This reagent is an alcohol (ROH), which is a weak nucleophile, so we expect an S_N1 reaction.

9.14 This reagent is an alcohol (ROH), which is a weak nucleophile, so we expect an S_N1 reaction.

9.15 The reagent is bromide (Br⁻), which is a strong nucleophile, so we expect an S_N2 reaction.

9.16 The reagent is hydroxide (HO⁻), which is a strong nucleophile, so we expect an S_N2 reaction.

9.17 The reagent is cyanide (N≡C⁻), which is a strong nucleophile, so we expect an S_N2 reaction.

9.19 A mesylate group (on the right side of the structure) is a good leaving group. Ethoxide (EtO⁻) is a very poor leaving group. So this structure has only one leaving group (a mesylate group).

9.20 This structure has two leaving groups (I⁻ and H_2O), both of which are very good leaving groups. Since HI ($pK_a = -11$) is a stronger acid than H_3O^+ ($pK_a = -2$), we can conclude that I⁻ is an even better leaving group than H_2O.

9.21 A tosylate group (the entire left half of the structure) is a good leaving group. H_2N^- is not a leaving group. So this structure has only one leaving group (a tosylate group).

9.22 Chloride is a good leaving group. Ethoxide (EtO⁻) is a very poor leaving group. So this structure has only one leaving group (chloride).

9.23 This structure has two leaving groups (Br⁻ and Cl⁻), both of which are very good leaving groups. Since HBr ($pK_a = -9$) is a

stronger acid than HCl ($pK_a = -7$), we can conclude that Br⁻ is an even better leaving group than Cl⁻, although they are both certainly very good leaving groups.

9.24 This structure has two leaving groups (Br⁻ and TsO⁻), both of which are very good leaving groups. Since HBr ($pK_a = -9$) is a stronger acid than TsOH ($pK_a = -3$), we can conclude that Br⁻ is an even better leaving group than TsO⁻, although they are both certainly very good leaving groups.

9.25 3-Iodo-3-methylpentane is a tertiary substrate with a very good leaving group (iodide) and is therefore expected to undergo an S_N1 reaction when treated with a nucleophile. In contrast, 3-methoxy-3-methylpentane is a tertiary substrate with a very poor leaving group (MeO⁻) and is therefore less likely to undergo an S_N1 reaction when treated with a nucleophile.

3-Methoxy-3-methylpentane   3-Iodo-3-methylpentane

9.26 The use of HCl will protonate the OH group and turn it into an excellent LG (H_2O), while also serving as a source of chloride ions for the desired substitution reaction.

9.28 Common examples of polar aprotic solvents include acetone, DME, DMSO, and DMF.

9.30 The substrate is secondary, the reagent (HS⁻) is a strong nucleophile, and the solvent is polar aprotic. All of these factors favor an S_N2 reaction.

9.31 The substrate is tertiary, and the reagent (H_2O) is a weak nucleophile. Both of these factors favor an S_N1 reaction.

9.32 The substrate is tertiary, although the leaving group is a poor leaving group. In the presence of acid, the poor leaving group is protonated to give an excellent leaving group (H_2O). In spite of the presence of bromide (a strong nucleophile), the tertiary substrate cannot undergo S_N2, so S_N1 is expected.

9.33 The substrate is secondary, there is no leaving group, and the reagent is a weak nucleophile. Because a leaving group is required for either S_N2 or S_N1 to occur, we expect neither reaction to occur.

9.34 The substrate is primary, the reagent (chloride) is a strong nucleophile, and the solvent is polar aprotic. All of these factors favor an S_N2 reaction.

9.35 The substrate is tertiary (LG = mesylate), and the reagent (CH_3OH) is a weak nucleophile. Both of these factors favor an S_N1 reaction.

CHAPTER 10

10.1 This alkyl chloride will react with a base to give two different alkenes, shown below. The more-substituted alkene is the Zaitsev product, while the less-substituted alkene is the Hofmann product:

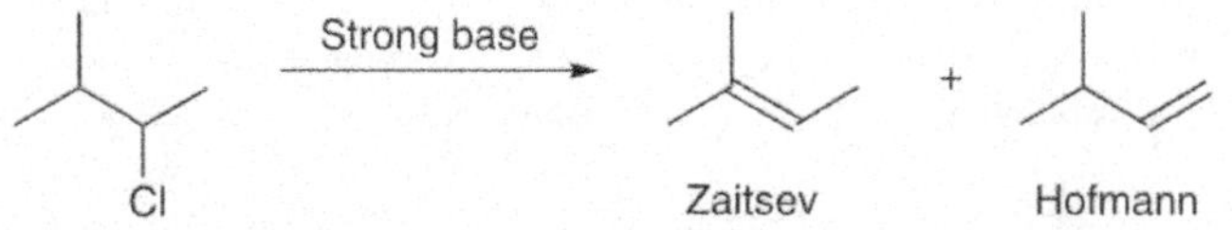

10.2 This alkyl chloride will react with a base to give two different alkenes, shown below. The more-substituted alkene is the Zaitsev product, while the less-substituted alkene is the Hofmann product:

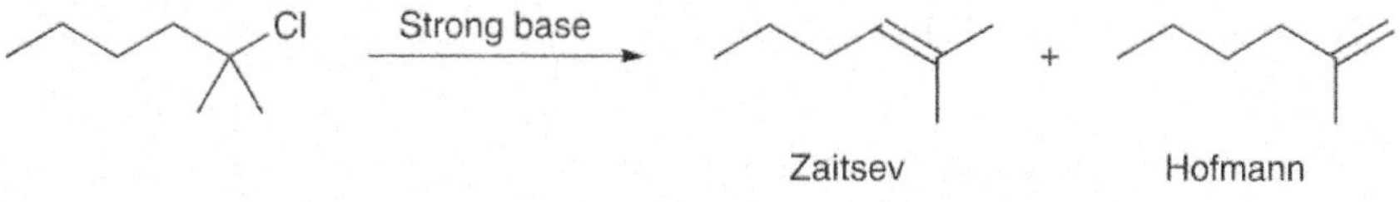

10.3 This cycloalkyl bromide will react with a base to give two different alkenes, shown below. The more-substituted alkene is the Zaitsev product, while the less-substituted alkene is the Hofmann product:

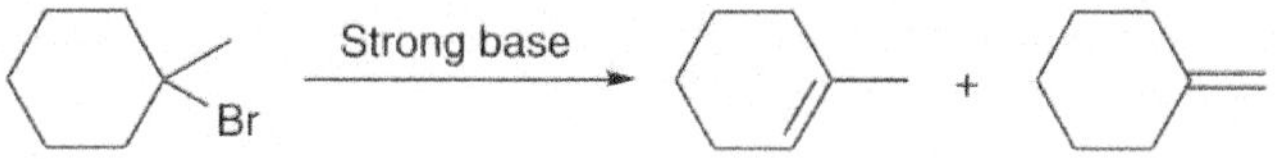

10.5 This substrate can react with ethoxide to give more than one alkene, although the Zaitsev product will be the major product because the base is not sterically hindered. To draw the Zaitsev product, we begin by drawing a Newman projection of the substrate, looking down the bond that will ultimately become a double bond. In the Newman projection, the beta proton (highlighted below) and the leaving group (also highlighted below) must be antiperiplanar, so we rotate about the central C—C bond of the Newman projection, in order to redraw the molecule in the conformation that is necessary for the reaction to occur:

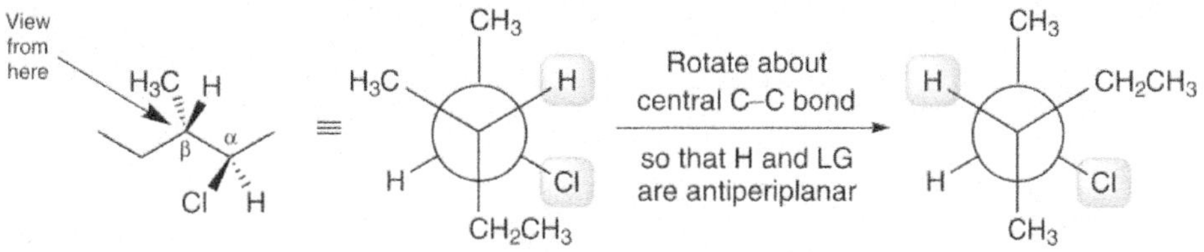

Finally, we remove the beta proton and the leaving group, and we draw a double bond between the alpha and beta carbon atoms. We use the Newman projection as a guide to determine which stereoisomeric alkene is obtained (E or Z), as shown here:

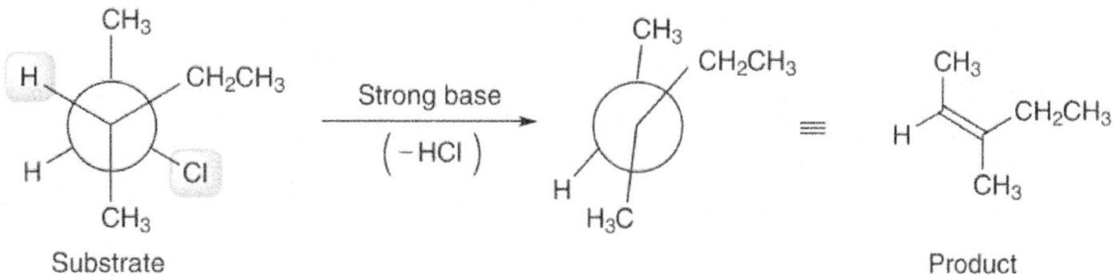

10.6 This substrate can react with ethoxide to give more than one alkene, although the Zaitsev product will be the major product because the base is not sterically hindered. To draw the Zaitsev product, we begin by drawing a Newman projection of the substrate, looking down the bond that will ultimately become a double bond. In this Newman projection, the beta proton (highlighted below) and the leaving group (also highlighted below) are antiperiplanar to one another, so we don't need to rotate about the central C—C bond of the Newman projection. The compound is already in the conformation that is necessary for the reaction to occur. So we remove the beta proton and the leaving group, and we draw a double bond between the alpha and beta carbon atoms. We use the Newman projection as a guide to determine which stereoisomeric alkene is obtained (E or Z), as shown here:

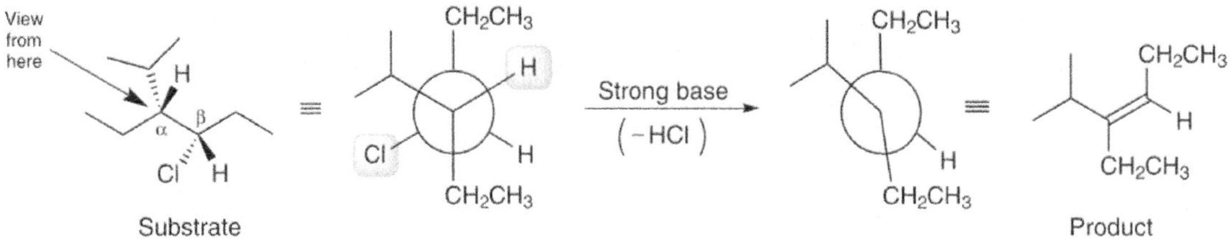

10.7 This substrate can react with ethoxide to give more than one alkene, although the Zaitsev product will be the major product because the base is not sterically hindered. To draw the Zaitsev product, we begin by drawing a Newman projection of the substrate, looking down the bond that will ultimately become a double bond. In the Newman projection, the beta proton (highlighted below) and the leaving group (also highlighted below) must be antiperiplanar, so we rotate about the central C—C bond of the Newman projection, in order to redraw the molecule in the conformation that is necessary for the reaction to occur:

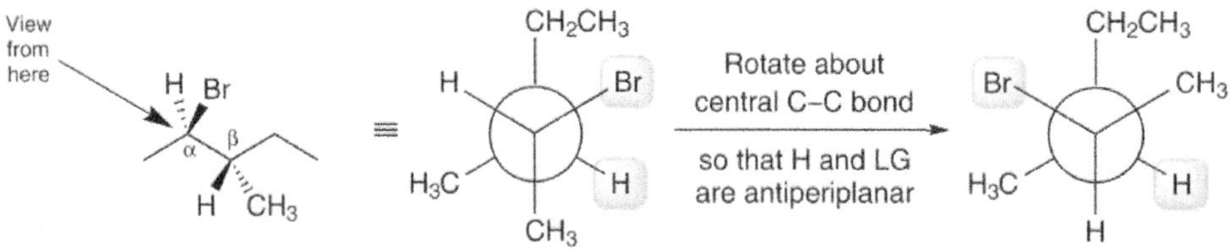

Finally, we remove the beta proton and the leaving group, and we draw a double bond between the alpha and beta carbon atoms. We use the Newman projection as a guide to determine which stereoisomeric alkene is obtained (E or Z), as shown here:

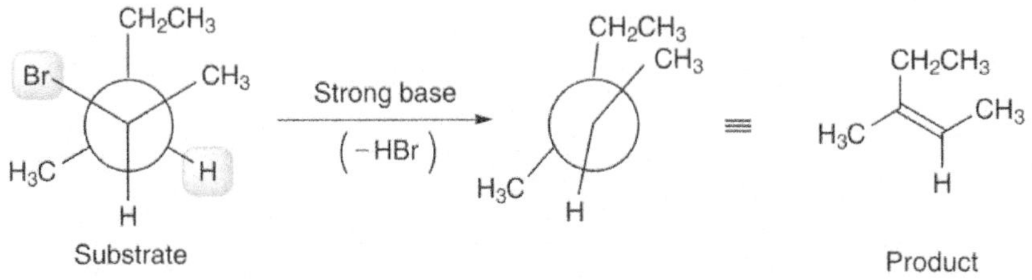

10.8 This substrate can react with ethoxide to give more than one alkene, although the Zaitsev product will be the major product because the base is not sterically hindered. To draw the Zaitsev product, we begin by drawing a Newman projection of the substrate, looking down the bond that will ultimately become a double bond. In this Newman projection, the beta proton (highlighted below) and the leaving group (also highlighted below) are antiperiplanar to one another, so we don't need to rotate about the central C—C bond of the Newman projection. The compound is already in the conformation that is necessary for the reaction to occur. So, we remove the beta proton and the leaving group, and we draw a double bond between the alpha and beta carbon atoms. We use the Newman projection as a guide to determine which stereoisomeric alkene is obtained (E or Z), as shown here:

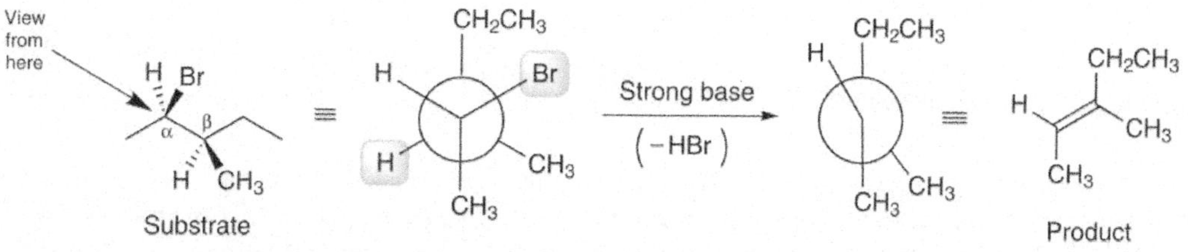

10.9 When a tertiary alcohol is heated in the presence of concentrated sulfuric acid, an E1 reaction occurs. In this case, there are two different alkenes that can be formed, as shown below. The major product is the more-substituted alkene (the Zaitsev product):

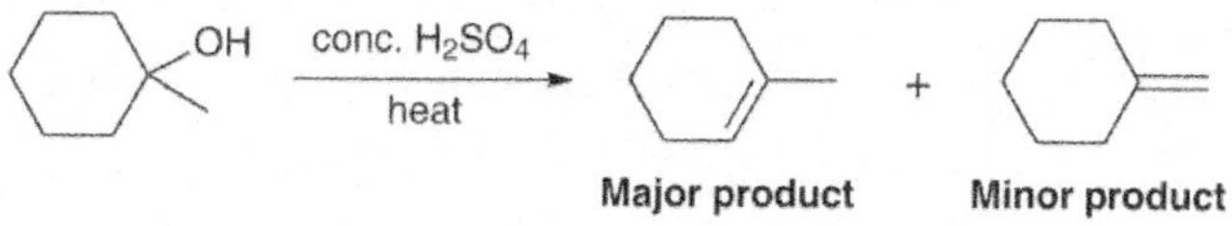

10.10 When a tertiary alcohol is heated in the presence of concentrated sulfuric acid, an E1 reaction occurs. In this case, there are two different alkenes that can be formed, as shown below. The major product is the more-substituted alkene (the Zaitsev product):

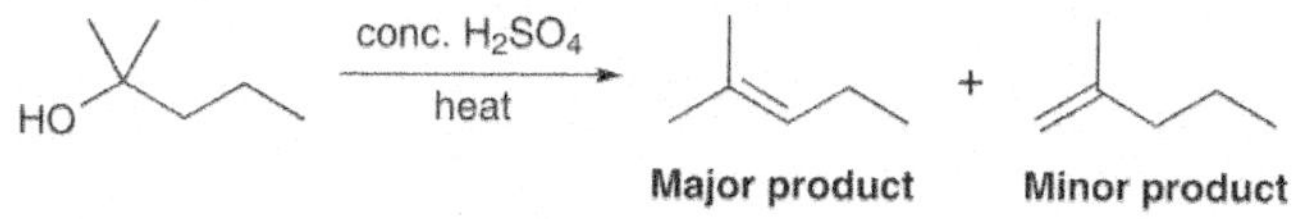

10.11 When a tertiary alcohol is heated in the presence of concentrated sulfuric acid, an E1 reaction occurs. In this case, there are two different alkenes that can be formed, as shown below. The major product is the more-substituted alkene (the Zaitsev product):

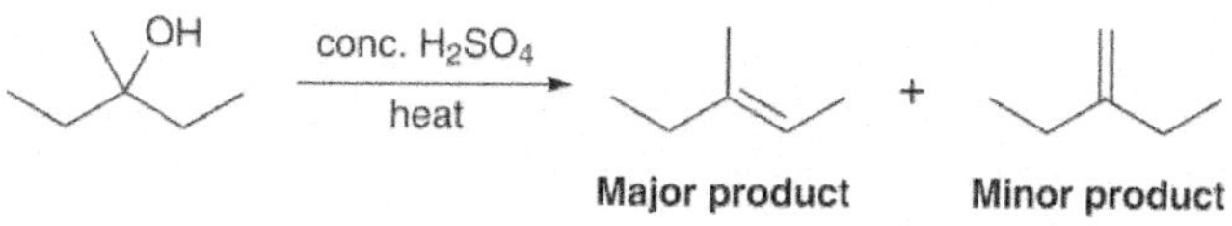

10.12 This compound is an alcohol (ROH), so it is a weak nucleophile and a weak base.

10.13 This compound is a thiol (RSH), so it is a strong nucleophile and a weak base.

10.14 Hydroxide (HO⁻) is a strong nucleophile and a strong base.

10.15 Chloride (Cl⁻) is a strong nucleophile and a weak base.

10.16 Water (H_2O) is a weak nucleophile and a weak base.

10.17 HS⁻ is a strong nucleophile and a weak base.

10.18 This compound is an alcohol (ROH), so it is a weak nucleophile and a weak base.

10.19 Iodide (I⁻) is a strong nucleophile and a weak base.

10.21 The reagent is ethoxide (EtO⁻), which is both a strong nucleophile and a strong base. The substrate is primary, so we expect both S_N2 and E2 products, although the S_N2 product is the major product.

10.22 When a tertiary alcohol is heated with concentrated sulfuric acid, an E1 reaction is expected.

10.23 The reagent is water, which is both a weak nucleophile and a weak base. The substrate is tertiary, so we expect S_N1 and E1 products.

10.24 The reagent is bromide (Br^-), which is a strong nucleophile and a weak base. The substrate is primary, so we expect only an S_N2 reaction to occur.

10.25 The reagent is bromide (Br^-), which is a strong nucleophile and a weak base. The substrate is tertiary, so we expect only an S_N1 reaction to occur.

10.27 We begin by analyzing the reagent. Hydroxide is both a strong nucleophile and a strong base. Next, we look at the substrate, which is primary. When a primary substrate is treated with a reagent that is both a strong nucleophile and a strong base, we expect S_N2 and E2 products, with the S_N2 product predominating. Because the leaving group is not connected to a chiral center, there is no inversion of stereochemistry to show:

Major product (S_N2) **Minor product** (E2)

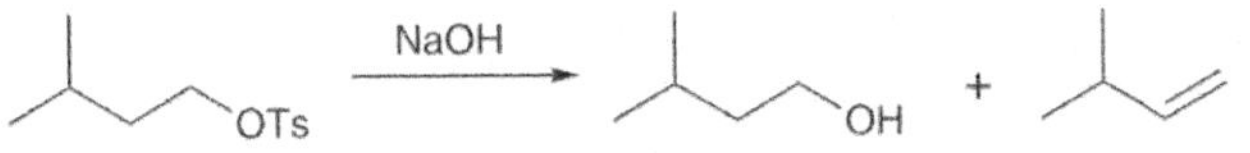

10.28 We begin by analyzing the reagent. Methoxide is both a strong nucleophile and a strong base. Next, we look at the substrate, which is primary. When a primary substrate is treated with a reagent that is both a strong nucleophile and a strong base, we expect S_N2 and E2 products, with the S_N2 product predominating. Because the leaving group is not connected to a chiral center, there is no inversion of stereochemistry to show:

Major product (S_N2) **Minor product** (E2)

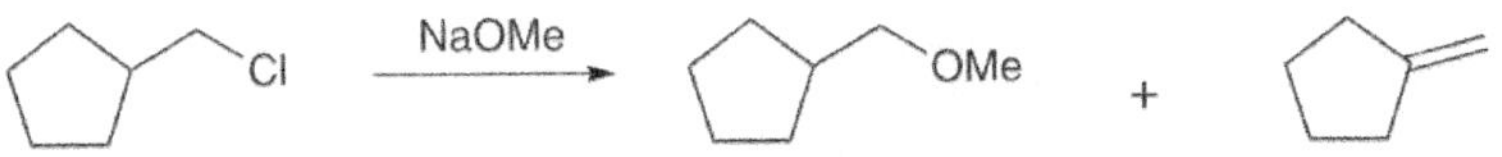

10.29 We begin by analyzing the reagent. Ethoxide is both a strong nucleophile and a strong base. Next, we look at the substrate, which is secondary. When a secondary substrate is treated with a reagent that is both a strong nucleophile and a strong base, we expect E2 and S_N2 products, with the E2 pathway providing the major product. The major product is the *trans* disubstituted alkene (the Zaitsev product). Minor products include the S_N2 product, the cis-disubstituted alkene, and the Hofmann product, as shown here:

Major product **Minor product** **Minor product** **Minor product**

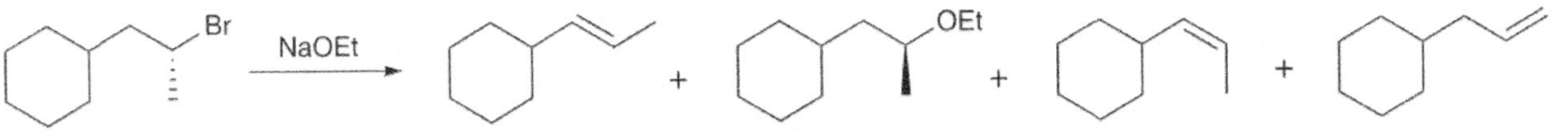

10.30 We begin by analyzing the reagent. Hydroxide is both a strong nucleophile and a strong base. Next, we look at the substrate, which is secondary. When a secondary substrate is treated with a reagent that is both a strong nucleophile and a strong base, we expect E2 and S_N2 products, with E2 giving the major product. The major product is the trisubstituted alkene (the Zaitsev product). Minor products include the Hofmann elimination product and the S_N2 product, as shown below. Notice the configuration of the double bond in the major product. This configuration is dictated by the requirement that the beta proton and the leaving group be antiperiplanar in order for an E2 reaction to occur.

Major product **Minor product** **Minor product**

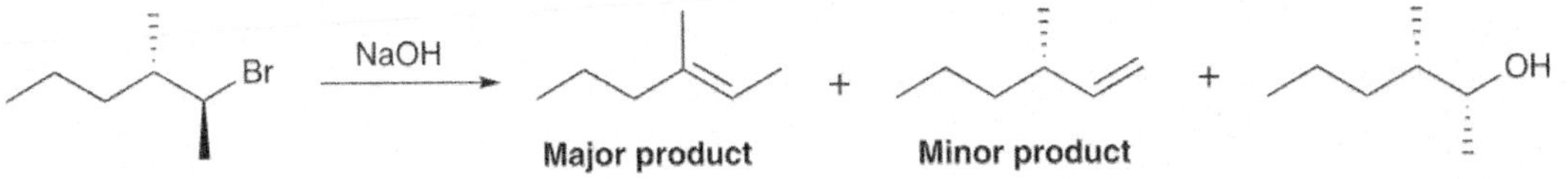

10.31 When a tertiary alcohol is heated with concentrated sulfuric acid, we expect E1 products. In this case, there are two possible regiochemical outcomes, and we expect both products, although the major product is the more-substituted alkene (the Zaitsev product):

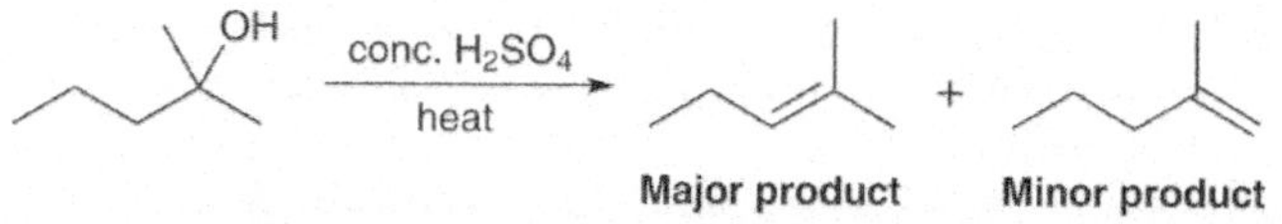

Major product   +   Minor product

10.32 We begin by analyzing the reagent. Water is both a weak nucleophile and a weak base. Next, we look at the substrate, which is tertiary. When a tertiary substrate is treated with a reagent that is both a weak nucleophile and a weak base, we expect S_N1 and E1 products. In a situation like this, it is often difficult to predict whether substitution or elimination will predominate, although the trisubstituted alkene (which is likely the major product) will certainly be favored over the monosubstituted alkene.

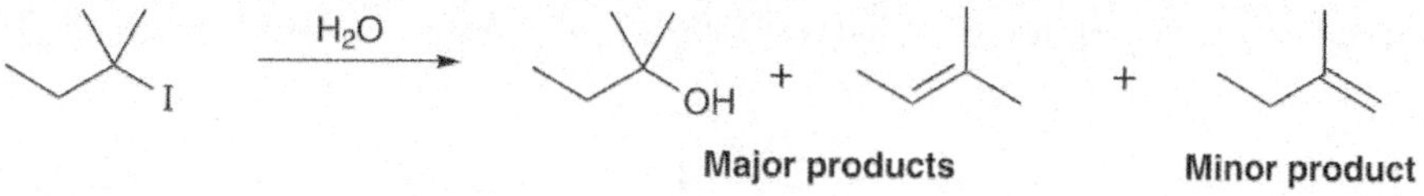

Major products   +   Minor product

10.33 We begin by analyzing the reagent. HS⁻ is a strong nucleophile and a weak base, so we expect substitution and not elimination. Next, we look at the substrate, which is secondary. When a secondary substrate is treated with a strong nucleophile, we expect the S_N2 product. The solvent is DMSO, which is polar aprotic, also suggesting an S_N2 process. In this case, the leaving group is connected to a chiral center, so we expect inversion of configuration:

10.34 We begin by analyzing the reagent. Ethoxide is both a strong nucleophile and a strong base. Next, we look at the substrate, which is tertiary. When a tertiary substrate is treated with a reagent that is both a strong nucleophile and a strong base, we expect only E2. There are two possible regiochemical outcomes (the Zaitsev product and the Hofmann product). Both products are obtained, although the Zaitsev product will be the major product, because we used a base that is not sterically hindered.

Major product   +   Minor product

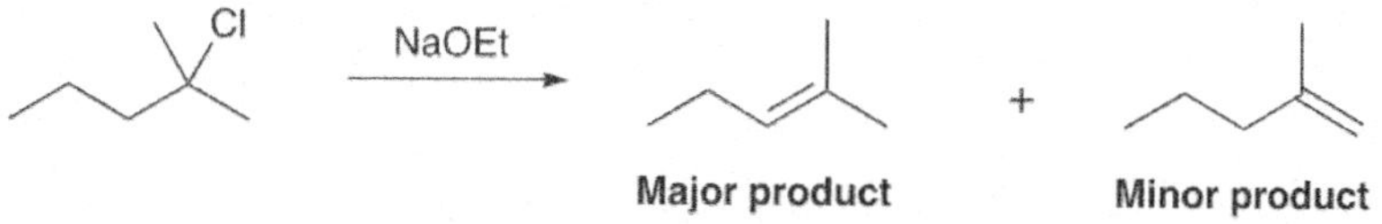

10.35 We begin by analyzing the reagent. *tert*-Butoxide is both a strong nucleophile and a strong base. Next, we look at the substrate, which is tertiary. When a tertiary substrate is treated with a reagent that is both a strong nucleophile and a strong base, we expect only E2. There are two possible regiochemical outcomes (the Zaitsev product and the Hofmann product). Both products are obtained, although the Hofmann product will be the major product, because *tert*-butoxide is a sterically hindered base.

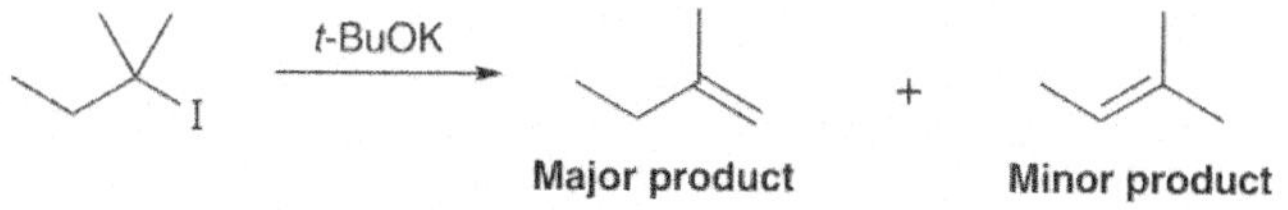

Major product   +   Minor product

10.36 We begin by analyzing the reagent. Methoxide is both a strong nucleophile and a strong base. Next, we look at the substrate, which is tertiary. When a tertiary substrate is treated with a reagent that is both a strong nucleophile and a strong base, we expect only E2. There are two possible regiochemical outcomes (the Zaitsev product and the Hofmann product). Both products are obtained, although the Zaitsev product will be the major product, because we used a base that is not sterically hindered.

Major product   +   Minor product

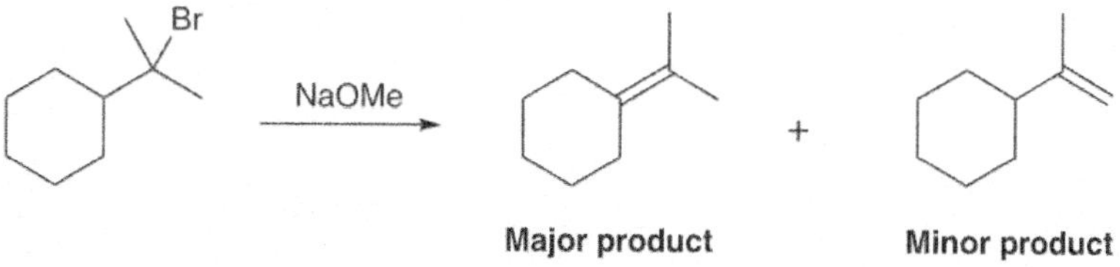

10.37 We begin by analyzing the reagent. HS⁻ is a strong nucleophile and a weak base, so we expect substitution and not elimination. Next, we look at the substrate, which is primary. When a primary substrate is treated with a strong nucleophile, we expect only the S$_N$2 product. Because the leaving group is not connected to a chiral center, there is no inversion of stereochemistry to show:

10.38 We begin by analyzing the reagent. MeS⁻ is a strong nucleophile and a weak base, so we expect substitution and not elimination. Next, we look at the substrate, which is primary. When a primary substrate is treated with a strong nucleophile, we expect only the S$_N$2 product. Because the leaving group is not connected to a chiral center, there is no inversion of stereochemistry to show:

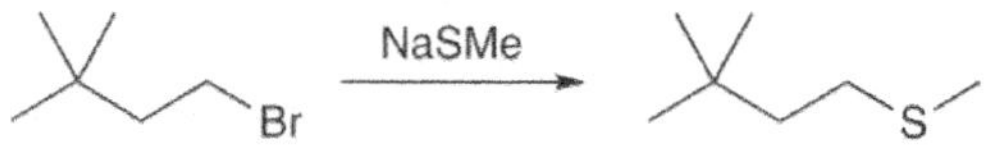

CHAPTER 11

11.2 The problem statement indicates an "*anti*-Markovnikov addition of H and Br," which is an addition reaction in which a bromine atom (highlighted below) is installed at the less-substituted position, and a hydrogen atom (also highlighted below) is installed at the more-substituted position:

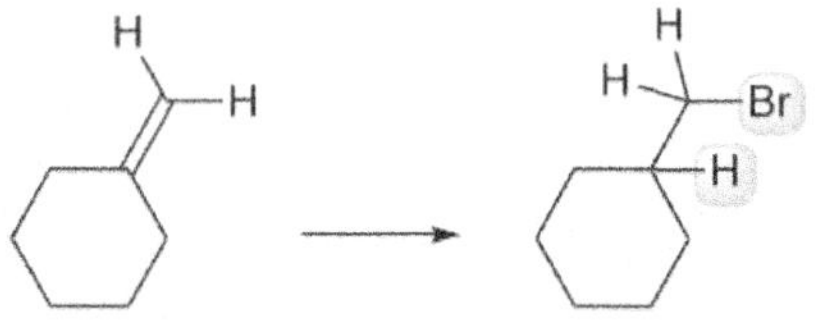

11.3 The problem statement indicates a "Markovnikov addition of H and Cl," which is an addition reaction in which a chlorine atom (highlighted below) is installed at the more-substituted position, and a hydrogen atom (also highlighted below) is installed at the less-substituted position:

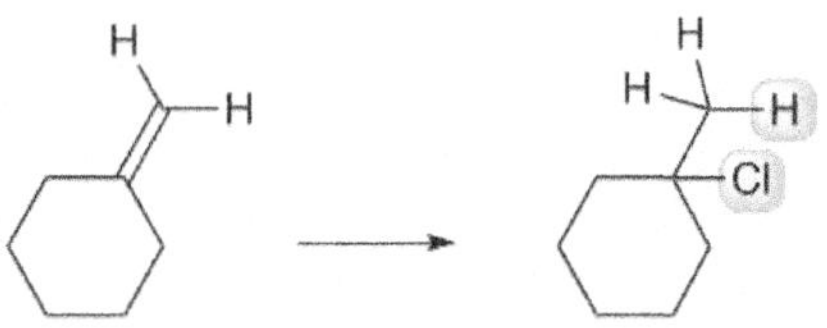

11.4 The problem statement indicates an "*anti*-Markovnikov addition of H and OH," which is an addition reaction in which an OH group (highlighted below) is installed at the less-substituted position, and a hydrogen atom (also highlighted below) is installed at the more-substituted position:

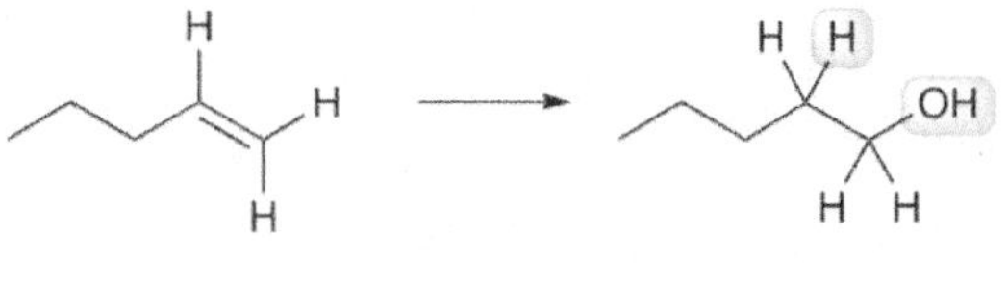

11.5 The problem statement indicates a "Markovnikov addition of H and OH," which is an addition reaction in which an OH group (highlighted below) is installed at the more-substituted position, and a hydrogen atom (also highlighted below) is installed at the less-substituted position:

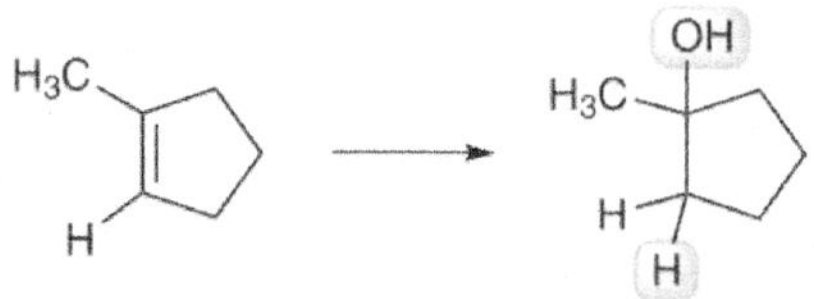

11.7 The problem statement indicates an "*anti* addition of OH and OH," which means that the two OH groups are installed on opposite sides of the double bond. Since the two groups are the same (both are OH groups), we don't need to consider regiochemistry (Markovnikov vs. *anti*-Markovnikov). In this case, a pair of enantiomers would be the result of an *anti* addition, as shown:

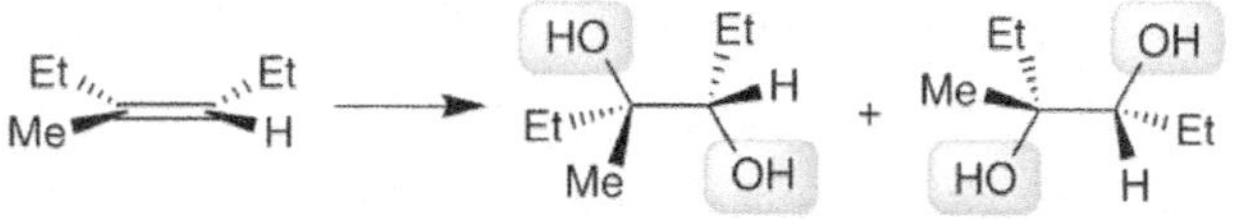

A pair of enantiomers

11.8 The problem statement indicates an "*anti*-Markovnikov addition of H and OH," which is an addition reaction in which an OH group is installed at the less-substituted position, and a hydrogen atom is installed at the more-substituted position. The problem statement also indicates that the reaction is a "*syn* addition," so H and OH are installed on the same side of the alkene, giving rise to a pair of enantiomers:

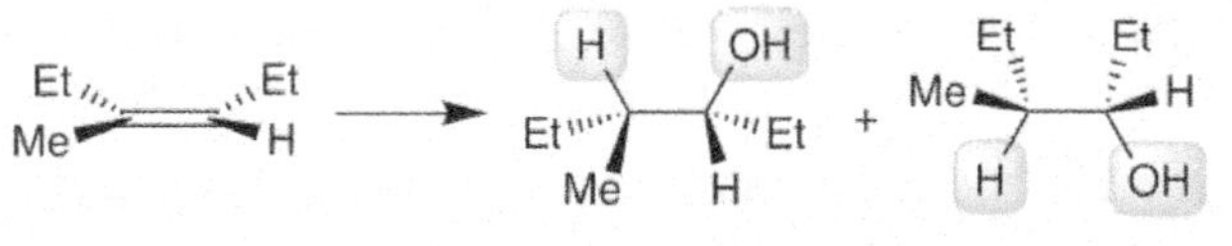

A pair of enantiomers

11.9 The problem statement indicates a "*syn* addition of OH and OH," which means that the two OH groups are installed on the same side of the double bond. Since the two groups are the same (both are OH groups), we don't need to consider regiochemistry (Markovnikov vs. *anti*-Markovnikov). In this case, a pair of enantiomers would be the result of a *syn* addition, as shown:

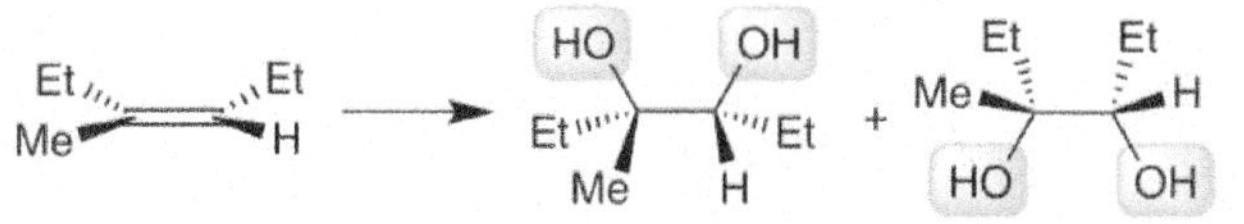

A pair of enantiomers

11.10 The problem statement indicates a "Markovnikov addition of H and OH," which is an addition reaction in which an OH group is installed at the more-substituted position, and a hydrogen atom is installed at the less-substituted position. The problem statement also indicates that the reaction is not stereospecific. A single chiral center is generated by the reaction, and we expect a pair of enantiomers:

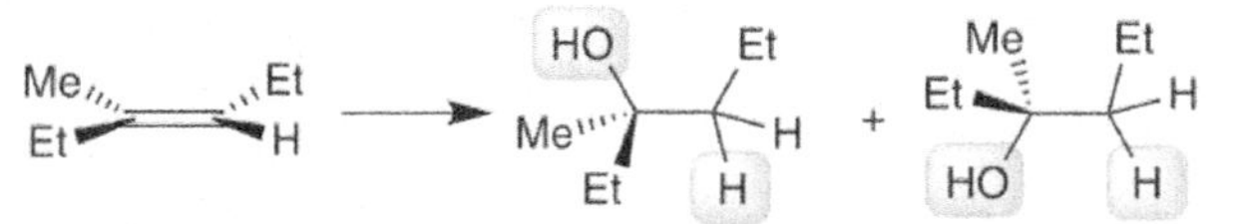

A pair of enantiomers

11.12 The problem statement indicates an "*anti*-Markovnikov addition of H and OH," which is an addition reaction in which an OH group is installed at the less-substituted position, and a hydrogen atom is installed at the more-substituted position. The problem statement also indicates that the reaction is a "*syn* addition," so H and OH are installed on the same side of the alkene, giving rise to a pair of enantiomers:

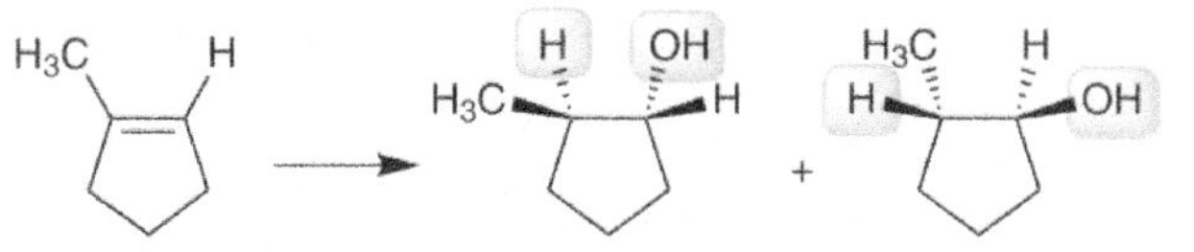

A pair of enantiomers

11.13 The problem statement indicates an "*anti*-Markovnikov addition of H and Br," which is an addition reaction in which Br is installed at the less-substituted position, and a hydrogen atom is

installed at the more-substituted position. The problem statement also indicates that the reaction is an "*anti* addition," so H and Br are installed on opposite sides of the alkene, giving rise to a pair of enantiomers:

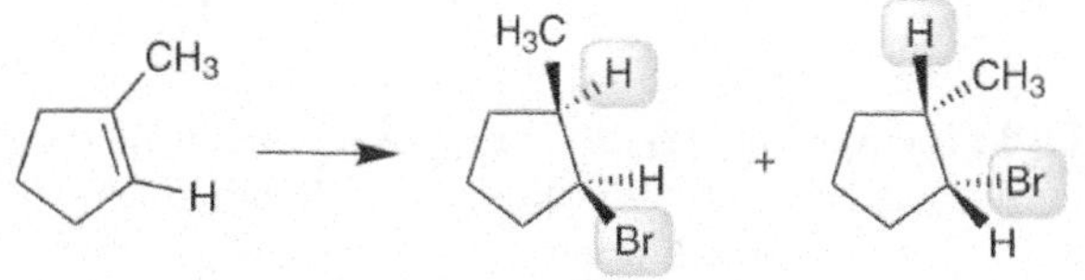

A pair of enantiomers

11.14 The problem statement indicates a "*syn* addition of OH and OH," which means that the two OH groups are installed on the same side of the double bond. Since the two groups are the same (both are OH groups), we don't need to consider regiochemistry (Markovnikov vs. *anti*-Markovnikov). In this case, a pair of enantiomers would be the result of a *syn* addition, as shown:

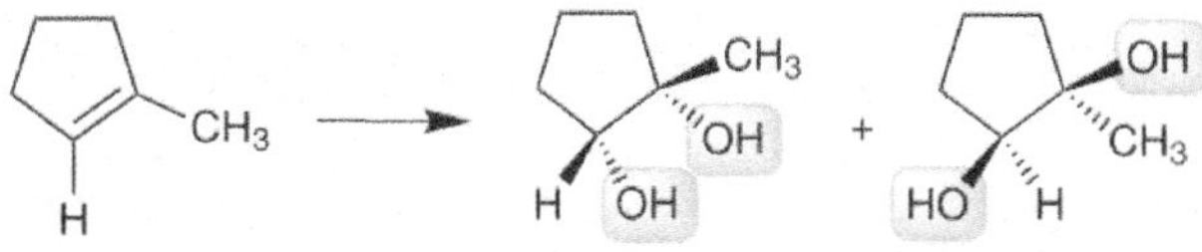

A pair of enantiomers

11.15 The problem statement indicates an "addition of OH and Br, with the OH group being installed at the more-substituted position," which means that Br is installed at the less-substituted position. The problem statement also indicates that the reaction is an "*anti* addition," so the two groups are installed on opposite sides of the alkene, giving rise to a pair of enantiomers:

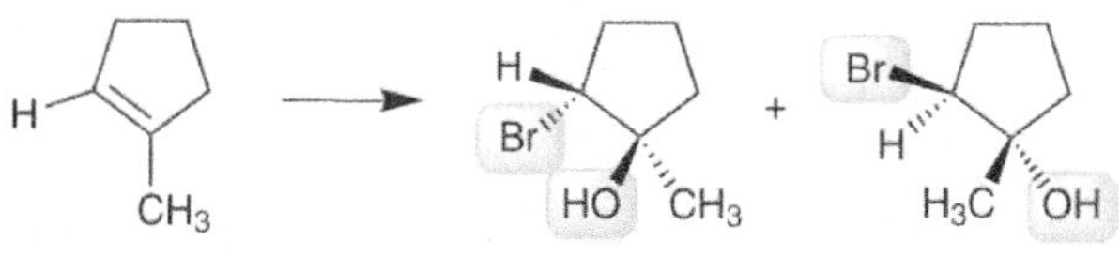

A pair of enantiomers

11.17 The problem statement indicates an "*anti*-Markovnikov addition of H and OH," which is an addition reaction in which an OH group is installed at the less-substituted position, and a hydrogen atom is installed at the more-substituted position. In this case, only one new chiral center is created, so the stereochemical requirement for a *syn* addition is not relevant in this case. If the reaction has been an *anti* addition instead of a *syn* addition, we would have obtained the same products, shown below. Indeed, even if the reaction had not been stereospecific, we would have still obtained the same pair of enantiomers below:

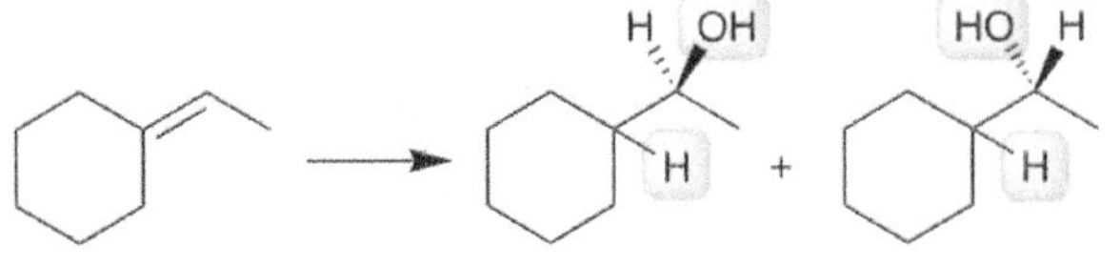

A pair of enantiomers

11.18 The problem statement indicates an "*anti* addition of Br and Br," so regiochemistry is not relevant. In this case, two new chiral centers are created, so the stereochemical requirement for an *anti* addition is relevant in this case. Specifically, we expect only the pair of enantiomers that would result from an *anti* addition:

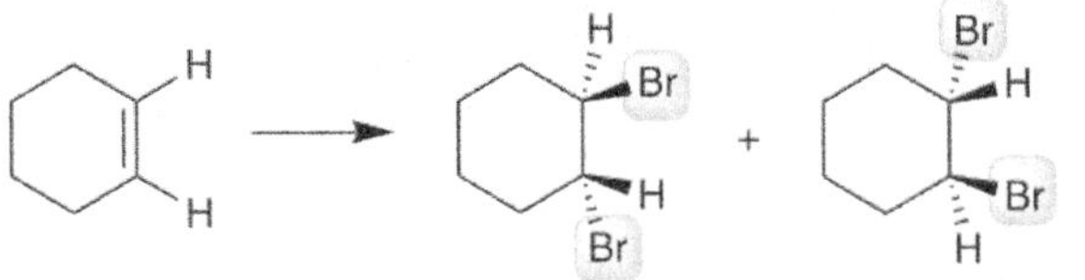

A pair of enantiomers

11.19 The problem statement indicates the addition of OH and OH (two of the same group), so regiochemistry is not relevant. In this case, two new chiral centers are created, so the stereochemical requirement for a *syn* addition is relevant in this case. Specifically, we expect only the pair of enantiomers that would result from a *syn* addition (not the pair of enantiomers that would have resulted from an *anti* addition):

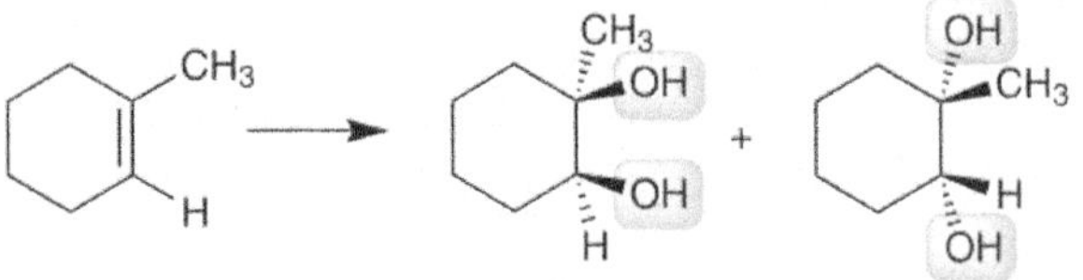

A pair of enantiomers

11.20 The problem statement indicates a "*syn* addition of H and H." Since the two groups have the same identity (H and H), we don't need to consider regiochemistry (Markovnikov vs. *anti*-Markovnikov). In this case, no new chiral centers are created, so the stereochemical requirement for a *syn* addition is irrelevant in this case. There is only one product:

11.22 The problem statement indicates an "*anti* addition of OH and OH," which means that the two OH groups are installed on opposite sides of the double bond. Since the two groups have the same identity (both are OH groups), we don't need to consider

regiochemistry (Markovnikov vs. *anti*-Markovnikov). In this case, two new chiral centers are being formed, and the result is a pair of enantiomers, as shown:

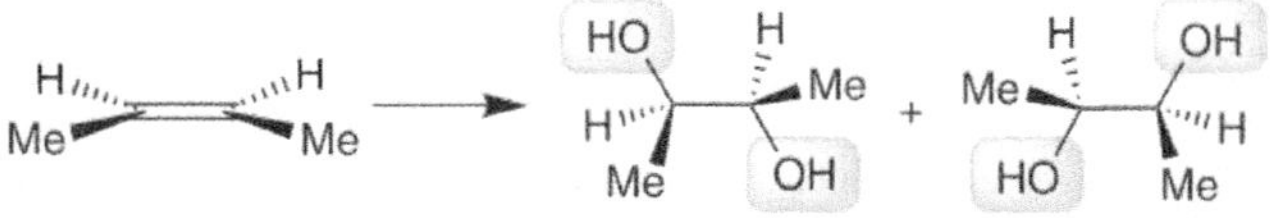

A pair of enantiomers

You might find it easier to draw the products in the following way, which also clearly shows an *anti* addition of OH and OH to give a pair of enantiomers:

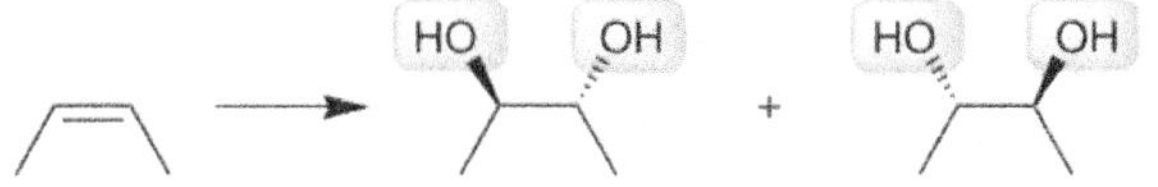

A pair of enantiomers

11.23 The problem statement indicates an "*anti* addition of OH and OH," which means that the two OH groups are installed on opposite sides of the double bond. Since the two groups have the same identity (both are OH groups), we don't need to consider regiochemistry (Markovnikov vs. *anti*-Markovnikov). In this case, two new chiral centers are being formed, but the result is a *meso* compound (rather than a pair of enantiomers), as shown:

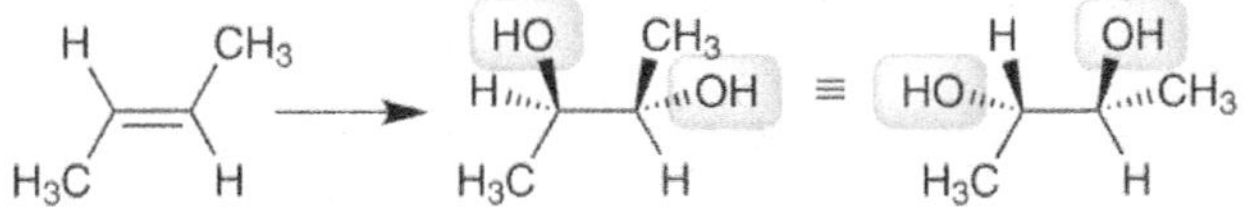

A *meso* compound

To see more clearly that this is indeed a *meso* compound, we can take either drawing above (let's look at the first drawing) and imagine rotating about the central carbon–carbon bond. The resulting conformation (shown below) clearly has an internal plane of symmetry, so this compound must be a *meso* compound. You should be able to do the same thing with the second structure above (change its conformation until it is clear that it is a *meso* compound).

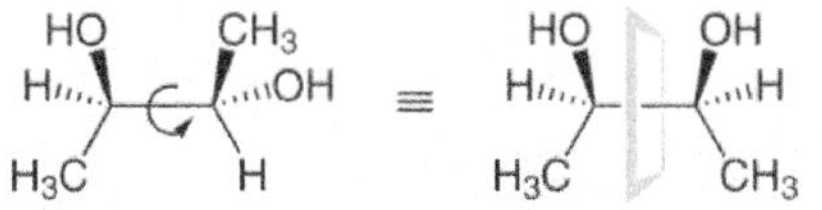

Internal plane of symmetry

11.24 The problem statement indicates an "*anti* addition of Br and Br," which means that the two Br groups are installed on opposite sides of the double bond. Since the two groups have the same identity (both are Br), we don't need to consider regiochemistry (Markovnikov vs. *anti*-Markovnikov). In this case, two new chiral centers are being formed, and the result is a pair of enantiomers, as shown:

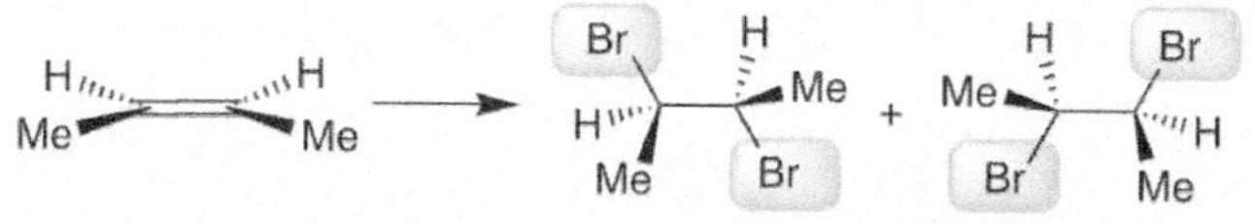

A pair of enantiomers

You might find it easier to draw the products in the following way, which also clearly shows an *anti* addition of Br and Br to give a pair of enantiomers:

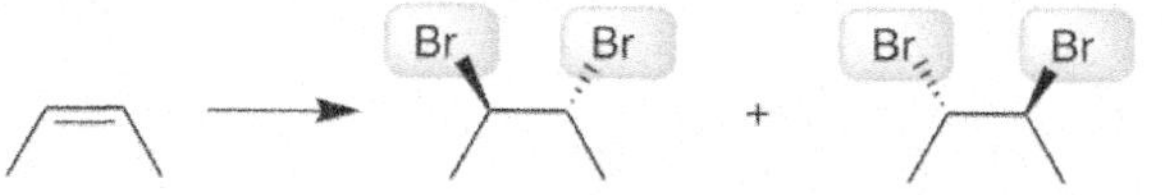

A pair of enantiomers

11.25 The problem statement indicates an "*anti* addition of Br and Br," which means that the two Br groups are installed on opposite sides of the double bond. Since the two groups have the same identity (both are Br), we don't need to consider regiochemistry (Markovnikov vs. *anti*-Markovnikov). In this case, two new chiral centers are being formed, but the result is a *meso* compound (rather than a pair of enantiomers), as shown:

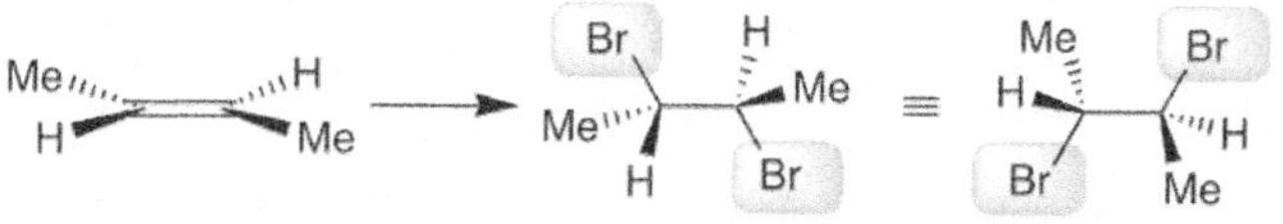

A *meso* compound

To see more clearly that this is indeed a *meso* compound, we can take either drawing above (let's look at the first drawing) and imagine rotating about the central carbon–carbon bond. The resulting conformation (shown below) clearly has an internal plane of symmetry, so this compound must be a *meso* compound. You should be able to do the same thing with the second structure above (change its conformation until it is clear that it is a *meso* compound).

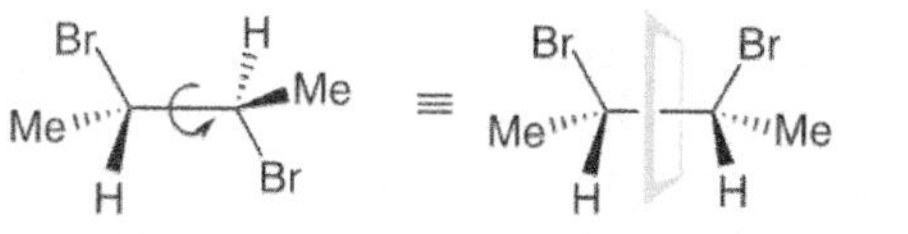

Internal plane of symmetry

11.27 The reagents indicate the addition of H and H across the double bond. Since the two groups have the same identity (both are H), we don't need to consider regiochemistry (Markovnikov vs.

anti-Markovnikov). In this case, only one new chiral center is created, so the stereochemical requirement for a *syn* addition is not relevant, and we expect a pair of enantiomers:

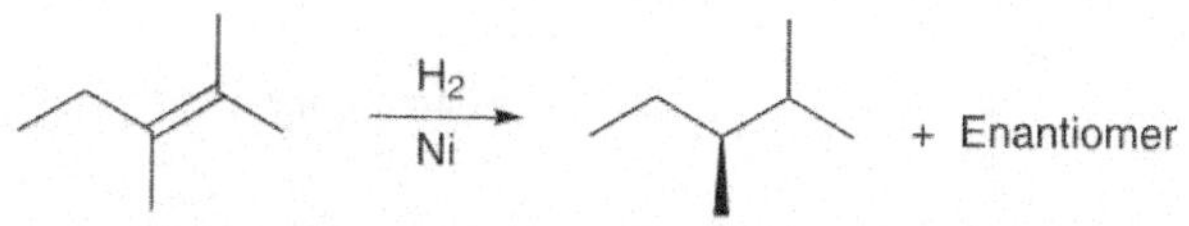

11.28 The reagents indicate the addition of H and H across the double bond. Since the two groups have the same identity (both are H), we don't need to consider regiochemistry (Markovnikov vs. *anti*-Markovnikov). In this case, only one new chiral center is created, so the stereochemical requirement for a *syn* addition is not relevant, and we expect a pair of enantiomers:

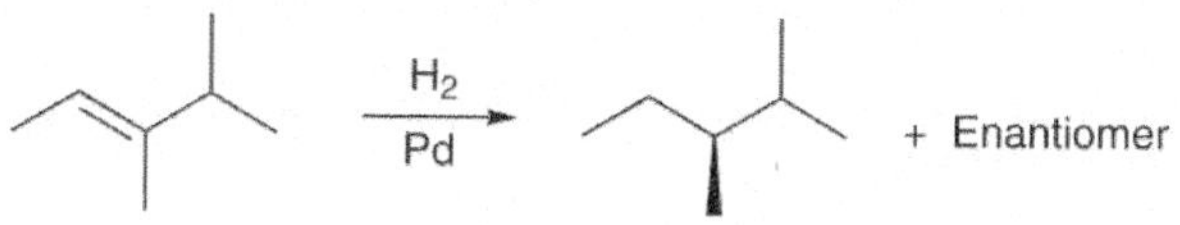

11.29 The reagents indicate the addition of H and H across the double bond. Since the two groups have the same identity (both are H), we don't need to consider regiochemistry (Markovnikov vs. *anti*-Markovnikov). In this case, no new chiral centers are created, so the stereochemical requirement for a *syn* addition is not relevant, and we expect only one product:

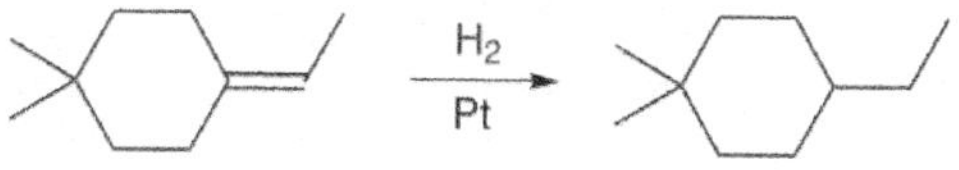

11.30 The reagents indicate the addition of H and H across the double bond. Since the two groups have the same identity (both are H), we don't need to consider regiochemistry (Markovnikov vs. *anti*-Markovnikov). In this case, two new chiral centers are created, so the stereochemical requirement for a *syn* addition is relevant, and we expect only the pair of enantiomers that would result from a *syn* addition (not the pair of enantiomers that would have resulted from an *anti* addition):

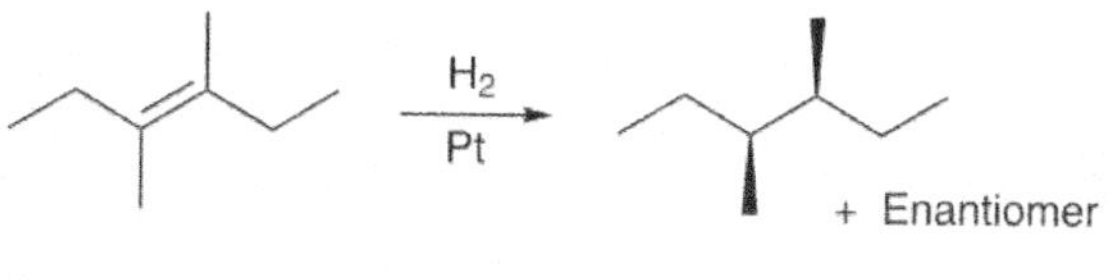

11.31 The reagents indicate the addition of H and H across the double bond. Since the two groups have the same identity (both are H), we don't need to consider regiochemistry (Markovnikov vs. *anti*-Markovnikov). In this case, two new chiral centers are created, so the stereochemical requirement for a *syn* addition is relevant. But the product is a *meso* compound, rather than a pair of enantiomers, because the product has an internal plane of symmetry:

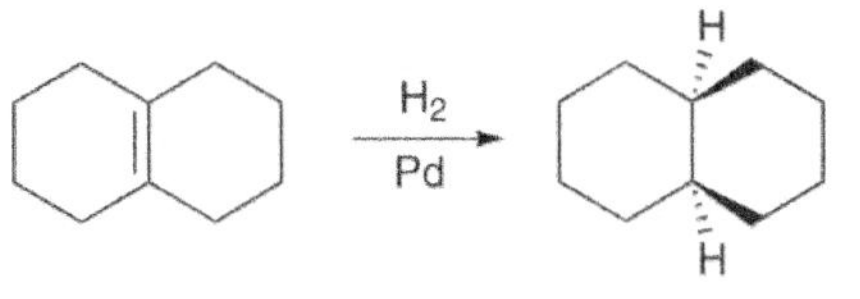

A *meso* compound

are H), we don't need to consider regiochemistry (Markovnikov vs. *anti*-Markovnikov). In this case, no new chiral centers are created, so the stereochemical requirement for a *syn* addition is not relevant, and we expect only one product:

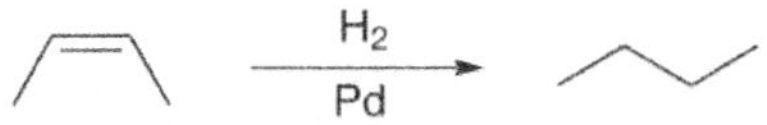

11.32 The reagents indicate the addition of H and H across the double bond. Since the two groups have the same identity (both

11.34 The first step has two curved arrows. The first curved arrow is drawn coming from the pi bond and pointing to the proton of HBr, while the second curved arrow is drawn coming from the H—Br bond and pointing to Br (generating a carbocation and a bromide ion). Then, in the second step, there is just one curved arrow, showing a bromide ion (formed in the first step of the mechanism) attacking the carbocation.

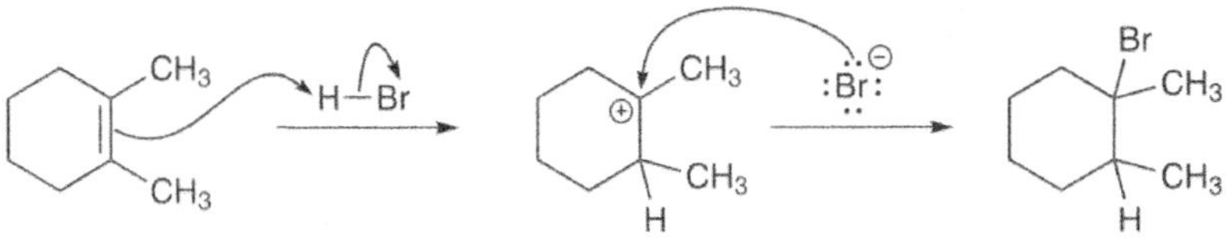

11.35 The first step has two curved arrows. The first curved arrow is drawn coming from the pi bond and pointing to the proton of HCl, while the second curved arrow is drawn coming from the H—Cl bond and pointing to Cl (generating a carbocation and a chloride ion). Then, in the second step, there is just one curved arrow, showing a chloride ion (formed in the first step of the mechanism) attacking the carbocation.

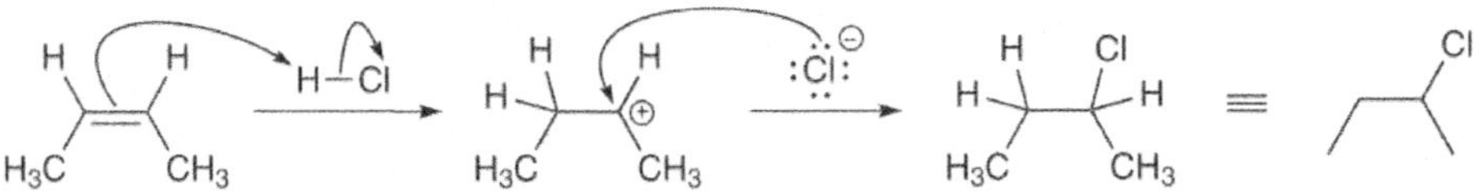

11.36 The first step has two curved arrows. The first curved arrow is drawn coming from the pi bond and pointing to the proton of HBr, while the second curved arrow is drawn coming from the H–Br bond and pointing to Br (generating a carbocation and a bromide ion). Then, in the second step, there is just one curved arrow, showing a bromide ion (formed in the first step of the mechanism) attacking the carbocation.

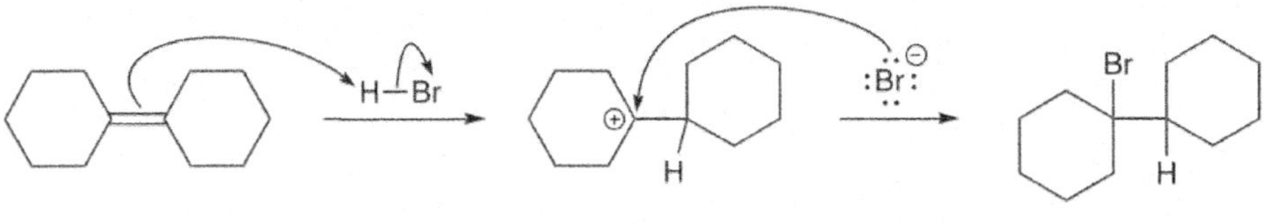

11.37 The first step has two curved arrows. The first curved arrow is drawn coming from the pi bond and pointing to the proton of HI, while the second curved arrow is drawn coming from the H—I bond and pointing to I (generating a carbocation and an iodide ion). Then, in the second step, there is just one curved arrow, showing an iodide ion (formed in the first step of the mechanism) attacking the carbocation.

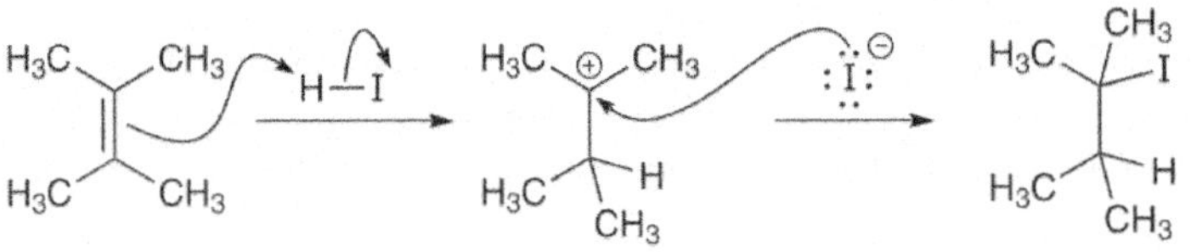

11.39 According to Markovnikov's rule, bromine is installed at the more substituted position. The product does not contain any chiral centers, so stereochemistry is irrelevant, and only one product is formed.

11.40 According to Markovnikov's rule, chlorine is installed at the more substituted position. The product does not contain any chiral centers, so stereochemistry is irrelevant, and only one product is formed.

11.41 According to Markovnikov's rule, bromine is installed at the more substituted position. The product does not contain any chiral centers, so stereochemistry is irrelevant, and only one product is formed.

11.42 According to Markovnikov's rule, iodine is installed at the more substituted position. The product has only one chiral center (not two), so we expect the following pair of enantiomers, regardless of whether the reaction proceeds via an *anti* addition or via a *syn* addition.

11.44 The first step has two curved arrows: the first curved arrow is drawn coming from the pi bond and pointing to the proton of HBr, while the second curved arrow is drawn coming from the H—Br bond and pointing to Br. The resulting secondary carbocation can then undergo a carbocation rearrangement to give a tertiary carbocation, and this step is illustrated with one curved arrow (showing a hydride shift). In the final step of the mechanism, there is just one curved arrow, showing a bromide ion (formed in the first step of the mechanism) attacking the carbocation to give the product.

11.45 The first step has two curved arrows: the first curved arrow is drawn coming from the pi bond and pointing to the proton of HCl, while the second curved arrow is drawn coming from the H—Cl bond and pointing to Cl. The resulting secondary carbocation can then undergo a carbocation rearrangement to give a tertiary carbocation, and this step is illustrated with one curved arrow (showing a hydride shift). In the final step of the mechanism, there is just one curved arrow, showing a chloride ion (formed in the first step of the mechanism) attacking the carbocation to give the product.

11.46 The first step has two curved arrows: the first curved arrow is drawn coming from the pi bond and pointing to the proton of HBr, while the second curved arrow is drawn coming from the H—Br bond and pointing to Br. The resulting secondary carbocation can then undergo a carbocation rearrangement to give a tertiary carbocation, and this step is illustrated with one curved arrow (showing a hydride shift). In the final step of the mechanism, there is just one curved arrow, showing a bromide ion (formed in the first step of the mechanism) attacking the carbocation to give the product.

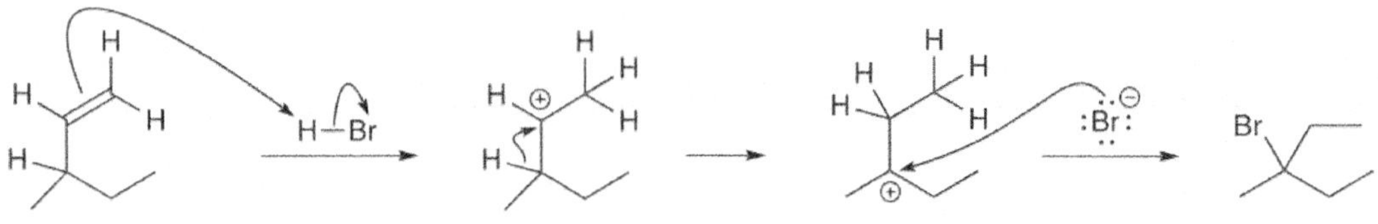

11.47 The first step has two curved arrows: the first curved arrow is drawn coming from the pi bond and pointing to the proton of HCl, while the second curved arrow is drawn coming from the H—Cl bond and pointing to Cl. The resulting secondary carbocation can then undergo a carbocation rearrangement to give a tertiary carbocation, and this step is illustrated with one curved arrow (showing a methyl shift). In the final step of the mechanism, there is just one curved arrow, showing a chloride ion (formed in the first step of the mechanism) attacking the carbocation to give the product.

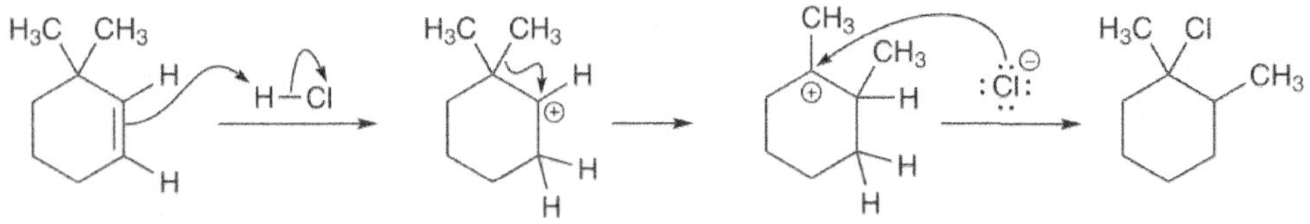

11.49 When treated with HBr in the presence of peroxides, an alkene will undergo an *anti*-Markovnikov addition reaction, in which bromine is installed at the less-substituted position. No chiral centers are created during this process, so stereochemistry is irrelevant, and one product is formed:

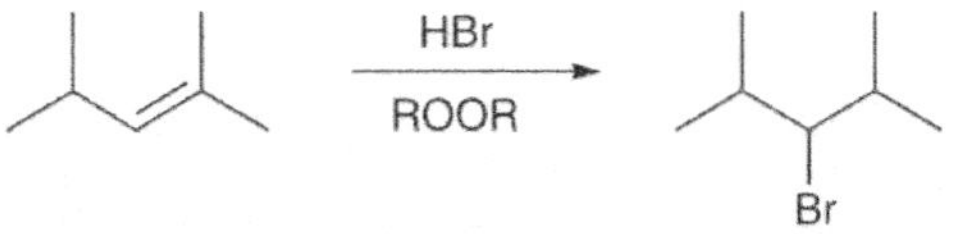

11.50 When treated with HBr in the presence of peroxides, an alkene will undergo an *anti*-Markovnikov addition reaction, in which bromine is installed at the less-substituted position. One chiral center is created during this process, so a pair of enantiomers is expected:

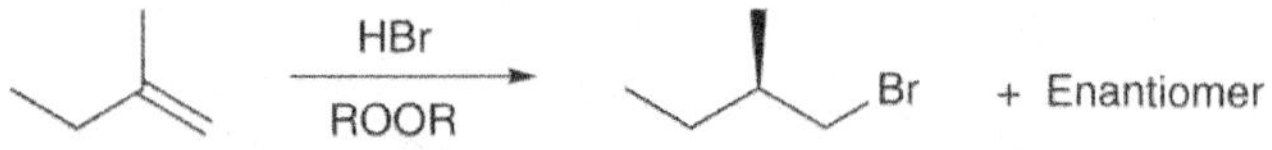

11.51 When treated with HBr in the presence of peroxides, an alkene will undergo an *anti*-Markovnikov addition reaction, in which bromine is installed at the less-substituted position. No chiral centers are created during this process, so stereochemistry is irrelevant, and one product is formed:

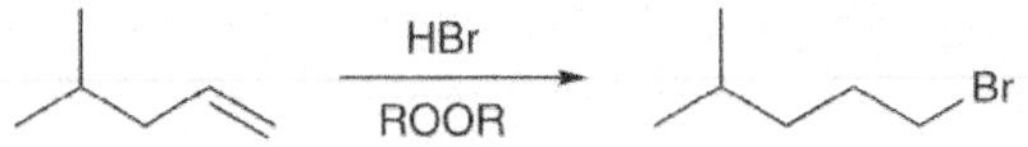

11.52 When treated with HBr in the presence of peroxides, an alkene will undergo an *anti*-Markovnikov addition reaction, in which bromine is installed at the less-substituted position. One chiral center is created during this process, so a pair of enantiomers is expected:

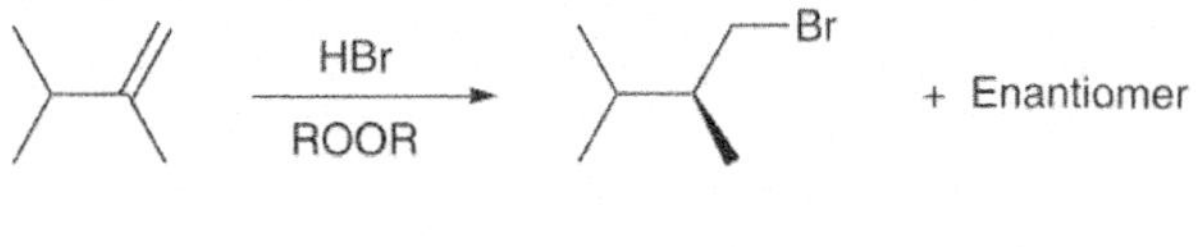

11.54 The desired transformation is an addition reaction in which a bromine atom is installed at the less-substituted position and a hydrogen atom is installed at the more-substituted position (*anti*-Markovnikov addition of H and Br). This can be accomplished by treating the alkene with HBr in the presence of peroxides (ROOR).

11.55 The desired transformation is an addition reaction in which a bromine atom is installed at the less-substituted position and a hydrogen atom is installed at the more-substituted position (*anti*-Markovnikov addition of H and Br). This can be accomplished by treating the alkene with HBr in the presence of peroxides (ROOR).

11.56 The desired transformation is an addition reaction in which a bromine atom is installed at the more-substituted position and a hydrogen atom is installed at the less-substituted position (Markovnikov addition of H and Br). This can be accomplished by treating the alkene with HBr only (no peroxides).

11.57 The desired transformation is an addition reaction in which a bromine atom is installed at the less-substituted position and a hydrogen atom is installed at the more-substituted position (*anti*-Markovnikov addition of H and Br). This can be accomplished by treating the alkene with HBr in the presence of peroxides (ROOR).

11.59 The starting material is an alkene, and the reagent is H_3O^+, which indicates an acid-catalyzed hydration. The result is Markovnikov addition of H and OH across the alkene to give an alcohol. Because no new chiral center is formed, we expect only one product.

The mechanism is believed to have the following three steps: 1) the pi bond is first protonated (which requires two curved arrows), and then 2) water functions as a nucleophile and attacks the carbocation, and finally, 3) the resulting intermediate is deprotonated by water to give the product:

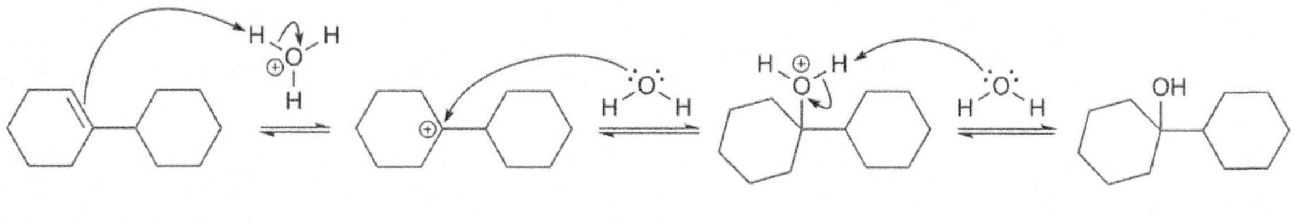

11.60 The starting material is an alkene, and the reagents are sulfuric acid and water, which are equivalent to H_3O^+ (because of the leveling effect). These reagents indicate an acid-catalyzed hydration. The result is Markovnikov addition of H and OH across the alkene to give an alcohol. Because no new chiral center is formed, we expect only one product.

The mechanism is believed to have the following three steps: 1) the pi bond is first protonated (which requires two curved arrows), and then 2) water functions as a nucleophile and attacks the carbocation, and finally, 3) the resulting oxonium ion is deprotonated by water to give the product:

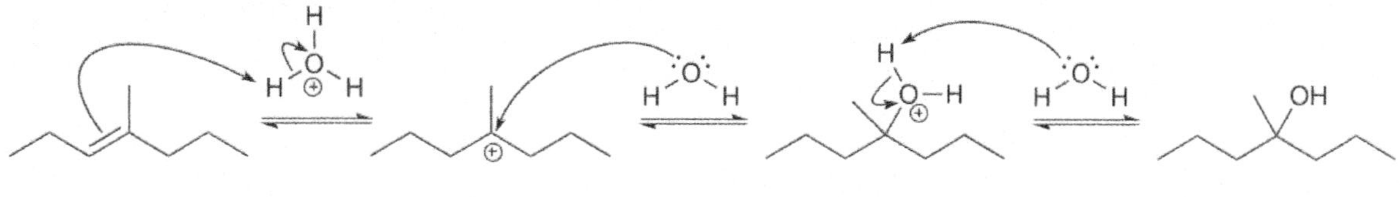

11.61 The starting material is an alkene, and the reagents are sulfuric acid and water, which are equivalent to H_3O^+ (because of the leveling effect). These reagents indicate an acid-catalyzed hydration. The result is Markovnikov addition of H and OH across the alkene to give an alcohol. Because no new chiral center is formed, we expect only one product.

The mechanism is believed to have the following three steps: 1) the pi bond is first protonated (which requires two curved arrows), and then 2) water functions as a nucleophile and attacks the carbocation, and finally, 3) the resulting oxonium ion is deprotonated by water to give the product:

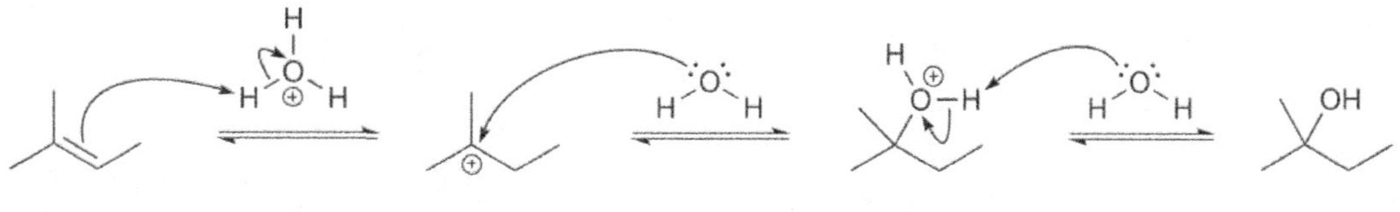

11.62 The starting material is an alcohol, and the reagent is concentrated sulfuric acid, suggesting an elimination reaction to give an alkene (the opposite of acid-catalyzed hydration). The mechanism should have the following three steps: 1) the OH group of the alcohol is first protonated (which requires two curved arrows) to generate a better leaving group, and then 2) water leaves as a leaving group to give a carbocation, and finally, 3) the resulting carbocation is deprotonated by water to give an alkene as the product:

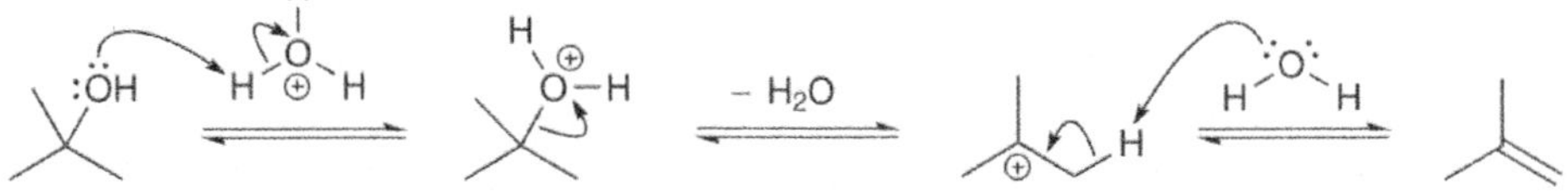

11.64 The starting material is an alkene, and the reagents indicate a hydroboration–oxidation process. The result is an *anti*-Markovnikov addition of H and OH across the alkene to give an alcohol (with the OH group being installed at the less-substituted position). This process achieves a *syn* addition of H and OH, and since two new chiral centers are generated during the reaction, we must be careful to draw only the pair of enantiomers that results from a *syn* addition:

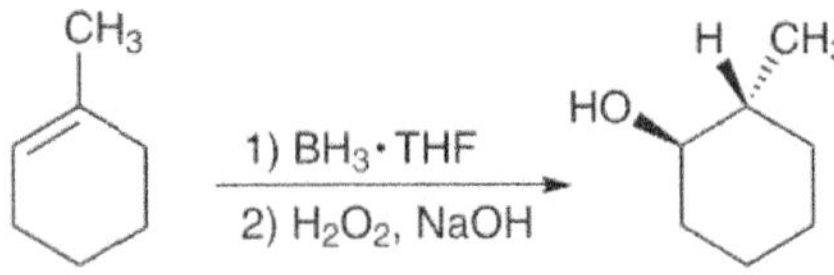

11.65 The starting material is an alkene, and the reagents indicate a hydroboration–oxidation process. The result is an *anti*-Markovnikov addition of H and OH across the alkene to give an alcohol (with the OH group being installed at the less-substituted position). The product does not contain any chiral centers, so we draw only one product:

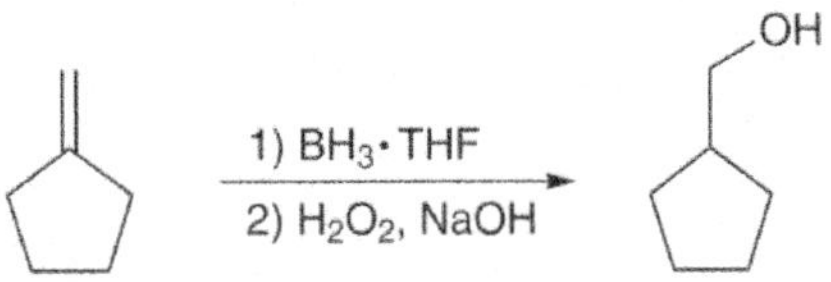

11.66 The starting material is an alkene, and the reagents indicate a hydroboration–oxidation process. The result is an *anti*-Markovnikov addition of H and OH across the alkene to give an alcohol (with the OH group being installed at the less-substituted position). This process achieves a *syn* addition of H and OH, but only one new chiral center is being generated, so the *syn* requirement is not relevant in this case. The process is expected to generate a pair of enantiomers, as shown:

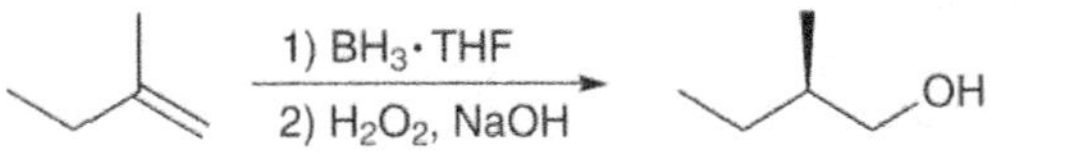

11.67 The starting material is an alkene, and the reagents indicate a hydroboration–oxidation process. The result is an

anti-Markovnikov addition of H and OH across the alkene to give an alcohol (with the OH group being installed at the less-substituted position). This process achieves a *syn* addition of H and OH, and since two new chiral centers are generated during the reaction, we must be careful to draw only the pair of enantiomers that results from a *syn* addition:

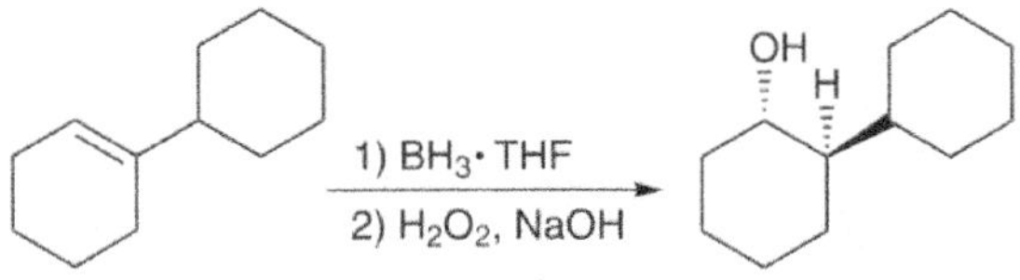

11.68 The starting material is an alkene, and the reagents indicate a hydroboration–oxidation process. The result is an *anti*-Markovnikov addition of H and OH across the alkene to give an alcohol (with the OH group being installed at the less-substituted position). This process achieves a *syn* addition of H and OH, but only one new chiral center is being generated, so the *syn* requirement is not relevant in this case. The process is expected to generate a pair of enantiomers, as shown:

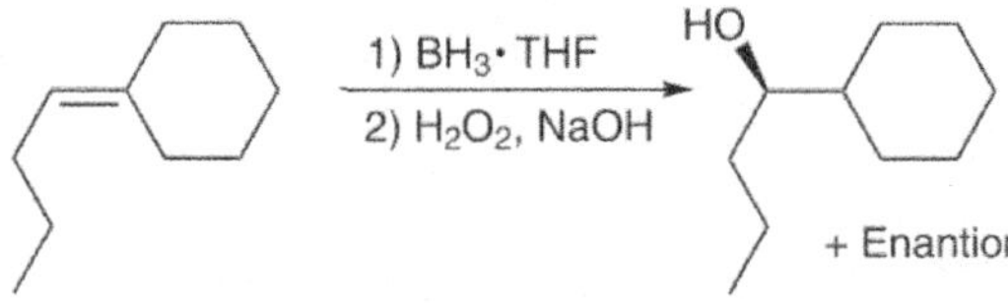

11.69 The starting material is an alkene, and the reagents indicate a hydroboration–oxidation process. The result is an *anti*-Markovnikov addition of H and OH across the alkene to give an alcohol (with the OH group being installed at the less-substituted position). The product does not contain any chiral centers, so we draw only one product:

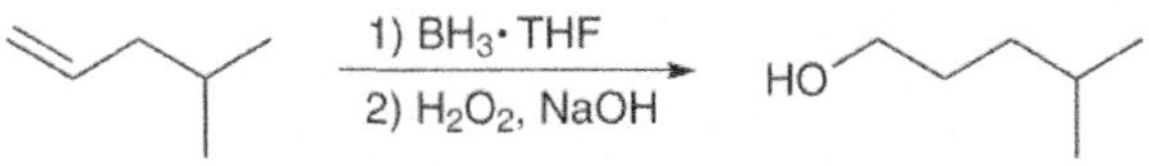

11.71 This transformation requires an *anti*-Markovnikov addition of H and OH, which can be accomplished via hydroboration–oxidation. The reagents for this process are shown here:

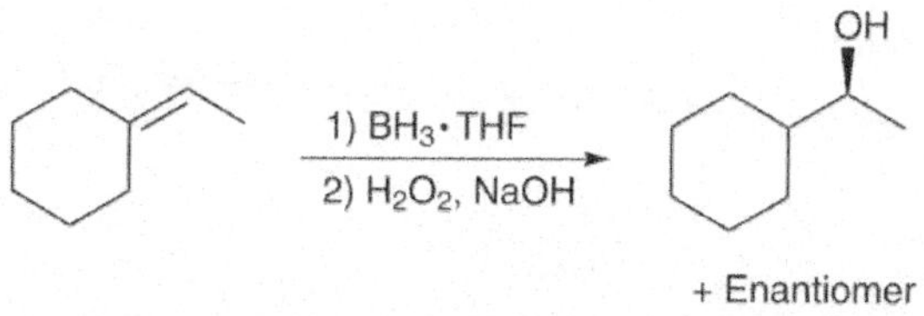

+ Enantiomer

11.72 This transformation requires a Markovnikov addition of H and Br, which can be accomplished by treating the starting alkene with HBr (without peroxides):

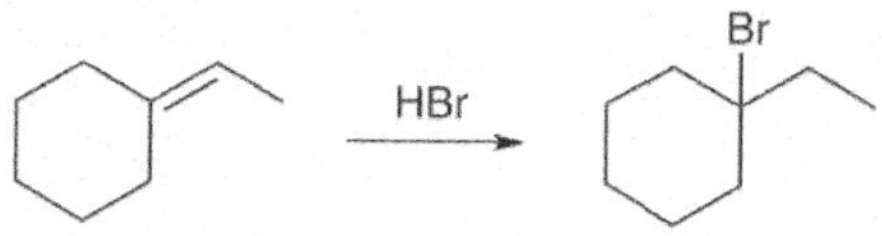

11.73 This transformation requires an *anti*-Markovnikov addition of H and Br, which can be accomplished by treating the starting alkene with HBr in the presence of peroxides:

11.74 This transformation requires the addition of H and H, which can be accomplished by treating the starting alkene with molecular hydrogen (H_2) in the presence of a suitable metal catalyst, such Pt, Pd, or Ni:

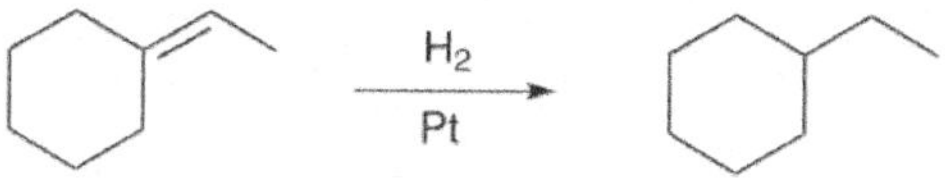

11.75 This transformation requires a Markovnikov addition of H and Cl, which can be accomplished by treating the starting alkene with HCl:

11.76 This transformation requires an *anti*-Markovnikov addition of H and OH, in a *syn* fashion, which can be accomplished via hydroboration–oxidation. The reagents for this process are shown here:

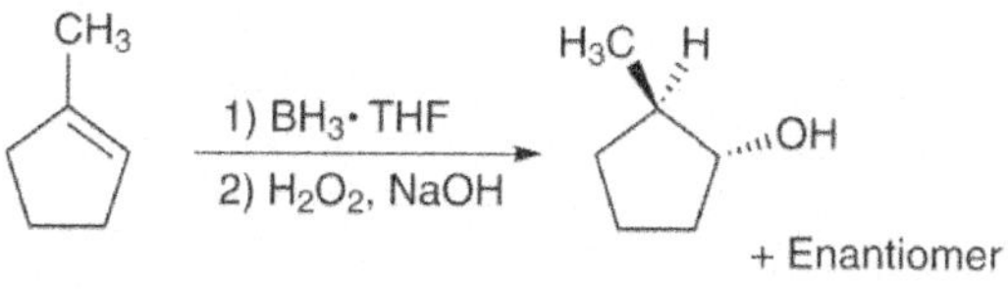

+ Enantiomer

11.77 This transformation can be achieved in one step, by treating the tertiary alkyl bromide with a strong base, such as NaOEt (or NaOMe, or NaOH), to give an E2 reaction. Note that we must use a strong base that is not sterically hindered because the Zaitsev product is the desired product (not the Hofmann product):

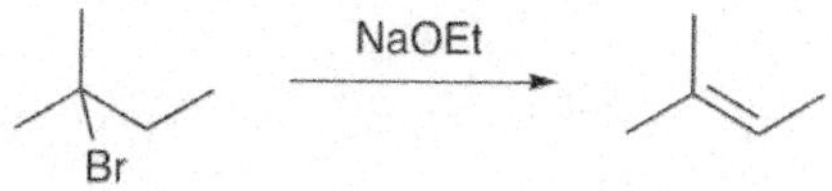

11.78 This transformation can be achieved in one step, by treating the tertiary alkyl bromide with a strong, sterically hindered base, such as *t*-BuOK, to give an E2 reaction. Note that we must use a sterically hindered base because the Hofmann product is the desired product (not the Zaitsev product):

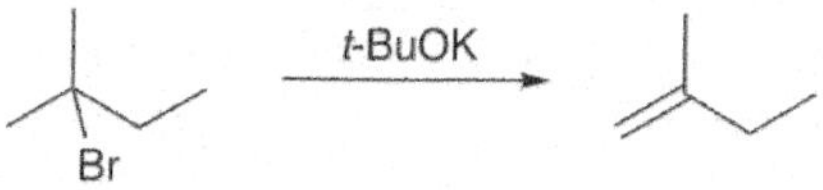

11.80 To solve this problem, we must change the location of the chlorine atom. This can be achieved via elimination, followed by addition. In the first step (elimination), we use *t*-BuOK (a sterically hindered base) to favor E2 over S_N2 because the substrate is primary (if we had used NaOH or NaOEt as the base, S_N2 would be favored over E2). The resulting alkene can then be treated with HCl to give a Markovnikov addition of H and Cl, resulting in the desired product:

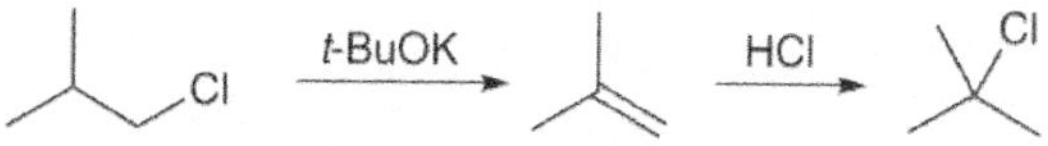

11.81 To solve this problem, we must change the location of the bromine atom. This can be achieved via elimination, followed by addition. In the first step (elimination), we use a strong base, such as NaOH or NaOEt, which gives only one E2 product. This product, an alkene, can then be treated with HBr in the presence of peroxides to give an *anti*-Markovnikov addition of H and Br, resulting in the desired product:

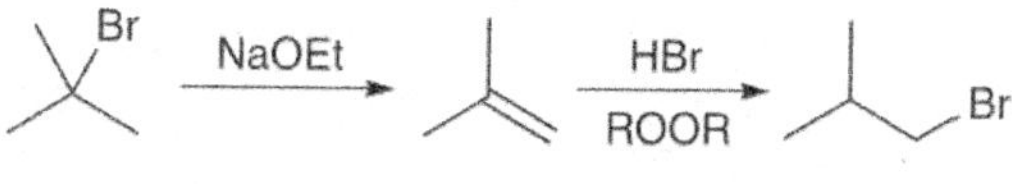

11.82 To solve this problem, we must change both the location and the identity of the functional group. Moving the location of the functional group can be achieved via elimination, followed by addition. And the identity of the functional group can be changed by considering which addition process to use. We can choose reagents that will install H and OH across the alkene. In the first step (elimination), a sterically hindered base must be used to give the Hofmann product. Then, hydroboration–oxidation can be performed to achieve *anti*-Markovnikov hydration, thereby installing an OH group at the less-substituted position, as shown here:

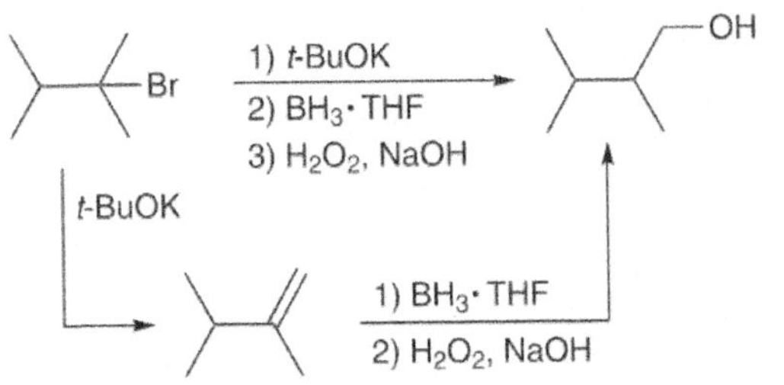

11.83 To solve this problem, we must change the location of the functional group. This can be achieved via elimination, followed by addition. But first, we convert the alcohol to a tosylate, which has a better leaving group (hydroxide is not a good leaving group, but tosylate is an excellent leaving group). After converting the alcohol to a tosylate, we treat the tosylate with a strong base, such as NaOEt or NaOH, to give an alkene (note that we do not use a sterically hindered base because we want the Zaitsev product, not the Hofmann product). Finally, hydroboration–oxidation can be performed to achieve *anti*-Markovnikov hydration, thereby installing an OH group at the less-substituted position. Since only one chiral center is formed in the process, we expect a pair of enantiomers (*syn* addition can occur on either face of the alkene, leading to a pair of enantiomers), as shown here:

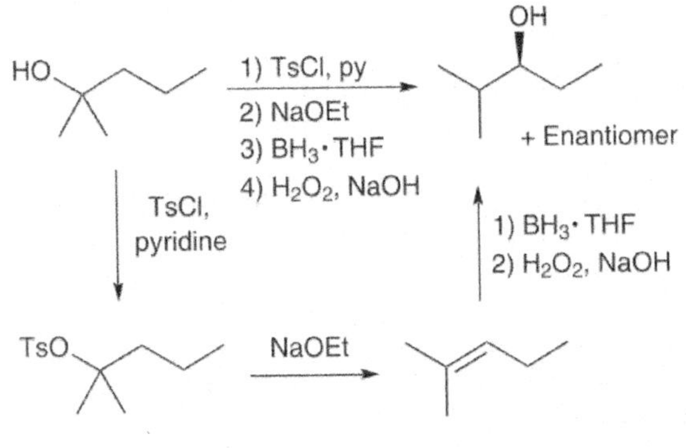

11.84 To solve this problem, we must change the location of the functional group. This can be achieved via elimination, followed by addition. But first, we convert the alcohol to a tosylate, which has a better leaving group (hydroxide is not a good leaving group, but tosylate is an excellent leaving group). After converting the alcohol to a tosylate, we treat the tosylate with a strong base, such as NaOEt or NaOH, to give an alkene (note that we do not use a sterically hindered base because we want the Zaitsev product, not the Hofmann product). Finally, hydroboration–oxidation can be performed to achieve *anti*-Markovnikov hydration, thereby installing an OH group at the less-substituted position. Two chiral centers are

formed in the process, and we expect the pair of enantiomers that result from a *syn* addition:

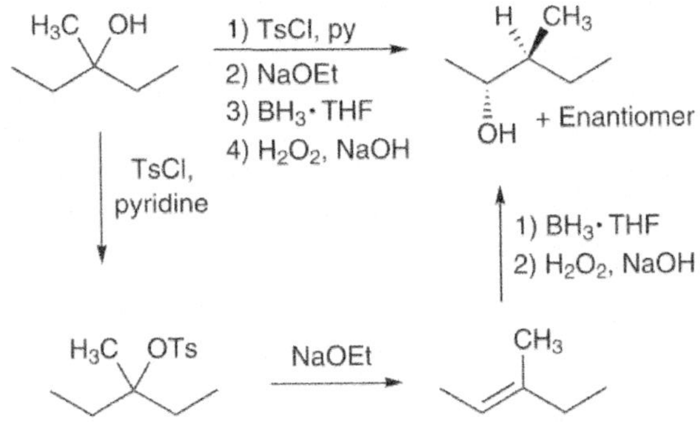

11.85 To solve this problem, we must change the location of the functional group. This can be achieved via elimination, followed by addition. In the first step (elimination), we use *t*-BuOK (a sterically hindered base) to favor E2 over S$_N$2 because the substrate is primary (if we had used NaOH or NaOEt as the base, S$_N$2 would be favored over E2). The resulting alkene can then be treated with HBr to give a Markovnikov addition of H and Br, resulting in the desired product:

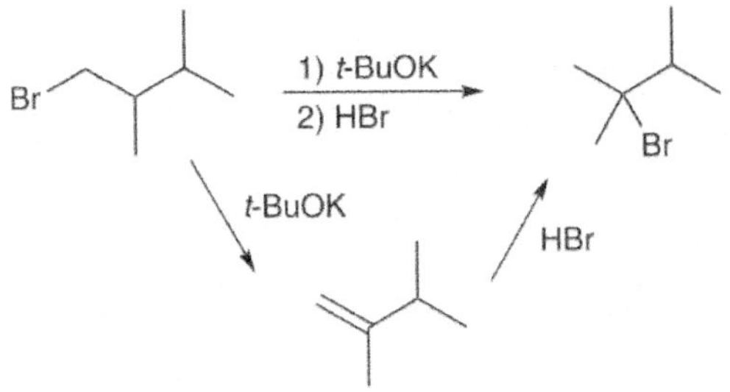

11.87 To solve this problem, we must change the location of the double bond. This can be achieved via addition, followed by elimination. In the first step, we use HBr with peroxides, thereby installing a bromine atom at the less-substituted position. This alkyl bromide can then be treated with a sterically hindered base to give the Hofmann product:

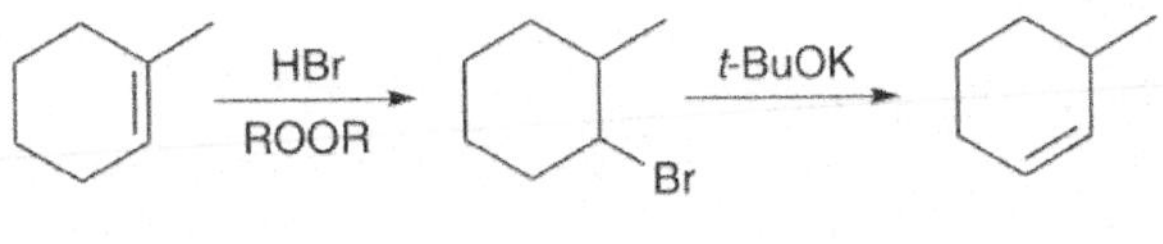

11.88 To solve this problem, we must change the location of the double bond. This can be achieved via addition, followed by elimination. In the first step, we use HBr (no peroxides), thereby installing a bromine atom at the more-substituted position. This alkyl bromide can then be treated with a strong base, such as NaOEt or NaOH, to give the Zaitsev product:

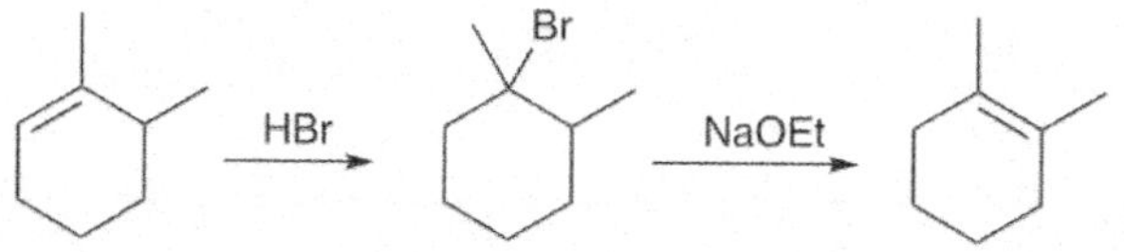

11.89 To solve this problem, we must change the location of the double bond. This can be achieved via addition, followed by elimination. In the first step, we use HBr (no peroxides), thereby installing a bromine atom at the more-substituted position. This alkyl bromide can then be treated with a strong base, such as NaOEt or NaOH, to give the Zaitsev product. This elimination process is stereoselective, affording the *trans* alkene as the major product.

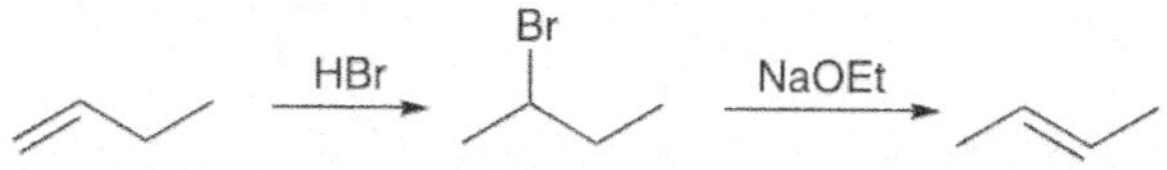

11.90 To solve this problem, we must change the location of the double bond. This can be achieved via addition, followed by elimination. In the first step, we use HBr with peroxides, thereby installing a bromine atom at the less-substituted position. This alkyl bromide can then be treated with a sterically hindered base to give the Hofmann product:

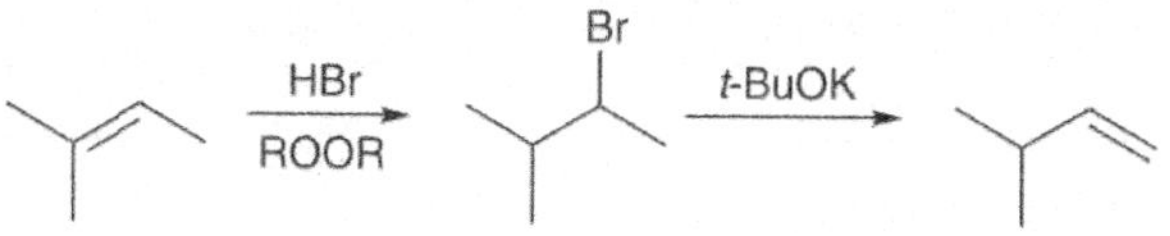

11.92 We are starting with a cycloalkane (no functional groups), which means that we must first install a functional group. This is best achieved via radical bromination, which selectively installs a bromine atom to give a tertiary cycloalkyl bromide. This cycloalkyl bromide can be converted into the desired product upon treatment with a strong sterically hindered base (an E2 reaction, giving the Hofmann product).

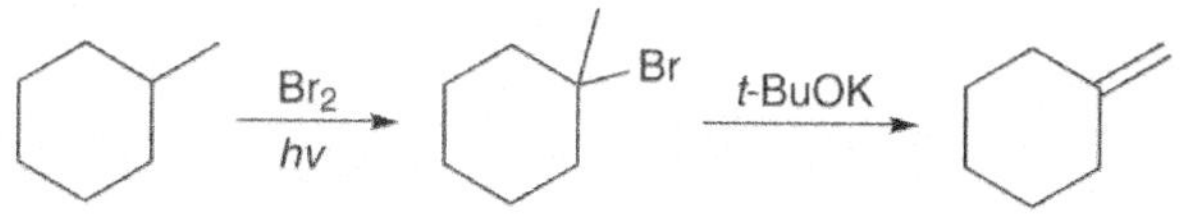

11.93 We are starting with a cycloalkane (no functional groups), which means that we must first install a functional group. This is best achieved via radical bromination, which selectively installs a bromine atom to give a tertiary cycloalkyl bromide. This cycloalkyl bromide can be converted into the desired product upon treatment with a strong base (an E2 reaction, giving the Zaitsev product).

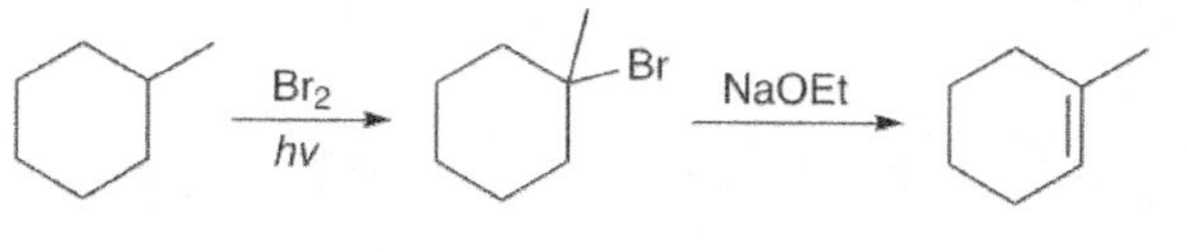

11.94 We are starting with an alkane (no functional groups), which means that we must first install a functional group. This is best achieved via radical bromination, which selectively installs a bromine atom to give a tertiary alkyl bromide. This alkyl bromide can be converted into the desired product upon treatment with a strong base (an E2 reaction, giving the Zaitsev product).

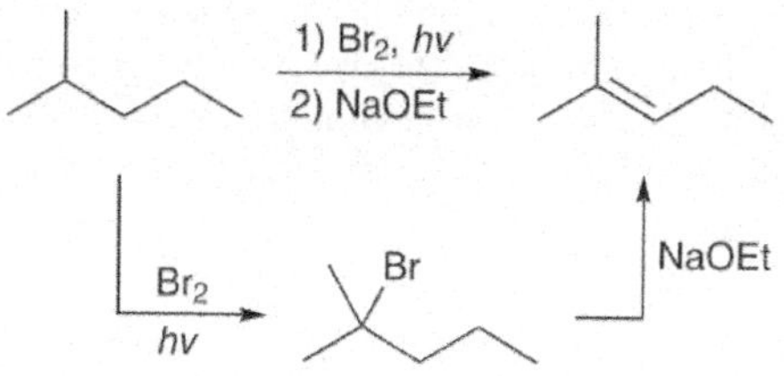

11.95 We are starting with an alkane (no functional groups), which means that we must first install a functional group. This is best achieved via radical bromination, which selectively installs a bromine atom to give a tertiary alkyl bromide. This alkyl bromide can be converted into the desired product upon treatment with a sterically hindered base (an E2 reaction, giving the Hofmann product).

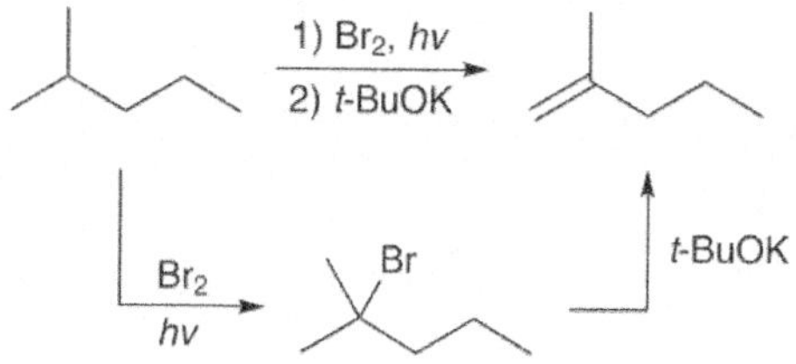

11.96 We are starting with a cycloalkane (no functional groups), which means that we must first install a functional group. This is best achieved via radical bromination, which selectively installs a bromine atom to give a tertiary cycloalkyl bromide. Converting this compound into the desired product requires moving the position of a bromine atom, which can be achieved via elimination, followed by addition. For the elimination step, we use *t*-BuOK (a sterically hindered base) to favor the Hofmann product. And for the addition step, we use HBr in the presence of peroxides to achieve an *anti*-Markovnikov addition of H and Br, resulting in the desired product:

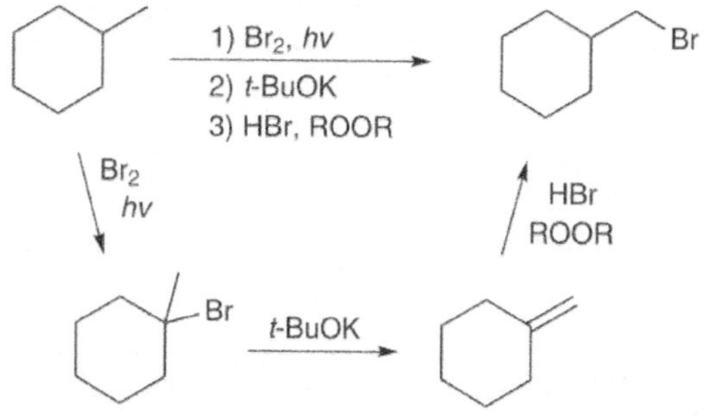

11.97 We are starting with an alkane (no functional groups), which means that we must first install a functional group. This is best achieved via radical bromination, which selectively installs a bromine atom to give a tertiary alkyl bromide. Treating this alkyl halide with a strong base (such as NaOEt or NaOH) affords the more substituted alkene. Converting this alkene into the desired product requires moving the position of the double bond, which can be achieved via addition, followed by elimination. For the addition step, we use HBr and peroxides to give an *anti*-Markovnikov addition of H and Br. And finally, in the elimination step, we use *t*-BuOK (a sterically hindered base) to favor the desired alkene (the Hofmann product), as shown here:

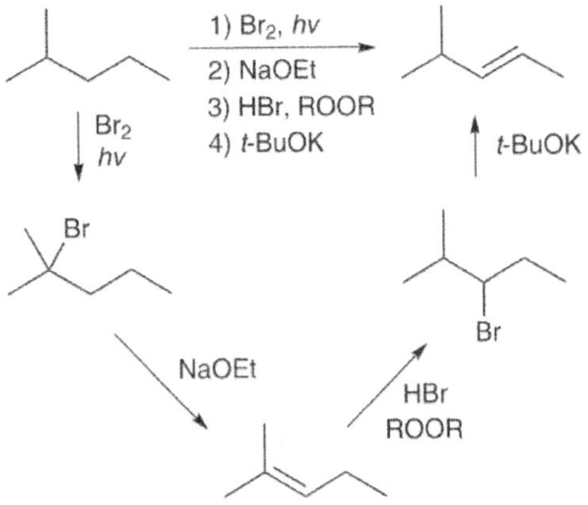

11.99 The starting material is an alkene, and the reagent indicates a bromination reaction. The result is the addition of Br and Br across the alkene. This process is an *anti* addition, but only one new chiral center is being generated, so the *anti* requirement is not relevant in this case. The process will generate a pair of enantiomers, as shown:

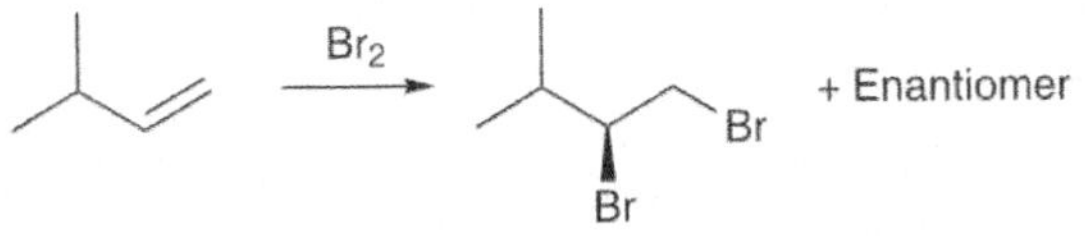

11.100 The starting material is an alkene, and the reagents indicate halohydrin formation. The result is the addition of OH and Br across the alkene, with the OH group being installed at the more-substituted position. This process is an *anti* addition, but only one new chiral center is being generated, so the *anti* requirement is not relevant in this case. The process will generate a pair of enantiomers, as shown:

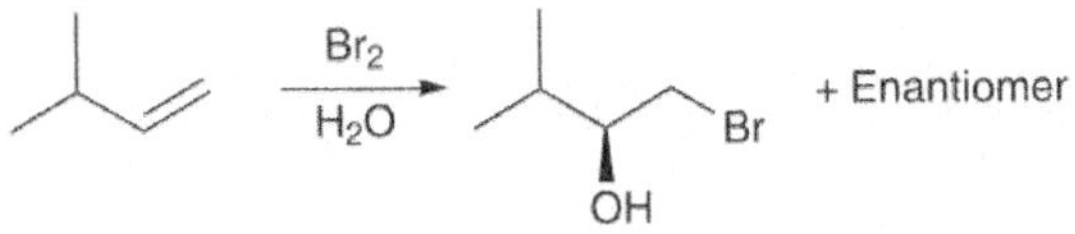

11.101 The starting material is an alkene, and the reagent indicates a bromination reaction. The result is the addition of Br and Br across the alkene. This process is an *anti* addition, and since

two new chiral centers are generated during the reaction, we must be careful to draw only the pair of enantiomers that result from an *anti* addition:

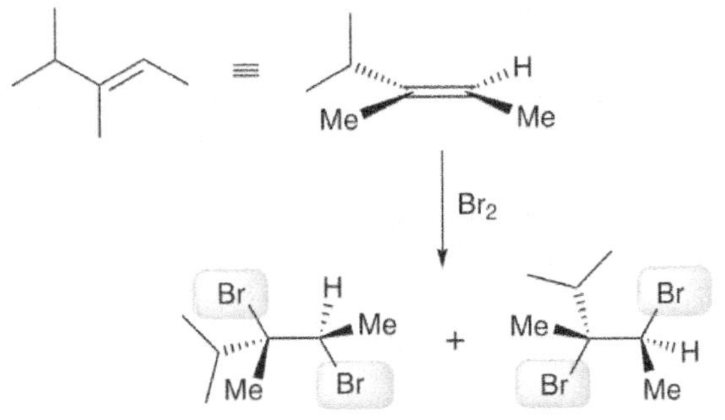

Alternatively, the products can be drawn in the following way:

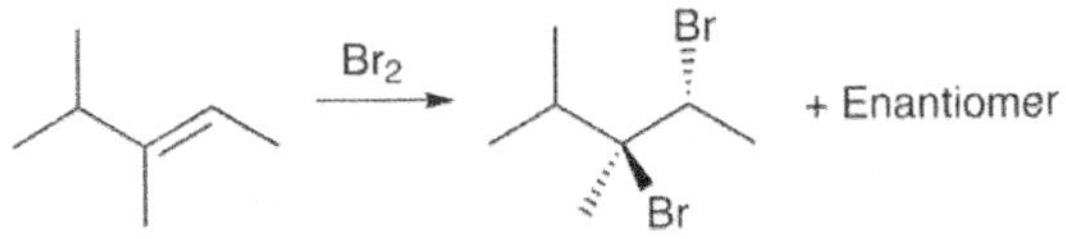

11.102 The starting material is an alkene, and the reagents indicate halohydrin formation. The result is the addition of OH and Br across the alkene, with the OH group being installed at the more-substituted position. This process is an *anti* addition, and since two new chiral centers are generated during the reaction, we must be careful to draw only the pair of enantiomers that result from an *anti* addition:

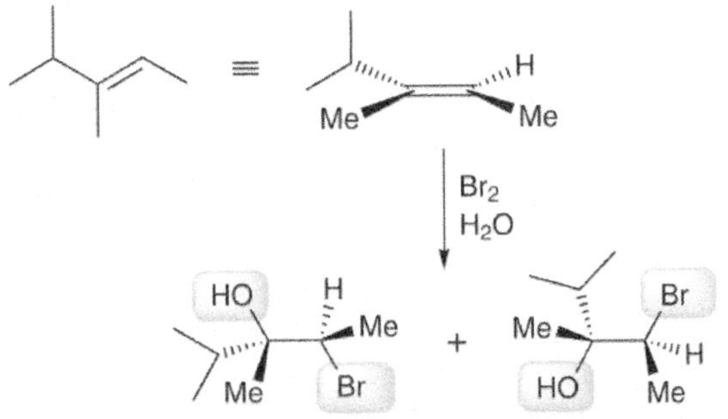

Alternatively, the products can be drawn in the following way:

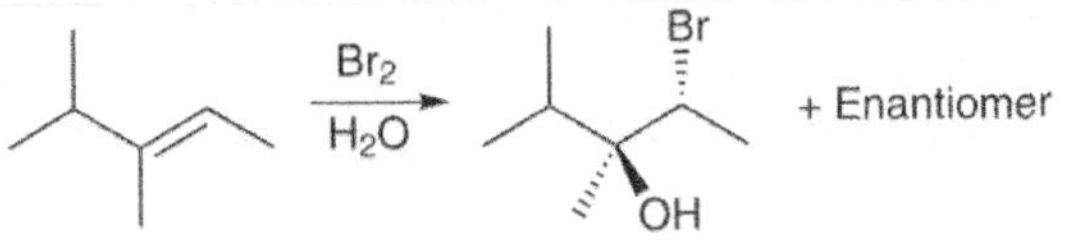

11.103 The starting material is an alkene, and the reagent indicates a bromination reaction. The result is the addition of Br and Br across the alkene. This process is an *anti* addition, and two new chiral centers are generated during this reaction. However, in this case, we don't get a pair of enantiomers, but rather, we get a *meso* compound:

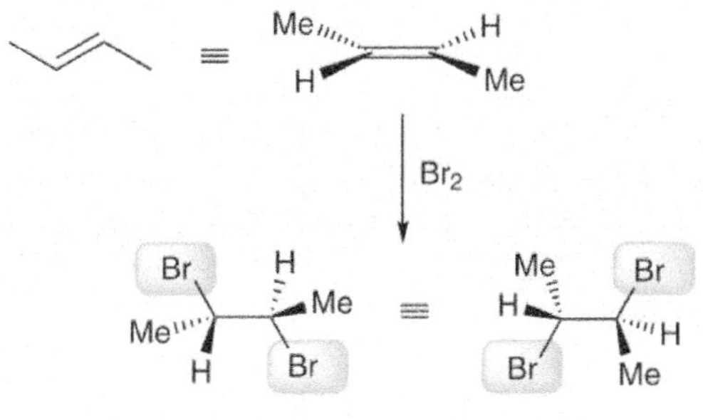

A *meso* compound

To see more clearly that this is indeed a *meso* compound, we can take either drawing above (let's look at the first drawing), and imagine rotating about the central carbon–carbon bond. The resulting conformation (shown below) clearly has an internal plane of symmetry, so this compound must be a *meso* compound. You should be able to do the same thing with the second structure above (change its conformation until it is clear that it is the same *meso* compound).

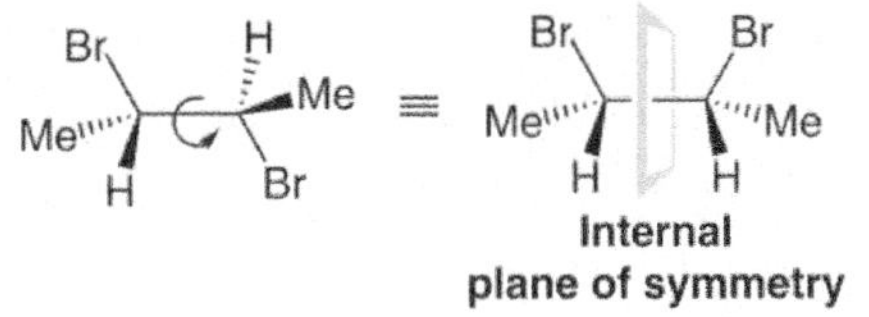

Internal plane of symmetry

11.104 The starting material is an alkene, and the reagents indicate halohydrin formation. The result is the addition of OH and Br across the alkene. The alkene is symmetrical, so we don't need to consider regiochemistry, but the stereochemistry is relevant. This process is an *anti* addition, and since two new chiral centers are generated during the reaction, we must be careful to draw only the pair of enantiomers that result from an *anti* addition:

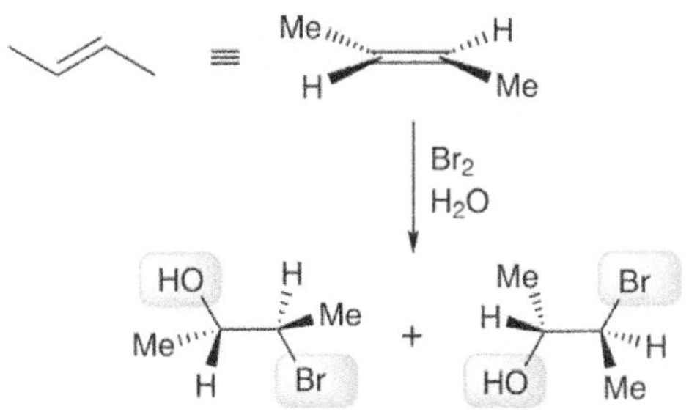

A pair of enantiomers

11.106 The starting material is an alkene, and the reagents indicate the addition of OH and OH across the alkene in an *anti* fashion. Since two new chiral centers are generated during the reaction, we

must be careful to draw only the pair of enantiomers that result from an *anti* addition:

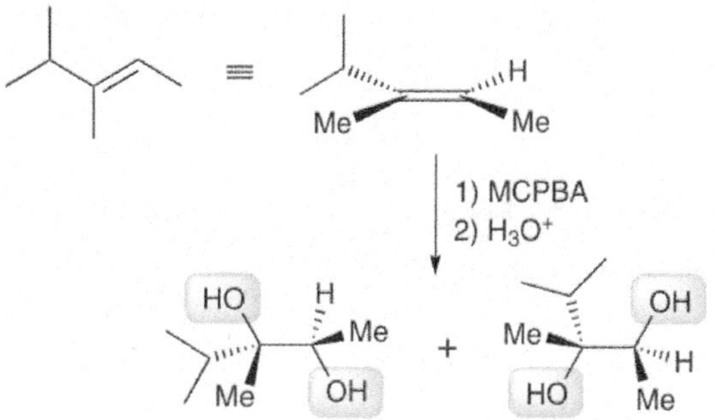

A pair of enantiomers

11.107 The starting material is an alkene, and the reagents indicate the addition of OH and OH across the alkene in an *anti* fashion. Since two new chiral centers are generated during the reaction, we must be careful to draw only the pair of enantiomers that result from an *anti* addition:

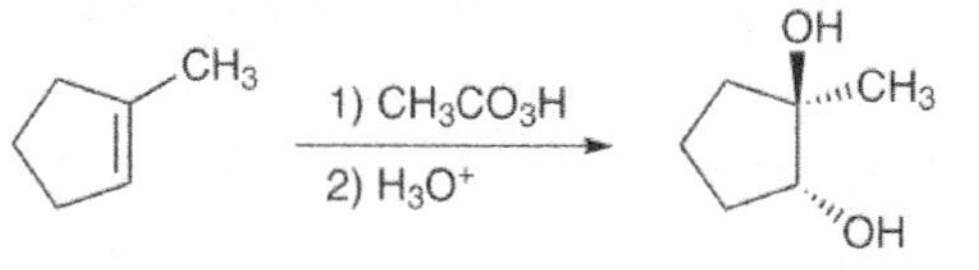

+ Enantiomer

11.108 The starting material is an alkene, and the reagents indicate the addition of OH and OH across the alkene in an *anti* fashion. Since no chiral centers are generated during the reaction, the *anti* requirement is not relevant, and there is only one product:

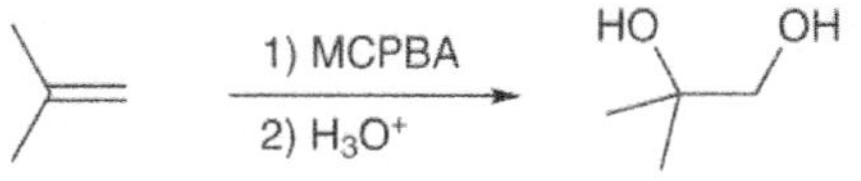

11.109 The starting material is an alkene, and the reagents indicate the addition of OH and OH across the alkene in an *anti* fashion. Only one new chiral center is being generated, so the *anti* requirement is not relevant in this case. The process is expected to generate a pair of enantiomers, as shown:

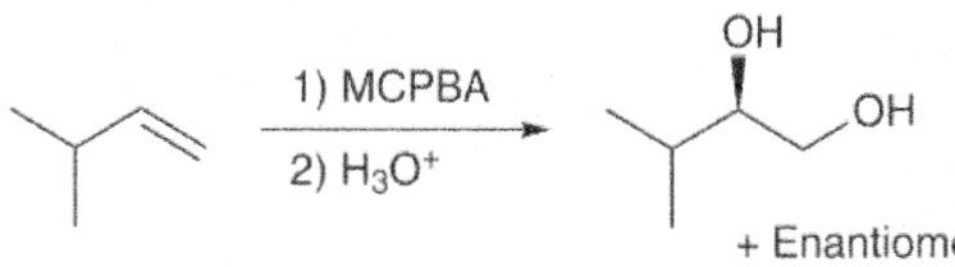

+ Enantiomer

11.111 The starting material is an alkene, and the reagents indicate the addition of OH and OH across the alkene in a *syn* fashion. Since two new chiral centers are generated during the reaction, we must be careful to draw only the pair of enantiomers that result from a *syn* addition:

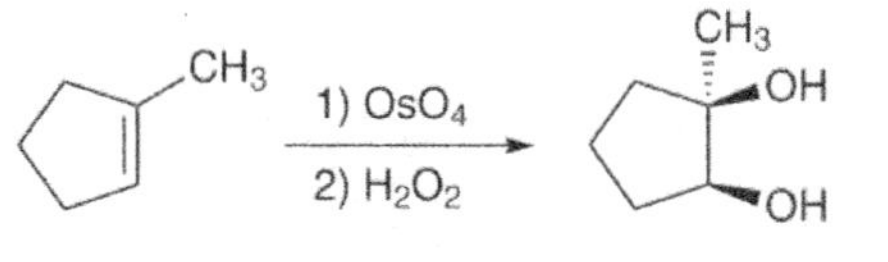

11.112 The starting material is an alkene, and the reagents indicate the addition of OH and OH across the alkene in a *syn* fashion. Two chiral centers are generated during this reaction; however, in this case, we don't get a pair of enantiomers, but rather, we get a *meso* compound:

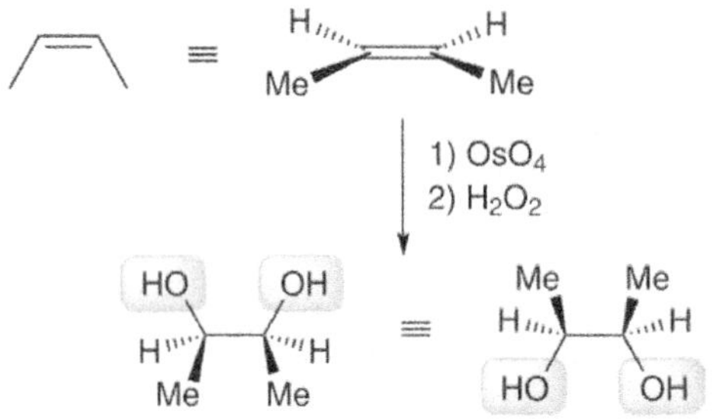

A *meso* compound

11.113 The starting material is an alkene, and the reagents indicate the addition of OH and OH across the alkene in a *syn* fashion. Since no chiral centers are generated during the reaction, the *syn* requirement is not relevant, and there is only one product:

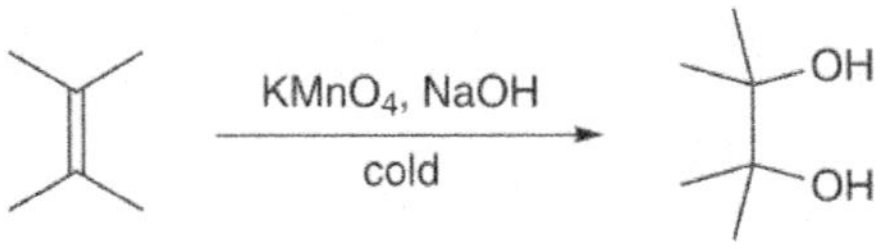

11.114 The starting material is an alkene, and the reagents indicate the addition of OH and OH across the alkene in a *syn* fashion. Since no chiral centers are generated during the reaction, the *syn* requirement is not relevant, and there is only one product:

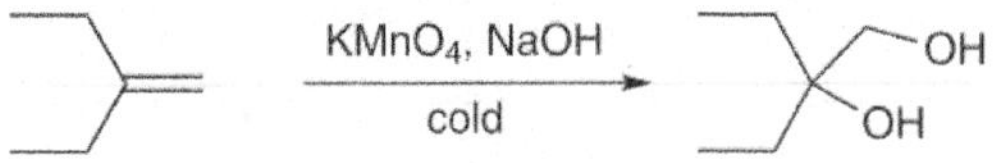

11.115 The starting material is an alkene, and the reagents indicate the addition of OH and OH across the alkene in a *syn* fashion. Only one new chiral center is being generated, so the *syn* requirement is not relevant in this case. The process is expected to generate a pair of enantiomers, as shown:

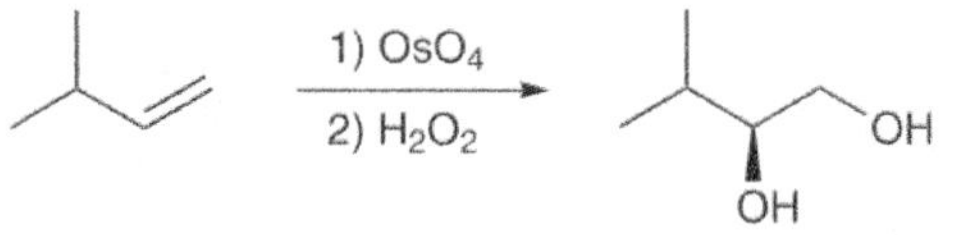

+ Enantiomer

11.116 The starting material is an alkene, and the reagents indicate the addition of OH and OH across the alkene in a *syn* fashion. Since two new chiral centers are generated during the reaction, we must be careful to draw only the pair of enantiomers that result from a *syn* addition:

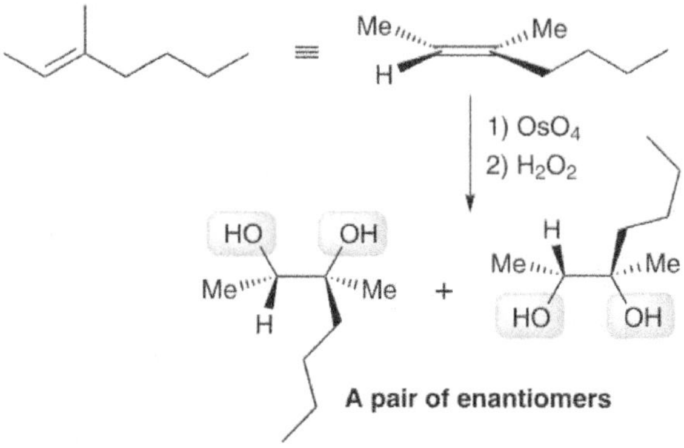

A pair of enantiomers

Alternatively, you might find it easier to draw the products like this:

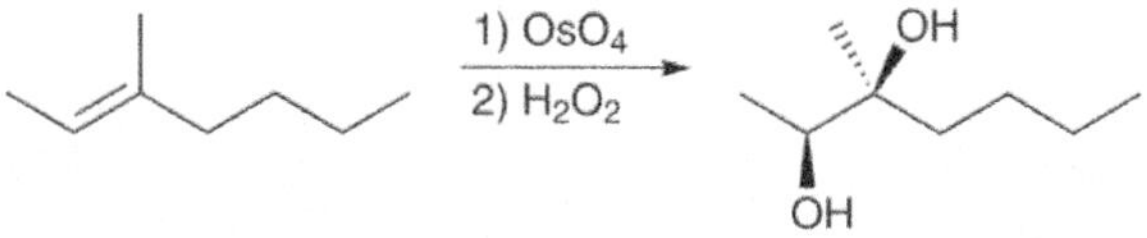

+ Enantiomer

11.118 The reagents indicate an ozonolysis process, which cleaves each C=C bond, and generates two C=O bonds in the place of each C=C bond, like this:

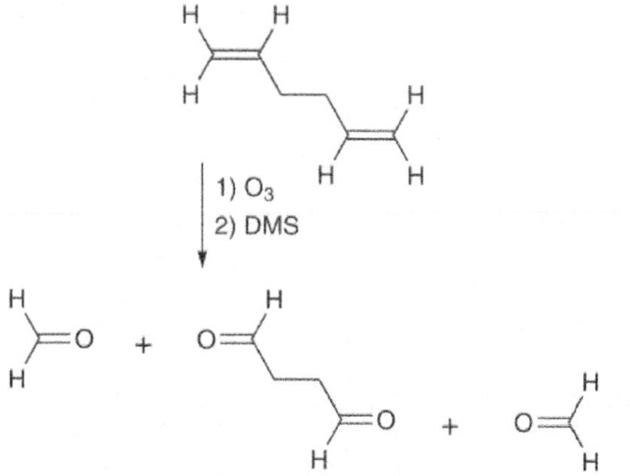

11.119 The reagents indicate an ozonolysis process, which cleaves each C=C bond, and generates two C=O bonds in the place of each C=C bond, like this:

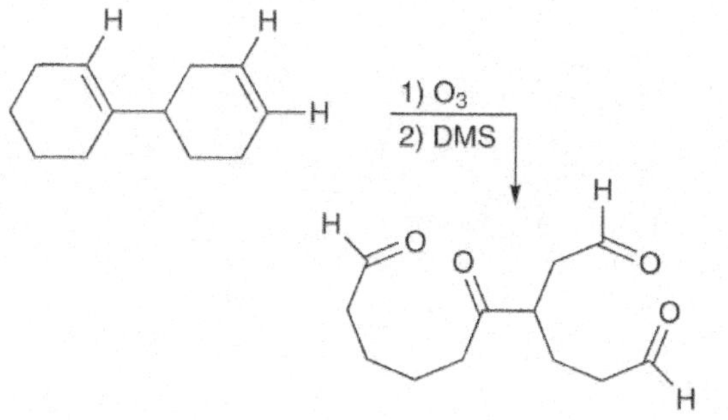

11.120 The reagents indicate an ozonolysis process, which cleaves the C=C bond, and generates two C=O bonds in the place of the C=C bond, like this:

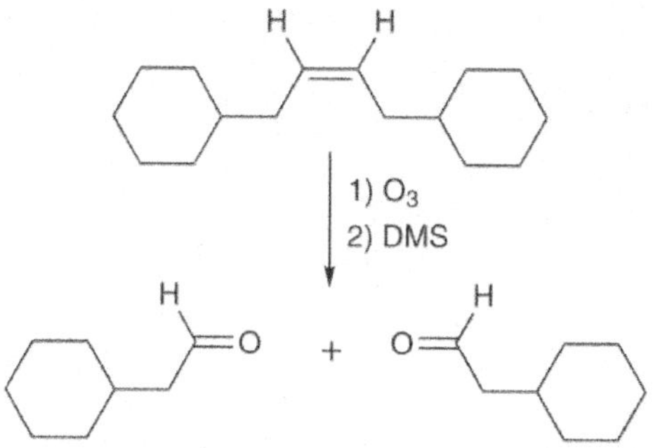

11.121 The reagents indicate an ozonolysis process, which cleaves the C=C bond, and generates two C=O bonds in the place of the C=C bond, like this:

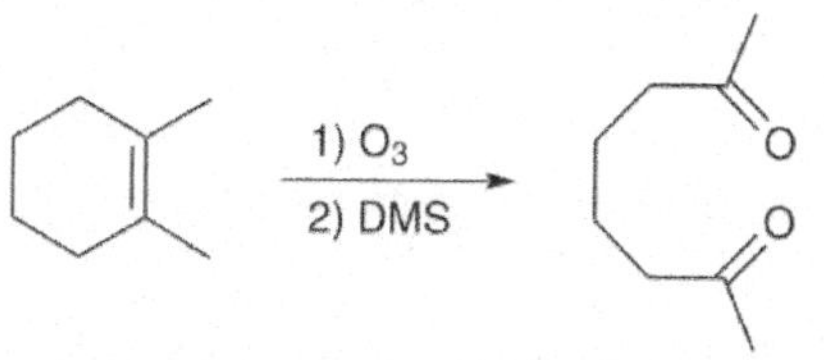

11.122 The reagents indicate an ozonolysis process, which cleaves each C=C bond, and generates two C=O bonds in the place of each C=C bond, like this:

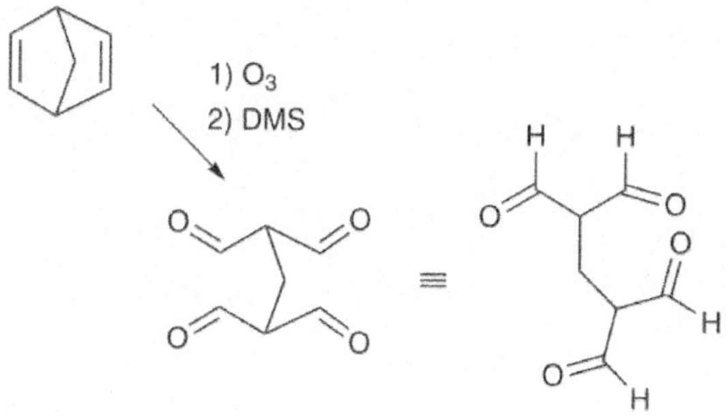

11.123 The reagents indicate an ozonolysis process, which cleaves the C=C bond and generates two C=O bonds in the place of the C=C bond, like this:

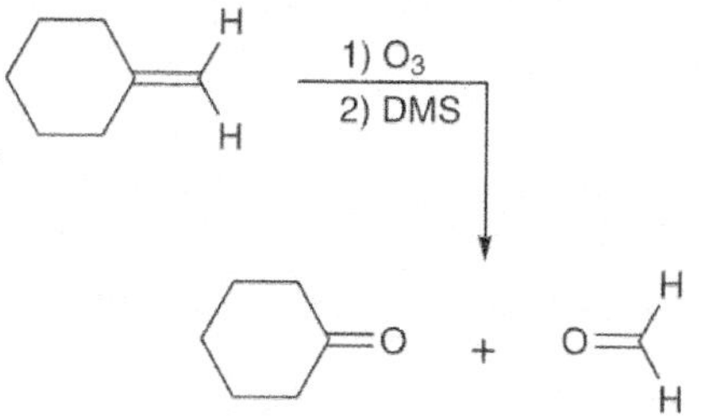

CHAPTER 12

12.2 The base has a negative charge on a nitrogen atom, so this base is strong enough to deprotonate the terminal alkyne, giving the following alkynide ion:

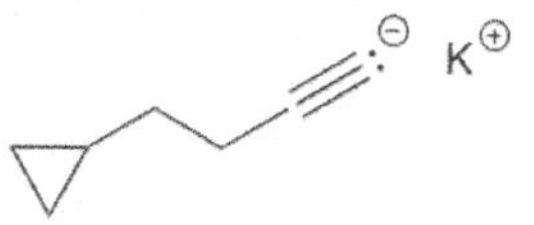

12.3 The base has a negative charge that is delocalized (via resonance) over two oxygen atoms. This base is not strong enough to deprotonate a terminal alkyne.

12.4 The base has a negative charge on an sp^3 hybridized carbon atom, so this base is strong enough to deprotonate the terminal alkyne, giving the following alkynide ion:

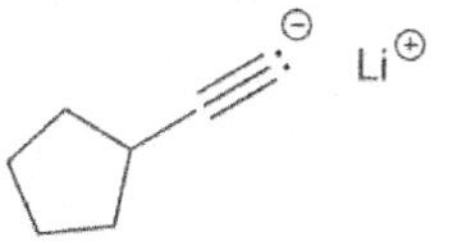

12.5 The base has a negative charge on an oxygen atom. This base is not strong enough to deprotonate a terminal alkyne.

12.6 The starting material is a geminal dibromide. When treated with excess $NaNH_2$, two successive E2 reactions occur, generating the terminal alkyne shown below. Under the strongly basic conditions, this terminal alkyne is deprotonated to give an alkynide ion. That is why water is used to work-up the reaction. Water serves as a weak acid that protonates the alkynide ion to regenerate the terminal alkyne.

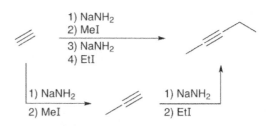

12.7 The starting material is a vicinal dichloride. When treated with excess NaNH$_2$, two successive E2 reactions occur, generating the terminal alkyne shown below. Under the strongly basic conditions, this terminal alkyne is deprotonated to give an alkynide ion. That is why water is used to work-up the reaction. Water serves as a weak acid that protonates the alkynide ion to regenerate the terminal alkyne.

12.8 The starting material is a geminal dichloride. When treated with excess NaNH$_2$, two successive E2 reactions occur, generating the terminal alkyne shown below. Under the strongly basic conditions, this terminal alkyne is deprotonated to give an alkynide ion. That is why water is used to work-up the reaction. Water serves as a weak acid that protonates the alkynide ion to regenerate the terminal alkyne.

12.9 The starting material is a vicinal dibromide. When treated with excess NaNH$_2$, two successive E2 reactions occur, generating the terminal alkyne shown below. Under the strongly basic conditions, this terminal alkyne is deprotonated to give an alkynide ion. That is why water is used to work-up the reaction. Water serves as a weak acid that protonates the alkynide ion to regenerate the terminal alkyne.

12.10 The starting material can be converted into the desired product via alkylation, using an appropriate alkyl halide, as shown below:

12.11 The starting material can be converted into the desired product via two successive alkylation processes. In the following reaction sequence, the methyl group is installed first and the ethyl group is installed last. Alternatively, the ethyl group could have been installed first.

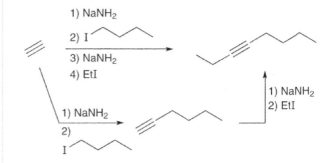

12.12 The starting material can be converted into the desired product via two successive alkylation processes. In the following reaction sequence, the butyl group is installed first and the ethyl group is installed last. Alternatively, the ethyl group could have been installed first.

12.13 The starting material can be converted into the desired product via two successive alkylation processes, each of which requires two steps. Even though we are installing two identical groups (two methyl groups), each group must be installed in its own separate alkylation process.

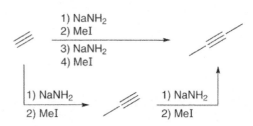

12.14 When an alkyne is treated with molecular hydrogen (H$_2$) in the presence of Pt, the alkyne reacts with two equivalents of H$_2$ to give an alkane, as shown:

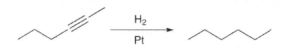

12.15 When an alkyne is treated with molecular hydrogen (H$_2$) in the presence of a poisoned catalyst, such as Lindlar's catalyst, the alkyne reacts with one equivalent of H$_2$ in a *syn* addition. In this case, the resulting alkene is not stereoisomeric, so the *syn* requirement is not relevant.

H₂
Lindlar's catalyst

12.16 When an alkyne is treated with sodium (Na) dissolved in liquid ammonia (NH₃), the alkyne is converted into a *trans* alkene.

Na, NH₃

12.17 When an alkyne is treated with molecular hydrogen (H₂) in the presence of a poisoned catalyst, such as Lindlar's catalyst, the alkyne reacts with one equivalent of H₂ in a *syn* addition. The product is a *cis* alkene.

H₂
Lindlar's catalyst

12.18 When an alkyne is treated with sodium (Na) dissolved in liquid ammonia (NH₃), the alkyne is converted into a *trans* alkene.

Na, NH₃

12.19 The starting material is a geminal dichloride, which is converted into a terminal alkyne upon treatment with excess NaNH₂, followed by water work-up. This alkyne undergoes alkylation upon treatment with NaNH₂, followed by methyl iodide, to install a methyl group. The resulting internal alkyne is converted into a *cis* alkene upon hydrogenation in the presence of a poisoned catalyst:

1) xs NaNH₂
2) H₂O

1) NaNH₂
2) MeI

H₂,
Lindlar's catalyst

12.20

(a) 1-Pentyne can be converted into pentane upon treatment with molecular hydrogen (H₂) in the presence of Pt, as shown:

H₂
Pt

(b) 1-Pentyne can be converted into 2-hexyne via an alkylation process that uses methyl iodide, as shown:

1) NaNH₂
2) CH₃I

(c) 1-Pentyne can be converted into 3-heptyne via an alkylation process that uses ethyl iodide, as shown:

1) NaNH₂
2) CH₃CH₂I

(d) 1-Pentyne can be converted into *cis*-4-octyne via an alkylation process (to install a propyl group), followed by reduction with a poisoned catalyst, as shown:

1) NaNH₂
2) CH₃CH₂CH₂I
3) H₂, Lindlar's catalyst

1) NaNH₂
2) CH₃CH₂CH₂I

H₂
Lindlar's catalyst

(e) 1-Pentyne can be converted into *trans*-4-octyne via an alkylation process (to install a propyl group), followed by a dissolving metal reduction to give the desired product, as shown:

1) NaNH₂
2) CH₃CH₂CH₂I
3) Na, NH₃

1) NaNH₂
2) CH₃CH₂CH₂I

Na, NH₃

12.21 The starting material is a terminal alkyne, and the reagents indicate an acid-catalyzed hydration reaction. Markovnikov addition of H and OH gives an enol, which rapidly tautomerizes into the methyl ketone shown:

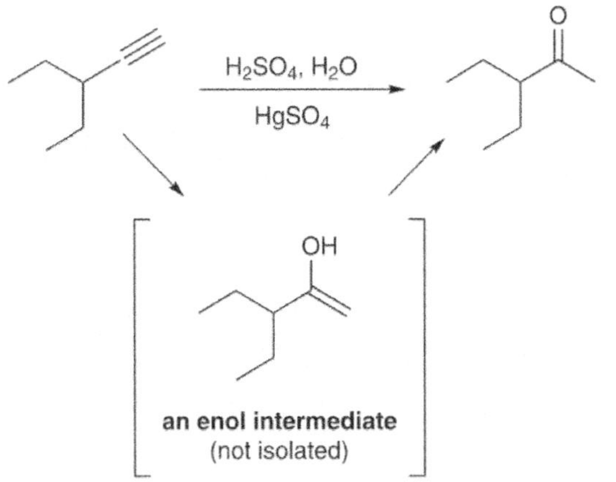

12.22 The starting material is a terminal alkyne, and the reagents indicate an acid-catalyzed hydration reaction. Markovnikov addition of H and OH gives an enol, which rapidly tautomerizes into the methyl ketone shown:

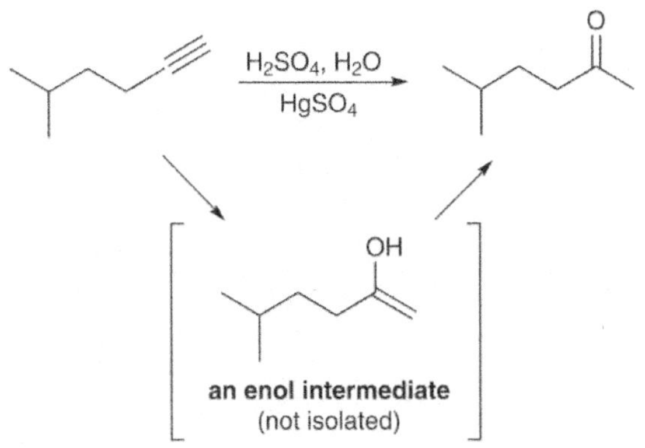

12.23 The starting material is a symmetrical alkyne, and the reagents indicate an acid-catalyzed hydration reaction. An enol is initially formed, but the enol cannot be isolated because it rapidly tautomerizes to give a ketone, as shown:

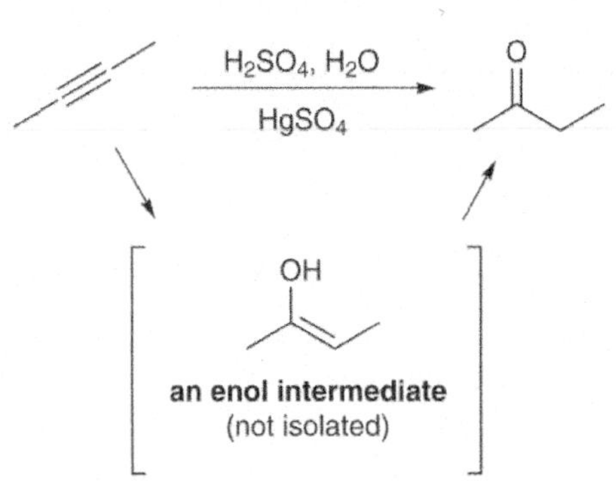

12.24 The starting material is a terminal alkyne, and the reagents indicate a hydroboration–oxidation process. An *anti*-Markovnikov addition of H and OH results in an enol, which rapidly tautomerizes into an aldehyde product, as shown:

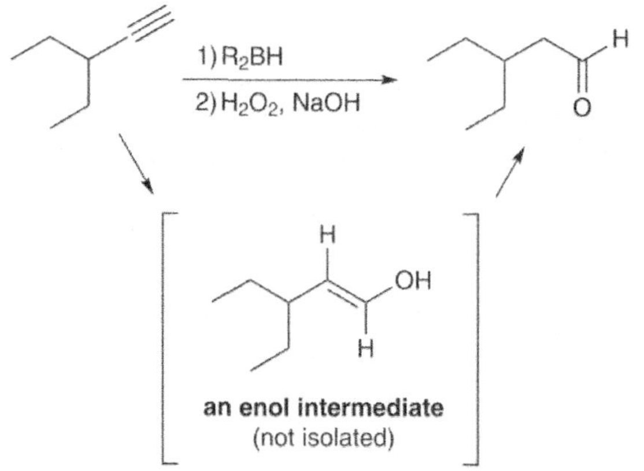

12.25 The starting material is a terminal alkyne, and the reagents indicate a hydroboration–oxidation process. An *anti*-Markovnikov addition of H and OH results in an enol, which rapidly tautomerizes into an aldehyde product, as shown:

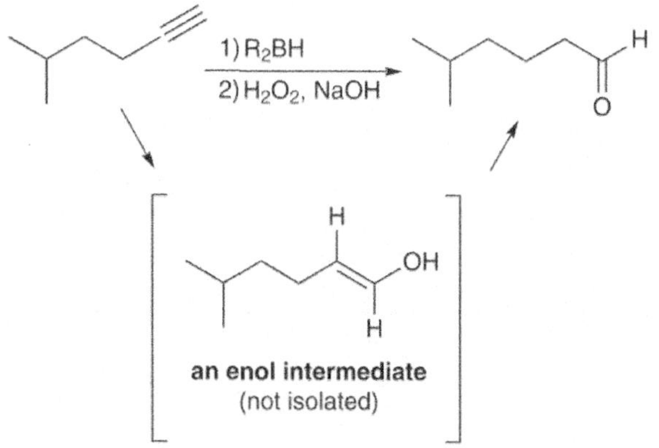

12.26 The starting material is a terminal alkyne, and the reagents indicate a hydroboration–oxidation process. An *anti*-Markovnikov addition of H and OH results in an enol, which rapidly tautomerizes into an aldehyde product, as shown:

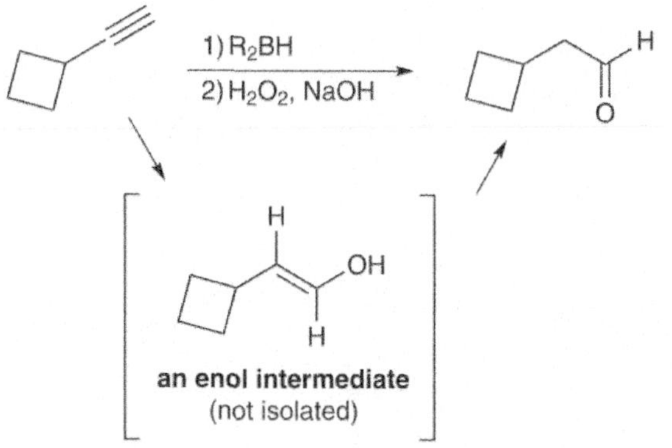

12.27 The desired transformation involves the conversion of a terminal alkyne into an aldehyde. This *anti*-Markovnikov hydration can be accomplished via hydroboration–oxidation. The reagents are shown below:

1) R₂BH
2) H₂O₂, NaOH

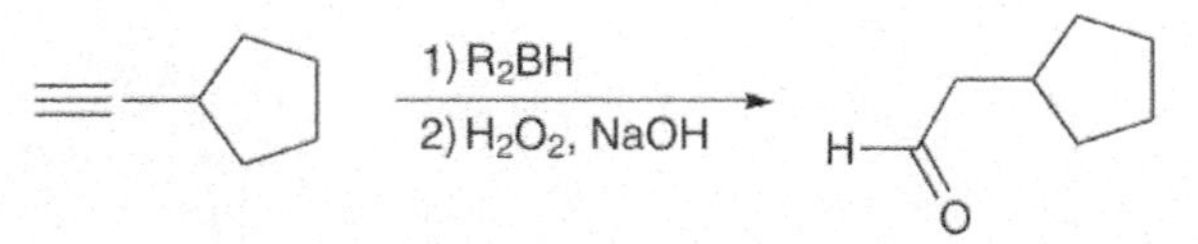

12.28 The desired transformation involves the conversion of a terminal alkyne into a methyl ketone. This Markovnikov regiochemistry can be achieved via acid-catalyzed hydration in the presence of mercuric sulfate (HgSO₄). The reagents are shown below:

H₂SO₄, H₂O
HgSO₄

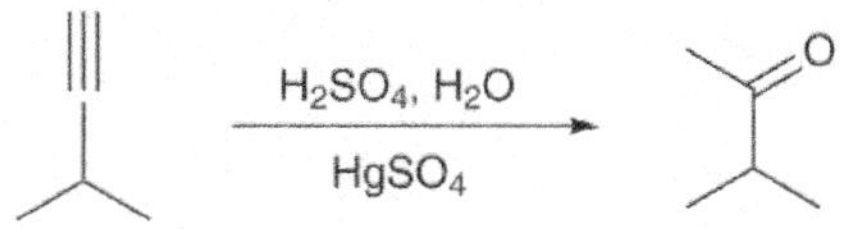

12.29 The desired transformation involves the conversion of a terminal alkyne into a methyl ketone. This Markovnikov regiochemistry can be achieved via acid-catalyzed hydration in the presence of mercuric sulfate (HgSO₄). The reagents are shown below:

H₂SO₄, H₂O
HgSO₄

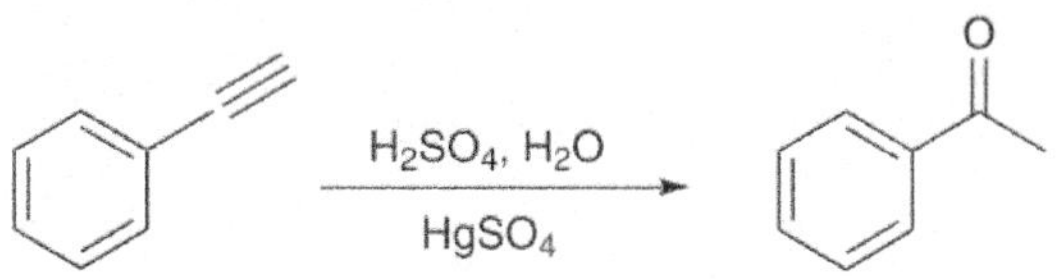

12.30 The desired transformation involves the conversion of a geminal dibromide into an aldehyde. This can be accomplished by first converting the geminal dibromide into a terminal alkyne and then converting the terminal alkyne into the desired aldehyde via hydroboration–oxidation. The reagents are shown below:

1) xs NaNH₂
2) H₂O
3) R₂BH
4) H₂O₂, NaOH

1) xs NaNH₂
2) H₂O

1) R₂BH
2) H₂O₂, NaOH

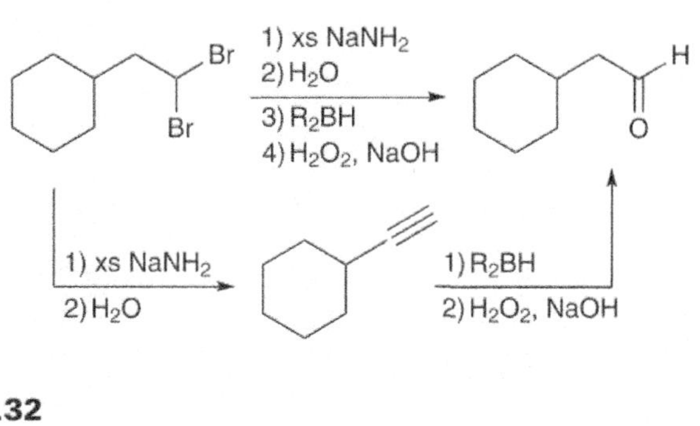

12.32

(a) In acidic conditions, the first step is protonation, using H₃O⁺ as the proton source. This step requires two curved arrows. When drawing the curved arrows for this step, make sure to protonate the pi bond, rather than protonating the OH group. The resulting, resonance-stabilized intermediate is then deprotonated by water to give the aldehyde:

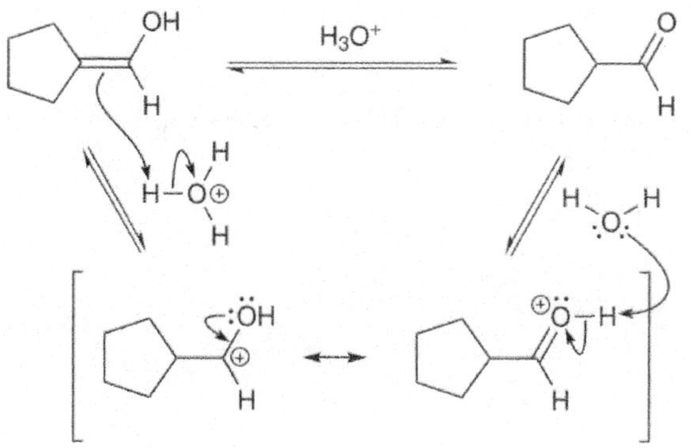

H₃O⁺

(b) In basic conditions, the first step is deprotonation, using hydroxide as the base. This step requires two curved arrows. The resulting, resonance-stabilized intermediate is then protonated by water (a weak acid) to give the aldehyde:

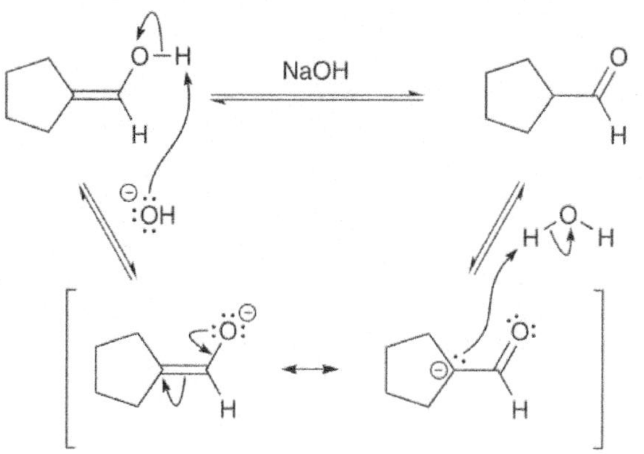

NaOH

12.33

(a) In acidic conditions, the first step is protonation, using H₃O⁺ as the proton source. This step requires two curved arrows. When drawing the curved arrows for this step, make sure to protonate the pi bond, rather than protonating the OH group. The resulting, resonance-stabilized intermediate is then deprotonated by water to give the ketone:

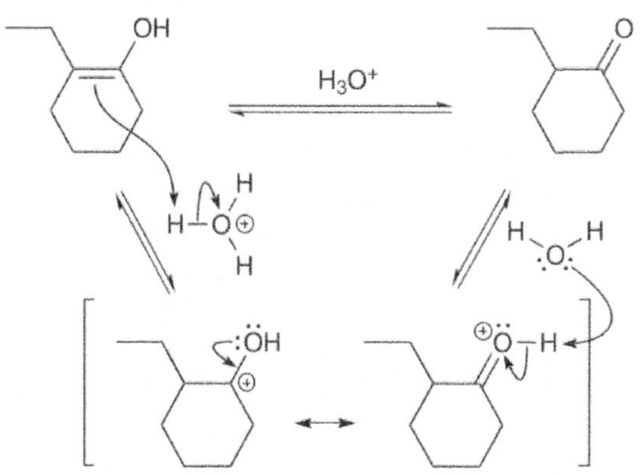

H₃O⁺

(b) In basic conditions, the first step is deprotonation, using hydroxide as the base. This step requires two curved arrows. The resulting, resonance-stabilized intermediate is then protonated by water (a weak acid) to give the ketone:

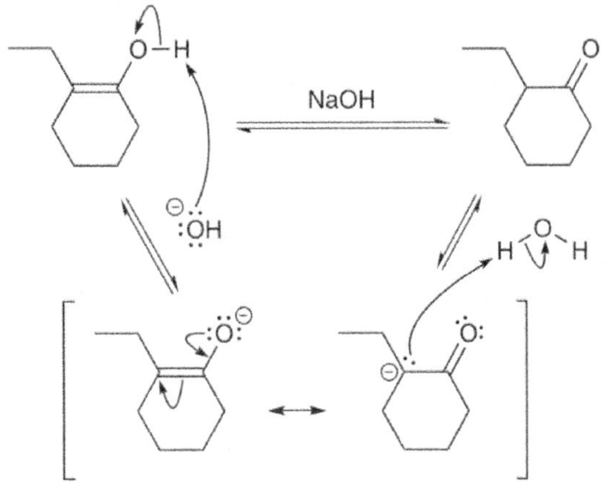

12.34

(a) In acidic conditions, the first step is protonation, using H_3O^+ as the proton source. This step requires two curved arrows. When drawing the curved arrows for this step, make sure to protonate the pi bond, rather than protonating the OH group. The resulting, resonance-stabilized intermediate is then deprotonated by water to give the ketone:

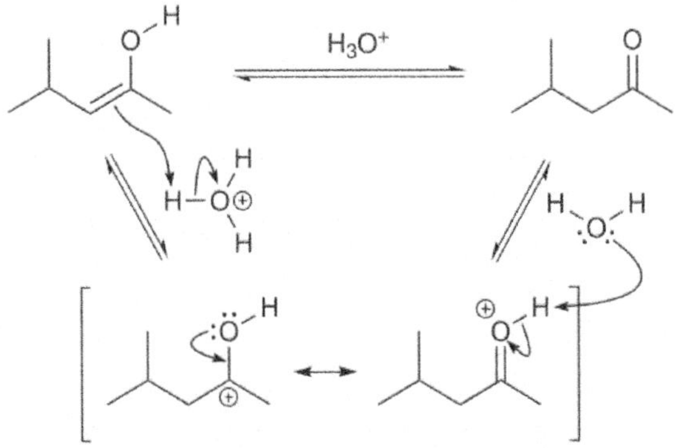

(b) In basic conditions, the first step is deprotonation, using hydroxide as the base. This step requires two curved arrows. The resulting, resonance-stabilized intermediate is then protonated by water (a weak acid) to give the ketone:

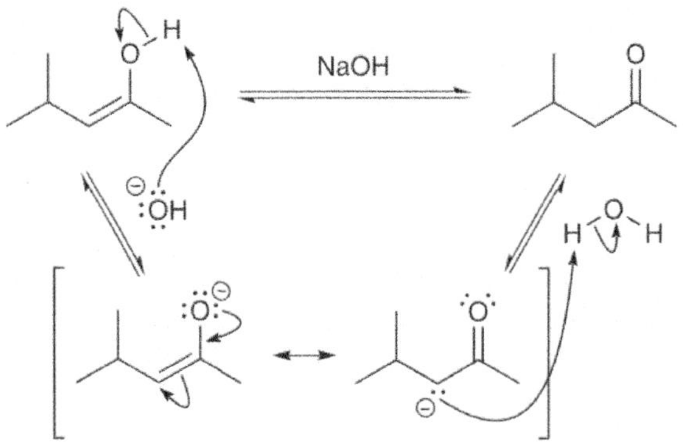

12.35 The starting material is an internal alkyne, and the reagents indicate an ozonolysis process, which cleaves the C≡C bond. This generates two carboxylic acids, as shown here:

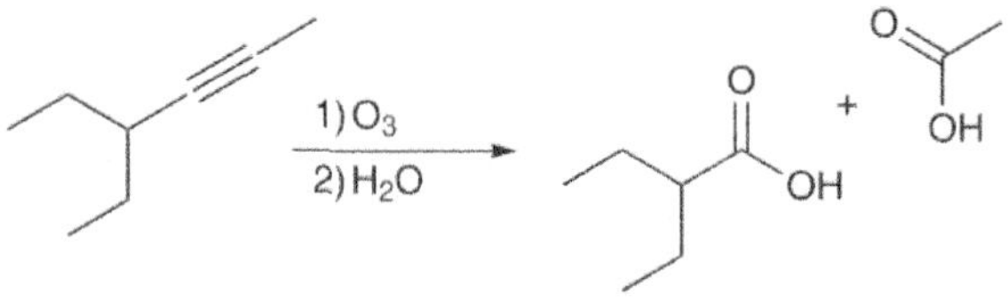

12.36 The starting material is a terminal alkyne, and the reagents indicate an ozonolysis process, which cleaves the C≡C bond. This generates a carboxylic acid and carbon dioxide, as shown here:

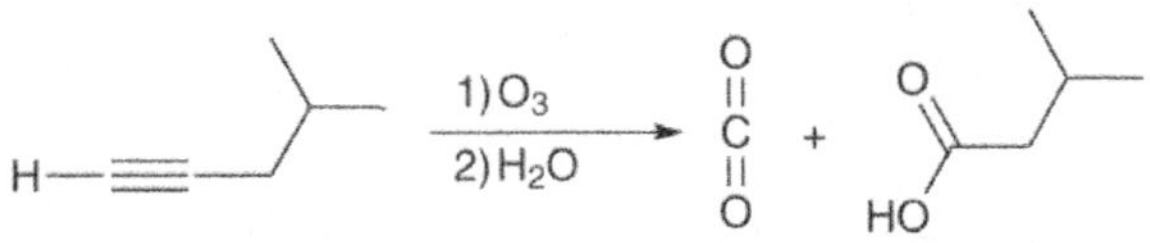

12.37 The starting material is an internal alkyne, and the reagents indicate an ozonolysis process, which cleaves the C≡C bond. This generates two carboxylic acids, as shown here:

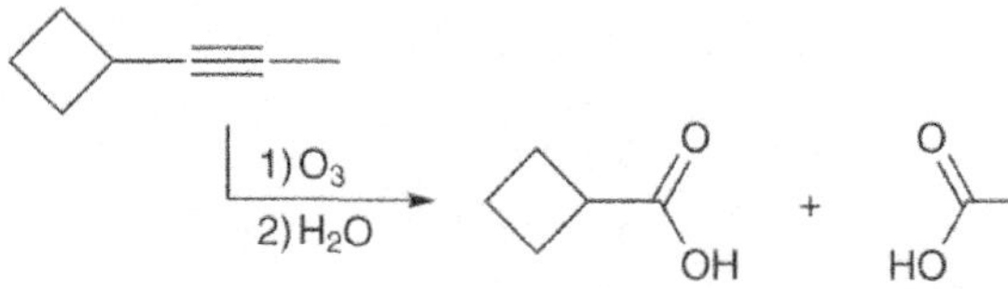

12.38 The starting material is ethylene, and the reagents indicate an ozonolysis process, which cleaves the C≡C bond and generates two equivalents of carbon dioxide:

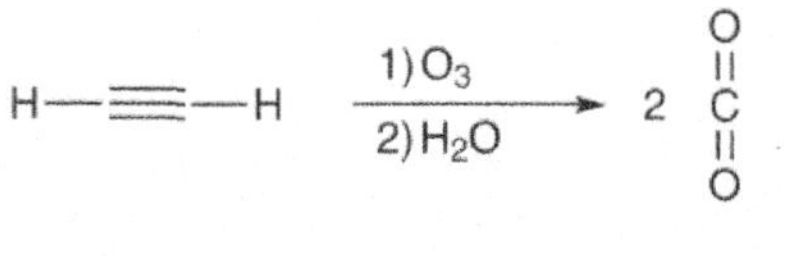

CHAPTER 13

13.2 The OH group is connected to a carbon atom that is connected to three alkyl groups (methyl, methyl, and butyl). Therefore, this compound is a tertiary alcohol.

13.3 The OH group is connected to a carbon atom that is connected to only one alkyl group. Therefore, this compound is a primary alcohol.

13.4 The OH group is connected to a carbon atom that is connected to only one alkyl group. Therefore, this compound is a primary alcohol.

13.5 The OH group is connected to a carbon atom that is connected to two alkyl groups. Therefore, this compound is a secondary alcohol.

13.7 This compound is an alcohol that has only two carbon atoms. Since this compound has fewer than five carbon atoms per OH group, the compound will be soluble in water. In fact, ethanol is miscible with water (soluble in all proportions).

13.8 This compound is an alcohol that has eight carbon atoms. Since this compound has more than five carbon atoms per OH group, the compound will be insoluble in water.

13.9 This compound is a diol (a compound containing two OH groups) that has five carbon atoms. Since this compound has fewer than five carbon atoms *per OH group*, the compound will be soluble in water.

13.10 This compound is an alcohol that has nine carbon atoms. Since this compound has more than five carbon atoms per OH group, the compound will be insoluble in water.

13.12 The alcohol shown below is the more acidic compound because its conjugate base has a negative charge on an oxygen atom, which is more stable than a negative charge on a nitrogen atom.

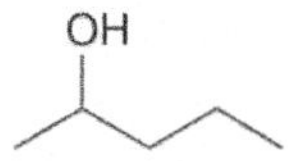

13.13 The alcohol shown below is the more acidic compound because its conjugate base has a negative charge on an oxygen atom, which is more stable than a negative charge on a carbon atom.

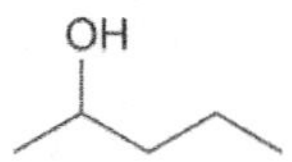

13.14 The alcohol shown below is the more acidic compound because its conjugate base is stabilized by the electron-withdrawing effect of the nearby fluorine atom.

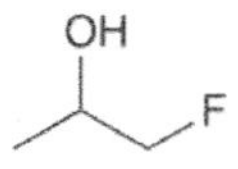

13.15 The compound shown below (called a carboxylic acid) is the more acidic compound because its conjugate base is stabilized by resonance.

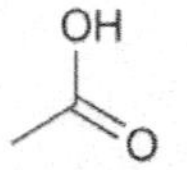

The conjugate base of the carboxylic acid is shown below. The negative charge is spread over two oxygen atoms.

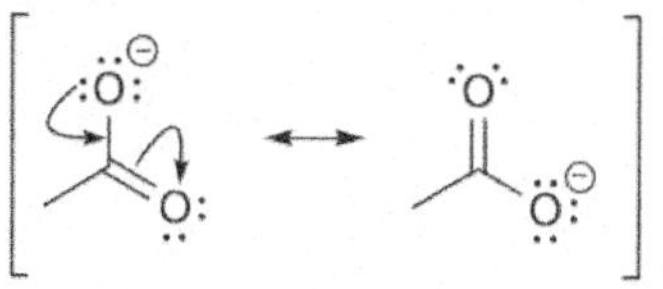

In contrast, the conjugate base of an alcohol is less stable because it has a localized negative charge.

13.16 The compound shown below (called phenol) is the more acidic compound because its conjugate base is stabilized by resonance.

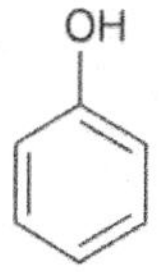

The conjugate base of phenol is shown below. The negative charge is spread over one oxygen atom and three carbon atoms.

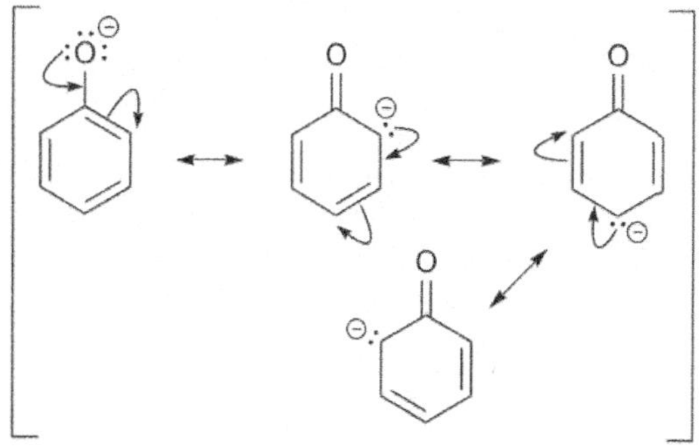

In contrast, the conjugate base of an alcohol is less stable because it has a localized negative charge.

13.17 The compound shown below is the more acidic compound because it has better resonance stabilization. Its negative charge is spread over three carbon atoms *and two oxygen atoms*.

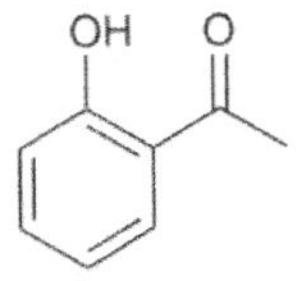

In contrast, the conjugate base of the other compound is less stable because it has a negative charge that is spread over three carbon atoms *and only one oxygen atom*.

The conjugate base of the more acidic compound is shown below:

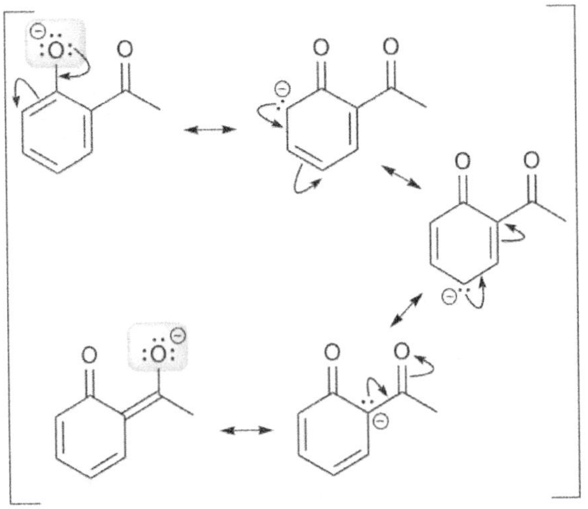

13.19 The starting material is an alkene, which can be converted into the desired alcohol via a Markovnikov addition of H and OH. This can be accomplished via acid-catalyzed hydration:

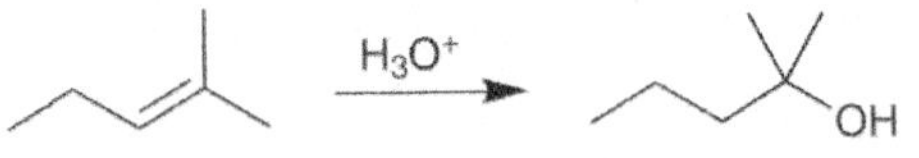

13.20 The starting material is an alkene, which can be converted into the desired alcohol via an *anti*-Markovnikov addition of H and OH. This can be accomplished via hydroboration–oxidation:

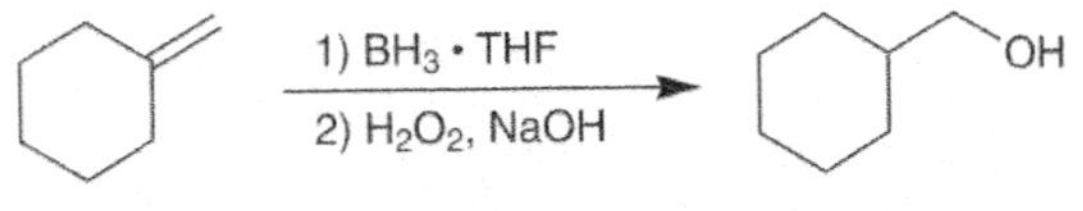

13.21 The desired transformation requires changing the identity and location of the functional group. This can be achieved via elimination, followed by addition. For the elimination process, a sterically hindered base must be used to favor the Hofmann product over the Zaitsev product. For the addition process, an *anti*-Markovnikov addition of H and OH is required, which can be achieved via hydroboration–oxidation, as shown:

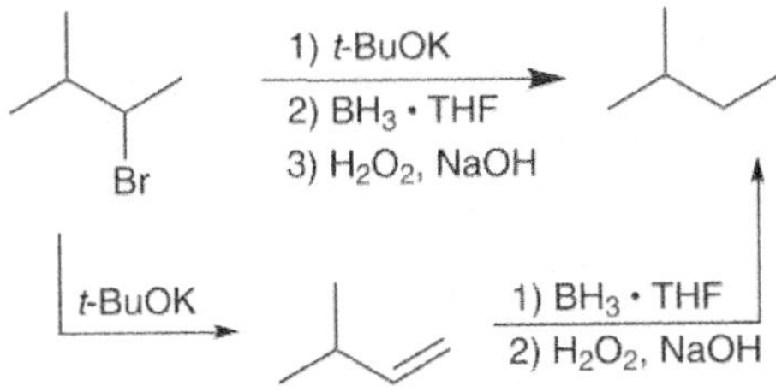

13.22 The starting material can be converted into the desired alcohol via an *anti*-Markovnikov addition of H and OH in a *syn* fashion. This can be accomplished via hydroboration–oxidation:

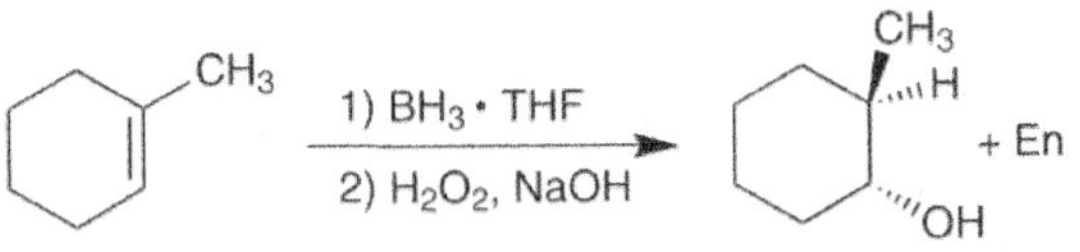

13.24 For each bond, we give both electrons to the more electronegative atom. For the C=O bond, oxygen is more electronegative, so oxygen gets all four electrons of the double bond. For the C—C bond, we treat the two electrons as being shared equally (one electron on one carbon atom, and the other electron on the other carbon atom). For the C—H bond, carbon is *very slightly* more electronegative than hydrogen, so carbon will get both electrons. In total, we count three electrons on the highlighted carbon atom, which means that it is missing one electron (it should have four valence electrons), and therefore, the oxidation state is +1.

13.25 For each bond, we give both electrons to the more electronegative atom. For the C=O bond, oxygen is more electronegative, so oxygen gets all four electrons of the double bond. Similarly, for the C—O bond, oxygen gets both electrons. For the C—C bond, we treat the two electrons as being shared equally (one electron on one carbon atom, and the other electron on the other carbon atom). In total, we count only one electron on the highlighted carbon atom, which means that it is missing three electrons (it should have four valence electrons), and therefore, the oxidation state is +3.

13.26 For each bond, we give both electrons to the more electronegative atom. For the C—O bond, oxygen is more electronegative, so oxygen gets both electrons. For each of the C—C bonds, we treat the two electrons as being shared equally (one electron on one carbon atom, and the other electron on the other carbon atom). In total, we count three electrons on the highlighted carbon atom, which means that it is missing one electron (it should have four valence electrons), and therefore, the oxidation state is +1.

13.27 For each bond, we give both electrons to the more electronegative atom. For each of the C—O bonds, oxygen is more electronegative, so oxygen gets both electrons. For the C—C bond, we treat the two electrons as being shared equally (one electron on one carbon atom, and the other electron on the other carbon atom).

And for the C—H bond, carbon is *very slightly* more electronegative than hydrogen, so carbon will get both electrons. In total, we count three electrons on the highlighted carbon atom, which means that it is missing one electron (it should have four valence electrons), and therefore, the oxidation state is +1.

13.28 For each bond, we give both electrons to the more electronegative atom. For the C—O bond, oxygen is more electronegative, so oxygen gets both electrons. For each of the C—C bonds, we treat the two electrons as being shared equally (one electron on one carbon atom, and the other electron on the other carbon atom). And for the C—H bond, carbon is *very slightly* more electronegative than hydrogen, so carbon will get both electrons. In total, we count four electrons, which is the number of electrons that a carbon atom is supposed to have, so the oxidation state is 0.

13.29 For each bond, we give both electrons to the more electronegative atom. For the C—O bond, oxygen is more electronegative, so oxygen gets both electrons. For each of the C—C bonds (and there are three of them), we treat the two electrons as being shared equally (one electron on one carbon atom, and the other electron on the other carbon atom). In total, we count three electrons on the highlighted carbon atom, which means that it is missing one electron (it should have four valence electrons), and therefore, the oxidation state is +1.

13.31 The desired product is a secondary alcohol, which can be made via reduction of the corresponding ketone. We can obtain the desired product by treating the ketone below with LiAlH$_4$ followed by water, or by treating the ketone below with NaBH$_4$ together with methanol, as shown:

13.32 The desired product is a secondary alcohol, which can be made via reduction of the corresponding ketone. We can obtain the desired product by treating the ketone below with LiAlH$_4$ followed by water, or by treating the ketone below with NaBH$_4$ together with methanol, as shown:

13.33 The desired product is a primary alcohol, which can be made via reduction of the corresponding aldehyde. We can obtain the desired product by treating the aldehyde below with LiAlH$_4$ followed by water, or by treating the aldehyde below with NaBH$_4$ together with methanol, as shown:

13.34 The desired product is a primary alcohol, which can be made via reduction of the corresponding aldehyde. We can obtain the desired product by treating the aldehyde below with LiAlH$_4$ followed by water, or by treating the aldehyde below with NaBH$_4$ together with methanol, as shown:

13.35 The desired product is a secondary alcohol, which can be made via reduction of the corresponding ketone. We can obtain the desired product by treating the ketone below with LiAlH$_4$ followed by water, or by treating the ketone below with NaBH$_4$ together with methanol, as shown:

13.36 The desired product is a primary alcohol, which can be made via reduction of the corresponding aldehyde. We can obtain the desired product by treating the aldehyde below with LiAlH₄ followed by water, or by treating the aldehyde below with NaBH₄ together with methanol, as shown:

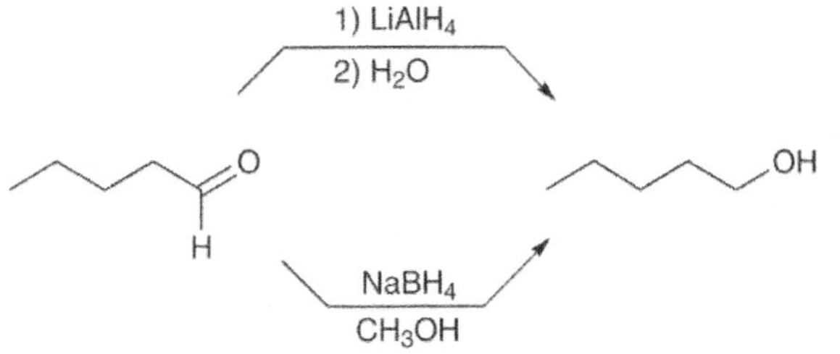

13.38 The desired alcohol can be made via a Grignard reaction between an aldehyde and methyl magnesium bromide, as illustrated in the following retrosynthetic analysis:

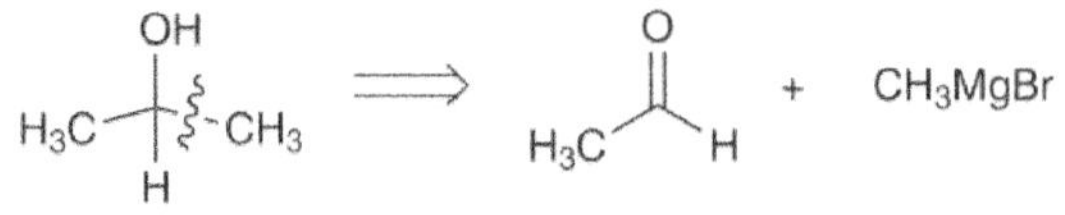

The forward process can be shown like this:

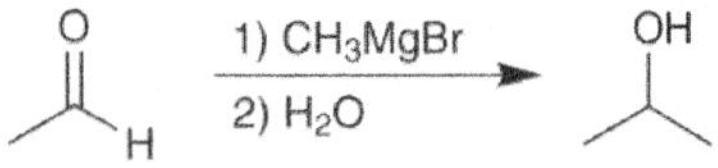

13.39 The desired alcohol can be made via a Grignard reaction between a ketone and methyl magnesium bromide, as illustrated in the following retrosynthetic analysis:

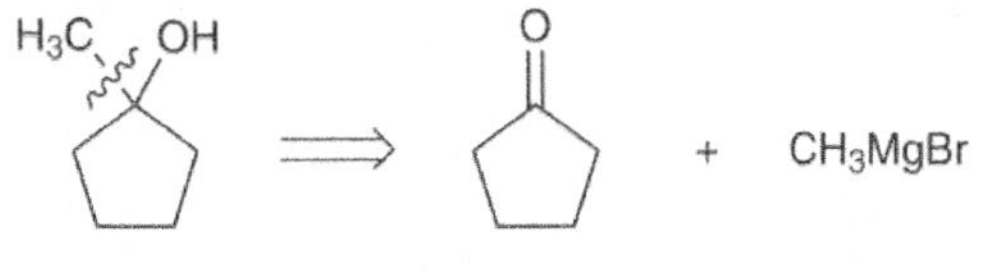

The forward process can be shown like this:

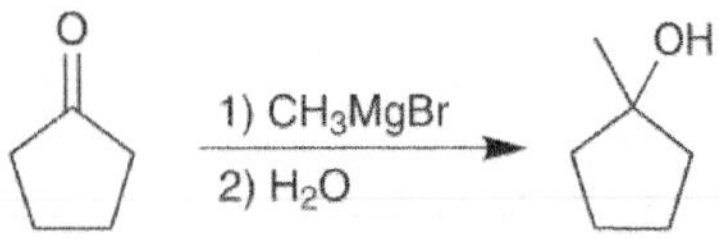

13.40 The desired alcohol can be made via a Grignard reaction between formaldehyde and a Grignard reagent, as illustrated in the following retrosynthetic analysis:

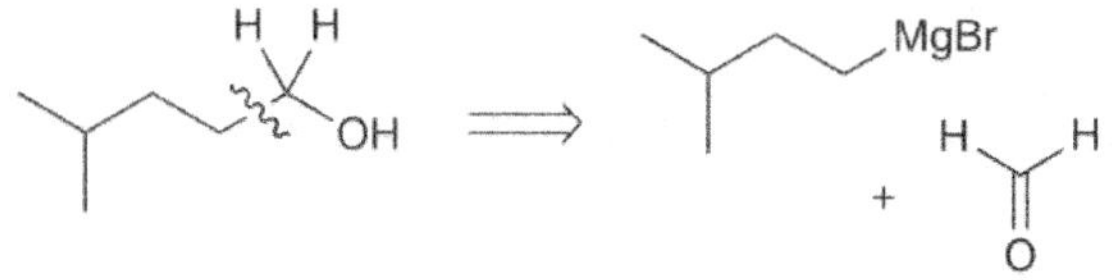

The forward process can be shown like this:

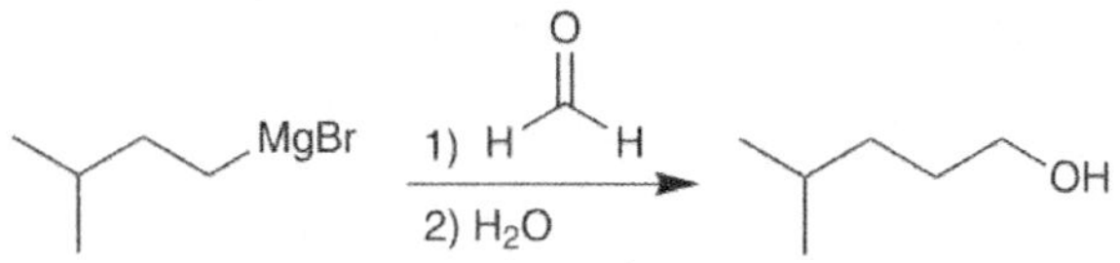

13.41 The desired alcohol can be made via a Grignard reaction between a Grignard reagent and formaldehyde, as illustrated in the following retrosynthetic analysis:

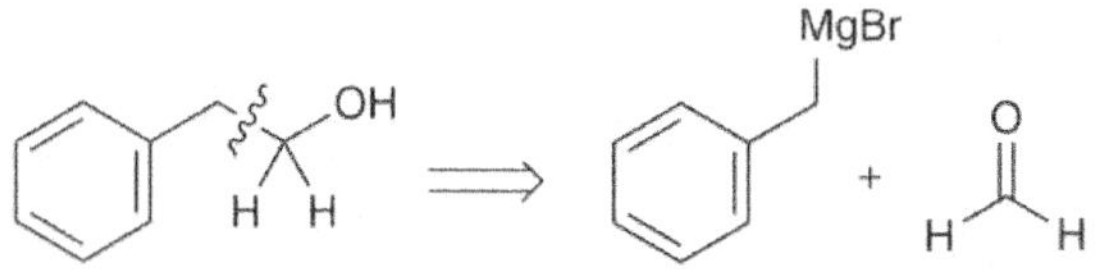

The forward process can be shown like this:

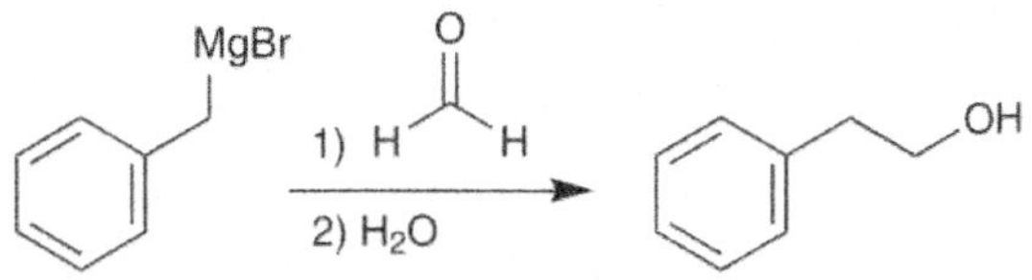

13.42 There are two different ways to use a Grignard reaction to make the desired alcohol, as shown in the following retrosynthetic analysis, showing both possibilities:

Possibility #1

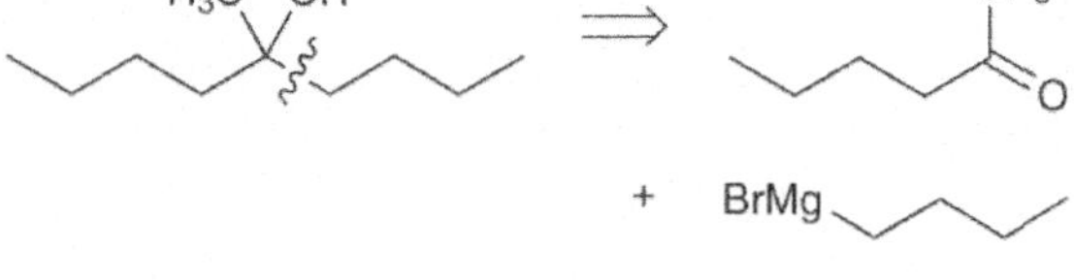

Possibility #2

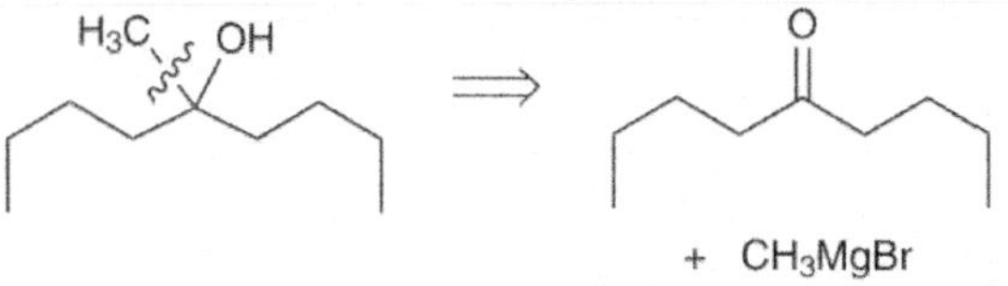

The forward processes can be shown like this:

Possibility #1

Possibility #2

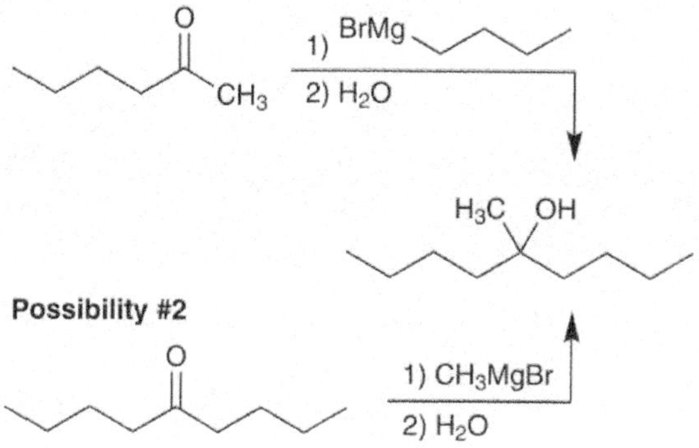

13.43 There are two different ways to use a Grignard reaction to make the desired alcohol, as shown in the following retrosynthetic analysis, showing both possibilities:

Possibility #1

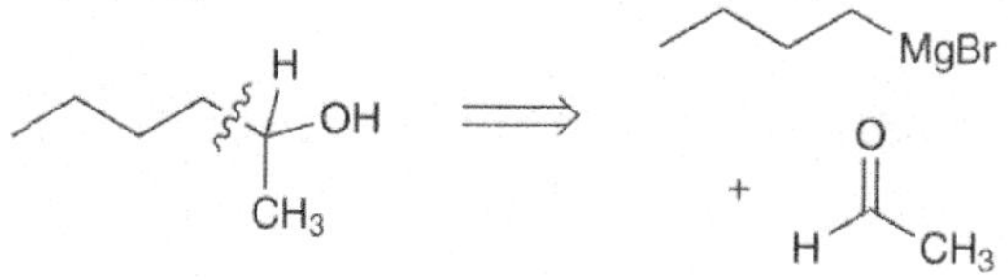

Possibility #2

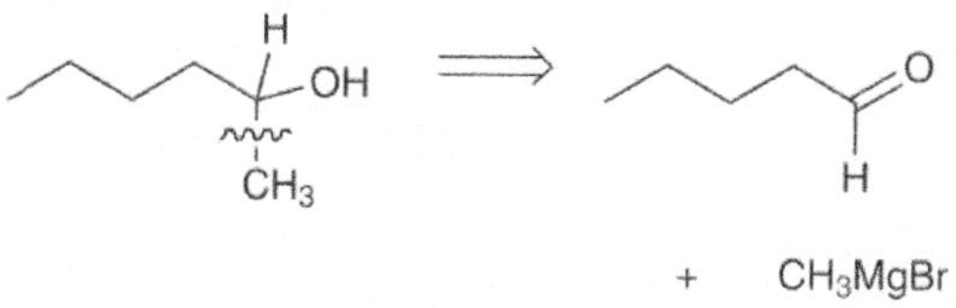

The forward processes can be shown like this:

Possibility #1

Possibility #2

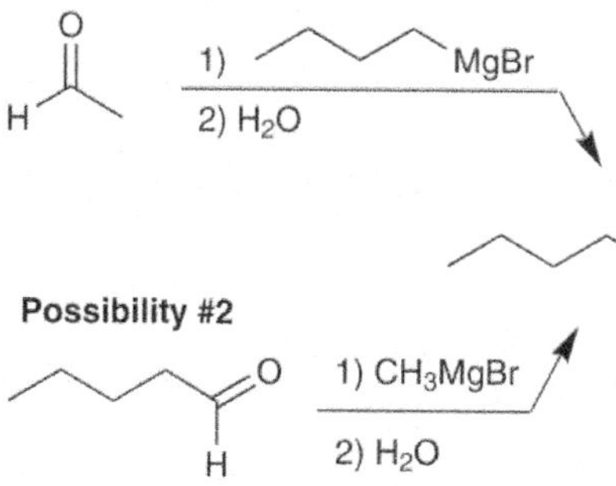

13.44 The starting ketone can be converted into the desired tertiary alcohol, via a Grignard reaction with methyl magnesium bromide, as shown:

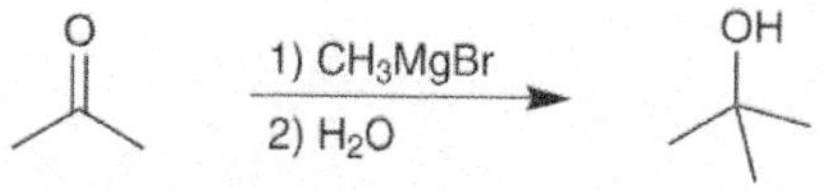

13.45 The starting ketone can be converted into the desired tertiary alcohol, via a Grignard reaction with ethyl magnesium bromide, as shown:

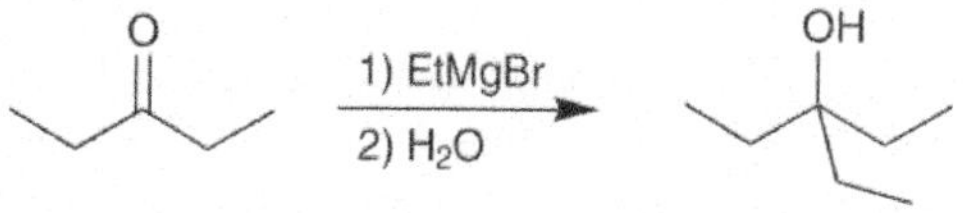

13.46 The starting ketone can be converted into the desired tertiary alcohol, via a Grignard reaction with phenyl magnesium bromide, as shown:

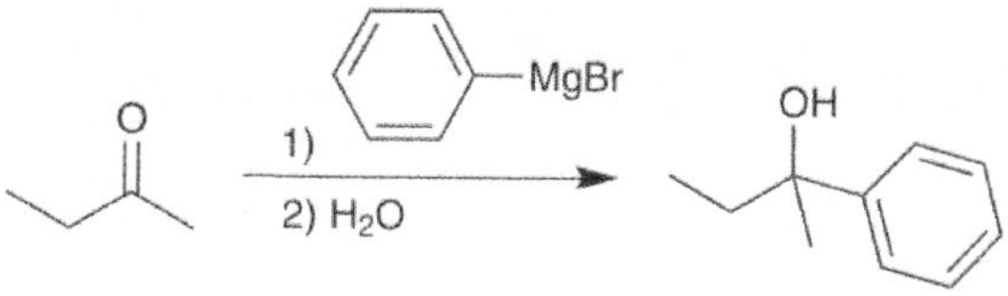

13.47 The starting aldehyde can be converted into the desired secondary alcohol, via a Grignard reaction with ethyl magnesium bromide, as shown:

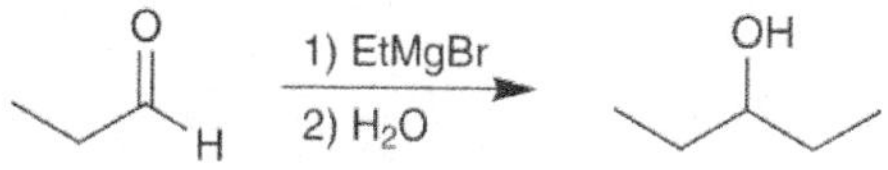

13.48 The starting material (a ketone) has three carbon atoms, and the product has five carbon atoms, so two carbon atoms (an ethyl group) must be introduced. This can be achieved via a Grignard reaction between the starting ketone and ethyl magnesium bromide, to give a tertiary alcohol. This alcohol can then be converted into the desired alkene via an elimination process, which can either be achieved via an E1 process (using concentrated sulfuric acid) or via an E2 process (requiring conversion of the alcohol into a tosylate and use of an unhindered base to favor the Zaitsev product), as shown:

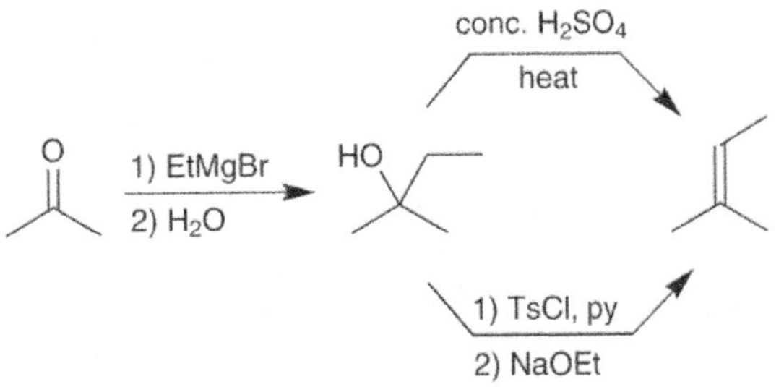

13.49 The starting material (a ketone) has six carbon atoms, and the product has seven carbon atoms, so one carbon atom (a methyl group) must be introduced. This can be achieved via a Grignard reaction between the starting ketone and methyl magnesium bromide, to give a tertiary alcohol. This alcohol can then be converted into the desired alkene via an elimination process, which can either be achieved via an E1 process (using concentrated sulfuric acid) or via an E2 process (requiring conversion of the alcohol into a tosylate and use of an unhindered base to favor the Zaitsev product), as shown:

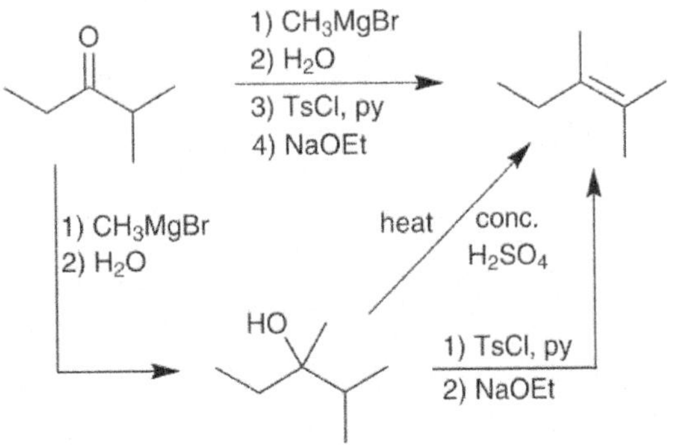

13.51 The desired alcohol can be made via reduction of a ketone, or via a Grignard reaction with an aldehyde:

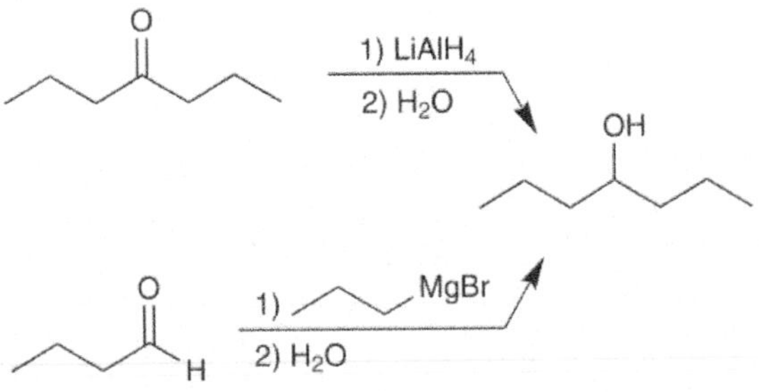

13.52 Below are four acceptable methods for making the desired primary alcohol. The desired alcohol can be made via 1) reduction of an aldehyde, or 2) hydroboration–oxidation of an alkene, or 3)

an S_N2 reaction with a primary alkyl bromide, or 4) a Grignard reaction:

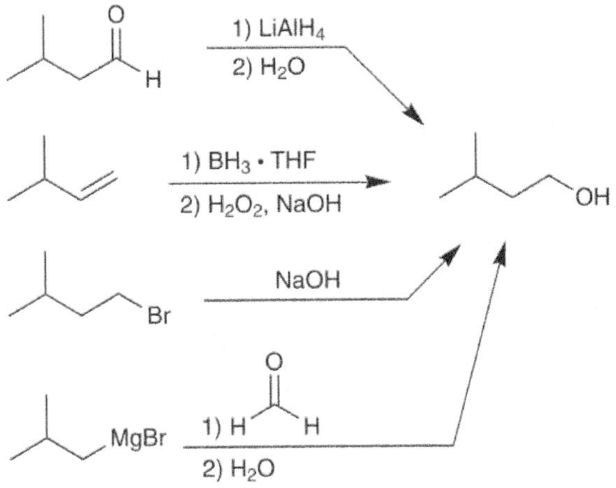

13.53 Below are five acceptable methods for making the desired tertiary alcohol. The desired alcohol can be made via 1) a Grignard reaction with an aldehyde, or 2) a Grignard reaction with a ketone, or 3) an acid-catalyzed hydration of a trisubstituted alkene, or 4) an acid-catalyzed hydration of a disubstituted alkene, or 5) an S_N1 reaction with a tertiary alkyl bromide:

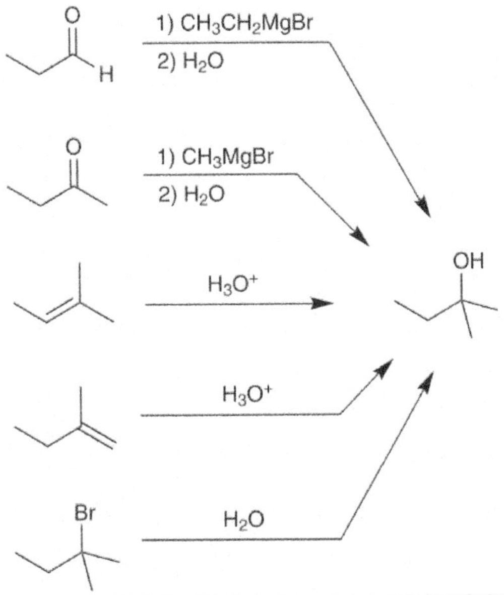

13.54 Below are four acceptable methods for making the desired alcohol. The desired alcohol can be made via 1) reduction of a ketone with LiAlH$_4$, or 2) a Grignard reaction between an aldehyde (called benzaldehyde) and methyl magnesium bromide, or

3) acid-catalyzed hydration of an alkene, or 4) an S_N1 reaction with a benzylic bromide:

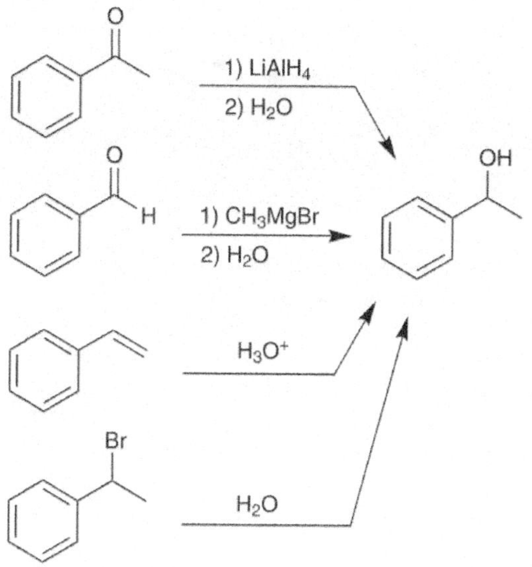

13.55 Below are three acceptable methods for making the desired alcohol. The desired alcohol can be made via 1) reduction of an aldehyde with $LiAlH_4$, or 2) hydroboration–oxidation of an alkene, or 3) a Grignard reaction with formaldehyde ($H_2C{=}O$):

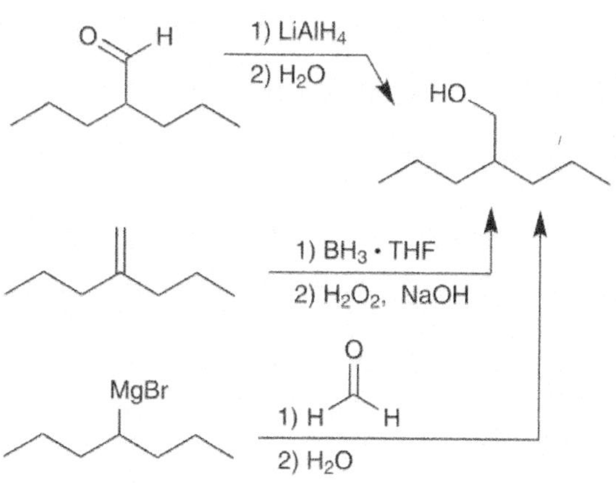

13.56 Below are three acceptable methods for making the desired alcohol. The desired alcohol can be made via 1) a Grignard reaction between a ketone and methyl magnesium bromide, or 2) acid-catalyzed hydration of an alkene, or 3) an S_N1 reaction with a tertiary substrate:

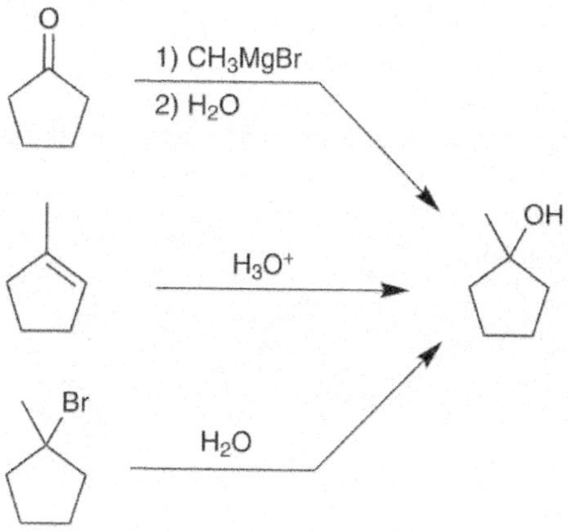

13.58 A tertiary alcohol can be converted into the corresponding tertiary alkyl bromide upon treatment with HBr (via an S_N1 process):

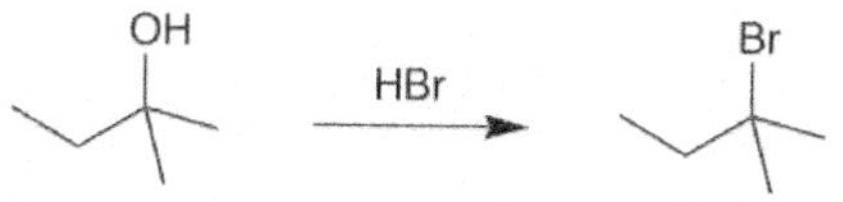

13.59 A primary alcohol can be converted into the corresponding primary alkyl chloride upon treatment with HCl and $ZnCl_2$ or upon treatment with thionyl chloride ($SOCl_2$) in the presence of pyridine:

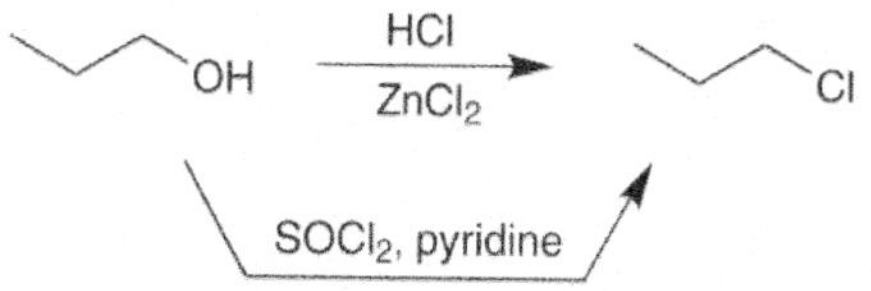

13.60 A tertiary alcohol can be converted into an alkene (the Zaitsev product) upon treatment with concentrated sulfuric acid (E1). Alternatively, the alcohol can be converted into a tosylate and then treated with a strong base that is not sterically hindered, to give the Zaitsev product (E2).

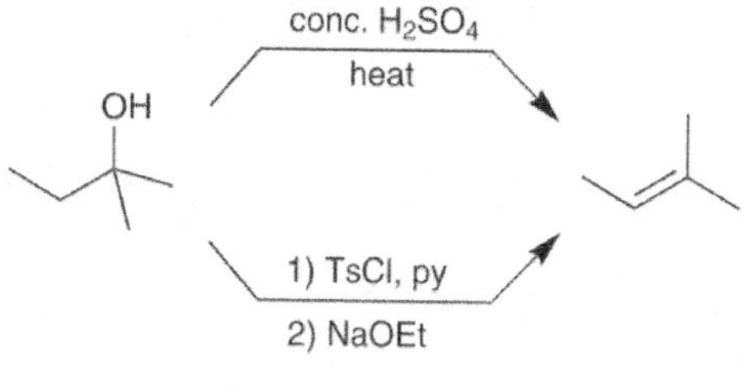

13.61 A tertiary alcohol can be converted into an alkene (the Hofmann product) by first converting the alcohol into a tosylate and then treating the alkyl tosylate with a strong, sterically hindered base, to give the Hofmann product (E2).

Note that we cannot use a strong acid to generate a better leaving group in this case because we want to treat the substrate with a strong base, so the use of an acid would be incompatible (we cannot use a strong acid and a strong base at the same time, as they will neutralize each other).

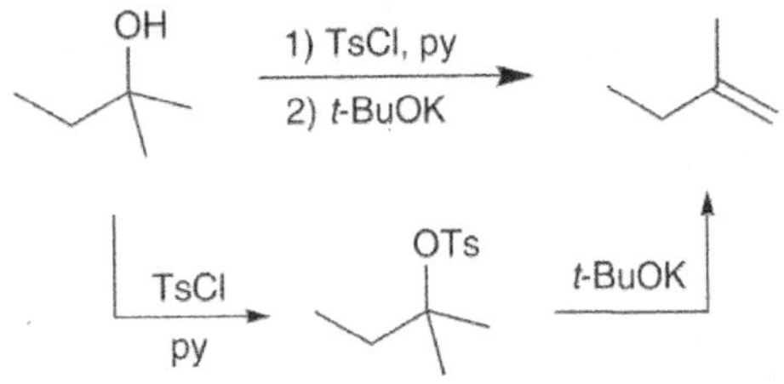

13.63 A secondary alcohol can be oxidized to give a ketone upon treatment with either chromic acid or PCC:

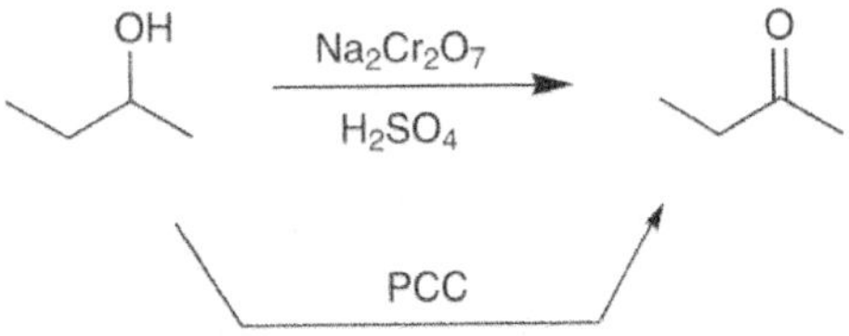

13.64 A primary alcohol can be oxidized to give an aldehyde upon treatment with PCC. Chromic acid cannot be used because oxidation with chromic acid would result in a carboxylic acid.

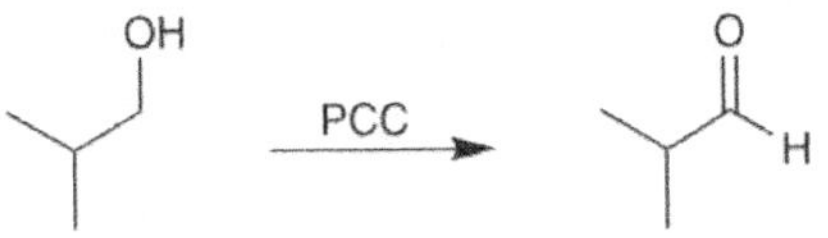

13.65 A primary alcohol can be oxidized to give a carboxylic acid upon treatment with chromic acid.

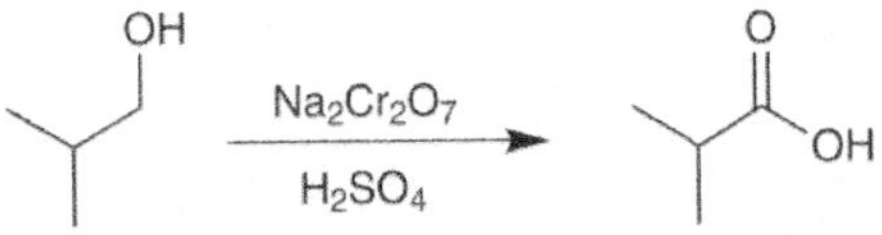

13.66 An aldehyde can be reduced to give a primary alcohol upon treatment with a reducing agent, such as lithium aluminum hydride or sodium borohydride.

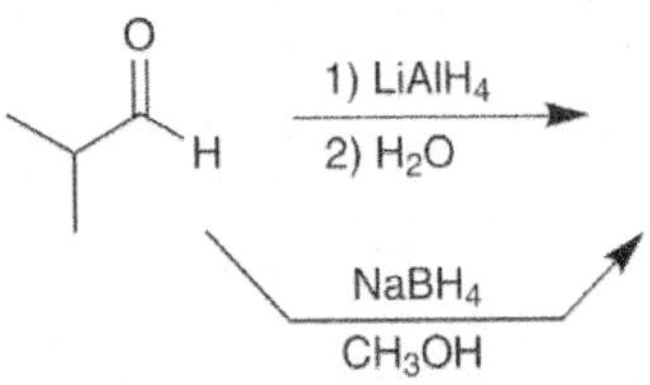

13.67 A ketone can be reduced to give a secondary alcohol upon treatment with a reducing agent, such as lithium aluminum hydride or sodium borohydride.

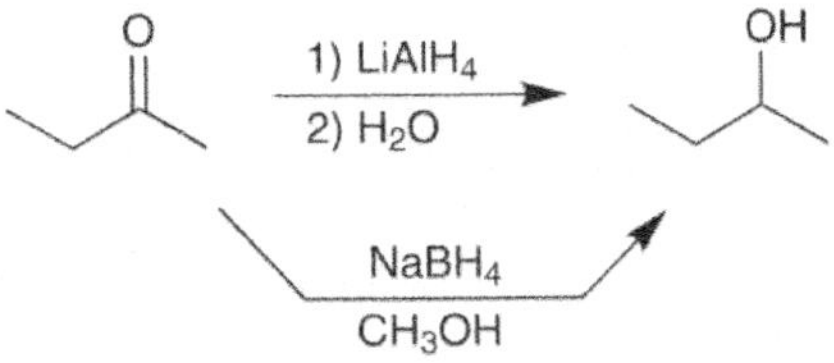

13.68 A secondary alcohol can be oxidized to give a ketone upon treatment with either chromic acid or PCC:

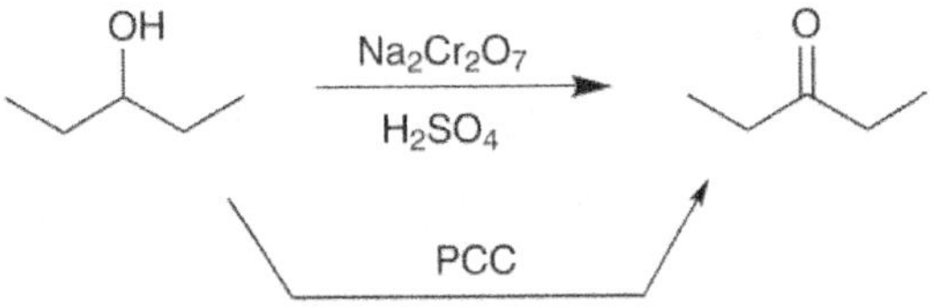

13.70 The desired ether can be made from 1-propanol via a Williamson ether synthesis, as shown in the following retrosynthetic analysis:

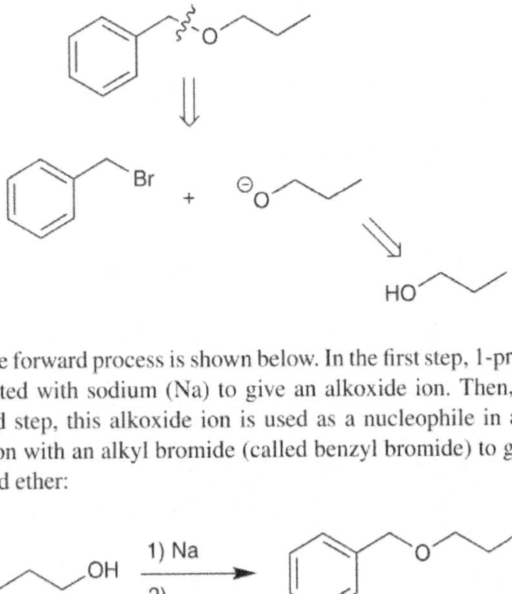

The forward process is shown below. In the first step, 1-propanol is treated with sodium (Na) to give an alkoxide ion. Then, in the second step, this alkoxide ion is used as a nucleophile in an S_N2 reaction with an alkyl bromide (called benzyl bromide) to give the desired ether:

13.71 The desired ether can be made from 1-propanol via a Williamson ether synthesis, as shown in the following retrosynthetic analysis:

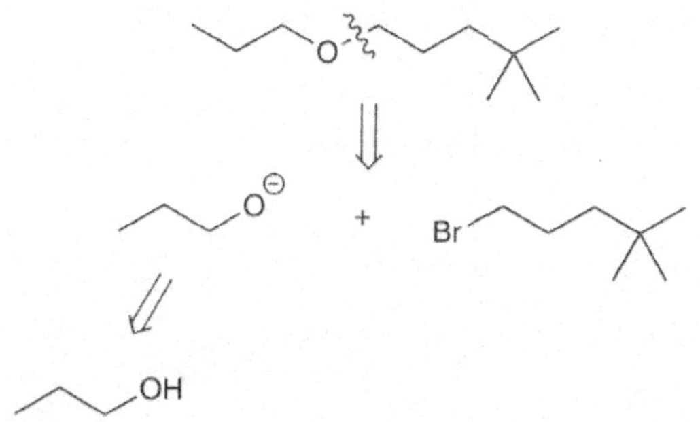

The forward process is shown below. In the first step, 1-propanol is treated with sodium (Na) to give an alkoxide ion. Then, in the

second step, this alkoxide ion is used as a nucleophile in an S_N2 reaction with a primary alkyl halide to give the desired ether:

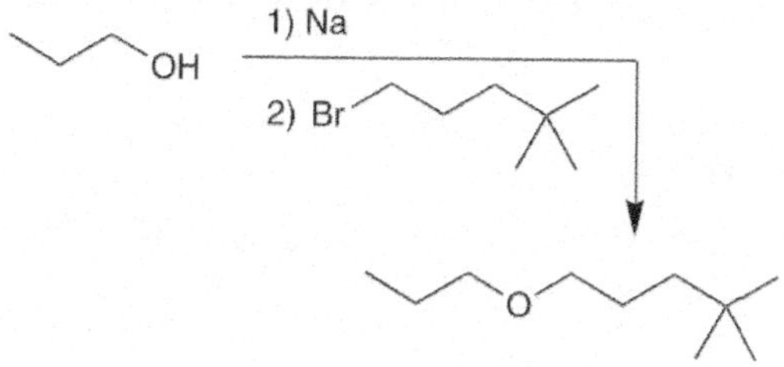

CHAPTER 14

14.1 This compound is an ether (R—O—R), where the two R groups are isopropyl and ethyl groups. Therefore, the common name of this compound is ethyl isopropyl ether.

14.2 This compound is an ether (R—O—R), where the two R groups are cyclobutyl and propyl groups. Therefore, the common name of this compound is cyclobutyl propyl ether.

14.3 This compound is an ether (R—O—R), where the two R groups are both isopropyl groups. Therefore, the common name of this compound is diisopropyl ether.

14.4 This compound is a crown ether with a ring size of eighteen atoms, six of which are oxygen atoms. Therefore, the common name of this compound is 18-crown-6.

14.5 When assigning a systematic name for an ether, the larger R group is named as the parent, while the smaller R group (together with its neighboring oxygen atom) is named as an alkoxy group. In this example, the parent is propane, and there is an ethoxy group connected to position #2 on the propane parent. Therefore, the name of this compound is 2-ethoxypropane.

14.6 When assigning a systematic name for an ether, the larger R group is named as the parent, while the smaller R group (together with its neighboring oxygen atom) is named as an alkoxy group. In this example, the parent is cyclohexane, and there is a methoxy group connected to the parent. By definition, the methoxy group is at position #1 of the ring, and it is not necessary to indicate the number "1" in the name because it is understood that the substituent is at position #1 (this is true for any monosubstituted cycloalkane). Therefore, the name of this compound is methoxycyclohexane.

14.7 When assigning a systematic name for an ether, one R group is named as the parent, while the other R group (together with its neighboring oxygen atom) is named as an alkoxy group. In this example, the parent is ethane, and there is an ethoxy group connected to the parent (at position #1 on the ethane parent, by definition, so the number is not included in the name). Therefore, the name of this compound is ethoxyethane.

14.9 We begin by considering both C—O bonds to determine which of them can be formed most readily during an S_N2 process. The following retrosynthetic analysis shows both possibilities, and the starting materials that would be required in each case:

Possibility #1

Possibility #2

Possibility #1 is not viable because it involves an S_N2 reaction at an sp^2 hybridized carbon atom. Possibility #2 is expected to be efficient because it involves an S_N2 reaction with a primary alkyl halide.

The forward process is shown below. In the first step, phenol (C_6H_5OH) is deprotonated to give a phenolate ion ($C_6H_5O^-$), which is then used as a nucleophile in an S_N2 reaction to give the desired product:

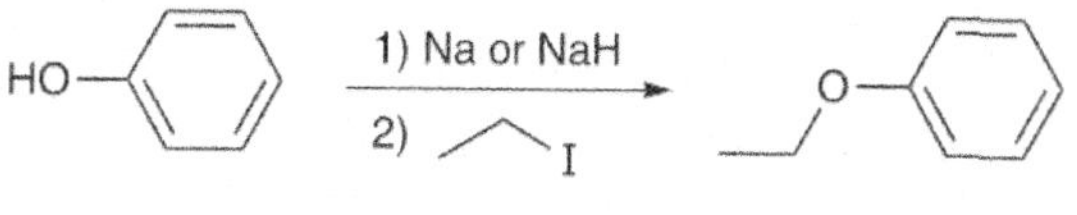

14.10 We begin by considering both C—O bonds to determine which of them can be formed most readily during an S_N2 process. The following retrosynthetic analysis shows both possibilities, and the starting materials that would be required in each case:

Possibility #1

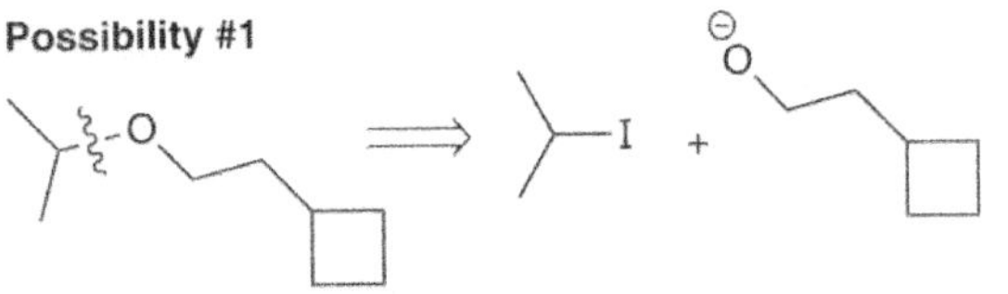

Possibility #2

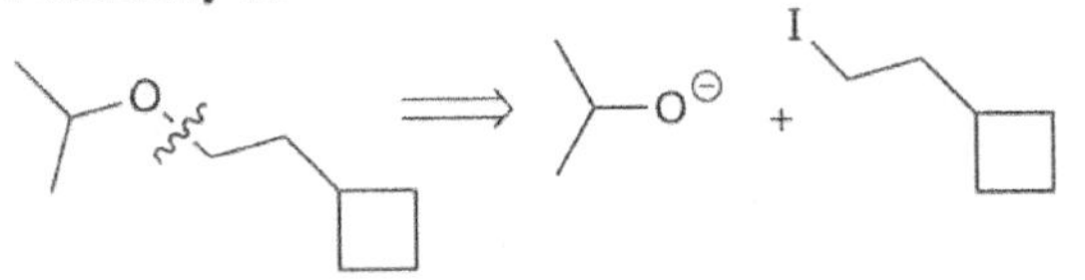

Possibility #1 involves an S_N2 reaction with a secondary alkyl halide, which is expected to favor E2 elimination. Possibility #2 is expected to be more efficient because it involves an S_N2 reaction with a primary alkyl halide.

The forward process is shown below. In the first step, the alcohol is deprotonated to give an alkoxide ion, which is then used as a nucleophile in an S_N2 reaction to give the desired product:

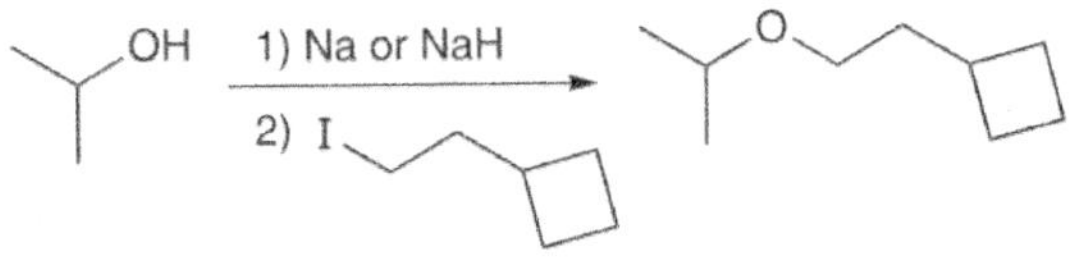

14.11 We begin by considering both C—O bonds to determine which of them can be formed most readily during an S_N2 process. The following retrosynthetic analysis shows both possibilities, and the starting materials that would be required in each case:

Possibility #1

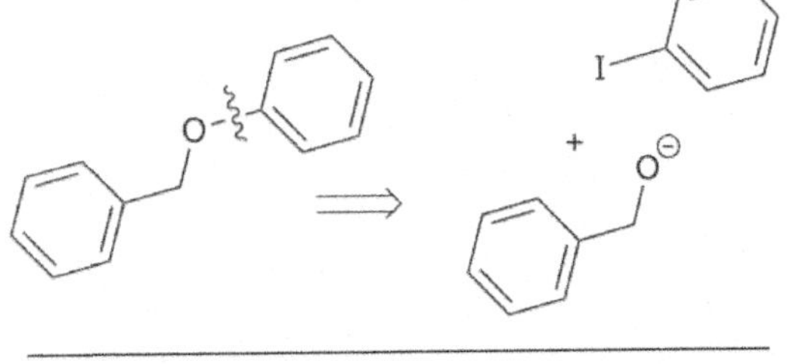

Possibility #2

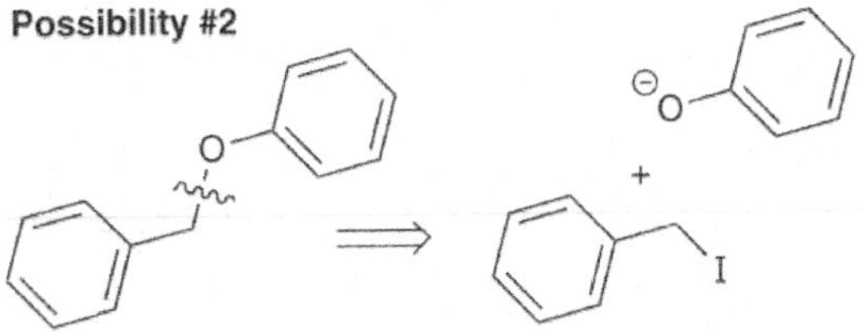

Possibility #1 is not viable because it involves an S_N2 reaction at an sp^2 hybridized carbon atom. Possibility #2 is expected to be efficient because it involves an S_N2 reaction with a primary (benzylic) halide.

The forward process is shown below. In the first step, phenol (C_6H_5OH) is deprotonated to give a phenolate ion ($C_6H_5O^-$), which is then used as a nucleophile in an S_N2 reaction with benzyl iodide ($C_6H_5CH_2I$) to give the desired product:

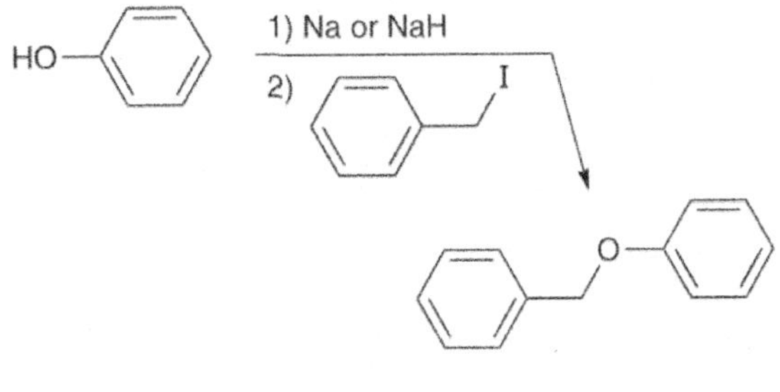

14.12 We consider the viability of forming each of the C—O bonds via a Williamson ether synthesis:

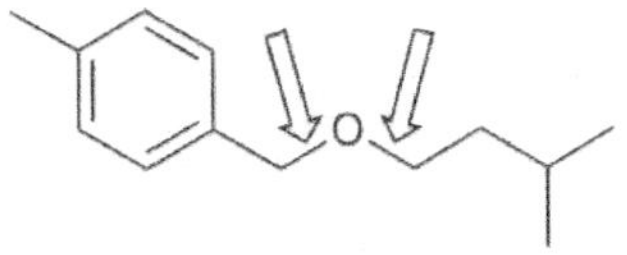

Indeed, either one of these bonds can been made via a Williamson ether synthesis. The C—O bond on the right can be made if we start with the following primary (benzylic) alcohol,

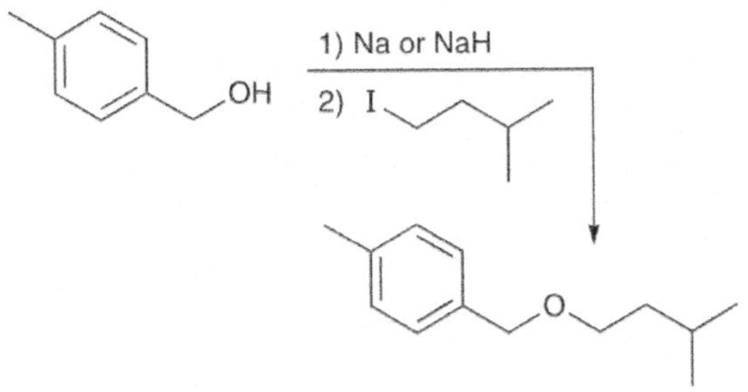

and the other C—O bond can be made if we start with the following primary alcohol:

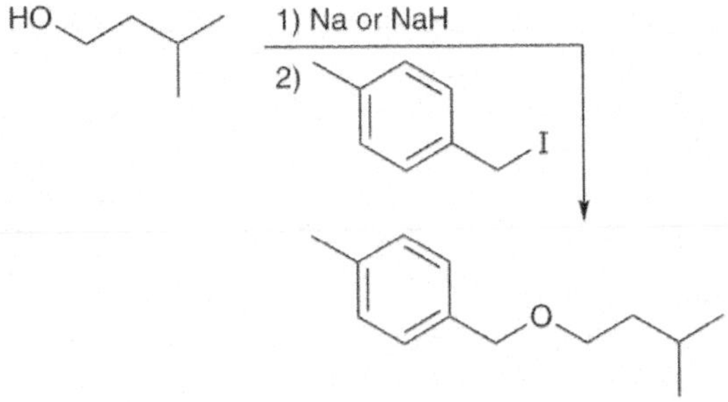

14.13 Treating this compound with NaH (or with Na) will result in deprotonation of the OH group to give an alkoxide ion.

This alkoxide ion can then undergo an intramolecular S_N2 reaction, expelling bromide as the leaving group, and giving the cyclic product shown.

14.14 This compound is an ether because the oxygen atom is connected to two carbon atoms. Each of these carbon atoms is sp^3 hybridized, so each of the C—O bonds will undergo acidic cleavage, giving two alkyl bromides and water:

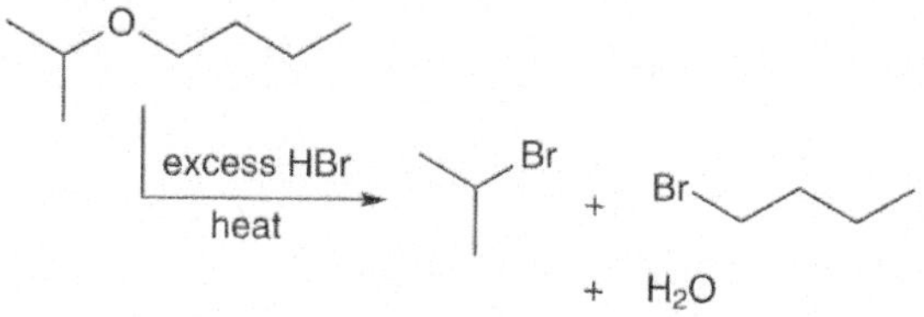

14.15 This compound is an ether because the oxygen atom is connected to two carbon atoms. One of these carbon atoms is sp^3 hybridized, while the other is sp^2 hybridized, so only one of the C—O bonds (the one with the sp^3 hybridized carbon atom) will undergo acidic cleavage, giving the following products:

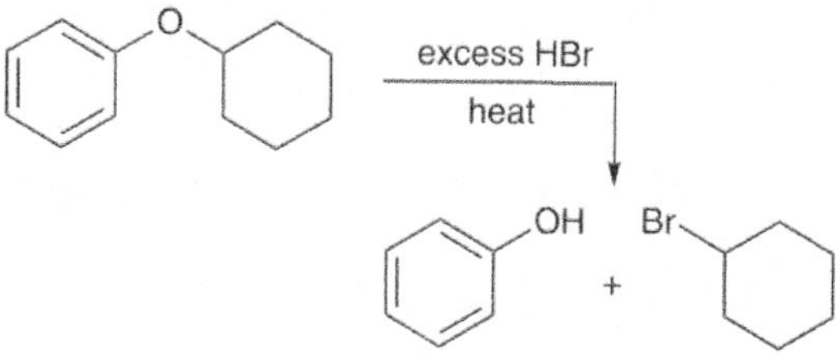

14.16 This compound is an ether because the oxygen atom is connected to two carbon atoms. Each of these carbon atoms is sp^3 hybridized, so each of the C—O bonds will undergo acidic cleavage, giving the following dibromide and water:

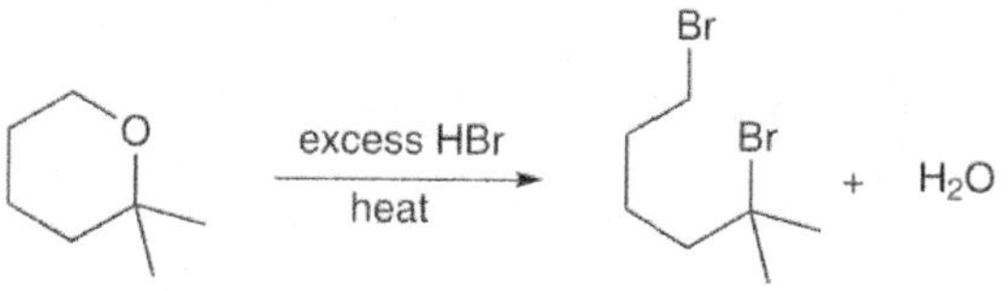

14.17 This compound has two ether groups because there are two different oxygen atoms, each of which is connected to two carbon atoms. Each oxygen atom is connected to one sp^3 hybridized carbon atom and one sp^2 hybridized carbon atom. So, for each oxygen atom, only one of the two C—O bonds (the one with the sp^3 hybridized carbon atom) will undergo acidic cleavage, giving the following products:

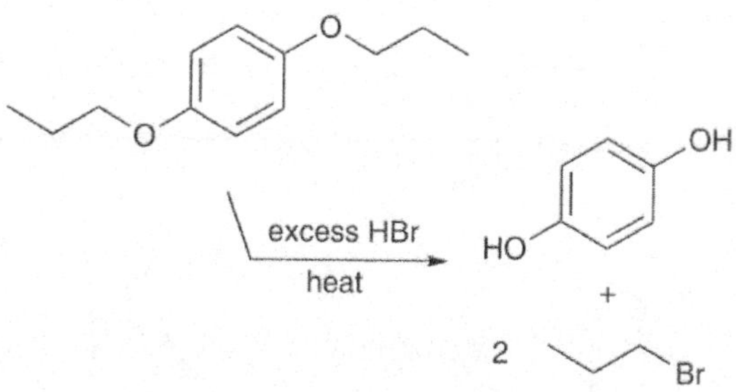

14.18 An alkene is converted into an epoxide upon treatment with a peroxy acid, such as MCPBA. In this case, the product has no chiral centers, so we don't need to consider the stereochemistry.

(no stereocenters)

14.19 An alkene is converted into an epoxide upon treatment with a peroxy acid, such as MCPBA. The starting alkene has a *trans* configuration, which is preserved in the product (the product has a *trans* configuration). Since epoxidation can occur on either face of the alkene, a pair of enantiomers is expected.

+ enantiomer

14.20 An alkene is converted into an epoxide upon treatment with a peroxy acid, such as MCPBA. The product is a *meso* compound, so we do not expect a pair of enantiomers in this case (epoxide formation on either face of the alkene will give the same product).

14.21 An alkene is converted into an epoxide upon treatment with a peroxy acid, such as MCPBA. Since epoxidation can occur on either face of the alkene, a pair of enantiomers is expected.

+ enantiomer

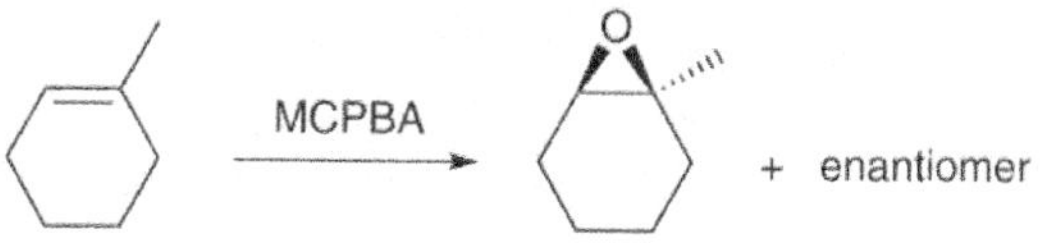

14.22 An alkene is converted into an epoxide upon treatment with a peroxy acid, such as MCPBA. The starting alkene has a *trans* configuration, which is preserved in the product (the product has a *trans* configuration). Since epoxidation can occur on either face of the alkene, a pair of enantiomers is expected.

+ enantiomer

14.23 An epoxide can be made from an alkene upon treatment with a peroxy acid, such as MCPBA. The starting alkene must have a *trans* configuration because the product has a *trans* configuration:

+ enantiomer

14.24 An epoxide can be made from an alkene upon treatment with a peroxy acid, such as MCPBA. To obtain the desired epoxide, cyclobutene must be used as the starting alkene.

14.25 An epoxide can be made from an alkene upon treatment with a peroxy acid, such as MCPBA. To obtain the desired epoxide, the alkene shown below must be used.

14.27 The starting epoxide is unsymmetrical. HS⁻ is a strong nucleophile, and the conditions are not acidic, so we expect the nucleophile to attack the epoxide at the less-substituted position. As a result of the reaction, the ring is opened, giving the following product. The product has no chiral centers, so we don't need to consider stereochemistry in this case:

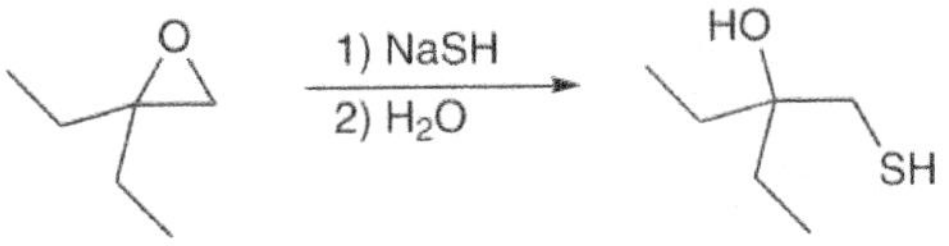

14.28 The starting epoxide is unsymmetrical. Br⁻ is a strong nucleophile, and the conditions are acidic, so we expect the epoxide to be protonated under these conditions, and the nucleophile will attack the epoxide at the more-substituted position. As a result of the reaction, the ring is opened, giving the following product. The product has no chiral centers, so we don't need to consider stereochemistry in this case:

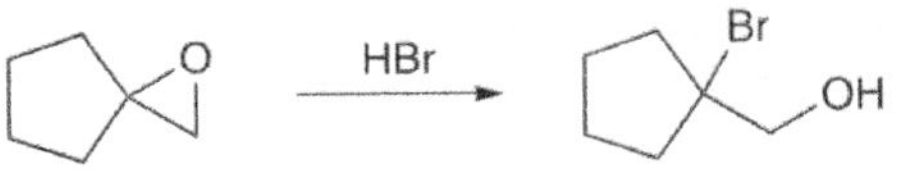

14.29 The starting epoxide is unsymmetrical and has a chiral center. Cyanide (N≡C⁻) is a strong nucleophile, and the conditions are not acidic, so we expect the nucleophile to attack the epoxide at the less-substituted position. As a result of the reaction, the ring is opened, giving the product shown below. Notice that the reaction does not occur at the chiral center, so the configuration of the chiral center is not affected during the course of the reaction.

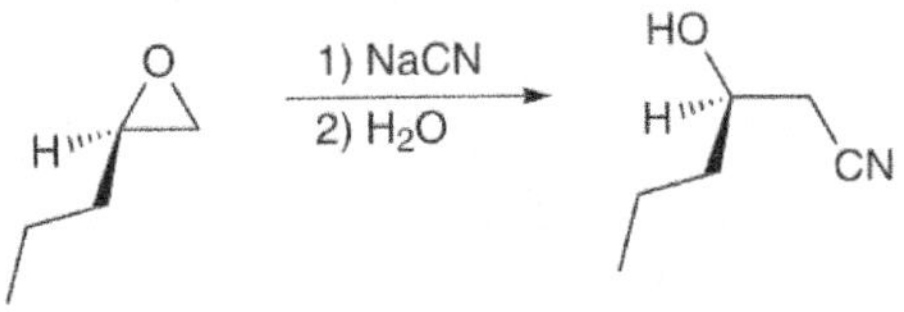

14.30 The starting epoxide is unsymmetrical and has a chiral center. Ethanol is a weak nucleophile, and the conditions are acidic. So we expect the epoxide to be protonated under these conditions, and we expect the nucleophile to attack the protonated epoxide at the more-substituted position. As a result of the reaction, the ring is opened, giving the product shown below. Notice that an S_N2 reaction is occurring at the chiral center, so the configuration of the chiral center is inverted during the course of the reaction. The starting epoxide had an *R* configuration, while the product has an *S* configuration.

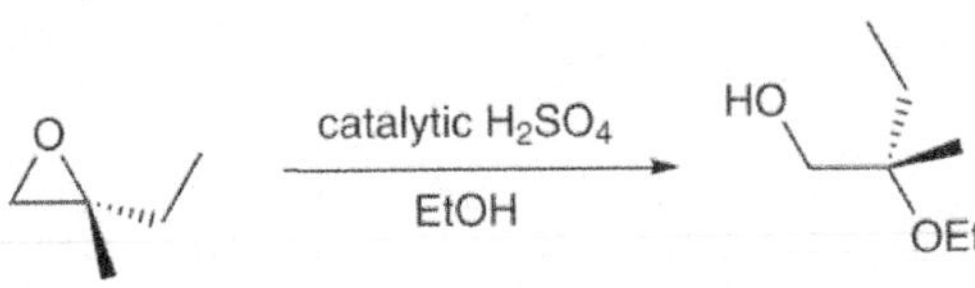

14.31 The starting epoxide is unsymmetrical and has a chiral center. Ethyl magnesium bromide is a Grignard reagent, which is a strong nucleophile. The conditions are strongly basic, so we expect the nucleophile to attack the epoxide at the less-substituted position. As a result of the reaction, the ring is opened, giving the product shown below. Notice that the reaction does not occur at the chiral center, so the configuration of the chiral center is not affected during the course of the reaction.

14.32 The starting epoxide is unsymmetrical. Methoxide is a strong nucleophile, and under these basic conditions, we expect the nucleophile to attack the epoxide at the less-substituted position. As a result of the reaction, the ring is opened, giving the following product. The product has no chiral centers, so we don't need to consider stereochemistry in this case:

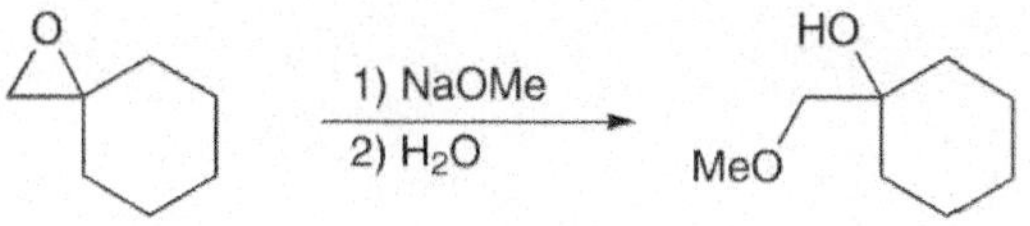

INDEX

Made in the USA
Middletown, DE
09 July 2022

68921107R00223